MORE ADVANCE PRAISE FOR *MOORE'S LAW*

"Finally, Gordon Moore gets the biography he deserves! One of the foremost pioneers of the digital revolution, he is a visionary, engineer, and revered leader. His 'law' defined and guided the growth of computing power, and his business acumen helped to create Silicon Valley. This is an inspiring and instructive tale of how brilliance and leadership can coexist with humility and decency in a truly extraordinary person."

—WALTER ISAACSON,
author of *Steve Jobs*

"*Moore's Law* is not only a definitive biography of a legendary figure in computing, but a fascinating account of the forces that triggered—and sustain—the digital revolution that has changed life for all of us."

—STEVEN LEVY,
author of *Hackers* and *In the Plex*

"If you think you know Moore's Law, prepare to be enlightened. If you think you know Gordon Moore, prepare to be enthralled. And if all of this is new to you, prepare for the ride of your life. This is the definitive story of the central theorem of the digital age, the man behind it, and its ongoing impact on us all."

—JOHN HOLLAR,
President & CEO, Computer History Museum

"With care and color, *Moore's Law* tells us how Gordon Moore, at the center of the IT revolution, applied his knowledge and insight in a quiet and effective way. When Gordon talked, everyone listened."

—GEORGE P. SHULTZ, former US Secretary of State and Thomas W. and Susan B. Ford Distinguished Fellow at the Hoover Institution, Stanford University

"Gordon Moore's story is one of disruptive innovation on the grandest scale, practiced by a brilliant technologist. Now at last we have the book that tells the story. *Moore's Law* offers a compelling, absorbing account of Silicon Valley, and its role in human progress."

—CLAYTON CHRISTENSEN,
author of *The Innovator's Dilemma*

"A remarkable book about a remarkable man, told with great style and refreshing candor."

—CARVER MEAD, winner of the National Medal of Technology and Innovation

"Almost everyone knows Moore's Law. Almost no one knows the Moore behind this law. Finally a book describing the quiet, unassuming technology godfather of Silicon Valley. A great read about a great man whose work truly changed the world."

—CRAIG R. BARRETT, Former CEO and Chairman, Intel Corporation

"Arnold Thackray and his co-authors integrate business history with the history of science and technology with great success, rendering this biography of Silicon Valley's most important revolutionary a captivating and deeply illuminating read. *Moore's Law* is also a signal contribution to the study of California history."

—DAVID A. HOLLINGER, Preston Hotchkis Professor of History, Emeritus, University of California, Berkeley

"This marvelous and well-written book about Gordon Moore captures his seminal role not only in Fairchild Semiconductor and Intel Corp, but also in the saga of Silicon Valley. The authors tell how Intel was managed into one of the great successes of all time. Gordon Moore in his quiet, non-threatening, and brainy manner created an atmosphere in which new ideas flourished and growth was encouraged. Moore's Law, his remarkable insight, has proved prescient. Woven into this story is the modest and loving relationship between Gordon and Betty, his wife of 65 years."

—ARTHUR ROCK, Co-Founder of Silicon Valley venture capital firm Davis & Rock and original investor of Fairchild Semiconductor

"A remarkable insight into the man who did so much to make Silicon Valley."

—DAVID MORGENTHALER, founder of Silicon Valley venture capital firm Morgenthaler Ventures

MOORE'S LAW

Moore's Law

The Life *of* Gordon Moore, Silicon Valley's Quiet Revolutionary

---

ARNOLD THACKRAY

DAVID C. BROCK

RACHEL JONES

BASIC BOOKS

A MEMBER OF THE PERSEUS BOOKS GROUP

New York

Published by Basic Books,

A Member of the Perseus Books Group

A catalog record for this book is available from the Library of Congress.

ISBN: 978-0-465-05564-7 (hardcover)

ISBN: 978-0-465-05562-3 (e-book)

10 9 8 7 6 5 4 3 2 1

For Betty

We are bringing about the next great revolution in the history of mankind—the transition to the electronic age.

— GORDON MOORE, 1976

CONTENTS

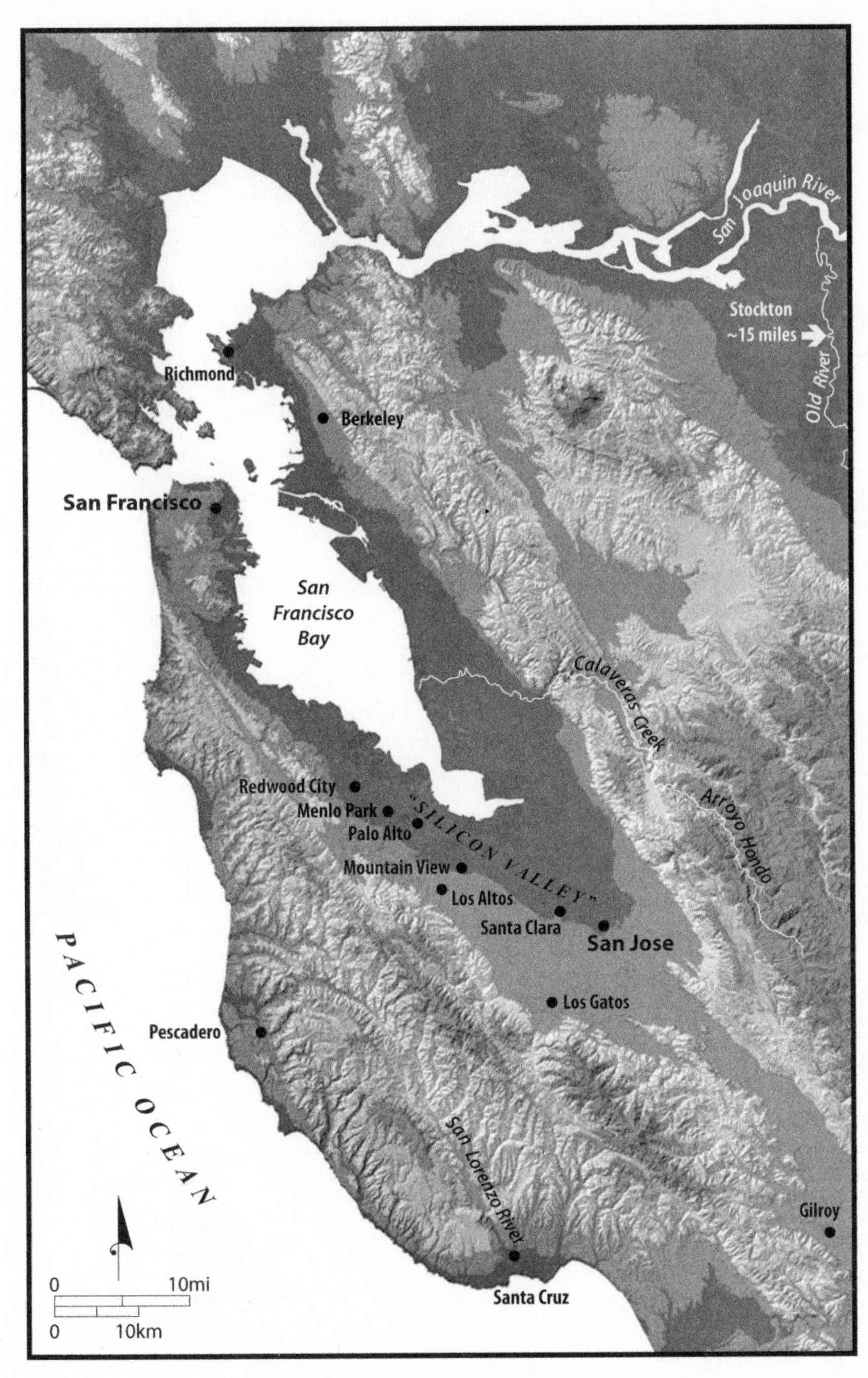

Silicon Valley, US.

COURTESY ROBERT SCHULTZ.

Gordon Moore.

SOURCE: COURTESY CAROLYN CADDES.

PRELUDE

"THIS IS SHOCKLEY"

The call proved fateful. At the time, it seemed merely unusual. In February 1956, not even scientific celebrities routinely telephoned long distance, especially if it was a simple matter of contacting an unknown young nerd.

The celebrity was William Shockley, renowned researcher and defense consultant, the nerd a postdoctoral chemist named Gordon Moore, in Silver Spring, Maryland. Moore had just arrived home from work at a government-funded, Cold War–oriented lab. He was tired and more than ready for the dinner his stay-at-home wife, Betty, had prepared. One-year-old son Ken was in his high chair. The cat was nosing its bowl. It was an ordinary end to an ordinary day.

The phone rang. Answering machines were unknown, so Moore perforce picked up: "Hello?" The response, deep and confident: "This is Shockley." Moore knew the name right away. William Bradford Shockley was a certified scientific god, revered as one of the world's leading solid-state physicists, coinventor of the era's most promising breakthrough in electronics—the transistor—for which he would shortly be awarded the Nobel Prize. The transistor was a wholly new type of on-off switch, a "semiconductor," offering tantalizing possibilities in military applications. Moore, who had recently heard Shockley lecture at the Cosmos Club in nearby Washington, DC, was shocked to realize a star was on the line.

But why was Shockley calling? The reason quickly emerged. His unabashed boast was that he was hiring the best and brightest young PhDs to join his venture in the little-known town of Mountain View, California. The aim? To perfect and mass-produce a novel silicon-based version of the transistor. This minute, solid device, no bigger than a fingernail, would be composed of silicon treated chemically in complex ways. It promised unprecedented standards of reliability to electronic devices. Shockley was counting on the reality that, in high-stakes defense markets, performance always trumped mere price. To accomplish his ambition, he urgently needed a competent chemist. He knew that Moore had recently turned down a position at a nuclear weapons lab in California. Might this young scientist be interested in the race to produce a reliable silicon transistor instead?

Moore *was* interested. After two years in the Washington, DC, area, he had become distinctly uneasy about both the direction of his career and the work of the government lab at which he labored. Shockley's call, implying that Moore might join his fledgling enterprise, offered an intriguing possibility—not least because Mountain View was close to Gordon's and Betty's families. As a fifth-generation Californian, languishing on the East Coast, he longed to return to his roots and to western living. Quickly agreeing to fly out to meet Shockley, he hung up and shared the startling news with his wife. "The window's opening," Betty responded. "We'd better make a beeline for it."

Gordon Moore was not someone to flaunt his talents. Nothing in his life to this point indicated that he was destined for greatness. Born on January 3, 1929, in remote Pescadero, three miles from the Pacific Ocean, he was the quiet second son to a steady, unexceptional local couple: Mira Moore and her husband, Walter, the town's part-time constable. On the day of Gordon's birth, Wall Street brokers a continent away were busy recalling staff from vacation. "As goes the opening, so goes the year," they gleefully remarked, savoring January's sharp rise. What happened in the months that followed, no one could have imagined.

The stock market's collapse in 1929 signaled the end of an era. Within two decades, British and European leadership—in politics, science, and business—would be in ruins. Through a major shift in the empires of the world, the United States would become the dominant superpower, locked with the Soviet Union in a tense Cold War embrace. As these new realities took hold, California would establish itself as a leading center of high-tech defense industries and an academic leader in the physical sciences. Large changes, indeed. Still larger, and only today becoming fully obvious, would be the changes that flowed from the silicon transistor and the call to Moore. Those changes have revolutionized our lives and continue to do so in accelerating fashion.

On that cold February evening in 1956, listening to Bill Shockley's pitch, Gordon Moore had little clue how fateful the moment was, but he and Betty saw clearly enough how important it could be for their own immediate future. Within three months, he was working for Shockley. Eighteen months on from the phone call, undaunted by Shockley's Nobel Prize, he would lead a revolt against his boss and with seven colleagues—"the Traitorous Eight"—form Fairchild Semiconductor, the breakaway start-up in what, as a direct result of their action, became Silicon Valley. Tom Wolfe would later remark, "Brainpower was the entire franchise. On that day was born the concept that would make the semiconductor business as wild as show business: defection capital." The eight, betting their future on the silicon transistor, quickly made good on their wager. They delivered their first hundred transistors to International Business Machines (IBM) in August 1958, at the handsome price of $150 each. Within two years, Moore and his colleagues had made their initial fortunes. Their company would go on to spawn more than four hundred spin-offs, the most successful of which, Intel, was cofounded by Moore himself.

At Fairchild a team in Gordon Moore's laboratory created a remarkable invention: the silicon integrated circuit, otherwise known as the "microchip." This was an entire electronic circuit, built from a host of transistors chemically printed onto a single sliver of silicon. In the microchip, Gordon Moore glimpsed an astonishing future. Trained as an experimental chemist, he first observed and then, through his work, fulfilled his prophecy for silicon transistors within these microchips: that they would double and redouble relentlessly—with ever-increasing use in an ever-proliferating array of products—even as their cost tumbled across the decades. This repeated doubling with plummeting price is known as "Moore's Law."

Moore's own deeper claim, arising as the clamor of the sixties subsided, was nothing if not bold: "We are bringing about the next great revolution in the history of mankind—the transition to the electronic age." At the time, hardly anyone took notice. Nevertheless, Intel would become the world's preeminent semiconductor manufacturer, as Moore, its chief strategist and largest shareholder, moved from research and development (R&D) head to longest-serving chief executive officer (CEO), board chair, and finally chair emeritus. By 1997, when he stepped down as board chair, Intel was in the top-fifty companies of the Fortune 500, Silicon Valley was the place to be, the electronic revolution was increasingly visible, and Moore himself—its theorist, quiet architect, and key facilitator—was closing in on a wealth comfortably in excess of $20 billion. Turning to become a major philanthropist, he would set records for the largest gifts ever made to conservation and to higher education. Gordon Moore transformed not only his own realities, but also those of his region, to say nothing of the global domains of research and business, the conduct of warfare, and the patterns of everyday experience.

The transistor has gone from a rare, exotic item of military hardware to the one essential ingredient in modern life. It underlies the electronic age and virtual reality: the realms of Twitter, Google, Facebook, and Amazon; of drones, government surveillance, "big data," "the cloud," and high-speed trading; of personal computers, Internet pornography, video gaming, smartphones, apps, tablets, and TVs—and soon of driverless automobiles, personalized medicine, fully automated surgery, and ubiquitous robots. All these wonders, enabling brainpower to revolutionize life, are simultaneously digital (wholly formed by an endless stream of ones and zeros) and material, handled and recorded by silicon transistors in microchips of staggering complexity.

Truth be told, many do not comprehend this new world or the fresh vistas it has opened to the imagination. Electronic reality allows us to be both present and absent, traveling in space and time via devices underpinned by billions of transistors. In 2016 well over 100 billion (*100,000 million*) transistors will be produced for every human being on the planet. This incredible profusion, springing directly from the ideas and work of Gordon Moore, is the key to Silicon Valley and to the altered dimensions of ordinary life. In digital electronics, the silicon transistor is the brick, the basic building block, yet Moore himself, the enabler of its availability and prophet of its role, today in his mideighties, remains little known. Why?

A clue is to be found in the paradoxes that characterize his life. He is one of the world's most exceptional achievers, yet he has consistently avoided opportunities to raise his profile. When Intel was named Electronics Company of the Year, his right-hand man, Andy Grove, beamed straight into the photographer's lens at the awards presentation. Moore—Intel's CEO—was mostly out of the frame, doing "something inscrutable in the margins." Internally driven and governed by the ticking of his watch, Moore believed his vision had global consequence yet worked quietly, within miles of where he was born and raised, eschewing the trappings of wealth and fame. His pursuit of revolutionary electronics brought extraordinary change, even as—with remarkable focus—he stuck to his knitting, doing one single important thing to the best of his ability. The logo "Intel Inside" speaks both of transistors and of Gordon Moore.

Whereas Larry Ellison, Andy Grove, Steve Jobs, Mark Zuckerberg, and a host of other immigrants to Silicon Valley command media attention, Moore has chosen to stay low-key. He has always known who he was, understood what he needed to do, and stayed on task. As far back as the mid-1970s, he was pointing to silicon electronics as "a major revolution in the history of mankind, as important as the Industrial Revolution." With his immediate colleagues, he was at its leading edge and foresaw how the transistor would leverage the power of human intellect. With a modesty that belied his passion, tenacity, and clarity of vision, Gordon Moore built one of the world's most successful companies, demonstrated the power of silicon technology, and established the relentless cadence of Moore's Law.

Today we know the truth of his perceptions and enjoy the fruits of his labors, even as we struggle to adjust to the scope of the novelties engendered by the transistor. With technical brilliance, focus, and unwearied assiduity, Gordon Moore has transformed our world in a way no political figure has done. On one level, Moore's story appears simple: his ability, drive, and persistence are, in hindsight, clear and remarkable. Yet at a deeper level, he provides a fascinating picture of a complex man—someone shaped by his pioneer heritage, small-town roots, and early familial experiences, a person characterized by avoidant tendencies, practical bent, and restless questing. Betty, his wife of sixty-five years, once asked, "What is he running from?" *Moore's Law* is an exploration of forces both obvious and hidden that drove Gordon Moore to find his rest in work that is transforming all our lives.

WHAT IS MOORE'S LAW?

Let's begin with a conclusion—Moore's Law is the product of human imagination. The phrase *Moore's Law* is known around the world as a technical observation, one that describes the development of digital electronics and computing. It is that, but it is also far more. It is an astonishing story of imagination, zeal, and world revolution.

The silicon transistor was initially created to serve the Cold War market of truly high-stakes poker, focused on nuclear annihilation, deterrence, and uneasy coexistence. Cost was a trivial concern compared with considerations of miniaturization, power consumption, and reliability. At Fairchild in 1959, Moore and his team achieved a breakthrough in the manufacturing of transistors, setting the stage for the microchip as we know it today. Innovations in chemical printing, to which Moore personally made several key contributions, replaced the old mode of making conventional circuits by wiring together individual components. In the early 1960s, the entire semiconductor industry adopted Fairchild's approach, but Moore went a stage further, prophesying both that the new methods meant microchips would be not only smaller, more reliable, and less power hungry but simultaneously cheaper and that this would change the world.

Fairchild's silicon microchips were, by 1963, already cheaper to make than the set of individual components required to build an equivalent conventional circuit. The microchip *was* the cheapest form of electronics. Moore, a deeply grounded and by then highly experienced chemist, believed that the industry could improve the chemical printing technology, without any fundamental barrier, almost indefinitely. With sufficient investment of effort and money, technology could be engineered to print finer features with great fidelity. Manufacturers would find their best competitive advantage lay in making microchips more complex, each incorporating an increasing number of transistors.

In April 1965, Gordon Moore's vision reached a larger audience: the tens of thousands of subscribers to *Electronics,* an industry magazine. In his article, Moore described how the chemical printing of microchips was open ended. If investment was made, technology would advance, and such investment would reward microchip makers handsomely. It was a win-win situation. By shrinking transistors, and putting more of them into individual microchips, everything became better: as chips became both better and less expensive, use would spread. Moore presciently envisaged the world we know today, "such wonders as home computers, automatic controls for automobiles, and personal portable communications equipment."

The *Electronics* article contained a fresh twist: a numerical prediction. Since 1959, when Fairchild Semiconductor made its breakthrough, the number of transistors on a chip had doubled each year, so that microchips now incorporated more than 50 transistors each. Moore predicted this dynamic would continue for the coming decade. By investing in chemical printing technology, doubling transistor counts each year, and shrinking cost—and with nothing on the horizon to deter either technology development or economics—manufacturers would in 1975 be making microchips containing not 50 but *65,000* transistors. This was the first formulation of Moore's Law, displaying its essence. Moore's research, management, and educational efforts within his industry over the next decade resulted in Moore's Law slowly becoming known, even as his home turf was dubbed Silicon Valley and the big world began to get curious.

Initially, Moore's article made little splash in the wider world, but a sprinkling of influential members of the electronics community such as Carver Mead, an electrical engineering professor at the California Institute of Technology, or Caltech, saw the potential of his predictions and helped to spread awareness. Meanwhile, Moore—a Berkeley graduate and a Caltech PhD in physical chemistry—left the seminal start-up he cofounded, Fairchild Semiconductor, to establish Intel Corporation. At Intel Moore pursued the fulfillment of his vision, championing the development of cutting-edge chemical printing and complex microchip technology to make memory chips and, in a groundbreaking development, microprocessors.

By 1975 Moore was CEO of Intel, and microchips did contain 65,000 transistors. No longer only a niche military product, they dominated the expanding field of "mainframe" business computers. Moore predicted that in the decade ahead, with mechanisms to develop the technology becoming more expensive, the "annual doubling law" would slow to a doubling every eighteen months. By 1985 microchips with 16 million transistors would represent the cheapest form of electronics. And so it went. Today, the transistor on a microchip has become the most manufactured object in all of history. Transistors now produced in a single year most likely exceed the proverbial grains of sand upon all the seashores of the world. The price of computing has fallen well over a millionfold, while the cost of electronics components has shrunk more than a billionfold.

Microchip complexity has increased at a metronomic pace for the past six decades, as Moore's Law is everywhere observed. That "law" is a social product, inspired by imagination, made possible through experience, and enforced through the cooperative and competitive efforts of the global semiconductor industry. The development of chemical printing and the design of complex microchips have required the investment of many billions of dollars and the coordinated effort of hundreds of thousands of people, through the organizing interventions of consortia, conferences, and "technology road maps."

In the history of technology, the silicon transistor within the microchip ranks alongside the steam railroad, the automobile, and the airplane in its revolutionary impact. The pace of its proliferation is unprecedented yet now barely marveled at, for that proliferation has become commonplace. Over the decades, consumers have found that electronic devices (from telephones to TVs, global positioning devices to video games) become better, for less, at a steady rate. Moore's Law is unique: the deliberate human creation of an unusually regular pace of unusually rapid change. We take this for granted and enjoy it. But it will not last.

"All good exponentials come to an end," observes Moore. He has long glimpsed the eventual emergence of fundamental barriers. On the technical side, it is impossible to print chemically a feature that is smaller than an atom (in 2015 some features of transistors on microchips are just tens of atoms thick). More significantly, Moore foresees disruption in the economic side of Moore's Law. The growing expense of ever more exacting manufacturing technology, in factories costing several billion dollars apiece, will erode economic incentives, slowing to a crawl the future career of the microchip. Moore's Law took a decade to change from a curiosity within a specialist community to a commonplace in the wider world. The prospect of the end of Moore's Law, today a niche debate in electronics and computing communities, will similarly transmute over the next decade into a discussion of profound consequence for our common future.

THE REAL REVOLUTIONARY

In 1973, as he saw his early predictions being fulfilled, Gordon Moore remarked to journalist Gene Bylinsky, "We are really the revolutionaries in the world today—not the kids with the long hair and beards who were wrecking the schools a few years ago." Undoubtedly, the technology is revolutionary, and Moore was its architect, but what of his own claim as a revolutionary?

Moore's Law frames Gordon Moore's deep understanding of silicon electronics, and he has spoken repeatedly about the myriad changes engendered by his work. Yet Moore did not originate the idea of electronics made from silicon or invent the silicon transistor, in either its original or its subsequent incarnations, nor did he invent an important kind of microcircuit or come up with the microchip, the microprocessor, the personal computer, or the smartphone. Was he simply, like his pioneer ancestors, a courageous bit player whose participation in exploring a territory—microelectronics—enabled other, more recognizable, figures to dig in, settle, and win a name? No: his contribution is fundamental, undergirding a transformation in human experience. Gordon Moore is the most important thinker and doer in the story of silicon electronics.

Gordon's first key step was his insight into the opportunity offered by Bill Shockley's out-of-the-blue call. Already in 1956, from his work in chemical research using electronic instruments, Gordon knew that transistors—invented barely eight years earlier—represented a technical and business opening. Working in Shockley's lab, he pioneered the chemical process for manufacturing novel diffused silicon transistors. He soon became convinced that chemical printing could enable mass production, that transistors would find a ready market in Cold War aerospace, and that, in time, they could compete more generally in electronics. By working with Bill Shockley's other young recruits, and by tackling diffusion—the way forward in chemical printing—he realized that the silicon "mesa," a new kind of transistor, had great promise. He championed it as a product and diffusion as a manufacturing technology.

Gordon's commitment to the transistor put him at odds with Shockley. As the cohesion of the lab dissolved, he took the lead, contacting investor Arnold Beckman on behalf of the dissidents and later assembling them in his own living room as they searched for a way forward. Moore's analyses of the promise of the transistor and of chemical printing by diffusion were critical to the success of Fairchild Semiconductor, the enterprise launched by Gordon and his colleagues in the Traitorous Eight. Their start-up quickly became the most important site of innovation in silicon electronics. Its buyout by its sponsor, Fairchild Camera and Instrument, earned Moore a small fortune. The heretical innovation of the planar process, by Moore's cofounder Jean Hoerni, paved the way for a further blockbuster: the silicon planar transistor. Following a concept devised by another cofounder, Robert Noyce, Moore oversaw a program to manufacture complete circuits of planar transistors on a single piece of silicon: the microchip.

In his writings and talks, Gordon worked to mobilize colleagues to make real the future he foresaw, creating graphs to show the exponential increase in the complexity of microchips and their regular decrease in cost: Moore's Law. His actions guided by this breakthrough insight, he became frustrated that at Fairchild Semiconductor he could not capitalize on his thinking or turn promising developments into manufactured products. Bob Noyce's determination to quit Fairchild Semiconductor proved a catalyst. Since Moore could not get the company to do what he thought best, it was time to go. His creation of Intel in 1968, with Noyce, was designed to make the most of his maturing insights.

Intel's structure, and the partnerships Gordon developed there, enabled the company to move quickly and efficiently along the tracks of his strategies. Key decisions and investments included the pursuit of DRAM memory chips and Gordon's support for EPROM, Intel's secret "cash cow." He championed the idea that digital electronics and computing should flow into every aspect of society, most especially the home. Moore's thinking set the agenda, positioning the firm to become the world's largest microchip company. He was not right about everything, nor did he directly invent seminal devices or products, but his strategy was overwhelmingly successful. Intel has followed it ever since, combining a championing of the microprocessor with the championing of Moore's Law.

In at least four distinct ways, Gordon Moore guided the "revolutionaries" he described to Bylinsky in 1973. First, he took a clear lead in the group of eight men who, in breaking from William Shockley to create Fairchild Semiconductor, established the blueprint for risk taking and innovation in what would become Silicon Valley. Not only did he innovate personally in manufacturing technology, but his resignation from Shockley helped establish "defection capital" as a key dynamic, making Silicon Valley the phenomenon it is today. Second, through his role as R&D director of Fairchild Semiconductor and in manufacturing the microchip, Moore was responsible not just for bringing to life individual innovations, but also for establishing the framework that would become central to the revolution. And third, in articulating and pursuing Moore's Law, he established the central dynamic of the global semiconductor industry over many decades, facilitating widespread, accelerating societal change. Finally, in his steady and tenacious focus on transistor technology, he sought continuous improvement in the basic building bricks of the information technology (IT) revolution that undergirds our daily lives through a cornucopia of products, services, data, and devices.

THE ENIGMA

In person, Gordon does not fit the stereotype of a revolutionary. He is not combative, colorful, or charismatic. Compared with contemporaries in the digital world, he is relatively unknown. Even Intel, the giant company he cofounded, is seen as the creation primarily of his long-term professional partners, Bob Noyce and Andy Grove. Moore has lived what is in most ways a conventional life, notable for its consistency and integrity. He has been married to the same woman, Betty Whitaker, for more than a half century and is the paterfamilias of a geographically and socially close clan. Moreover, despite decades of leading a large and hard-nosed company, and being at one moment the richest person in California, he has drawn little opprobrium: few colleagues or competitors have an ill word to say of him. He is universally liked and perceived as calm, insightful, technically brilliant, and an intellectual powerhouse. So who is this man, Gordon Moore?

First and foremost, he is a man apart. From childhood, his apparent surface ease has been undergirded by an abiding detachment and reserve. Born into an inexpressive family with highly traditional values, of a lawman father and a quiet mother, Moore early formed powerful defenses and inner strengths. He became both self-contained and adaptive, emulating the example of his father, Walter, who provided a solid model of how to suppress emotions, restrain oneself physically, and appear calm. Stoicism fed Gordon's detachment, as did his life in Pescadero, a tiny, remote settlement, frequently fog bound, on Northern California's far Pacific Coast. In this no-frills environment, Gordon became accustomed to operating alone and relying on his own resources. As pioneers, his ancestors had crossed the American wilderness; like them, he used what he had and turned it to advantage.

His experience was shaped by the history and position of his family as bastions of the community: the Moores of Pescadero. From his earliest days, he experienced a sense of rootedness: his place in the natural order was secure, no matter what happened. His first schoolteachers saw his quiet self-reliance and containment as a developmental problem; later on, in marriage and business, more subtle issues emerged, even as his style of relating—dismissing or minimizing the need to engage—gave him a huge capacity to focus on both practical and strategic matters.

Gordon would retain a lifelong orientation toward practical pursuits. Fishing and hunting were activities that combined outdoor solitude with strategic thought, offering victory through mastery and control. He was in the game to catch, kill, and even (as a young boy) sell his spoils. Gordon's father and maternal grandfather, through their daily work in law enforcement and in storekeeping, would reinforce the importance of business and solid, rational strategy. One had to earn, track, and save money carefully, requiring long hours of dedicated hard work, with few diversions. Feelings were displaced into doings. Long-term goals and the practical problems of the moment were what mattered.

Gordon's discovery of chemistry as a boy of eleven would be a game changer. Quiet, but not a loner, his need to connect was real, if suppressed. Explosions could literally make others sit up and take notice; even his reticent mother could not ignore his setting the neighbor's house on fire. Through careful application of his intelligence to chemical experiments—understanding and controlling the "bang"—Gordon could announce his identity and connect, while simultaneously developing his skills and remaining detached. Through his prowess as a chemist, he could turn material things into language, expression, and engagement.

Partnership was central to Gordon's success, in business and in private life, but warm relationships were rare. His low-key courtship of Betty evolved into a marriage that for Gordon was characterized by careful maintenance of boundaries. It was a union of two solitary people who would become strongly reliant upon and devoted to each other. Betty was not only a "safe" person, but also one he could "do business" with. A pattern quickly emerged: affection and intimacy were expressed by doing things together rather than talking about feelings. This transactional mode meant Gordon could remain in a trusting partnership—one that gave him crucial, steadying ballast—while keeping part of himself off limits. ("The Moores have no emotions," concluded Betty, resignedly.) His avoidant tendencies were problematic, but partnership allowed him to be simultaneously absent and connected. It would be an effective armor.

Gordon insists that he became an entrepreneur by chance; mere happenstance, born of necessity. "I had no training in business. After my sophomore year of college I didn't take any courses outside of chemistry, math, and physics. My career happened quite by accident. And it ran counter to early predictions." However, he did come from a long line of entrepreneurs. In 1847 his ancestor Eli Moore, born in North Carolina, had become a "pre–Gold Rush" adventurer in the arduous trek to California; once there, from necessity, Eli and his sons turned to multiple moneymaking endeavors, all highly practical. Ancestors exemplified for Gordon what it meant to be a pioneer, providing role models for his plunge first into Shockley's venture and then to play a leading role in the Traitorous Eight. Others, less willing to strike out, did not sign the legendary dollar bills that launched Fairchild Semiconductor.

Gordon is a man apart, an accidental entrepreneur whose success arises from an inborn pioneering spirit, steady application, and a particular combination of talents enabling him to master diverse roles: chemist, strategist, technologist, investor, businessman, leader, and industry statesman. Within semiconductor electronics, Moore was shrewd, driven, focused, apart, open to taking risks, and ruthless. With characteristic modesty, he remarks, "I was completely lucky to be there at the beginning." He was in the right place at the right time—but Gordon Moore was also the right man.

CALIFORNIA LOCAL

In 1998 Betty, suffering from arthritis and wishing to avoid cold winters, arranged for a partial move to Hawaii. Gordon now spends a good portion of his time at their large house on the Kona Coast of the Big Island, but in important ways he has never left California, the place of his ancestors, childhood, success, and identity. He maintains his official status as a Californian. His address and main office are at Mountain Meadow, his home in Woodside, California, where his mail still accumulates. A part-time helper sifts it, while the on-site caretaker oversees a never-ending stream of maintenance work on the house and grounds.

When Gordon flies in alone from Hawaii to spend brief periods in Woodside, piles of nonurgent mail greet him. His house is remote yet connected, set within the redwood-covered Santa Cruz mountain range but only minutes from Highway 280—and a half hour north into downtown San Francisco or south into Silicon Valley. At Mountain Meadow, he is very much at home. Disconcertingly, as one arrives at his door, he is quick to greet. You might wish to linger, enjoying the flowers or the day's weather, but he is leading the way to his study. The house is noticeably silent, save for the ticking of a wall clock. Occasionally, from outside comes the sound of hammering or a leaf blower's whine.

Inside, the man and the setting are perfectly matched. In his mideighties, Gordon is tall, close to six feet. His early days hauling oyster-shell sacks, playing football, and tumbling in the school gymnasium are long gone, but he retains broad shoulders and a straight frame, full but trim. In comfortable slacks and shirt, casual walking shoes and eyeglasses, and with a balding top and discrete hearing aids, he has the look and air of a retired chemistry professor. At Intel, compared with the slick Noyce and peremptory Grove, Moore was seen as "folksy" in style, "a reluctant oracle for the chip industry." Unpretentious, he comes across as "a charming uncle, more prone to wisecracks than grandiose statements," or "a true engineer, who loves to make astonishing calculations."

His study is entirely of a piece. Books accumulated across six decades, most with technical titles, fill the bookcases and are piled on his large desk and on side chairs; they rise in stacks at the corners of the room. A bank of files overflows into a blizzard of strewn papers. Fishing books, bits of memorabilia, and pieces of fishing reels and lures overlay the whole. An unremarkable older computer sits on another desk. It is up to the visitor to move aside a cardboard box, filled with mail still to be sorted, in order to sit down. In a swivel chair, Gordon leans back and fixes his interviewer with a steady gaze, his countenance inexpressive, listening closely to each question. He shows little inclination for small talk; it is not his job to keep the conversation going, and he is not moved by the need to please, although when his hearing aid acts up and requires attention, he details the latest struggles with his microchip-powered model. He also makes reference, in passing, to a recent trip to Costco.

Newer acquaintances, at first, strain to hear the ends of his sentences. Gordon is surprisingly soft-spoken for a titan of the digital age. Subtle cues betray his varying levels of engagement: improved posture, eyes growing wider, a hint of a smile, and answers that come louder and quicker. Technical and scientific queries elicit this more enthusiastic response, but when the conversation turns to questions concerning emotion or psychological motivation, the pauses become longer and the replies shorter. "Well, people are people" is a favorite. Asked about his own feelings, "I remember the event" is often all he will say. When he does not deem himself competent to answer a technical or factual question, he remains silent. Sometimes the pause becomes uncomfortable, and the conversation moves on.

After an hour or two of discussion, Gordon will confess there is little in the house to eat. At a nearby restaurant, he orders a cup of soup and a half-sandwich combo, with a diet Coke. The discussion turns to events in the news, including a recent proposition for the colonization of Mars; he rips it apart with surprising passion, describing a litany of technical reasons that make the proposition ridiculous. As he lights up, animation dissolves his quiet introversion. Away from his home environment, the image of a retired, slightly shabby professor from Central Casting subsides; even at Mountain Meadow, contrasts are everywhere. His cluttered study is filled with exquisite, understated woodwork and paneling; the technical jottings and printouts scattered on his desk conceal letters and invitations from America's richest and most powerful citizens. The house itself, today infrequently occupied, sits on acres of pristine land, situated in one of the wealthiest communities on earth.

These contrasts are part and parcel of Gordon Moore—the self-contained chemist from Pescadero, the longest-serving CEO of the largest and most successful semiconductor company in the world, the quiet prophet of the electronic revolution. Personally mild, he is an aggressive competitor who has built one of America's largest fortunes. On one level inarticulate, he is the writer, speaker, and organizer who worked consistently to build the social consensus required to follow Moore's Law. To the coming generation—even to young scientists and engineers studying at the University of California (UC) at Berkeley—Moore, the long-retired leader of Silicon Valley's premier company, is all but unknown. He likes it that way. Gordon's is not simply a rags-to-riches tale of a man with great talent and relentless focus, but a paradoxical story of self-realization, framed by an enduring need for self-concealment. At the heart of Moore's psyche is a desire to create an impact while minimizing attention to himself on a personal level.

Moore is a Californian through and through, shaped by the experience of pioneer generations in an unsettled land of promise, forever on the "edge of novelty." To understand him, and to appreciate Silicon Valley, it is first necessary to examine some very traditional elements in the American (and above all the California) story. Those elements, lionized, bowdlerized, and mistranslated, have been represented in a myriad of Hollywood movies about the westward migration that spilled out of the original thirteen colonies in the early to mid-nineteenth century. It is there, as the Moore family prepares to head west by wagon train, that our story starts.

1

THE MOORES OF PESCADERO

INTO THE UNKNOWN

Eli Moore Hears the Call

Eli Moore and his wife, Lizbeth, were bit players in the westward migration, making an initial move from Tennessee to Missouri in 1835. In their late twenties, they already had a growing family. Eli was a man of parts: a farmer and a hunter and skilled in working with wood. Intensely practical, adaptable, simple in his tastes, he had what it took to survive and prosper in the harsh, unforgiving business of heading west. He was a pioneer. Alexander, his eldest son, shared many of Eli's traits, not least the discernment that would one day characterize even more strongly Alex's great-grandson Gordon Moore.

Missouri was the stepping-stone to the Far West, to the little-known regions of Oregon and California, with their virgin lands. Jedediah Smith and fellow trappers had been the first Americans to penetrate overland from the East in 1826. A decade later, John Marsh set out from Missouri. He wrote many articles for eastern newspapers, praising the West's bounteous resources and hyping the ease of travel through the mountains into the Mexican-held province of Alta California. Only a small scattering of American citizens had previously reached that fabled territory, via a perilous sea voyage around South America. In the early 1840s, growing numbers set out overland from Missouri in wagon trains, prepared to face an equally long and risky journey, but at least on dry land.

Alta California was lightly settled, with fewer than ten thousand "Californios." These Spanish-speaking people, having moved north by land from Mexico, had large ranchos and governed through alliances among leading families. The years of independent Mexican rule from 1821 had been chaotic, as the earlier Spanish system collapsed. Under both, the population of indigenous Californians was badly diminished. The arrival of American settlers served only to further destabilize an already unsettled region. Manifest

Destiny was in the air. The term itself was coined in 1845 to denote the belief that all the land between the Atlantic and the "natural" western boundary given by the Pacific belonged to the United States and was there to be settled by Americans. The theory was used to support the expansion plans of President James K. Polk's administration and to justify war with Mexico. Many who went west were inspired by these views. Others, of Catholic faith like the Moores, were attracted by the scent of opportunity and by the idea of life in a Catholic culture.

Ambitious and enterprising, Eli Moore envisaged Missouri as his jumping-off point, but exactly to where was not clear. He found a mentor in Charles Hopper, who had been in the pioneering Bartleson-Bidwell Party that journeyed by land to Oregon and Alta California in 1841. The group faced extreme hardship, going for two days without water while crossing the Salt Lake flats and at one point being forced to abandon wagons as they pressed on through the Sierra Nevada before the engulfing winter snows. The Donner Party, using a variant route a little later, was not so well prepared in fall 1846, or so lucky. Its fate in the mountains, involving starvation, death, and cannibalism, became legendary.

Eli listened to Hopper's stories with keen interest, imbibing the pioneer creed: "Do not follow where the paths may lead, but rather lead where there is no path, and leave a trail for others." In January 1847, he resolved to head west. In anticipation, his eldest son, Alex, now twenty-one, married Adaline Spainhower, also from Tennessee. There was no time for the newlyweds to settle down. The Donner Party's demise made it plain that travelers must reach and conquer the Sierra Nevada before the snows came in early November. Assuming they covered fifteen miles a day, the journey would take four to six months; May was the latest possible safe departure. Hopper and his family would also be among this party of four hundred, most bound for Oregon.

On May 9, 1847, the entire Moore family—Eli and his wife, son Alex and daughter-in-law Adaline, and Alex's five younger siblings—joined the wagon train assembling in Independence, Missouri. No matter that Adaline, Alex's new wife, was well into the second month of her first pregnancy, nauseous and fatigued. In the dawn air, families loaded wagons and readied livestock. Even heavy items, including Eli's beloved weight clock, would travel with them. Two of Adaline's sisters were also in the group. Adrenaline masked her anxiety: this was a fresh start, offering possibilities of greater prosperity. The hardships were unknown and unknowable.

By July 4, after roughly two months, the wagon train reached Independence Rock at the edge of the Rocky Mountains (see map "The Journey West in 1847"). Weeks later, its travelers drank from the naturally carbonated water bubbling from the ground at Soda Springs. At nighttime, when

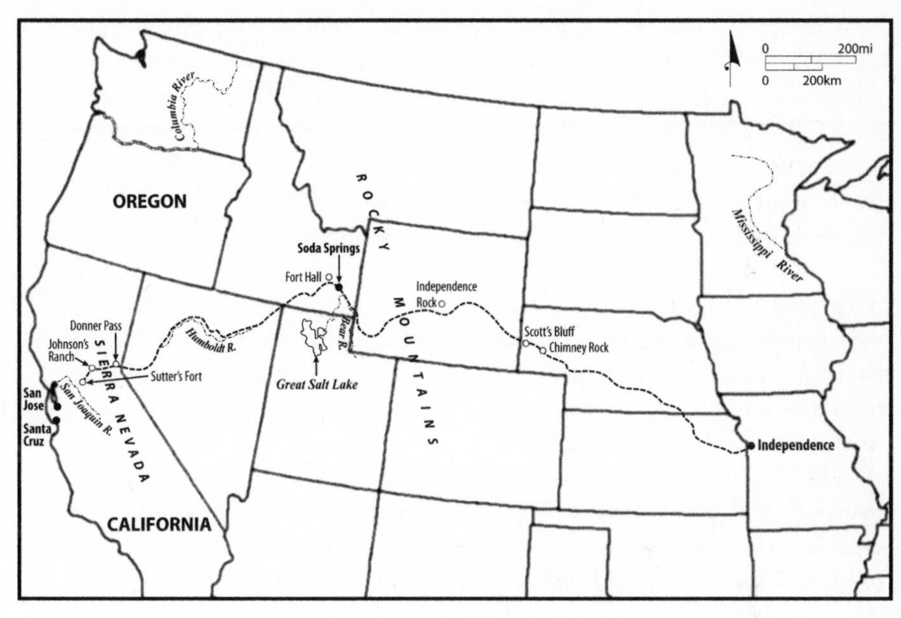

"The Journey West in 1847."
(Modern state boundaries are shown for convenience.)
COURTESY ROBERT SCHULTZ.

they rested, they parked the wagons in a circle, with Alex Moore's younger brother Tom shooting at targets "to show his ability with a gun, and to scare the Indians away." At Fort Hall (in today's Idaho), the group encountered Lieutenant John C. Frémont, who that January had accepted the surrender of Mexican military commander Andrés Pico. Frémont gave the travelers the word that hostilities with Mexico were over. The future belonged to the United States. California was there for the taking.

California: Land of Opportunity

The wagon train split at Fort Hall. The larger party headed for the Columbia River and Oregon Territory, while Eli Moore and his family joined the remnant that was intent on California. Headed by Hopper, this group comprised fifteen families and twenty wagons. They struck off south for the Humboldt River, traveling past recent notches in trees, above the snow level, that had been made by members of the Donner Party. On October 2, 1847, after a further two months of travel, the wagons finally reached the Pacific Coast delta, at Johnson's ranch on the Bear River, fifty miles from Sacramento. The wagons continued past Sutter's Fort, where, a year later, the California Gold Rush would begin.

German-born captain Charles Weber, who had arrived in California with the 1841 Bartleson-Bidwell Party, by now enjoyed an interest in a Spanish land grant and was busy enticing travelers with free land to create a hoped-for settlement (Tuleberg, now called Stockton). Weber, an acquaintance of Hopper, offered each in the group a square mile of land. "Myself, Capt. Hopper and five or six of our party were willing to accept the offer," Alex recorded. His father had more promising things in mind. Eli had heard that the best land and climate were on the coast. With three or four other families, Hopper among them, he declined Weber's offer and journeyed on, using makeshift means to ford the San Joaquin River. Here as elsewhere, Eli displayed pioneer strengths. His determination to push on to fresh opportunity and live on the edge of novelty would become a leitmotif in his life, as in California's culture.

The diminished wagon train finally approached San Jose. There, its members fell in with Isaac Branham, an experienced hunter, who in Missouri had run a sawmill, gristmill, and distillery. Eli and Alex Moore agreed to erect a sawmill in Branham's settlement of Lexington, in the mountainous area en route to the coast and the Santa Cruz Mission (today a ghost town under the Lexington Reservoir). The party headed into the hills and fixed up cabins, with the intention of wintering there. Despite primitive conditions, it was better than the constant sickening motion of the wagons. In particular, the pregnant and postpartum women needed rest.

The agreement with Branham proved short-lived. Appealing vistas continued to emerge, and the Moore men could not resist moving forward. This time it was Alex who stoked the appetite for risk, opportunity, and adventure. He struck out toward Santa Cruz and—with another pioneer, John Doak—dropped in on two older men well versed in the opportunities afforded by California: Isaac Graham and Joseph Ladd Majors. Graham, five years older than Eli Moore, was in one description a "swashbuckling soldier of fortune, trapper, hunter, rifleman, ranchero, lumberman and litigant." In 1840 he had been involved in a coup that led to the Graham Affair, a diplomatic crisis involving the rival territorial ambitions of Mexico, the United States, and Great Britain. To the young Alex Moore, he was a glamorous figure. Majors, a mountain man from Tennessee, had come to California with Graham in 1834, via the Santa Fe Trail. Alex returned to his father "very impressed and reported favorably. I liked Santa Cruz better than any portion of California I had yet seen." Eli was "well pleased with the prospect and climate." Back in the Lexington settlement, the women in their party—including Hopper's wife, pregnant with her eighth child—learned the group would head to Santa Cruz around the mountains, a roundabout journey "by way of San Jose and Gilroy, the only route then open for wagons."

Santa Cruz, Entrepreneurship, and the Gold Rush

Santa Cruz Mission had been secularized in 1834. Secularization became a license for plunder and for exploiting the land, labor, and resources of the Native American population. As Mexican landholdings were traded, split, and acquired by "gringos," the balance of power shifted to the new arrivals. As early as 1812, Russians had established a base at Fort Ross. In Santa Cruz, Jose Bolcoff (formerly Osip Volkov), a Siberian-born fur trader, assimilated himself into the local culture. By the mid-1830s, he was the administrator of the secularized Santa Cruz Mission. Bolcoff exchanged movable assets and land for liquor with the dwindling remnant of Native Americans, while they in turn raided settlements for horses and livestock, sometimes selling them back to American settlers.

American ways superseded the rancho system, as nimbleness in adaptation and a fearless entrepreneurial spirit were rewarded. The existing social order collapsed, and Eli and Alex Moore were among those well placed to exploit the situation. Arriving before the Gold Rush, they saw an opportunity to make a much better life, if not a fortune. To reach the West and survive, pioneers had to be practically minded, physically robust, innovative, and tenacious. They had to hunt for food, find water, repair broken wagons, lead oxen, and deliver babies. A century later, in an equally open-ended context, the physical and mental traits that kept Eli and Alex in good stead would nourish Gordon Moore, as he dove into the rough-and-tumble of opportunity on the electronic frontier.

On November 15, 1847, the Moore family was among the first Yankee settlers in Santa Cruz itself. They set up camp on the plaza by the mission church and soon moved into an adobe building owned by Bolcoff. Alex and Adaline Moore were content to winter there, planning to build their own cabin the following spring. Their main focus was the arrival of their firstborn son, two weeks before Christmas. They named him Eli Daniel, proud he was "the first male child born in this neighborhood of American parentage."

Alex's acquaintance with Joseph Ladd Majors paved the way for Eli to buy from Bolcoff part of Rancho El Refugio, the mission's former grazing lands. Eli quickly completed the construction of a log cabin (on today's Front Street) directly east of the adobe home occupied by Alex and Adaline. The first wooden building of any consequence in Santa Cruz, Eli's redwood cabin was nevertheless primitive, with small windows and low rooms, including one occasionally rented to wayfarers.

Pioneering, risk taking, and entrepreneurial, Eli and Alex—and Eli's younger sons William and Thomas—looked to enter any potentially profitable venture. They kept several business irons in the fire, as did their descendant Gordon many years later. Time would tell which was best. One strategy was to target the booming lumber market. Early in 1848, they formed a partnership with Doak and Bolcoff to build a sawmill. Bolcoff would own half and underwrite its cost; the others would contribute labor. But as Alex later recalled, "When we were busy getting ready to finish our mill, gold fever broke out, and the help all left us." The fabled California Gold Rush had begun.

In January 1848, James Marshall found a gold nugget on the South Fork of the American River, at Sutter's Mill. He tried to keep his find secret, but a San Francisco newspaper, the *Californian,* published the news. That summer another prospector, Benjamin Woods, discovered gold on a branch of the Tuolumne River. A few mining camps sprang up; lacking a railroad connection, however, California retained its isolation. Even so, within a year, tens of thousands of men were flocking in from Oregon, Hawaii, Latin America, and the East. At first, nuggets could be picked up. Later, prospectors panned endlessly in streams and riverbeds. In all, gold worth tens of billions of dollars was recovered. A few men became rich beyond their wildest dreams. Most did not.

The Gold Rush was a transformative event, sparking a massive surge in population together with the development of commerce and transportation and a boom in ranching and agriculture. These were "headlong, heedless, prodigal years," notable for the spontaneous social organization that would propel California into statehood in record time. Within six years, San Francisco grew from a small settlement of two hundred people to a town of forty thousand residents. As Kevin Starr, California's preeminent historian, puts it, "Not for California would there be—or would there ever be, as it turned out—a deliberate process of development. California would, rather, develop impetuously through booms of people and abrupt releases of energy."

CALIFORNIA POPULATION, 1845–1860

Year	Number of Non-Native Residents
1845	17,900 (~150,000 Native Americans)
1848	26,000
1849	50,000
1850	93,000
1860	380,000 (~30,000 Native Americans)

Sources: US Census Bureau; Andrew Rolle, *California: A History* (Arlington Heights, IL: Harlan Davidson, 1987); James Rawls, *Indians of California* (Norman: University of Oklahoma Press, 1986).

The turbulence of the Gold Rush brought with it an era of mining innovations that helped establish California's reputation in technology. Combining pragmatism, ingenuity, and urgency, the thousands of hopeful miners adapted old, even ancient means to the California landscape, pushing tools such as rockers and sluices to the limits of efficiency. These self-taught entrepreneurial seekers—a breed quite unlike the pedigreed mining engineers of Europe—made novel contributions, notably the breakthrough innovation of hydraulic mining. Many years later, the semiconductor industry in Silicon Valley would enjoy similar booms and releases of energy driven by a company known for always being in a hurry: Gordon Moore's Intel. "We were too darned busy," Moore would say ruefully of his company's feverish activity, pursuing the edge of novelty. "A lot could have been learned more efficiently."

Without the labor to build their sawmill, Eli and Alex scrambled to improvise. Alex went prospecting himself, while Eli invested in the Eldorado Mining Company. Finally, they completed Bolcoff's mill and began to supply timber to build Long (Central) Wharf in San Francisco and homes in Santa Cruz. As the local population skyrocketed, Alex won a contract to provide eighteen thousand feet of timber for a county jail. Eli became involved in a flour-mill venture on Santa Cruz Creek. The Moores also began to farm their land. Their eastern knowledge impressed the Californios; their grain cradle and a steel plow manufactured in Peoria, Illinois, became objects of wonder. Alex recalled how locals would come "day after day to see our plow, proclaiming in Spanish 'very good.'"

The coastal land around Santa Cruz was rich with possibilities. Potatoes soon became the new gold. In the fall of 1852, an extraordinary harvest drove men from mines to farms. With spuds selling for high prices, cultivating potatoes became more rewarding than digging for gold. The rush was on again, and Eli headed north along the coast in search of more land. Coming to the Pescadero Creek, he marveled at finding wild oats growing higher than his horse. To him, this remote, unsettled place was not to be grazed, Native style; rather, its rich alluvial soil cried out for cultivation. It was a classic "Eureka" moment, akin to James Marshall's discovery of gold. "Openly and without shame," Eli and his fellow Americans "coveted what they saw. They felt California's present possessors unworthy of their inheritance." Eli soon purchased an eight-hundred-acre tract for the not inconsiderable sum of six thousand dollars. His great-granddaughter Louise Williamson, Gordon Moore's aunt, would recall the family's pride in how Eli sought to buy "as much land as he could survey" in the area's first sale to a white settler.

Through Eli's gifts and influence, his sons came to own land in and around the Pescadero area. Eli himself never lived in Pescadero. Instead, in

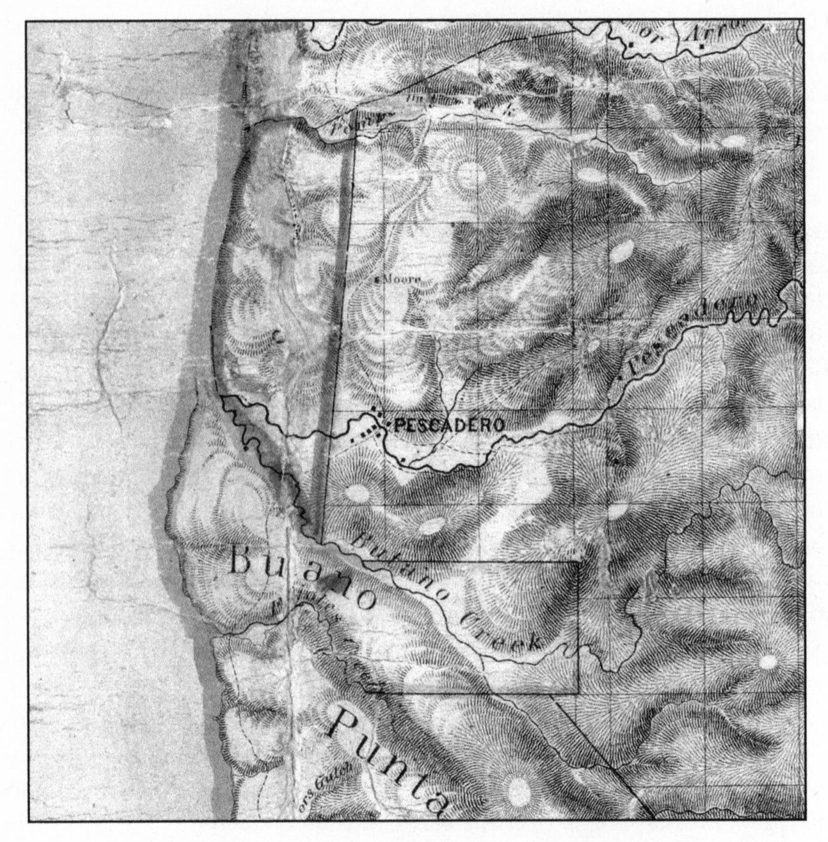

Pescadero's place on the map, including original Moore purchase, 1867.

SOURCES: CHARLES HOFFMAN, *MAP OF THE REGION ADJACENT TO THE BAY OF SAN FRANCISCO* (1867); DAVID RUMSEY MAP COLLECTION, WWW.DAVIDRUMSEY.COM.

Santa Cruz he was at the center of a growing web of relationships, his status as a father of the town foundational to his reputation and to the security of his family. Since immigrants could display few trusted credentials, preexisting ties—"blood and baptismal relationships"—played a crucial role in cementing business deals. Wagon-borne American families with children began to settle the area in significant numbers. More women appeared as "schools and churches outnumbered saloons and whorehouses," and Santa Cruz evolved from frontier outpost to settled community.

The town owed much to Eli. He served as president of the new town council and became a county supervisor, engendering a tradition of civic engagement and respect for law and order that continued to characterize the Moore family eight decades later, in Gordon's youth. Eli died in his midfifties, in June 1859. His widow, free to inherit his property in her own name by dint of California's novel state constitution, lived on for another quarter century, outlasting four of her own children. Eli's log-frame house remained in use under other owners. Later, his family donated the earlier adobe dwelling to form part of the site for the county's first courthouse. Today, Moore Creek flows through the old ranch and through the Moore Creek Preserve.

Pescadero Pioneers

Alex and Adaline had settled with their firstborn son in Santa Cruz in 1847. A second son was born there in March 1849 and named Joseph Ladd Majors Moore, with an eye to their influential friend's further favor. A third son, William, arrived in 1851. When Eli Moore promised Alex part of the acreage in the Pescadero Valley and urged him to move there, it was time to take another calculated risk. "In 1853 I came to Pescadero," Alex recalled later, "and have lived here ever since." The family had to pay further moneys to the state when a board of land commissioners finally assessed the validity of Mexican and Spanish land-grant claims, as California settled into statehood. Few grantees emerged with intact holdings. Still, Eli's prescience and investment established the Moores' claim.

The little town was close to, but not quite on, the rocky, windswept coast. Sentinel hills kept the stronger winds from the valley, if not the ubiquitous fogs; the climate was rationalized as a "happy mean between the heat that parches and the cold that chills," with mean temperatures rarely deserting the midfifties Fahrenheit. The dense redwood forests, fresh-scented canyons, and abundant fishing creeks that characterized the coastal range of the Santa Cruz Mountains would provide later tourists with magnificent picnic and camping grounds. Already in the 1860s, a visitor saw the scene as "more lovely than any we had ever looked upon." Such real estate puffery—an American and California trope—said nothing of the area's utter remoteness.

If California was "splendidly isolated," Pescadero was its outermost outpost, barred by the rocky coastline from booming San Francisco to the north and by forests and mountains from the Bay Area flatlands to the east. A special breed lived there, "hardened men, hearty farmers forced to travel miles over impossible roads to get their produce to market, and self-sufficient women willing to forgo ordinary conveniences and comforts." Adaline, a midwife, traveled by horse along the trails to attend women in labor, sometimes accompanied by one of her children. "We had no roads and little communication with the rest of the world," reported Alex Moore.

Until the transcontinental railroad opened in 1869, California remained remote; in 1857, from St. Louis, it took twenty-five days of continuous

Adaline Moore, wife of Alexander Moore.

Alexander Moore, 1847.

Alexander Moore (*left*) in a hunting party.

SOURCE: PESCADERO HISTORICAL SOCIETY ARCHIVES.

The Moore home in Pescadero.

SOURCE: PESCADERO HISTORICAL SOCIETY ARCHIVES.

travel to reach San Francisco by Wells Fargo's fastest stagecoach. Pescadero's still further seclusion led to an intermarried community. The Moores prospered as no-nonsense settlers, attuned to the unsentimental realities of pioneer rural life. Alex and Adaline had their oxen drag lumber along the coast to build the valley's first frame house, styled on homes in a New England village. This L-shaped, light-filled residence was substantial and attractive, "built for gracious living," with eight bedrooms, a two-story veranda, and pierced porch columns that would be widely copied in the town. It made a statement of confidence in the future of Pescadero. Among the furniture Alex installed was the still-functioning weight clock brought west by his father. More than a century later, the house was derelict but upright on its solid foundation.

By 1860 Alex and Adaline had three more sons—George, John, and David—and a daughter, Ida Jane. A school was sorely needed, and Alex built one at the corner of his orchard. Four of the seven pupils were his children. In 1864 the couple celebrated the birth of their last child, Walter Henry, but tragedy soon followed: siblings George and John died in quick succession. Quietly grieving, their father, Alex, continued his community commitments, helping in the construction of St. Anthony's Catholic Church in 1868. He was also successful in a campaign to remove Pescadero from Santa Cruz County and join it to San Mateo County, with jurisdiction based in Redwood City, over the coastal range to the east. The transfer meant that the long horseback ride to Santa Cruz was replaced by a coach ride from Half Moon Bay to Redwood City, a less taxing if still tedious journey. The San Mateo County Board of Supervisors appointed Alex as trustee for the Fifth Township, their freshly acquired precinct. He and his family were the undisputed leaders of small but prospering Pescadero, a community of three hundred souls in the fertile but secluded valley.

BOOM YEARS AND FAMILY SENESCENCE

Settling Down

For all practical purposes, the pioneer days were over. The task now was not to search for opportunities over the hill, but rather to develop and pursue immediate commercial realities. One option was milling, which became big business as the population increased. The coast redwood, *Sequoia sempervirens,* provided a rich if onetime bounty of ideal material. Logging camps along Gazos Creek yielded timber to build the local towns, and San Francisco became "a product of Pescadero's ancient redwoods." Farming was also lucrative: the rich soil of the valley nourished dairy, potato, and grain products. In the 1870s, California was America's leading wheat producer, and half the "ground fruit" (potatoes) consumed in San Francisco came from this part of the coast.

The children of Alex Moore began to make their way. By 1880 only four—William, Ida Jane, David, and sixteen-year-old Walter Henry—were living in the big frame house with their parents. By 1890 all were married and were characterized by the *Register* of San Mateo County as farmers. Eli Moore's success, and the bounty of game, fish, woods, and soils in the Pescadero region, enabled his family to be fruitful and multiply. By the twentieth century, seventeen grandchildren and twenty-five great-grandchildren constituted stable networks that framed the third and fourth generations, family members who surrounded Gordon Moore, the grandson of Walter Henry and great-grandson of Alex, in his early years. Eli's perceptiveness, tenacity, and business sense had established the possibility for their secure life in Pescadero and surrounding San Mateo County.

As the decades progressed, the risk-taking spirit of the pioneer generation transmuted into the settled ways of an established, insular clan. The particular combination of fertile valleys, fecund waters, and redwood forests, coupled with Pescadero's inaccessible remoteness, led to a consistent set of folkways. If each new and larger generation diluted available assets, basic security and family ties were a given. That fundament of rootedness would become a hallmark of Gordon Moore's life, as would a rekindling

The sons of Alexander Moore. Gordon Moore's grandfather Walter Henry Moore sits second from the left.
SOURCE: PESCADERO HISTORICAL SOCIETY ARCHIVES.

of the pioneer drive displayed by Eli Moore. To these were added an unswerving, persistent attention to business deriving from Josiah Caldwell Williamson, Gordon's maternal grandfather, who in the 1860s also traveled to Pescadero to make his fortune.

A Sea Passage to Pescadero

In May 1869, ten years after the death of Eli Moore, California governor Leland Stanford swung a silver sledgehammer to drive the last spike and create a transcontinental railroad link, reducing the journey from the East Coast to less than a week. Six months later, eighteen-year-old Josiah Caldwell Williamson journeyed West by sea instead, probably for reasons of financial expediency. This blue-eyed young man of average height possessed an uncommon appetite for risk and adventure, a determined entrepreneurial drive, and pioneering traits to rival those of Eli Moore.

Born in Marshfield, Massachusetts, in 1851, Josiah had grown into a hardy, self-reliant young fellow. He was eager to succeed as a businessman, a man of enterprise. Like Eli, he looked west. An older friend had already ventured forth from Marshfield and ended up in Pescadero. Josiah planned to join him. On November 19, 1869, he bid his parents and sister farewell, leaving Boston by steamship. He arrived at New York's Pier 39 early the following morning. By afternoon he was on board the SS *Alaska,* bound for eastern Panama, more than 2,000 miles away. After crossing overland to the Pacific, the sea voyage from western Panama to "Frisco" was a further 3,250 miles.

Josiah kept a meticulous journal of the voyage. He enjoyed measuring and analyzing as well as noting details and statistics. Recording his observations also helped deflect anxiety about the unpredictable elements of the journey. (Many years later, grandson Gordon Moore would adopt a similar meticulous note-taking habit to cope with business uncertainties.) The ship carried four hundred passengers, "the greater number being in steerage." Josiah was impressionable enough to enjoy seeing his own name on the passenger list, but in other ways he was wise beyond his years, coping stoically with privations while keeping his own counsel. A shrewd observer of people and their differences, he disapproved of the Panamanians he saw lounging in the shade: "I do not see what they do to support themselves." With a strong Puritan ethic, he longed to be in productive work. Above all, he was practical. His grandson Gordon would display those same qualities.

Baja (Mexican) California appeared to Josiah "the most desolate place I ever saw." Two weeks before arrival in San Francisco, he noted his own mounting impatience: "Want to get on shore and be at work." San Francisco failed to make a good impression: "It is not so much of a city as I supposed, being hilly and the streets narrow and muddy, and overrun with people who have no work and cannot get any." After an overnight rest, Josiah rose early to take the train along the Bayside to San Mateo, a stop on the newly built railroad to San Jose, and then traveled onward through the mountains for eight hours by stagecoach. It was a lonely, anxious day, but by early evening he was in Pescadero, his final destination, with a room at the local inn. It was time to turn his dreams into reality.

The country at large was uneasy. The US government was printing "greenbacks" to defray the cost of the recent Civil War. That September there had been a crash on the New York Gold Exchange and a run on the banks. The West would experience its own panic as the Bank of California failed, but to boat-weary Josiah his new location seemed to offer welcome promise. Pescadero was a "pleasant agricultural town and summer resort, at a similar distance from San Francisco and Santa Cruz," its equable climate ensuring that "those who go there to spend a few days seldom come away again as soon as they intended." Josiah initially worked at a dairy and in 1873 leased a parcel of land to launch his own farming venture. It was bad luck, or bad timing. That year America entered the "Long Depression,"

Josiah Williamson's Pescadero general store.

SOURCE: PESCADERO HISTORICAL SOCIETY ARCHIVES.

the worst in its young history. Eventually, Josiah was forced to settle for a position in the leading general store as clerk, druggist, and telegraph operator—not exactly the high enterprise he had imagined. In his spare hours, he became a correspondent for the *San Mateo Times & Gazette*.

Five years later, he was secure enough financially to marry Hattie Honsinger, a prosperous dairyman's daughter, and to start a family: Nellie, who died young, and Frank born in 1884. The following March, he finally opened the store that was to be his future. Williamson's became a long-enduring fixture on the Pescadero scene and a place in which Gordon Moore would spend many childhood hours, imbibing basic business realities. In 1887 Josiah also became the town's postmaster, a task he fulfilled from his store. Year upon year, with a steady focus, he built his enterprise. In April 1890, the *Redwood City Times & Gazette* reported an extensive addition to the store, with two additional rooms for residential purposes. By 1893 he had two more daughters, Ella Gladys and Florence Almira, known as Gladys and Mira. Mira would become the mother of Gordon Moore.

Like other Pescadero settlers, the Williamsons looked toward the twentieth century with optimism. The local population had grown to about a thousand. Josiah served on the town council with Alex Moore, who offered part of his land for a half-mile racetrack with a grandstand. The town had an enthusiastic brass band, two blacksmiths, a freight line, two livery

Alexander and Adaline Moore on their Pescadero porch with some of their children and grandchildren.
SOURCE: PESCADERO HISTORICAL SOCIETY ARCHIVES.

stables, a boot maker, a butcher, and a flouring mill. The business directory listed two physicians, a constable, and a justice of the peace, and the new public school boasted ninety students.

A half century later, sociologist C. Wright Mills would characterize entrepreneurs such as Alex and Josiah as members of the "old middle class," self-employed farmers, merchants, and producers who might also employ other family members. "An individual established a farm or an urban business and expanded it, rising up the scale of success as he expanded his property." A smaller order of capitalists lay above; below came laborers and the dispossessed. The small-town, small-business orientation of the middle class engendered civic spirit that, "in the American form, involved a widespread participation in local affairs on the part of those able to benefit a community by voluntary management of its public enterprises." Alex, the founder of Pescadero, and Adaline Moore were just such bastions of the community.

In 1898 they celebrated their golden anniversary. Alex had continued deer hunting in the hills, in accord with family practice, well into his seventies: "Being a very good shot, he took only three shells with him." In January 1902, Adaline died. That August, tended by his sons and daughter, Alex died. His simple Pescadero gravestone would state proudly: "Pioneer of 1847." In the ensuing decades, the Moore family had become integral to the evolving town and region, not as rich rentiers or speculators, but through their participation as civic figures and working farmers, ranchers, storekeepers, lumbermen, and hunters. Both Alex and Josiah Williamson were among those who "remain for years in one spot, forming the mass of the settled population, and giving a tone to the institutions of the country."

Historians of the American West point to the existence of both "movers" (a transient minority who farmed for a time before pushing on) and "stickers" (those who persisted in the area, intermarried, and passed farmland down to their descendants) as integral to the formation of community. As early as 1830, an observer had noted the reality of those "left behind, who cling permanently to the soil, and bequeath their landed possessions to their posterity." Posterity and land played their roles, along with opportunity and migration, in shaping the character of Pescadero. Through ties of marriage with other long-established families, the Moore kinship group would come to constitute a significant strand, even the backbone, of the town's population.

Children of a New Era

Within the Moore clan, and in Pescadero, life followed familiar and predictable routines. As the years went by, the town crested and then slid into slow decline, in part because its lumber industry disappeared as the last reachable redwoods fell to the ax. In the early years of the twentieth century, tourism hopes revived, as local entrepreneurs pushed forward the idea of a coastal railroad. Work began on the double-track Ocean Shore Electric Railway, aimed at a route that would halve the five-hour travel time from Pescadero to Santa Cruz. The grand opening was set for early 1907. Instead, on the morning of April 8, 1906, disaster struck: the San Francisco earthquake, among the worst natural disasters in US history. Pescadero was twelve miles from the line of the San Andreas Fault and counted itself lucky to escape with minor damage. "Williamson's store and stock were damaged about $300; the Catholic Church is off its foundation and pretty badly wrecked; the school will need extensive repairs, before it is safe."

If railroad plans had kindled the possibility that Pescadero could revive as a tourist destination, the earthquake tolled the death knell. The motto "Ocean Shore reaches the beaches" could now be taken literally. "More than 4,000 feet of right-of-way slipped into the ocean, along with most of the construction equipment for the northern end. Of greater significance than any physical loss were the nearly complete financial losses of most of the backers of the railroad." As the twentieth century progressed, Pescadero would become a modest, dwindling coastal enclave, described not as a destination but simply as an agricultural and recreational area. Hunting and fishing continued in nearby forests, while farming and ranching took place on the plains and terraces. The coast offered commercial and recreational fishing and provided a link to the outside world via the shipping station at Pigeon Point.

PESCADERO REGION POPULATION, 1880–present

Year	Population
1880	1,113
1890	1,015
1900	999
1910	1,016
1920	1,125
2015	643

Source: US Census.

The Moore family, too, settled into low-key maturity. Pescadero had its family empires, such as the vibrant Steele dairies, but the five surviving sons of Alex and Adaline (Eli, Joe, Bill, Dave, and Walter Henry) displayed little of the driving, risk-taking attitude so evident in their grandfather Eli and their father. These brothers grew up in a stable community, bearing the name of a highly respected pioneer family, enjoying a life of comforting security. Walter Henry, like his brothers, embarked on a farming career, living close to the land. He foraged for abalones, mussels, and other seafood around Pigeon Point. His son Walter Harold, Gordon Moore's father, would recall a childhood of playing ball and hunting—squirrels, rabbits, and deer. He would speak of start-of-the-century Pescadero as "a lot busier then," with sawmills, bars, and hotels.

Walter Henry well understood the power of the Moore name and connections as well as the importance of the practical. He was a local figure, "an organizer who could do most anything." As the key player in the area's Fourth of July picnic in 1906, after the earthquake's devastation he was "busy on a great celebration, and harnessed contributions from all over the county. The day of the picnic, there were firemen, and a parade that went clear around the block." The Chamarita, a Pentecost festival organized in conjunction with the Catholic churches, was also a highlight of each year. "It was such an important event, we bought new clothes for it," recalled Walter Harold. "All the young people looked forward to that dance."

A far greater and more permanent disruption than the earthquake befell the Moore family when Walter Henry died just before Christmas 1909, after an agonizing three-year battle with cancer. Only forty-five, he left five children behind. Barely fifteen, his son Walter Harold perforce became the breadwinner. "My father had to leave school and start supporting the family," explains Gordon Moore. "He didn't get any more formal education." Walter Harold traveled widely around the region to find work: "Any place

I hung my hat was home sweet home to me. The rest of the family always lived in Pescadero. Until my mother died, I gave her $20 every month to help the family." At first, Walter Harold was based in Watsonville; later, he drove a team of horses for the Campbell Lumber Company, hauling timber from Gazos Creek. In 1912 at the age of eighteen, he was back in Pescadero, working for a lumber company.

Over the Top

By then Walter Harold was a seasoned lumberman. He would travel back and forth from Pescadero over the hills into what is now called Silicon Valley. The round trip from Pescadero to Palo Alto was a distance of about twenty miles, as the crow flies (see Map 1). Six horses pulled loaded wagons from the valley up the steep grades and then down to Palo Alto to unload. He later recounted how "we would leave nearby La Honda, the loading point, at 2 am. There was a hotel and a barn in which we could leave the horses. We had two wagons loaded with lumber. We left one and would go to the top of the mountain, which took seven hours: seven miles, seven hours! Then we had to come all the way back for the trail wagon, then on into Palo Alto. The third day we came back. The horses were rested and it was easier on the return trip."

Then in 1917, the outside world at last impinged directly on the Moores of Pescadero, as Walter Harold registered for the draft, despite an attempt by his mother to claim his exemption for family support. "I went in the army," he remarked laconically. "I was in the 363rd Infantry Regiment, 91st Division, which was supposed to be California's own." (A son of Walter Henry's brother Bill joined the same division, while another cousin, Ida Jane's son Charles Steele, also enlisted.) By the start of April 1918, Walter was at Camp Lewis, Washington. Five days later, the United States entered World War I, and by the end of June the division, now twenty-seven thousand strong, was at Camp Merritt, New Jersey, with each man receiving a steel helmet and two pairs of hobnailed trench shoes.

By the end of July, they were in France, entraining for the interior "in small side-door Pullmans known to every Allied soldier as 8:40 trains. (The box cars were stenciled '40 Hommes-8 Chevaux.') Two nights and a day gave the men all the 'chevauxing' they desired." The activities of previous months—close drill, bayonet combat, and mock gas attacks—signaled to these raw recruits that they were about to enter a very different universe. An outdoorsman since early youth, Walter, now twenty-three, was well equipped to withstand war's physical demands, yet he and his colleagues could scarcely prepare for, or even imagine, the mental impact of what lay ahead. Fresh young American blood was needed for the endless carnage.

Last orders were issued to the division on September 25: to go over the top, in what was billed as yet another drive for victory. First came a vast, stunning German bombardment: "Those who lay under it during the hours before the 'jump-off' will never forget. The noise was so overwhelming that no one could grasp the whole." Emerging, the 91st Division found battered enemy frontline trenches scattered with the debris of a hasty evacuation. The division advanced through the city of Very to northerly high ground, where they dug in. Fierce fighting ensued, with artillery fire becoming steadily more severe. Conditions worsened, with heavy rain and no blankets for overnight protection. On the morning of October 4, they finally withdrew. By then, the division had a 20 percent casualty rate, with more than one thousand slaughtered, a far cry indeed from quiet Pescadero.

Other offensives and counterattacks followed in short order. Despite the danger, discomfort, and distress, Walter escaped both physical mutilation and serious illness. That November, still in the trenches where life was nasty, brutish, and short, he had little idea that his period of active service was about to end. On November 3, preparing to cross the Scheldt River, the 91st Division received orders to withdraw. The Armistice was signed eight days later, on November 11, a date so significant to him that he would choose it as his wedding day six years later. Caught in the long, slow demobilization process, Walter landed in New York exactly a year after he had arrived at Camp Lewis. Finally discharged, with the rank of corporal, "I lacked five days of being in the army thirteen months." At last, he was free to return to his hometown and family.

Normalcy, Prohibition, and Lawful Life

When Walter came back to Pescadero, it must have seemed to him that time had stood still. There was the familiar sweep of pasture, dotted with whitewashed wood-frame buildings and small church spires, reflecting the eastern and midwestern origins of those who, across one lifetime's span, had built the community. Pescadero's surface seemed unruffled. Josiah Williamson continued to tend his store on San Gregorio Street, the town's main thoroughfare, where it jostled for customers with McCormick's Merchandise Store. Nearby were Duarte's Tavern, the Elkhorn Saloon, and Ginola's, "the wildest bar in town." Josiah remained deeply involved in local affairs, perhaps in an attempt to keep busy after the death of his beloved wife, Hattie, in 1917, at fifty-nine, from chronic heart disease. "I didn't know my maternal grandmother," recalls Gordon Moore. "I just know her name."

Josiah's son Frank joined his father in the store. He had married into the Moore clan in 1909, wedding Walter Harold's sister Louise in the months before the early death of his father-in-law, Walter Henry Moore. Louise was a vivacious person and became a fixture at the store and in town. The couple's support enabled Josiah to diversify and to become a surrogate banker and counselor within the wider community. Pescadero had seen the growth of a sizable Portuguese colony, farm laborers and largely illiterate. They came to rely on Josiah to protect their gold pieces in his large vault and to read and write letters for them. This innovative protobanking business included interest-free loans to those whom Josiah trusted. Willingness to extend credit made the store enduringly popular. Williamson's was "a place where the owner sits by the cash register with a personal word for everyone who comes in."

For Walter Harold, nothing—and everything—had changed. At first, he operated a small ranch, then "ran a tractor for the county," and worked for a company called Shoreham Properties, spending several unsettled years trying to find his place. As a young teenager, he had seen his father die under painful circumstances. He had roamed the area, living a life of unrelenting labor, with no time to think about formal education or bettering his own lot. "Being a man" meant working hard, saying little, and taking responsibility for others' welfare. The war was a nightmare interlude, best met with silence. Pescadero was safe—home, the locus of his mother, sisters, and wider family, including his aunt Ida Jane in the Alexander Moore house with her daughter. (That house, "sway-backed from settling but still handsome," remained occupied by Moore descendants until Ida Jane's son Grover died in the 1960s.)

The passing of the Volstead Act in 1919 launched the United States on the thirteen-year era of Prohibition. The consumption of alcohol declined, but organized crime rose dramatically. Thanks to its remote coastal location, Pescadero became "a center of serious smuggling operations involving rumrunners of all different ranks." Walter Harold's main chance for a steady occupation arose when, drawing on his family's reputation and following in his uncle's footsteps, he was elected part-time constable for the San Mateo County coast in 1923, as Prohibition fed a steady increase in the need for law enforcement. Angling for and securing the constable's position required, as a political necessity, that he belong to the Native Sons, the American Legion of Honor, and the Benevolent and Protective Order of Elks. Revealing his real, much more reserved self, he later admitted, "I belonged to the Elks for twenty years and the only time I went was the night I was initiated. I'm not much of a joiner and don't attend anything regularly." Nevertheless, in accord with the family's place in the community, and like his father before him, he served on the committee for the Fourth of July celebrations, along with his uncle Eli and cousin James.

Walter slid back into small-town life. As part-time constable, he drew a seventy-dollar monthly salary and became deeply involved in trying to contain bootlegging operations. Initially, he used his own horse and buggy to police the bounds. Automobiles were still rare; Josiah Williamson was the first in town to get one. "They called it E.M.F," recalled Walter. "'Every morning, fix it!'" Walter in his turn became the proud driver of an official Dodge and journeyed frequently to Santa Cruz (the county line), with red lights blazing, "to fool the rumrunners into believing that the cops were everywhere." On the contrary, it was liquor that was everywhere, from the cannery at Pigeon Point to the silos on the farms. One resident recalled seeing even Josiah's roadster "filled with so much liquor that it sank low to the ground."

Despite his official Dodge, Walter would often drive in his horse and buggy to La Honda for Saturday-night dances. Horses were his great love, and he was known as a very fine hand, with the ability to drive eight in hand. He also loved dancing and benefited when Pescadero—adopting the frivolous spirit of the 1920s—began Saturday-evening dances at the community hall. Mira Williamson had been at school with Walter, and her brother, Frank, was married to Walter's sister Louise. When Walter returned from the war, Mira saw him frequently at dances and family gatherings. "In Pescadero, everybody knew everybody," Gordon Moore explains. Walter himself simply noted, "We had known each other all our lives." Even so, it was not until 1924, following Walter's securing of his position as constable, that the couple finally tied the knot, on Armistice Day.

Mira was already in her thirties. Gladys, her surviving sister and last close female relative, had died that January, helping to precipitate Mira's agreement to Walter's proposal. The newly married couple invited widower Josiah—a blind eye turned to his bootlegging—to live with them. The following December, Mira gave birth to the couple's first child, Walter Elsworth Moore. The birth marked the start of another generation, while also signaling the continuance of the Moore presence in Pescadero. Less than three years later, Mira was again pregnant, at the age of thirty-six.

PESCADERO PASTORALE

Good Times in Bad Times

Gordon Moore was born on January 3, 1929. That same year, "Black Tuesday" heralded the start of the Depression, which would bring prolonged hardship to the entire country. To make matters worse, one of the most severe droughts in history struck in 1930. During the next several years, crops vanished and livestock died of thirst on a half-million farms. Hunger was widespread. Nearly half of American children did not have adequate food, shelter, or medical care. Myriad young people, believing themselves a burden on poverty-stricken households, took to riding the rails. In California a fifth of the population (some 1.25 million individuals) went on public relief.

Migrants from drought-stricken states fled west on "the Mother Road" (Route 66). These "Okies," immortalized in John Steinbeck's novel *The Grapes of Wrath,* became the butt of derogatory jokes and the scapegoats for a shattered economy. In this grim era, novel forms of escapism offered by early electronic media such as radio and the movies were increasingly appealing. Record numbers of people went each week to watch such "talkies" as *King Kong* (1933). Robert Naughten, who would become a schoolmate of Gordon Moore, recalled: "My father had a small gold mine in the Sierras that helped us through the Depression. He and my mother's father would heat the gold in a frying pan to drive off the mercury. That was our income. We grew most of our own stuff, vegetables and fruit. We had a radio, and we heard some of the Fireside Chats. A Will Rogers movie came to town. Nobody could afford to go."

Gordon Moore was more fortunate. His father, Walter Harold, had the advantage of what in 1933 became a full-time government position under the county sheriff. Not only that, but his grandfather Josiah Williamson owned the general store (today's Pescadero Country Store; Duarte's Tavern—the Duartes had settled in Pescadero in the 1890s—is still just down the way). Walter and Mira's modest bungalow was right across the street; in it Gordon shared a bedroom with his older brother, Walt Jr.

Some Moore relatives had substantial property, but Gordon's father possessed little in the way of land. "He owned a house in town, and a lot across the street, but no ranch or farmland," recalls Gordon. However, "we were never short of food, because we had the store. The family also had a cow, kept behind the store. We had all the milk and cream you could imagine. Childhood days certainly weren't dark." Gordon's father and Josiah, his maternal grandfather, were the central male figures in his young life. Known as "Mr. Cy," Josiah, in his ninth decade, "was a regular member of the family. He lived with us, ate meals with us."

Williamson's was by now the only store left. It carried "everything from picks to women's underwear. It didn't have much of a meat selection, but it had dry goods, groceries, hardware: the works." Farmers came from surrounding areas to buy supplies. The clientele was mostly local residents, "but anybody going through town who wanted something had to buy it at that store. Granddad was there all the time. He'd have breakfast, go to the store, and be there until dinnertime. He was the storekeeper in a very real sense. He kept pretty busy at it." Josiah was joined by his son, Frank, and by Frank's wife, Louise. Daughter Mira—Gordon's mother—stayed home, keeping house for the family.

The Sins of the Fathers

As all parents eventually discover, they unwittingly bequeath to their children far more than a set of genes. In their earliest years, infants absorb cultural signals—most especially from parents—that deeply influence their manner of being in the world. Walter Harold, Gordon's father, sent clear if unstated messages about how to be a man. His experiences in adolescence and young adulthood had confirmed and sealed his own beliefs. The awful experience of watching his middle-aged father die from cancer could hardly be acknowledged, since he was required to remain strong for his mother and sisters, leave school, and roam to find work. He then faced death directly, repeatedly going "over the top" in wartime France. How little he chose to share these experiences with Mira and his family may be glimpsed in a much later comment of Gordon's: "My father was in the middle of the action. Nine months was a long time, at the front. He didn't speak of it much. I didn't realize that the war was such an important experience for him." In very different circumstances, Gordon would be equally laconic in describing to his own sons the daily battles and triumphs of his life in business. Moore men were stoic—"'Nuff said!"

Mira Moore had her own unspoken sadness. She had lost her mother and her last surviving sister in the decade preceding her marriage and, once married, found herself caring for three males instead: husband Walter, young son Walt, and father Josiah. On learning that she was pregnant once more, Mira hoped for a daughter. Instead, her second child was another boy, Gordon. Her unsatisfied yearnings were expressed in his early years, as she dressed him in girl's clothes for his first eighteen months. "Gordon made the cutest darn girl," according to May Knapp Moore, cousin-in-law of Walter Harold. This response to young Gordon's arrival was memorable—one of the few elements of his early life that both Gordon and his wife, Betty, still mention—but while it became a lasting family story, made light of at times and somehow still puzzling after all these decades, whatever lay at the root of this situation was not explored or dealt with.

Both Walter and Mira were quiet figures, with few words and limited discourse, living frugally as part of a small-town community, conforming to its social norms. They simply did not know what to do with complex feelings, turning instead to physical pursuits. Hunting, fishing, policing, and house- and storekeeping all helped keep emotions at bay. Even had there been a desire to explore their own interior lives, Pescadero afforded practical, not psychological, help. Gordon, too, would learn to channel his anxieties and his energies into outdoor pursuits and to the quantifiable certainties of the laboratory notebook and the business ledger. When

Gordon Moore (*left*) with his older brother, Walter, early 1930s.

SOURCE: GORDON MOORE.

there is little language for dealing with grief and strong emotions, things may come out sideways or turn inward, to great advantage or to great cost.

Adaline Moore, pioneer wife of Alex, had ridden the trails to help other women give birth. Subsequent generations of Pescadero women were more directly home centered. Life took place within gendered spheres: Gordon's father and grandfather represented authority and the male world outside; mother Mira was the incarnation of femininity, a gentle homemaker whose life revolved around her menfolk: "If I picture my mother," recalls Gordon, "she's cooking and washing dishes, taking care of us kids. I don't have memories of her telling us off. She was more likely to be comforting me." In contrast, "When I picture my father, he is doing something. At the dinner table, he would be speaking—him or my younger brother. I don't remember him laughing a lot."

The settled local connections and structures of the Moore family would foster Gordon's quiet self-assurance as he matured. The flip side, perhaps related to his mother's earliest hopes in a gender-bifurcated world, is his self-containment and his emotion- and conflict-avoiding nature. When very young and highly impressionable, Gordon received strong signals: Moore men go out and do practical things; women stay home and provide support. Women were essential but mysterious, perhaps able to turn boys into girls, both embodying and controlling the essential divide. At an inward preverbal level, Gordon learned that it was necessary "to be a man," to master one's emotions and one's expressions—to absorb the codes of masculinity and avoid feminine modes and manners. Self-containment was all, but the corollary was an inability openly to connect with others on the plane of feeling.

In August 1934, when Mira Moore delivered her third son, Francis Alan (Fran), she finally accepted that it was her lot to raise boys. In time she became close to daughter-in-law Betty, Gordon's wife, who fostered family links and took Mira out for "lunch and lady shopping." "She loved what I did for her. I would find very feminine lingerie for all of her gifts. She was very appreciative of the things I purchased," Betty recalls.

Huntin', Shootin', and Fishin'

Walter Harold was not a joiner. The outdoors and action, not the church, saloon, or social interchange, formed the warp and weft of life. His real loves were horses and hunting. Walter's most prized possessions were his Arabian pinto and a mounted silver saddle. "My father always had horses," Gordon recalls. "He had a fair amount of ranch experience growing up, so he liked rodeos. We'd go to them regularly." One such rodeo was the annual event at Salinas, which in the 1930s hosted Hollywood stars such as Will Rogers and Gene Autry. Once again, sex segregation was the natural order: "My mother didn't ride. This was my father's thing." Conversely, as time went on, Walter "never went to see films," and Mira would go by herself. Gordon, in turn, failed to appreciate the "heartwarming" movies that his wife, Betty, loves. "There's something about the Moore genes that doesn't allow you to be all human," Betty quips.

Mira's distaste for riding was confirmed in 1934, when five-year-old Gordon broke his arm riding a bull calf in a rodeo: "They found me a very small bull and told me that you were supposed to hold on—like *this*—on the rope. Evidently, I didn't get my hand out when I fell off. I still have a crooked arm; my left arm sticks up in the air!" He needed to travel through the hills to San Mateo for physical therapy twice a week. Mira took him to these appointments, cementing his detached role. Gordon had disappointed his father's implicit expectation. Walter Harold continued to frequent rodeos, eventually taking younger brother, Fran, along. "Fran became interested," says Gordon, "but breaking my arm took all the cowboy out of me!"

With the Pescadero Creek running by his backyard, Gordon Moore developed an early and lifelong passion for fishing, a traditionally male and frequently solitary occupation. His childhood friend Ron Duarte recalls: "We spent time in the creek. We started out with crayfish. All you do is put a piece of meat on the end of a string and throw it out, have a net to put underneath, and pull them in. Steelhead fishing came later." In another male Moore tradition, Walter Harold would go hunting on weekends, whenever deer season came around. In 1938, as he turned nine, Gordon went too. Like fishing, hunting would become a lifelong pursuit. "My father hunted for quite a while. Later, it became a time to see my older brother, Walt. We'd hunt in a family group. I went every year, right through the decades until I was in my sixties. We ate a lot of venison!"

Outdoor life fostered self-reliance and self-containment, with activity driven by necessity, tradition, practicality, and facts—rather than by emotional impulse. In long hours spent fishing and hunting, Gordon would develop an appreciation for wild places and a respect for the natural world. His accompanying introversion, apparent from his earliest years, became part of the affect that would characterize his life and enable his remarkable ability to maintain his focus on a long-term agenda.

As a boy, Gordon had two close friends: Frank Huglin, a little older, and Ron Duarte, a little younger. With them he spent time "building paths through the bushes and making little forts." Older brother Walt was "mild-mannered, easygoing, a homebody," while younger brother Fran was a more entrepreneurial type. The three brothers became close later in life, but in childhood Gordon did not count them as peers. "They were closer to each other than I was to either. We got along fine, but didn't have much in common."

Small-town life was familiar and safe. Minor adventures stood out: "I used to take my BB gun out and hunt birds. There was an old Italian shoemaker who would buy our dead birds. We received a nickel for a robin and a penny or two for a sparrow. He would cook them, after paying us to shoot them." A generation later, writer Rob Tillitz described the unchanging nature of local life: "Years sometimes went by that I did not leave the Pescadero Creek watershed. No reason to. For many it was that way. Grow your vegetables, buy what you need at Williamson's Country Store and that was that. The grammar school showed movies on Friday nights—old ones!

There was an occasional dance. We did most of our playing in Pescadero Creek. Summers, we'd go to the beach. We would go fishing around Pigeon Point. I remember hearing the foghorn, six miles away."

For Gordon, talent and ambition were as yet deeply buried. Instead, childhood days were filled with practical activity and bounded by secure routines. It was only later, in the light of experience, that he realized that "in Pescadero, there wasn't a heck of a lot to do."

Walter: The Local Moore's Law

Gordon's father was reelected twice as part-time constable, demonstrating "the confidence and esteem in which he was held by the people of Pescadero." "That job was my earliest memory of him," says Gordon. "He was well known and respected in a very small community." In 1933, a decade after his first election, a revised charter form of government made his position permanent, as part of the staff of the county sheriff. Walter became a full-time employee, with a monthly salary raised to $165. As an undersheriff, he now worked directly for James McGrath, the sheriff of San Mateo County.

In an era of furtive drinking and big-time gambling, the county had become a place "where if you couldn't get away with it in San Francisco, you could get away with it 'down in the country.'" McGrath was an old-style politician, attuned to the corruption and hypocrisy that accompanied Prohibition, and adept at survival. Close contacts included Emilio Georgetti, owner of the Hollywood Turf Club, in which McGrath later invested, and Horace Amphlett, publisher of the *San Mateo Times*. Despite having no background in law enforcement, McGrath rose to the top job. "There was lots of opportunity for aggressive law enforcement, but this was not McGrath's style."

For most of the 1930s, Walter Moore was on his own. He was the only deputy on the Pacific side of the coast range in San Mateo County; from the Santa Cruz vicinity to the edge of San Francisco, an eighty-mile-long region, its width stretching from ocean to ridgetop, he was *the law*. The county itself was dubbed "the most lawless" in California, and he was completely outnumbered by those engaged in illegal activities. "That he succeeded as a lawman through the boozy 1920s and the gambling 1930s and 1940s is a tribute to his dedication and honesty." The end of Prohibition somewhat eased the pressures, but working for a corrupt boss was never easy, a fact he concealed from his sons.

Walter was a natural for his job: physically strong, accustomed to outdoor life, at ease with horses and guns, and—from his time in France—au fait with life's harshest realities. He had relatives and connections throughout the region, and he well understood the topography of the hills and valleys. He had the Moore family's strong sense of civic duty and a practical, unemotional temperament, not given to panic. Walter Harold did his work without fuss, bragging, or complaint. Like his own father, he stood just shy of six foot, weighing between 180 and 200 pounds. Gordon saw him as an authority figure: "He only had to holler at us, and he had our attention in a hurry." Robert Naughten remembered him as "imposing" in uniform. "He definitely maintained order," says Ron Duarte, whose mother was raised with the Williamsons and Moores.

Ken Moore, Gordon's older son, who knew Walter well toward the end of his life, found him very practical, "a salt-of-the-earth guy, hands-on and mechanical," a man who occasionally looked after his young grandsons and enjoyed building things for them. Walter smoked a pipe, but more often chain-smoked or chomped on inexpensive cigars. Gordon inherited his father's acceptance of prevailing realities and his practical aptitude, not least the ability to remain calm in a crisis and to suppress emotions. "Lying awake at night worrying doesn't help much," he says.

Gordon's father maintained order in the everyday world, operating quietly and shrewdly in a chaotic, shifting environment. Gordon himself would aspire to and thrive in a culture of order, discipline, and control at Intel, while navigating an ever-changing industry on the equally lawless electronic frontier. In business he would take pains to avoid direct conflict, using others—notably Andy Grove—to handle confrontations for him. In family life, he would display a similar stance, depending on his vivacious, outspoken wife, Betty. Grove would amplify, simplify, communicate, and implement his technological insights and agendas, while Betty handled the conflicts and politics of family life in a noticeably autonomous manner. Gordon made wise choices in enlisting and co-opting these two very willing people, even as they in turn learned how to facilitate his agenda. In very different ways, they would become his most important partners.

School Days

In the fall of 1935, Gordon Moore started school. Pescadero had eight grades with three teachers, each overseeing up to thirty students in mixed-grade rooms. Not much had changed since his parents' time: "There were half a dozen boys of my age and half a dozen girls, and everybody knew everybody." Walter Harold had dropped out in seventh grade, while Mira had actually finished high school. Only two members of the Moore clan had any real academic grounding—Gordon's double cousin Harriet (daughter of Louise Williamson) and his father's sister Ione (who had sufficient training to gain a technology-related job in the Bell system, in San Francisco).

Certainly, Gordon's forebears had overseen the decision to create buildings for primary and religious education, and, pragmatically, Walter Harold kept an eye on his sons' early schooling. But the simple truth was that male members of the Moore family, with their outdoor orientation, saw small value in education. Practical pursuits, even a senior position in the sheriff's office, required little formal knowledge. The skills and orientation to hunt or to run a ranch or general store successfully were not academic in nature. As Betty later saw, "His family didn't have an academic dimension. Libraries didn't mean much to them. They loved TV!"

In the Moore household, daily conversation revolved around mundane, tangible realities, with a focus on goings-on at the store. Josiah, the self-made shopkeeper, who lived in the home and ate his meals with the family, could always be counted on to counsel thrift, patience, and hard work. Gordon fitted right in. He was a quiet child, exploring the outdoors, shooting birds, and fishing in the creek. Yet when he entered school, his self-contained, unsocial manner was seen as remarkable, even for Pescadero.

At the end of first grade, he brought home a report card saying he was to be kept back: the school was concerned about his social development. Detachment and inarticulacy—introversion—were the issues. Walter Harold, not usually one to make a fuss, raised a ruckus. "He went to the elderly teacher and had me promoted. He didn't want his kid spending another year playing in the sandbox. I'd have hated to be a year behind, though I would have been a better football player in high school."

Allowed into second grade alongside his buddy Frank Huglin, Gordon responded well to a new young teacher and began to pay attention to his schoolwork. Ron Duarte, a year behind, testifies that "Gordon was always a tremendous student. He was reserved; he didn't have a lot to say, but enough. He and Frank, another smart guy, started a year before I did. By lunchtime they were already in the second grade!" In junior high school, Gordon would again suffer the stigma of his detachment problem. His parents completely failed to intervene on this second occasion. Education was not that high a priority.

Outside school, life followed a predictable routine. Gordon would join his father and older brother to go hunting. With other kids, he played softball alongside the store. "Even the adults played in the evening," says Duarte. At home life was quiet. The family—except for his father—rarely used the telephone, which required cranking to operate. "I wouldn't have been calling anybody out of town," says Gordon. "I didn't know anybody out of town." If he were going to push forward to new frontiers, the drive would have to come from within, as it had with Eli and Gordon's great-grandfather Alex and grandfather Josiah.

Looking Back, Looking Forward

Two people who had early influences on Gordon were his aunt Louise and Harriet, her daughter. These women had followed wholly different paths: Louise—at the center of Pescadero community life—was a "sticker," while Harriet, fifteen years older than Gordon and the only family member to pursue a higher education, was a "mover." "Harriet stimulated my interest in the University of California," says Gordon. "She graduated from Berkeley in the mid-1930s and became an English teacher." Harriet's trajectory, via public university to a schoolteaching career, was modest—another example of the family's practical bent—but it demonstrated to Gordon that a life beyond the coastal valleys, and a focus on ideas, were possible: "Later I got into the Cal group because of her influence."

With the exception of Louise, busy in the family store, all of Walter Harold's sisters—Ida, Ione, and Bernice—would eventually move to San Francisco, as Pescadero declined and they married. Ida and her husband became foster parents, and both Ione and Bernice married men from the San Francisco Police Department. "There's a strong law-and-order streak in the family. It's amazing I am ever chosen for a jury, with all the relatives I've had in law enforcement," quips Gordon.

Paternal grandmother Frances also eventually left Pescadero to live with daughter Bernice in the metropolis. As a child, Gordon was taken once or twice a year to visit her—"a hazy figure"—and his San Francisco aunts. Occasionally, the trip was augmented with a visit to the zoo. "It was an all-day trip up there and back. We didn't go often. Sometimes, we'd get together for a picnic in the summer. We'd have a modest turnout; it wasn't a big event." When his grandmother Frances died in 1937 of coronary thrombosis, still in her seventies, Gordon scarcely registered the loss.

Family links within Pescadero were renewed most wholeheartedly at golden wedding celebrations. Alex and Adaline Moore had set the tone with a lavish party in 1897; their daughter, Ida Jane, reached the milestone with her husband, Charles Steele, in 1930, the year after Gordon's birth. Great-uncle Bill and his wife, Hattie, celebrated three years later. These were the last hurrahs of the pioneer age. In their infancy, Gordon's great-uncles Bill and Joe Moore had been among Pescadero's first white settlers, directly connecting Gordon and his family with the pioneer generation. Both men lived well into their eighties; both died in 1936.

A far greater loss was that of his maternal grandfather, Josiah Williamson, who was eighty-four when he died. He had run his store in Pescadero for exactly fifty years. The *San Mateo Times* called him a "Yankee-born California pioneer, merchant, banker and public figure in San Mateo County since 1869." Of all Gordon's forebears, it was with Josiah that he had the closest relationship. Josiah had launched out in late adolescence with great courage and enterprise, leaving his home and family on the East Coast to restart life on the other side of the United States, where he used his drive and acumen to begin a venture of his own. Against local competition, he proved resourceful and adaptable, adjusting strategies to suit the times while also looking out for those less fortunate. Josiah possessed a dry humor and an ambition that pushed for change and improvement in multiple arenas. For Gordon, he exemplified what it meant to be a successful, self-sufficient, and respected man of business.

On Josiah's death, his children Frank and Mira—Gordon's mother—inherited equal halves of the store. Frank remained in charge. "My uncle ran the store," remembers Gordon. "He would spend his waking hours there. The store was all he was concerned about." His wife, Louise, was still "very outgoing and knew everything happening in town" and was "at the store a good portion of the time, talking to everyone who came in." Mira had little interest in the business. She found fulfillment as wife, mother, and homemaker. What she shared with her brother and her sister-in-law was a settled approach to life. Charles Jones described people who have lived in Pescadero all their lives in his book *A Separate Place* (1974). "They don't need change for the sake of change. That is not their character. They are not for or against change; they are simply indifferent to it." Gordon's own older brother, Walt Jr., displayed just such an approach, choosing to pass his whole adult life in Pescadero after a brief hiatus in World War II.

From the mid-1930s onward, the town became further isolated. The extension of Route 56—California Highway 1—rerouted traffic, leaving the shrunken settlement two miles off the beaten track. "The main coast road used to go right through the middle of town, and then they rebuilt it out on the shore," explains Gordon. Pescadero thus became one of the few places in coastal California that escaped mid- and late-twentieth-century development almost entirely. "The weather is not very attractive, and it's a long commute. It's the only town I know of in California that's smaller now than it was sixty years ago."

Already by 1939, Pescadero was reverting to a backwater, home to storekeepers, lumbermen, hunters, ranchers, and Portuguese and Japanese vegetable farmers. It was also the domain of Eli Moore's descendants, several of whom still remained as stalwarts of the town. The effects of what Gordon's family created were manifest in other ways. The 3.6-mile Honsinger Creek, named for his maternal relatives, is a tributary of the main stream where Gordon used to fish, and the numerous Moore tombstones in the Pescadero graveyard and the fine house that Alex Moore built remain token enough of family claims.

Walter Harold, proud and secure as one of the Moores of Pescadero, slowly made his own way forward. Despite his fractured adolescence, he matured into a family man of settled existence, not a pioneer or entrepreneur, but rather a survivor. As a full-time employee of the sheriff, in 1939 he made the momentous decision to accept further promotion and leave declining Pescadero for the "big-time" life of nearby but far different Redwood City and the sheriff's own office. Though there was no way to see it at the time, Redwood City and Bayside life would serve to push Gordon to his future, for, unlike brother, father, mother, grandparents, aunt, and uncle, he would never be content to settle down to the class-based confines and unchanging existence of a remote enclave.

As he headed west, Eli Moore's actions propelled his family into the launch of California's unfolding drama, where in due course his descendants became an enduring presence. Eli possessed the discernment to play a significant role on the frontier of an expanding nation. Gordon Moore, Walter and Mira's second son, would also be a man of parts, possessing a determination that led him to fresh frontiers. Both a "sticker" and a "mover," securely rooted yet restless to conquer unknown territories, he would go on to mix constancy and certainty with revolutionary change, as he pursued mastery of the electronic frontier. Indeed, Gordon's eventual life as the silicon revolution's most important thinker and critical entrepreneur echoed the pioneer creed: "Do not follow where the paths may lead, but rather lead where there is no path, and leave a trail for others."

2

THE CHEMISTRY OF ROMANCE

OUT OF THE BACK WOODS

Worlds Aborning

When Eli Moore arrived in California in 1847, there were comfortably fewer than five hundred white settlers in the whole San Francisco Bay Area, including the Pacific coastal lands down to and through the Santa Cruz Mountains. Fifty years later, Santa Cruz alone had twenty thousand residents, and San Francisco—thanks to the Gold Rush—had surged to more than three hundred thousand. Remote Pescadero—at what would turn out to be its apogee—boasted around a thousand inhabitants, secure in their preindustrial world.

The twentieth century would see Pescadero dwindle, as improving roads, reliable automobiles, and fresh forms of communication such as the telephone and the radio put California itself ever more firmly on the map. The Southern Pacific, developed under the direction of Leland Stanford, now provided railroad links both to the east and along the Bay from San Francisco to San Jose.

Following the death of his son and heir, Leland Stanford gave a massive gift of land and money to found Stanford University "for the children of California." Stanford opened its doors in 1891. Initially, it was seen as little more than a playground for the rich. Throop University in Pasadena, in Southern California—the forerunner to Caltech, where Gordon Moore would gain his PhD—was established the same year. By 1917, under the leadership of astronomer George Ellery Hale, Throop had a department of aeronautics and its own wind tunnel; four years later, with the arrival and help of chemist Arthur Noyes and physicist Robert Millikan, Hale renamed Throop the California Institute of Technology and launched it as a center of scientific excellence. Stanford University and the University of California at Berkeley were also changing, if more slowly, from local to national centers.

Pescadero was but a stone's throw from leading examples of California's changing realities. A score of miles east-northeast lay the newly launched Stanford University, in Palo Alto. A half-dozen miles to its north, along the rich soils of the Bayside flatlands, Redwood City had become a thriving manufacturing and shipping center. North again was San Francisco, where finance, communications, and military industries were developing. To the south of Palo Alto lay the Santa Clara Valley, stretching for some twenty miles to San Jose and beyond. Renamed "the Valley of Heart's Delight" by hopeful boosters of its benign climate, bountiful orchards, and abundant blossoms, the area had become a major fruit production region. Growth was everywhere, except the far mountain valleys of the Pescadero coast, though populations were far less than they would later become. Palo Alto in 1920 had five thousand residents, and Redwood City's population was somewhat smaller. Los Gatos, nestled in the Santa Clara Valley, reported fewer than five hundred people. But growth, development, and in-migration were very much in evidence.

By 1940 San Francisco's population was more than six hundred thousand, and rural San Mateo County had grown to one hundred thousand. The automobile, or "horseless carriage," surely helped this cascading growth, but in Pescadero's particular case it served rather as a means of egress. At last there was an easy way to negotiate the daunting mountainous valleys to the metropolis of San Francisco or to the much more agreeable climate and more modern life that characterized the towns on the Bayside plain.

Electric Magic

Fateful in its implications for the nation and the world was the creation of electric power. Legendary inventor-entrepreneurs such as Thomas Edison, with his General Electric, and George Westinghouse, with his eponymous Westinghouse Electric, spurred system building in the production, distribution, and use of electricity. The earlier industrial age morphed into an electrically lit and driven era over the four decades to 1929. Soon long-distance electrical power was being transmitted to Bay Area towns from dams and hydroelectric plants on the edge of the Sierra Nevada. Stanford University, recognizing the potential of electricity, taught the subject of electrical engineering from its earliest days. The Bay Area's power companies used the Stanford High Voltage Laboratory to study the challenges of long-distance transmission. That cooperation fostered a generation of engineers who could match the know-how of the East Coast.

Alongside electrical power, a novel technology that was reliant on it took hold in the Bay Area: "wireless" electronic communications. Using electric power, information could now be communicated through the air. Wireless telegraph messages could be sent from shore to ships and back, much to the delight of the US Navy fleet centered on the San Francisco Bay. Voice transmissions soon followed. Beyond two-way exchange of information came one-way mass communication by radio broadcasting. Voice and music pulsed across the region from some of the earliest radio stations. Key to both forms of wireless was a singular invention that defined the new world of "electronics"—the vacuum tube. Akin to complex lightbulbs, vacuum tubes could produce electrical signals by acting an on-off switches and could also amplify these signals. "Tubes" were the key to transmitting and receiving information in the radio world.

Tubes were power hungry. They worked by producing an electrical current from a heated delicate filament and controlling its flow. They were as troublesome as they proved indispensable. Reducing their thirst for electric power and mitigating their frailty became important as uses proliferated. With tubes and radio introduced in the Bay Area, fresh phenomena arose: a fledgling electronics industry and a community of radio amateurs. The Bay Counties Wireless Telegraph Association formed in 1907. As teenagers Charles Litton, a child prodigy, and Fred Terman, a precocious, brilliant "faculty brat" from Stanford, were key figures in setting the stage for what would ultimately become Silicon Valley, as they ran their own amateur radio sets, sending and receiving messages—Litton in Redwood City and Terman in Palo Alto.

At ten Litton had built a radio set, and he was soon blowing glass to make vacuum tubes. His home in Redwood City grew two one-hundred-foot antennae towers, with which he managed to establish communication with radio stations in Australia and New Zealand. Meanwhile, Terman would head to the Massachusetts Institute of Technology (MIT), where he saw the importance of building links from science and engineering to industry. On his return to Stanford in the mid-1920s, as a credentialed electrical engineer, he sought to sharpen the university's focus by making the rich boys' playground into a significant center for the field of radio and communications electronics.

The Bay Area made other early steps in the production and use of electronics. Stanford student Charles Herrold founded the West's first, and the world's second, radio station, in San Jose in 1909. Another alumnus, Cyril Elwell, started the Poulsen Wireless Telephone and Telegraph Company in Palo Alto, working from a cottage on the corner of Channing Avenue and Emerson Street. Renamed the Federal Telegraph Corporation, his company began to rival East Coast corporations. One of his employees, Lee de Forest, made the accidental discovery that if he fed the outgoing current from a vacuum tube back into the input—the "feedback principle"—the signal swelled to such an extent that "if one dropped a handkerchief a few inches from a telephone transmitter, there was a loud thud in the earphones."

The impact of this finding was consequential and rapid. The vacuum tube's newfound ability to amplify enabled the creation of devices that made long-distance telephone and "wireless" communications much more satisfactory. As a result, de Forest's work helped usher in the age of electronics, the name increasingly given to vacuum tube–dependent technologies such as telephone transmission, telegraphy, and radio.

A couple of blocks away from de Forest, Gordon Moore's father-to-be, Walter Harold, still in his teens, was working equally hard in Palo Alto's Forest Avenue lumberyard. Arriving regularly with logs from over the hills, on three-day round-trips, he was part of an older world. Palo Alto also became home to three-year-old William Shockley, who in 1913 moved from England to 959 Waverley Street, as his parents sought the comfort of the area's benign climate. A little more than forty years later, Shockley would return from the East Coast to Palo Alto and to his mother. By then he was the inventor of the junction transistor, a rival to the vacuum tube as the basic building block of electronics.

One example of the fresh possibilities flowing from vacuum tubes was how American Telephone and Telegraph (AT&T), in New York, used them to amplify the first coast-to-coast telephone signals in 1915. During World War I, the US military made prodigious use of tubes for electronic communications. Afterward, civilian uses of these developments came to the fore.

Firms on the far more populous East Coast came to predominate in most electronics technologies. RCA, the Radio Corporation of America, in New Jersey, became the dominant player in the as yet unnamed field of consumer electronics. On the West Coast, "Federal was doomed and other Bay Area–based radio companies survived only on military applications." If Federal was doomed, it had been triumphant in making the vacuum tube the one essential ingredient in ever-multiplying telephone connections, in this new world of radio, and soon, in the novel technology of television.

US PRODUCTION OF VACUUM TUBES, 1921–1939

Year	Units
1921	1,000,000
1923	4,500,000
1924	12,000,000
1925	20,000,000
1926	30,000,000
1927	41,200,000
1928	50,200,000
1929	69,000,000
1939	98,500,000

Source: Electronics Industries Association Electronic Yearbooks and Market Data Books.

Electronics Enters Ordinary Life

Vacuum tube production in the United States mushroomed from the 1920s to the 1940s—from 1 million to nearly 100 million per year. Telephones steadily spread throughout the nation, thanks to the vacuum tube's improvement by the Bell Telephone Laboratories—inaugurated in 1925 as the R&D arm of mighty "Ma Bell" (AT&T). By 1940 nearly 40 percent of US homes had telephone service. Radio took the nation by storm. By that same year, almost three-quarters of US households had a radio, up from only five thousand homes in 1920. Radio provided a new layer of reality: the mass communication of information by electronics. Music, comedy, adventure series, and national news emerged miraculously from a box in one's living room. The Radio Manufacturers' Association was founded in 1926, becoming the Electronics Industry Association in 1957 (today's Consumer Electronics Association, with an annual trade show attended by 150,000 purveyors of electronic reality).

Most Americans had little concern for the interplay of physics, chemistry, business, and military concerns that lay behind vacuum tube electronics. What they experienced was the novelty of affordable radio "receivers" that connected them, as ordinary citizens, with a new realm of electronic information that stretched far beyond their individual farms, houses, towns, and cities. For the Moore family, isolated in Pescadero, listening to the radio became a part of daily life. *Little Orphan Annie* was a fifteen-minute show that went national in 1931, with two separate casts: one in San Francisco and the other in Chicago. *Jack Armstrong, the All-American Boy* was an adventure series sponsored by Wheaties cereal. Walter Winchell was a controversial newspaper columnist read by millions every day. Gordon's whole family would sit together to hear his radio show on Sunday evenings. Gordon remembers, "There was no daily newspaper, so the radio was the principal source of news. Walter Winchell was an important part of the whole culture."

Like other children, Gordon sent away for "secret decoder rings" featured in national radio advertising campaigns: "I would drink my Ovaltine and eat my Wheaties." The technical feats behind them—radio and tubes—did not capture his imagination. In Pescadero he lived in an older, more bounded world. "I fiddled around and built a crystal set, but I never was into that very much." Even when Gordon entered high school and his younger peers—his brother Fran and schoolmate Ray Dolby—took vacation work in electronics companies, Gordon spent his high school summers in manual labor. One whole summer was spent working for his uncle Frank in the store, sweeping the floor, washing the windows, and stocking the

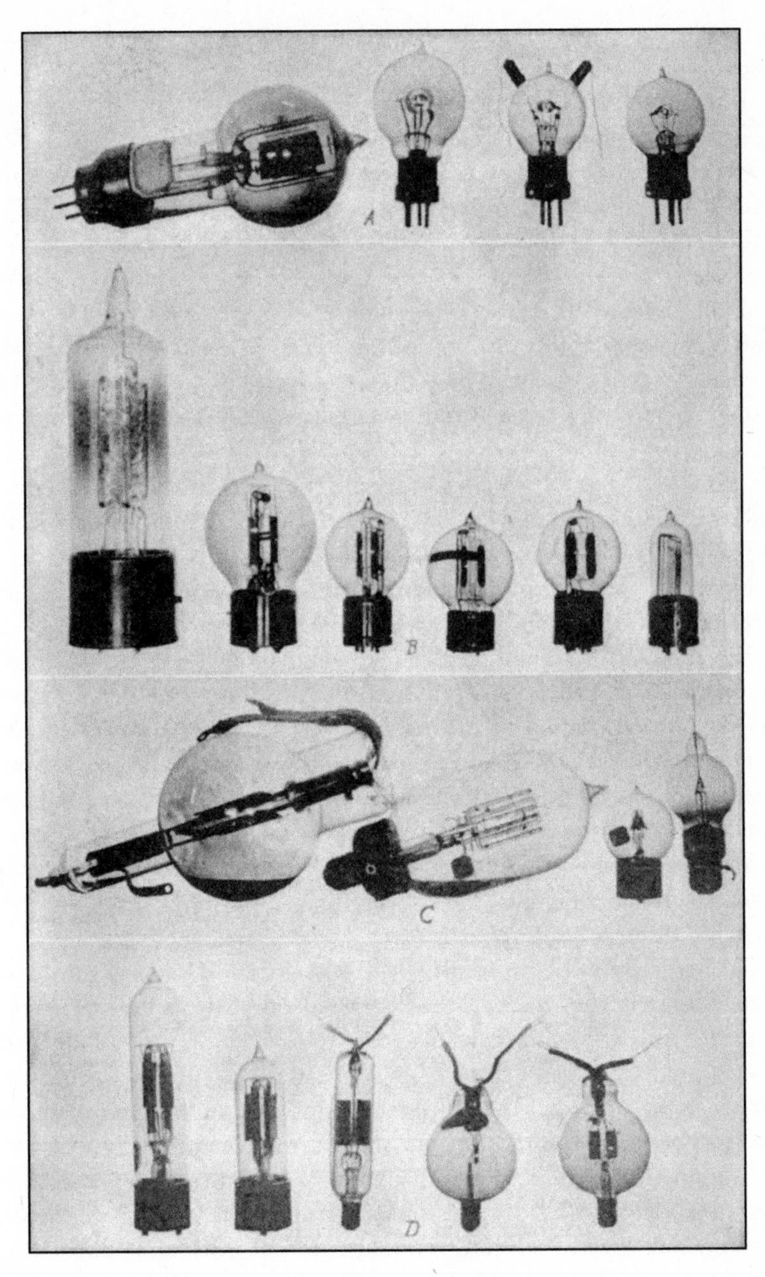

Early vacuum tubes.

SOURCE: US ARMY/WIKIMEDIA COMMONS.

shelves. In subsequent years he hauled sacks of shells at Redwood City's cement plant, before chemistry became an option.

Through the 1930s on the San Francisco Peninsula, a small community was devoted to improving vacuum tube electronics. Stanford physicist William Hansen worked closely with brothers Russell and Sigurd Varian—a physicist and a pilot, respectively—on a powerful vacuum tube called the klystron. Capable of generating "microwaves," like radio waves but able to pass through the air with greater ease, klystrons proved vital in World War II, in radar and atom smashers. The Moores, secure in their Pescadero bungalow, knew nothing of such possibilities. Their interests lay closer at hand or—when listening to the radio—in news of the conflicts in Europe. Like most Americans, they viewed European developments from a comfortable distance. They had little wish to risk lives and livelihoods for foreign causes. The actual outbreak of war in Europe did not alter the popular desire to avoid international entanglements. Instead, public opinion slowly shifted toward limited US aid to the Allies.

Electronic information compressed Americans' experience of time and space, as geographies and populations were more tightly woven by the threads of communication. Markets became national. The radius of trade and exchange expanded. News flowed into and out of centers of power. The demand for vacuum tubes from large organizations allowed manufacturers to achieve economies of scale. Prices dropped dramatically. The radio became the public's primary experience of electronics as a mass medium. Tubes further transformed cultural reality, as they were used in the creation and playback of phonograph records. Studios employed tubes to add audio content to motion pictures. Here again, tubes created a mass culture of shared information and experiences.

For nine-year-old Gordon in late 1938, electronics and war were undiscovered territories. Instead, the big news was that his father had been promoted. The whole family would move to Redwood City, so that Walter Harold could work from Sheriff McGrath's headquarters on an increased salary of $190 a month. The move meant not only a higher salary and the possibility of further promotion, but also a better climate.

Gordon remembers, "My brother and I had bronchitis almost continually on the foggy side of the hill, so my parents wanted to get over here on the sunny side to see if our health would improve." Mira was reluctant, but Walter was firm. Policing the coast was lonely work. While headquarters would bring its own challenges, he had little reason to stay in Pescadero. Both of Walter's parents were now buried, and all but one of his sisters had left. Once again, opportunity called. Redwood City would be "better." It was, for Gordon would fall in love with chemistry in Redwood City, leading him to a life he could never have imagined.

IMPACT OF ELECTRONICS, 1930–1940

	US households with:			
Year	**Telephone**	**Radio**	**Refrigerator**	**Total US electronics sales**
1930	41%	39%	20%	$220,000,000
1940	37%	73%	60%	$340,000,000

Year	**Vacuum tubes produced**	**Value**
1930	40,200,000	$37,900,000
1940	108,000,000	$27,600,000

Sources: Electronics Industries Association Electronic Yearbooks and Market Data Books; US Census Bureau.

New Home, New Life

Having swelled to more than twelve thousand residents by 1939, Redwood City was many times bigger than Pescadero. By 1950 its population would double again, to more than twenty-five thousand. "It was a pretty good size with a real commercial center," remembers Gordon. "There were lots of streets and a much larger school." Father, mother, and the three young boys settled at 196 Westgate Street, just off Alameda De Las Pulgas, in a standard two-bed, one-bath cottage, costing $5,600. Over the years Walter and Mira would do little to change it. It would be home for the rest of their lives.

A major port expansion had occurred two years earlier, and Redwood City was booming. The town's roots went back to an eighteenth-century grant of seventy thousand acres of land between the San Mateo and San Francisquito Creeks, and by the mid-nineteenth century the Gold Rush and the demand for redwood lumber were feeding a rapid increase in population. The location was known for its exceptionally deep channel flowing into San Francisco Bay, a water highway ideal for moving redwood logs onward in their journey from the hills. By 1859 the town already had its own courthouse, which served as a jail, dance hall, meeting room, school overflow, and recruitment headquarters for the Civil War.

In 1918 Redwood City witnessed the launch of the world's first cement-hulled oceangoing vessel, *Faith,* built for fighting German U-boats. The accumulation of clam, oyster, and mussel shells in the channel "represented a reserve of tens of millions of tons of good-grade limestone, and the other ingredients for cement." The interest of the Pacific Portland Cement

The Moore family home in Redwood City, 196 Westgate Street.
SOURCE: GORDON MOORE.

Company was aroused. Patented a century before, Portland cement was formed from chemicals such as calcium carbonate, silicon dioxide, and alumina. The company's new plant, opened in 1924, added an important chemical factory to Redwood City. Expanding onto thirty thousand acres, on both sides of the channel, the cement company became one of the area's principal businesses. Soon, plans were developed to deepen and widen the shipping channel. Four months after the opening of the Golden Gate Bridge in San Francisco in May 1937, Redwood City folk gathered to celebrate their improved port.

Progress of a different sort was unfolding near what would become Gordon's new home. Charles Litton, the onetime child prodigy, had founded Litton Engineering Laboratories at the residence of his parents, a half mile from the site of the Moore family's cottage on Westgate Street. In 1938, in a collaboration with Stanford University on a fresh form of vacuum tube, Litton welcomed a young graduate student, David Packard, into his home-based laboratory. A year later, joined by William Hewlett, Packard founded Hewlett-Packard (HP) to produce electronic instruments and equipment. As Stanford students, both were favorites of Fred Terman, who had taken them to local vacuum tube companies.

The Varian brothers, with Stanford physicist Hansen, developed their microwave tube into an electronic device that could "sweep the sky with an invisible beam, then recapture and project that beam as it echoed off any object." Stanford University provided the brothers with $100 to develop their invention, which would provide the heart of radar systems—vital to aerial warfare. Litton, for his part, went on to help Raytheon develop the magnetron, a tube that further strengthened US efforts in World War II by increasing the range of radar. Throughout the 1940s Litton Industries would grow to rival the large companies of the East Coast and help secure the technological and industrial foundations for the electronic revolution that would transform the San Francisco Peninsula in succeeding decades.

Walter Harold, at forty-six, now earning $2,280 a year as a deputy sheriff, worked forty-four hours a week. His neighbors had rank-and-file jobs: a project clerk in construction, a hairdresser in a beauty salon, a certified public accountant, a private secretary, a freight clerk, a carpenter for a building contractor, a tire salesman, a landscape gardener, a third mate on a tanker ship, a plumber, and an insurance agent. Gordon's father, a child of the old middle class, was now a white-collar worker. In time his sons would hark back to "old middle-class" occupations: Gordon would start his own company, and his brother Fran, like his grandfather Josiah, would own a small business. Walter Harold, marked by a fractured adolescence and war, preferred the security of America's newer class of salaried employees. His eldest son, Walt Jr., would also work as an employee, running his father-in-law's ranch.

If Walter Harold appreciated his increased status and security, his wife was less happy. Relatives, including her brother, remained in Pescadero. Walter had his social network at the office, but Mira—a housewife coping with three sons—was cut off from familiar family circles and found it hard to adjust. Gordon recalls the move as "traumatic" for his mother, despite frequent trips back to Pescadero on weekends to soften the blow. For Gordon himself, the change was easy. Not one for introspection, he focused on fresh activities, while enjoying his returns to Pescadero. "You couldn't do outdoor activities much, not in Redwood City. There was no fishing nearby, and I was through shooting birds with my BB gun. There was nobody to buy the birds anyway. Instead, I'd head over to the Pescadero area and go hunting. I couldn't do those things in Redwood City."

A handful of times that first year, Ron Duarte, his Pescadero friend, made the journey to stay with the Moores and play with Gordon. Duarte recalls, "We'd do things like come down a street with both of us in a wagon; it was a paved street with a little slope on it. There weren't many cars on the road, luckily, but his father—holy criminies—we'd catch hell over that." For Gordon, outdoor activity continued to appeal. "We did the things kids used to do at that age. I rode a bike all over the place. I couldn't get around the corner without taking my bike. We played baseball in the middle of the street, the Alameda De Las Pulgas, and would have to get out of the way when the occasional car came by."

The main drawback to the move was the school: McKinley, on Duane Street. Gordon was placed in fifth grade, where teachers once again mistook his quietness for developmental impairment. Soon he was "stuck in a remedial speech class, the real bottom one with the first and second grade kids." In a fresh, unfamiliar setting, a certain level of caution on his part was only natural, yet his self-containment, reserve, and avoidant behavior once again became a concern rather than being interpreted as a sign of high intelligence. Unused to conversational discourse and still a child, Gordon had no way to explain. Seven decades on, the matter still irks him. "I was quiet, reasonably studious, maybe a baby in the class, but I don't know what I did to convince them I couldn't talk. I never felt I was completely inarticulate, but I was put in this class where I didn't belong. I didn't believe there was any problem or issue with my quietness. I didn't understand it at the time, and looking back, I can't figure out why the heck I was stuck in that class. I was really out of place."

Lacking parental intervention, Gordon continued in the class for two years. It was a distressing escalation of his first grade experience in Pescadero, where but for his father he might have remained in the sandbox. By now, however, it was not the simple matter of failing a test, which might be retaken; this was a deeper issue, relating to language, expression, and interaction. Games, math, school-yard scuffles, and other nonverbal activities did not faze him, but it seems the experiences of family life had reinforced or even helped create his limited capacity for and ease with interpersonal communication. Gordon was the opposite of a gregarious, people-pleasing middle child: instead, he was a boy with exceptional concentration and focus, oriented not toward words and emotional engagement, but toward practical results—with or without companions.

The Romance of Chemistry

Things were about to get better. In January 1940 Gordon Moore turned eleven and discovered his first true love, one that would determine the course of his life. He might be quiet in a school setting, but he did interact with his next-door neighbor on Westgate Street. Donald Blum, born a few months before him, was the only child of a traveling confectionery salesman, Edwin Blum, and his English wife, Cecilia, who worked in a department store. Donald executed the fateful move, introducing Gordon to his future: "He was given a chemistry set for Christmas. I started playing with the set with him. In those days you used to have some really nice chemicals in these sets, like potassium chlorate, with which you could get fascinating results. Explosives were what caught my attention. I became interested in a variety of the experiments. I decided, 'Gee, I want to be a chemist.'"

Today, potassium chlorate is a favorite of terrorists and suicide bombers. Gordon describes his own experiences with it, creating colorful flames, fuel for toy rockets, and small explosions, as "different from anything I'd done before." He was hooked. Flashes, bangs, and transformations: here were both excitement and control, a source of pleasing intellectual puzzles and a satisfying means of releasing inner emotions and tensions. By following the instructions of a chemistry set, one could gain astonishing capabilities. For years he had fished in the creek; he was well versed in the conditions for success in the natural world. The chemistry set, with its transmutations and smells, provided a far more dramatic take on nature, at once similar yet remarkably different. It offered startling change that could be domesticated into reproducible effects amenable to cataloging and study.

These realities met a deep need in young Gordon: "It really got my attention." He explains, "Hands-on is important." It allows the experimenter "to experience the excitement." He had found his vocation. Diving in, Gordon Moore focused in a manner that would later characterize his response to the transistor. He built a small laboratory in one corner of the shed attached to the back of his family's garage and bought his own reagents. "I started collecting a bunch of stuff, sending off to the supply houses for beakers, flasks, and miscellaneous chemicals."

The Blums soon moved to 531 Alameda De Las Pulgas, a short walk away: "I still had interaction with Donald, but not as close as when he was my next-door neighbor." The fresh neighbors at 190 Westgate Street were Fred and Anna Linsteadt, who had previously lived in Cook, Illinois, where their elder son, another Donald, was born in April 1929. Fred was, like Walter Moore, a war veteran, and he worked as an automobile mechanic. The couple's two sons, at eleven and six, were exact contemporaries of Gordon and his younger brother, Fran. Don Linsteadt, also a student at McKinley, became a useful partner in crime. "He was interested in my chemistry experiments. Donald liked to blow things up, too. So we did that together."

Gordon might be young, but he was utterly serious about his laboratory. Symbolizing that this was no passing fancy, he poured concrete to create a solid floor to the shed. He embraced a work habit of thinking and experimenting alone, in his own space. "There was continuity from the chemistry set to my much more complete laboratory: it was exponential growth in very small increments." Walter and Mira did nothing to veto their son's hobby. "My parents didn't know exactly what I was doing, but I'm sure they knew I was making things that exploded. I don't know how they could have missed it. They were not discouraging; it was just, 'Be careful.'" Contemporaries at school were also experimenting. "Another friend blew a couple of fingers off—not in my laboratory, fortunately. I don't know if my mother knew about it."

Ron Duarte remembers Gordon "tinkering with some experiments in the garage" and causing the Redwood City firemen such concern that "the fire department had to talk to his father. 'Don't let him do any more of this stuff.'" Gordon's own words, characteristically, are much more low-key: "I had minor mishaps. For example, when you get through making nitroglycerine, you separate it off from the acids. You obtain a big beaker full of whatever is left behind. If you don't take care, it suddenly starts a very vigorous reaction, with brown fumes of nitrogen dioxide pulling off it. I had that happen a couple of times. It made a heck of a mess." Once he came close to serious injury: "I did burn myself relatively severely. When you mix potassium permanganate and aluminum powder, though the mixture doesn't ignite easily, it eventually goes 'poof!' I thought we had a bad batch, so I lit it with a match. I scorched my whole hand. I fried it. That was the closest thing to a real accident—the most I ever hurt myself."

Moving through adolescence, Gordon began to haunt the secondhand bookstores of Palo Alto in search of chemistry manuals. In part, his passion was fueled by the notion that he could control what was happening. If he carefully approached and appropriately studied his subject, he could catch the tiger by the tail. The name of the game was focus, understanding, and consistency. Responding to an inner drive, Gordon in his teenage years found a subject that simultaneously offered voyages into abstract thought and the most obvious, immediate, and astonishing of practical results. Chemistry was a game he could and would play—often as a solitary hand, but also as a member, then leader, of a team.

Athletics and War

Unlike Pescadero, Redwood City offered the able player an opportunity to join in many other activities. In the fall of 1942, thirteen-year-old Gordon entered Sequoia High School, named for the local redwoods. Sequoia High already had a rich history. First organized at the end of the nineteenth century, it gained its own forty-acre campus in 1920. When Gordon became a student, Sequoia was well established. In addition to the main curriculum, subjects included drama, art, journalism, homemaking, shop, business courses, and gardening. Sequoia even had its own theater. It was an "excellent school with good equipment, a beautiful and functional campus, a trained and devoted faculty, and an excellent curriculum, which became nationally known." Gordon's overwhelming impression was that the school was "pretty big." He was part of a class of four hundred, which included many from McKinley, though not Don Linsteadt or Don Blum.

At Sequoia Gordon had a fresh start, no longer among those with language difficulties: "I did pretty well in English, with the grammar. I could diagram sentences better than anybody." Even so, his analytical orientation, methodical approach, and preference for the patterns and rules of mathematics and chemistry made him a slow study. His family was not one to read or discuss literature, and Gordon was not equipped to decode the emotional messages of books and plays. He could not diagram a character's motivations, or the vagaries of human nature, quite the way he could a sentence. "I was a slow reader, and still am. I'd seldom finish the books they assigned in English class. I wasn't that big on literature. Someday I'm going to go back and read *A Tale of Two Cities*!" As he matured, Gordon would develop a Skinnerian approach to human psychology, an understanding of behavior based on cause and consequence, with a preference for studying observable behavior rather than internal mental events. At his main company, Intel, he would be a devotee of stock options and performance-based pay, measurable motivations for staff. In the fullness of time, the Gordon and Betty Moore Foundation (GBMF) would seek the elusive goal of "measurable results" in its philanthropic work.

Sequoia High was outstanding not least for its provisions for physical education: two large gymnasiums, a swimming pool, athletic fields, and tennis courts. A history of the school describes its strong staff of physical education teachers and that, as a member of the Peninsula Athletic League, the Sequoia varsity won six championships, for each of eight consecutive years in Gordon's era. Given his outdoor upbringing, good physical coordination, and quiet but approachable manner, it followed easily that "I was on football every year, and track, and swimming. I was a diver. I did gymnastics. I was out for athletics all year long, too. I was in good shape in those days."

On the football team, the Cherokees, he started out playing halfback: "My brother [Walt Jr.] had been a guard, so the coach decided I ought to be, too. I was really too small, at 160 pounds. Even in those days, most linemen were bigger. I didn't get to play as much as I would have liked, but I dove in." He played three years on varsity as a right guard. "His blocking and tackling have earned him a starting position," reported the *Sequoia Times*. Gordon recalls, "We were champions one year and strong contenders the others. We did pretty well." Later in school, he became captain of the gymnastics team: "I never became truly expert, but I was always near the top of my gym class. We did free-floor tumbling and the horizontal bar and parallel bars. We used trampolines. Sequoia was not competitive with

Gordon, the high school gymnast, mid-1940s.

SOURCE: KEN MOORE.

schools that focused on gymnastics, but we had a lot of fun. It was all about physical agility."

Gordon was also an accomplished diver. Running and swimming came to him less naturally: "People would go in the pool, lay back, and float. I'd sink to the bottom. My body fat was pretty low." By his early adolescence, his competitive drive was well developed. Failing at anything was disappointing. In childhood he had spent many hours standing in the creek, fighting to land a fish or pursuing deer with his father and brothers. Gordon had become a "player," athletic and studious, distinguished by his self-possession, disarming smile, and unusual passion for chemistry. Athletics and chemistry framed his life. In keeping with his reserved nature and Pescadero background, he pursued this repertoire without attention to the wider world. Focusing on, and succeeding in, impersonal, tangible action—fishing, experimenting, sports—gave comfort and satisfaction.

In actuality, the wider world was about to be transformed. In December 1941 the attack on Pearl Harbor jolted America awake. The nation was at war. In February 1942 President Franklin Roosevelt signed an executive order to move 110,000 people of Japanese descent to internment camps: the US Army was soon busy enforcing evictions of "enemy aliens." The effects were dramatic in Pescadero and throughout San Mateo County. Gordon's wife, Betty, who in the 1940s was growing up with her mother and grandparents on a ranch in nearby Los Gatos, remembers, "Japanese were transported from the area, never to come back. In high school we helped save the berries and crops that they would normally have been out there to collect."

There was strong anti-Japanese sentiment in Redwood City. Bob Naughten, a fellow student of Gordon's at Sequoia High, says that many of the people killed at Pearl Harbor came from the Bay Area. "There was a lot of animosity and bad feeling. We were so angry at these people; I'm afraid there may have been a mess if they'd stayed. That's not nice to say, but this was what was going on." Fear of Japan was widespread. Betty recalls being anxious after hearing whispers that a Japanese submarine had "lobbed some explosives" onto the ranch of a relative's property in Big Sur. "The government hushed that up, but it made us think that they were right off our shore." (A Japanese submarine had in fact torpedoed and sunk an oil tanker off the Big Sur coast, two weeks after Pearl Harbor.) She also remembers one particular afternoon at Los Gatos High School in 1943. "Army vehicles were parked outside, and there was word that they might come through the Santa Cruz Gap to get to the valley: somebody had sighted submarines off the coast. There were soldiers and vehicles right in front of our eyes as we were having class. It made an impact; at least it did for me."

For ordinary Americans, the war affected daily realities by making supplies scarce and travel at night impossible during blackouts. Walter Harold, supremely practical, engineered a set of plywood inserts for the windows at 196 Westgate Street, "so we could keep the lights on." In World War I, the family had been ignorant of his daily tribulations in France. Now, with Pearl Harbor putting the Pacific Coast firmly in the spotlight, everyone was drawn into the wartime drama, one way or another. "People learned how to garden," recalls Betty. "The emphasis was on victory gardens, and everything was victory." Gordon remembers "rationing of gasoline, sugar, meat—enough to keep us all involved at some level." At nights on Betty's Los Gatos ranch, attempts were made to steal gasoline allotted to her grandparents for farm use. "People were out there, siphoning gas out of this big Shell tank. We had this great old gun. My grandmother—she was such a kick—would get out there and shoot this thing up in the air. You could hear people scrambling away, because they thought some tyrant was coming after them; they didn't know it was an old woman."

One stark reality was the loss of close friends in the conflict itself. "Some enlisted. Those who graduated and were eighteen were called up," says Betty. "I lost a couple of my friends. They were killed immediately. It was a frightening time." Nearby, Dibble General Hospital (now the site of the Stanford Research Institute and Menlo Park Civic Center) cared for soldiers wounded during Pacific operations, specializing in plastic surgery, blind care, neuropsychiatry, and orthopedics. "We would lose friends, brothers, only a year or so older than us," recalls classmate Naughten. "People wore uniforms to graduate from high school and then disappeared into the conflict."

When Gordon's older brother, Walt Jr., graduated in 1943 and was drafted, war once again came home to the Moore family. "I'm sure my father didn't discourage him," says Gordon. "He was a young man, thinking of the future. He went in as a private, came out as a sergeant." Walt was in the ordnance division in time to follow the June 1944 invasion to France and worked as a mechanic, repairing trucks. The family received letters periodically. "A lot were censored in those days. You worry." One immediate effect of his older brother's absence was that Gordon took possession of his car. "That was a great deal for me. I was in high school and suddenly had a vehicle. Trouble was, there was no gasoline. Four gallons a week didn't go very far."

Gordon's age allowed him to escape the direct impact of hostilities. War did not shape his life as it did that of his older brother and of his father. Walter Harold's brief but intense experience cast a long shadow, one acknowledged obliquely by Gordon when he commented, "The war was a short period in my father's life, so I was really surprised when he wanted to be buried not in Pescadero, but in the veterans' plot much closer to Half Moon Bay." War also changed the life of Walt Jr. Returning from Europe in 1946—already a "secretive" youth who communicated little of how he felt—he struggled to regain his foothold. Gordon remembers, "My brother had a very high IQ, but didn't have the sense of direction that I had. He started at San Jose State thanks to the GI Bill, but soon dropped out to drive a truck." From there, it was only a hop, skip, and a jump to marrying a Pescadero girl and settling down to local life.

Gordon and Walt Jr. were both intensely practical and highly intelligent, yet differences of circumstance and volition led them onto wholly different paths. Walt developed no passion comparable to Gordon's for chemistry. Gordon—unscathed by war, inner directed, and fueled by the desire to experiment and learn—quickly entered San Jose State, moving on

NAME MOORE, Gordon Earle

ADDRESS

COUNSELOR E. WOOLDRIDGE

ADVISER

GROUP 1-1

ACE Psych 1942 [illegible]
Ter. M-A-C- [illegible]
Prog. Math Adv. B [illegible] 90%ile
OTIS [illegible]
NEW STANFORD ACH. — Adv. Bat. [illegible]
HIGH SCHOOL CONTENT
Date of birth: 1-3-29 [illegible]
Place: San Francisco

PHOTO 9-42

MARKS USED
TRANSFER-GREEN MAKE-UP-RED
A. B—COLLEGE RECOMMENDING
C —PASSING
D —BARELY PASSING
E —FAILURE
INC.—INCOMPLETE

YEAR 1942-43

SUBJECT	1 QTR.	2 QTR.	SEM.	3 QTR.	4 QTR.	SEM.	SEM. PER	REC. PER
Wooldridge	C+	C-	C-	C+	C+	C+	20	
Soc. Liv. I	B-	C	C	B-	C+	C+		
H.&G.S. Anderson	B	B+	B+	A	B+	A-	10	
Span. I Hughes	B-	C+	B-	B+	B+	B+	10	
Alg. Kauffman	B-	B	B	C+	B	B-	10	
PHYSICAL EDUCATION [illegible]	B	B	B	C	B	B	OK	
TOTAL UNITS							50	
C. S. F.								
ATTENDANCE ABSENT TARDY								

YEAR 1943-44

SUBJECT	1 QTR.	2 QTR.	SEM.	3 QTR.	4 QTR.	SEM.	SEM. PER	REC. PER
	C+	B+	B	A-	A-	A-	20	
Soc. Liv II Wooldr.	B-	A	A	A-	A-	A-		
Geom. Kauff	A-	C+	B	B	B-	B-	10	
Biology Bigler	B-	B	B	B	B+	B+	10	
German Dusel	C	C+	C+	B-	C+	C+	10	
PHYSICAL EDUCATION Dw	B	A	A	A	A	A	OK	
TOTAL UNITS							50	
C. S. F.								
ATTENDANCE ABSENT TARDY								

YEAR 1944-45

SUBJECT	1 QTR.	2 QTR.	SEM.	3 QTR.	4 QTR.	SEM.	SEM. PER	REC. PER
	B+	B+	B+	B+	B	B	10	
ENG III WOOLDRIDGE	B+	A-	A-	A	A	A		
US HIST BLASE	B	B+	B+	B	B+	B+	10	
GERM II DUSEL	B-	B+	B+	B-	B-	B-	10	
MATH III KAUFFMAN	A	A-	A-	A-	A-	A-	10	
CHEM KYLE	A-	A	A	A	A	A	10	
PHYSICAL EDUCATION GRIFFIN	A	A	A	A	C	B	OK	
TOTAL UNITS							50	
C. S. F.			✓			✓		
ATTENDANCE ABSENT TARDY								

YEAR 1945-46

SUBJECT	1 QTR.	2 QTR.	SEM.	3 QTR.	4 QTR.	SEM.	SEM. PER	REC. PER
	D	C	C	B	B	B	10	
Eng IV Melton	C	C	C	B	B	B		
Math IV Kauffman	A	A-	A-	A-	A-	A-	10	
Ger III Walters	B+	A-	A-	A-	A-	A-	10	
Physics Kyle	A	A	A	A	A	A	10	
Typ I Aubrey (Garcia)	B	A	A	A-	A	A	10	
PHYSICAL EDUCATION [illegible]	B	A	A	A	A	A	OK	
TOTAL UNITS							50	
C. S. F.			✓					
ATTENDANCE ABSENT TARDY								

YEAR

SUBJECT	1 QTR.	2 QTR.	SEM.	3 QTR.	4 QTR.	SEM.	SEM. PER	REC. PER
PHYSICAL EDUCATION								
TOTAL UNITS								
C. S. F.								
ATTENDANCE ABSENT TARDY								

IOWA H. S. CONTENT EXAM.
Date taken 5/28/46
No. in class 354
Score 97 %ile
Rank 10

RATINGS IN SOCIAL CHARACTERISTICS

YEARS	1	2	3	4	5
APPEARANCE - MANNER	1122	11231	221222	21221	
DISPOSITION	1121	11221	212121	21111	
HONESTY	1121	11222	122111	11111	
ACCURACY	1142	13442	222221	31212	
APPLICATION	1132	12332	222121	31211	
INITIATIVE	2243	12332	222221	22311	
COOPERATION	2123	12322	212121	11211	
LEADERSHIP	[illegible]	14342	313223	23211	

GRADUATED 6-14-46
NO. GRADUATING CREDITS 200 G.W.W.
DIPLOMA NOMENCLATURE Col. Prep
RANK IN CLASS 10 in class of 354
TRANSCRIPT TO: San Jose State - 7/13/46 Uni. Cal. 2-18-[illegible]
[illegible] 12-27-87
Juliane Walters
Vice-prin.

NAME MOORE, Gordon Earle

Gordon's high school transcript.

SOURCE: CHEMICAL HERITAGE FOUNDATION.

to Berkeley and then to research for a PhD with world-leading professors. Chemistry first fired his imagination and then meshed with his focused concentration, leading to a life lived not simply in the Moore family's ancestral haunts, but simultaneously as one that transformed world realities through focus, innovation, and vision.

The Sheriff and the Blind Eye

Walter Harold had moved to Redwood City as a deputy sheriff. Four years later his promotion to chief deputy confirmed his earlier intimation of opportunity. He was now the highest nonelected officer under Sheriff McGrath. "With the sheriff being a politician, my father was the chief operating officer. He ran things," recalls Gordon. The laconic observation conceals a complex truth.

The county of San Mateo, born in corruption in 1856, remained notorious. It spanned more than five hundred square miles, including mountainous valleys, fog-draped hills, rural farmland, and densely populated settlements, all lying between the San Francisco Bay and the Pacific Ocean. Its courthouse, jail, and county seat were in Redwood City. "Big Jim" McGrath, who weighed in at 300 pounds, would become its longest-serving sheriff, elected by voters for four terms, despite his implication in county corruption. "In the late 1920s, he inherited a community where a powerful

Deputy Sheriff Walter Moore, Gordon's father.

SOURCE: SAN MATEO COUNTY SHERIFF'S OFFICE.

gambling network permeated every shadowy political corner. An equally potent, more positive influence during McGrath's tenure was the dramatic growth of the county, thanks to World War II. Industry moved in, home building was on the rise, and immigration soared." The county's 1910 population of twenty-five thousand people quadrupled within three decades and by 1970 was more than a half million.

As Walter Harold established himself in Redwood City, peninsula life included "dog racing, lush gambling casinos, slot machines, and big-shot gamblers with political influence." Card rooms for draw poker and horse racing were legal, and vice operators could locate their businesses conveniently just over the county line from San Francisco. Gambling clubs were widespread. If attempts were made to close them, "attorneys for the owners would produce evidence to show their clients operated innocent social organizations." Many San Mateo grand juries launched investigations into the county's gambling, yet the most these produced were discreet temporary closings. "When the heat was on, the gambling houses would shut down—only to quietly reopen when the pressure subsided." McGrath's approach was to accommodate, while prioritizing the payment of his own political debts. Well known for his involvement with gambling organizations, he was reelected because of his success in keeping them in line. Rather than fight a crusade, he managed corruption.

As McGrath's right-hand man, Walter Harold was well aware of political realities. He had a ringside seat to the complex maneuvering. Compartmentalizing his world, he chose to focus on high-level administration, assignment of personnel, and courthouse duties, while also remaining closely involved with the investigation of violent crimes. He was not a man to avoid reality, nor was he squeamish. His daily work could be grim—finding the frozen bodies of two men electrocuted while working on a radio transmitter, for example—yet he was calmly practical, able to pursue procedures and direct others. In 1944 he investigated the case of the sister of one of Gordon's schoolmates, whose body had been disinterred in Menlo Park. He also investigated the disturbing case of two little girls murdered by their father. High-profile inquiries included the ice-pick slaying of socialite Carola Hartog.

In 1950, a decade after Walter and Mira Moore moved their family to Redwood City, Sheriff McGrath met his downfall at the polls. He had won plaudits for his work preparing peninsula communities for a predicted atom bomb attack by the Soviet Union, but less salubriously, he had been named publicly in a gambling controversy, involving his sale of shares of Hollywood Turf Club stock at premium prices to an underworld friend. The influx of postwar suburbanites was dissatisfied with business as usual. A holdup at the Cabbage Patch club involved a gunshot death, propelling the issue of illegal gambling back to the top of the county agenda.

McGrath sought one last reelection. His opponent, Earl B. Whitmore, was a handsome young Redwood City police sergeant and a graduate of Sequoia High School. Ambitious, dashing, innovative, and an expert at public relations, Whitmore was a modernizer, promising a shake-up in law enforcement and meritocracy in the sheriff's office, replacing seniority as the basis for promotion. He "swept the veteran sheriff out of office" by three votes to one. Walter Harold rolled with the punches, adapting to the changing order and to an office that grew to ninety employees. He was confirmed as chief deputy sheriff and later recognized for his role in combating organized crime on the peninsula. Whitmore himself faced corruption allegations in the 1960s, but by then Walter had retired, after nearly forty years in law enforcement.

In an age before the popularization of psychotherapy, Walter Harold kept his own counsel, even within his immediate family. "He would disappear and come back, and occasionally we'd get accounts of what went on, but usually not," says Gordon. Walter was highly successful as the "inside man" at the sheriff's office, playing a role complementary to the "outside" political maneuverings of McGrath and later of Whitmore. Gordon, who famously became the inside man at Intel, witnessed these roles and—consciously or not—took his father as a model. The message was clear: by avoiding the messy, compromising entanglements that went with political work, one could actually get things done and build an unshakable reputation within an organization through steady, pragmatic action. As an adult, Gordon was tight-lipped, opaque, and reluctant to disclose details of his work when home with family.

Big Bangs

While his father was a steady presence in the lawless outside world of San Mateo County, Gordon, in his self-equipped cement-floor laboratory, was blowing things up in style. Teaching himself from advanced textbooks, he steadily expanded his repertoire of bombs and rockets, paying for equipment out of his earnings. "Chemistry was a self-financed hobby. I may have had a small allowance, but I worked summers as soon as I could, in menial jobs."

After getting his start in the family store, Gordon labored for local chemistry-based industry. He lugged sacks of bark and hides at a tannery. He graduated to moving sacks of shells at the cement plant: "At the beginning of the summer, lifting a hundred-pound sack was tough; by the end, you were throwing them up seven high in the railroad cars. It was a good bodybuilding exercise; we were ready for football season!" His paychecks went straight to materials for and texts on explosives. In the days before the advent of global terrorism, "you could buy anything." Today such activity by a young person would be highly suspect, but all Gordon had to do was put in an order, and reagents would be shipped freight, no questions asked. "To make nitroglycerine you had to have concentrated acids. You had things like picric acid. All you had to do was dry it out and detonate it! It was like TNT. We used metallic sodium, which spontaneously ignites on contact with water. Potassium was a lot more expensive and didn't do much more than sodium."

By the time Gordon took his first chemistry lesson at Sequoia High, in late 1944, he was—not surprisingly—way ahead of the class. A new Donald would now be significant: his science teacher Donald Kyle. "It was before they had advanced-placement courses," but Kyle enjoyed and encouraged the three or four students who had already taught themselves. At home Gordon was into serious endeavors, using *Nitroglycerine and Nitroglycerine Explosives*, "a great big, thick book which was lying around in one of the used bookstores

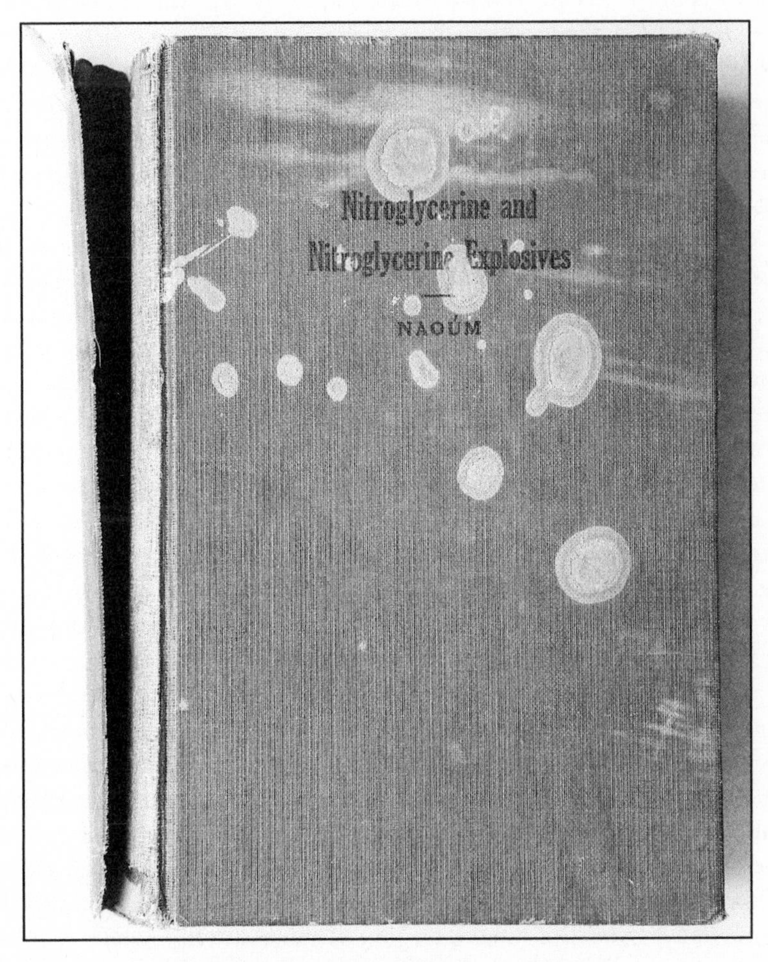

Gordon Moore's well-used book on explosives.

SOURCE: GORDON MOORE.

in Palo Alto—Stanford fallout. I was reasonably cautious, following directions and everything. The only difference was, I couldn't buy fuming nitric acid or fuming sulfuric acid. The recipe called for both acids to be fuming, but I could only get the concentrated stuff, so I had to diddle the proportions." Such was Gordon's fondness for the *Nitroglycerine* book, soon with "multiple acid burns on it," that he treasured it throughout the decades. A manual for making chemical warfare agents also became so important that he paid to keep it rather than return it to the public library. "It dealt with various tear gases and had descriptions of a whole bunch of explosives. I wasn't especially interested in the tear gases, but it told you how to detonate the explosives."

The deepening engagement with chemistry and the growing ambition to become a chemist linked easily to a powerful current in that era. The DuPont Company, using the slogan "Better things for better living, through chemistry," was riding high, successfully introducing nylon and other polymers as scientific breakthroughs. Further chemical wonders of the 1940s included synthetic rubber and the miracles of penicillin and effective antibiotics. Catalytic cracking, enabling the efficient production of gasoline from copious domestic oil supplies, was also making news.

Chemists' understanding of the molecular world was simultaneously being transformed by electronic instruments—new eyes on the world—and by novel physical theories. Linus Pauling, a brilliant young academic, had recently decoded the chemical bond, a key to chemical reactions and behaviors in terms of shared electrons and their quantum mechanical properties. By August 1945, as the atom bomb was dropped, Gordon's own knowledge had—with the help of a piece in *Time*—developed to a point at which he felt he could explain to his friends "how you could get energy out of uranium." The destruction wrought by the atomic bomb was stunning: "phenomenal," he remembers. "All these battles, one after the other, and all of a sudden, complete cities disappearing. It was a dramatic change." It was also his first real glimpse of how applied chemistry had the power—quite literally—to transform the world.

At school, encouraged by his teacher, Gordon steadily gained confidence in his lab skills. "By the age of sixteen, I was well advanced. Mr. Kyle was aware of what I was up to. You couldn't get by without making nitrogen tri-iodide and spreading it around the laboratory. I'm sure we bragged about it." Neither, at home in Westgate Street, could his parents entirely ignore the fact that he was building and even distributing bombs and rockets. "I was testing my skills, trying different things. It was more a manufacturing operation than a learning operation. When I gave my superfirecrackers to some friends, I found out they blew up people's mailboxes with them. After that, I didn't think I ought to give them away."

As deputy sheriff Walter Harold was intimately familiar with guns, explosives, and violence. Gordon remembers his parents as "relaxed" about his hobby: "The only time they became involved was when one of the rockets had a piece of smoking paper come down on the neighbor's roof. We ran over there with the hose and squirted it down. The neighbor wasn't very happy, so he had the local police come. The policeman knew my father well, so for me it was not quite the bawling out you'd ordinarily get."

Having been cut a little slack, Gordon soon found an opportunity to repay the favor. "My father had come upon this big bunch of what looked like safecracking tools—drills and pullers and a little bottle of yellow liquid. He asked me, could I tell if this was nitroglycerine." Being old enough to drive, Gordon grabbed a small anvil and headed to the sheriff's office to conduct a test: "I put a drop of the liquid on a piece of filter paper on the anvil and hit it with a hammer. That gives a beautiful bang with nitroglycerine. And it did! My father didn't know what to do. I said, 'Oh, I'll take care of it.' I'd been making nitroglycerine in larger quantities than that at home for quite a while. So I took the nitroglycerine and made it into dynamite." This was classic Gordon: quiet assurance, a dash of humor, then a big bang. In his own words, "Most people who knew me then would have described me as quiet, except for the bombs." At sixteen he possessed a sophisticated understanding and calm confidence in executing his ideas. These traits, combined with aptitude and determination, set him apart from his peers.

His brothers were both of lesser intellectual desire. Having quickly dropped out of San Jose State, Walt Jr. turned back to familiar haunts and was soon courting a young Pescadero woman, Darlene Cabral, while working for her father on the Cabral ranch, Pacific Acres. Walt and Darlene married at a candlelit ceremony in Half Moon Bay in 1948. Gordon was the best man. The couple would spend the rest of their days in Pescadero, raising three sons—Walter, Merritt, and Russell—while growing barley and flax and raising cattle. Over the years Gordon would remain in touch with Walt through the behaviors with which they both grew up. "I saw my brother eight or ten times a year. I would go over there hunting, or he'd come on my boat fishing."

Fran, five years younger than Gordon, left high school in the early 1950s. Like Walt, he had little interest in college. For a time he worked at the Redwood City electronics firm Ampex. A natural salesman, he soon joined a car dealership on the peninsula. Later, he established a distributorship for machine tools, "not the big tools, but the accessories that go along with them," explains Gordon, "the measuring equipment, drills, bits, end mills, nuts, and bolts." Fran married twice and had two daughters by his first marriage. The tool business was modestly successful, with up to twenty employees, including one of Darlene and Walt's sons. Fran's annual barbecue became another family gathering point for the Moores.

Ready to Move On

After a slow start, Gordon found high school "pretty easy. I got by without spending a lot of effort." As he embarked on his final year, he began to achieve better grades. His math teacher became an important influence. "While I got into chemistry by myself, I would never have dug into math without help. I toyed with switching to being a math major, as I was impressed with the logical simplicity of elementary calculus." Mathematics—with its reproducibility, clarity, and precision—had an obvious appeal to someone of Gordon's analytical bent, but with an eye on the future, he decided, "There weren't many good opportunities for mathematicians." Chemistry had the connection to doing, to the practical world, to exciting and useful capabilities; it would remain his love: he had "self-defined as a chemist. The chemistry-set experience really set me going. I enjoyed it; that was the gist of it."

Sequoia possessed well-equipped laboratories for physics, chemistry, and biology. It also had a comprehensive industrial arts program, but, despite his affinities for mechanical tinkering and for physical labor during school vacations, Gordon (together with other academically inclined students) was forbidden to take shop, "because it would ruin our hands." Sequoia High turned out many outstanding athletes, but his interest in sport was waning. With remarkable clarity, Gordon made an unsentimental decision that would prove essential to future success: he would consciously avoid the investment of time in activities in which he could not excel: "At college I fiddled very briefly with diving, but decided I wasn't good enough. Success lay on a different level." He dropped formal sports in favor of the lab from that point on.

Notwithstanding the lack of precedent in his nuclear family, Gordon "kind of assumed" he would go to college. Walter Harold had briefly tried to push his obviously gifted middle son toward medicine, but discovered that Gordon had no interest in being a doctor. Conversely, Bob Naughten, also a gifted student, had determined early on that he would be a physician. As a child, he had witnessed his father suffering a fracture dislocation of the neck and how the prompt actions of a young doctor prevented enduring paralysis. "I thought, 'That's a neat way to live.' I decided to be a doctor, and I wanted to go to UC Berkeley.

Gordon, having heard positive stories from double cousin Harriet Williamson, also had an interest in Berkeley, some forty miles away. He and Naughten were both offered enrollment, but realized it would be easier to study locally. In 1946 floods of war-returned, college-bound veterans were putting schools under enormous pressure. "All of these guys came back, and the priority of colleges and universities was to get them through school on the GI Bill," explains Naughten. "We'd been accepted at Cal, but housing was impossible. We'd be commuting the whole day to and from classes." Gordon, always practical, saw that starting at nearby San Jose State would be "the easiest route. I could continue to live at home." Gordon and Naughten both planned to transfer to Berkeley when the dust had settled and they had tackled their introductory classes.

His immediate future decided, he bid farewell to his high school peers on Friday, June 14, at a graduation ceremony today notable for its lack of interest in Gordon himself. Both the Scholarship Medal and the Science Medal were awarded to Walter Stanley Scheib Jr., who went on to study chemistry at Stanford and chemical engineering at MIT. With twenty others, including Naughten, Scheib was granted lifetime membership in the California Scholarship Federation; Gordon was one of four to earn the lesser "novitiate pin." At this stage he was perceived as a steady if quiet performer, giving little indication of the technical brilliance—combined with clarity of vision, shrewd decision making, and unusual doggedness—that would be fully apparent in less than a decade.

About the electronic activity around him, he remained largely ignorant. Redwood City and its neighboring regions were humming amid the longest and most rapid economic expansion in the history of the United States, as California's population passed 10 million. This explosion of activity was paralleled by an exponential rise in defense-contracting dollars flowing into California, both south and north. Vacuum tube manufacturing for radios, communications, and radar was growing steadily, thanks to the war-induced boom in defense industries. San Carlos, just north of Redwood City, was home to the vacuum tube efforts of Charles Litton, Dalmo, Eitel-McCullough, and the Varian brothers, among others. Yet Gordon was aware "only at a superficial level that things like radar were becoming important. Seeing things in the press didn't change my direction. I was determined to be a chemist."

With a place at San Jose State lined up, Gordon packed away most of his chemicals. The days of his intense work with explosives were over. "When I was at college, I probably made a bomb now and then, for old times' sake, but I was pretty well out of the production business." The lab, with its concrete floor, remained a fixture at his parents' home: "Eventually, my father figured he had to get rid of the chemicals and made the mistake of hauling them all down to the dump and throwing them away. I wish he had enlisted my aid. Lucky he didn't kill somebody. There were bottles of cyanide, bottles of sulfuric acid. Whoa! He did ruin a pair of shoes!"

BOY MEETS GIRL

San Jose State

In the fall of 1946, Gordon Moore—quietly studious, fond of explosions, and not yet eighteen—became a freshman at San Jose State College, today's San Jose State University. He acquired his own wheels in the form of an ancient Ford flatbed truck purchased for twenty-five dollars and commuted daily with Bob Naughten. Naughten recalls, "We would take off every morning, down to San Jose and back. Four of my friends from San Carlos went to San Jose State, but only Gordon and I commuted in his truck." Before long, the convenience of the Southern Pacific train triumphed over the glamour of the truck. Gordon enjoyed the challenge of "a pinochle game on the train on the way down, as eight or ten of us commuted together. The student fares were really cheap. I'd drive to the station in the morning and take the train to school and back." Tuition was low, though students had to purchase their own books, and science students were required to buy a slide rule, which, remembered Naughten, was "terribly expensive."

Gordon, a Pescadero boy at heart, was impressed by the busy campus: "It was big and impersonal. All the veterans were back. There was a line way down the hall for using the men's room. Nobody received much in the way of attention." San Jose State has the distinction of being the oldest public institution of higher education on the US West Coast. Founded in San Francisco in 1857, it grew to be the California State Normal School, moving to its permanent home in San Jose in 1871. The main buildings, with a central grass quadrangle, were located on a sixty-two-acre site downtown. A southern arm, home to many of the school's athletic facilities, lay just over a mile away. The city itself still had a population under 100,000, compared to today's 1 million plus.

As a chemistry major, Gordon knew what he had to do. His matter-of-fact report stresses, "There was the usual math: analytic geometry and a couple of kinds of calculus. The chemistry was easiest because I'd done a lot of that on my own. The physics class separated the people who ought to go into technical careers from those who shouldn't. I had to learn a lot of stuff in a hurry, stuff I had never seen before." He couldn't resist displaying his chemical expertise. Nitrogen tri-iodide was one staple. Naughten remembers, "Gordon loved explosions. He put together some stuff in the chemistry labs. He'd coat it on things. It would go pop, pop, pop. The instructor was a little upset about that."

College was to Gordon's liking: a place he could get down to business and deal with necessary Mickey Mouse "more German and the usual California State requirements." He moved "all of those out of the way in two years," clearing the decks to pursue upper-level courses in his preferred areas—chemistry, math, and physics. "The first year at San Jose was intense. I had never done much homework before. I took tough courses, and homework kept me busy." The summer of 1947 was equally busy, thanks to his need to earn cash. He returned to the cement plant in Redwood City, but with enhanced status as a college man. This time he was in the quality-control laboratory. "I was running around collecting the samples. About the most complicated thing I did was measure how much water there was in the slurry."

Gordon Moore enjoyed his adolescence, despite being naturally quiet: "Hormones were strong! I could just get up enough nerve to ask girls to go out with me, although it was always a chore. There were dances every week in high school, and I went—at least after the first year or two." That September, uncharacteristically without focus, and untypically on a whim, he attended a presemester student government conference at Asilomar, at the time a YMCA retreat on the Monterey Peninsula overlooking the sea, some sixty miles to the south of San Jose: "I don't know why I went to that meeting. I had a week with nothing else to do, so I happened to go." It was a momentous choice, for in Asilomar Gordon met the woman who would become his wife and lifelong partner.

Enter Betty Whitaker

Betty Irene Whitaker was a journalism major at San Jose State, one year ahead of Gordon and six days short of a year older. She lived at home on her family's fruit ranch in semirural Los Gatos, thirty miles down the peninsula from Redwood City and near San Jose. In high school Betty had been active as a staffer on her school newspaper. Given her opposition to the influence of cliques in the school and a corresponding interest in student governance, she quickly persuaded her buddy Barbara Kinney to come with her to the conference. "Asilomar was very rustic," she recalls. "They had a great glass meeting hall that looked out over the ocean. It was beautiful."

In September 1947 Betty Whitaker was a vivacious nineteen years old, a confident, lively, and determined young woman with a direct gaze, dark eyebrows, a full mouth, dimples, and midlength hair, waved in the fashion of the time. She was outgoing and enjoyed dancing. Gordon was immediately struck by her manner, very different from his own: "She was pretty and had a lot of spirit. I was attracted to her. There was a dance at the meeting. I saw this pretty little girl and started to dance with her." Betty, for her part, was drawn to Gordon's personal composure and quiet confidence: "Some of the people I'd been around were pushy. He didn't seem like he was hassling me. He looked like somebody I really wanted to know." They

Betty Whitaker and Gordon Moore, at San Jose State.
SOURCE: KEN MOORE.

traveled back together on the conference bus. As they jointly recall, a flirtation quickly developed.

> Betty: "Gordon was throwing pennies. I remember how lucky you were at tossing those pennies."
> Gordon: "We started matching pennies, and I won everything. I had pockets full of pennies by the time we arrived back. She was afraid I was a big gambler."
> Betty: "I didn't know what kind of a guy you were. I was thinking, 'What the heck is going on here?'"
> Gordon: "I never had won like that, gambling, in my life."
> Betty: "Everything he touched turned to gold."

With Betty by his side, Gordon Moore would continue to be blessed with the golden touch. As they traveled home on the bus, he had little idea that he had found his life companion, a young woman who was in some ways his opposite, yet shared his values, understood and mirrored his need for privacy and space, and—overcoming the legacy of her own parents' divorce—gave him the stability, space, and family support that would enable him to drive remorselessly forward, over many decades.

Like Gordon, Betty had strong local roots, though hers were more recent, cosmopolitan, and affluent. Her paternal grandparents had settled in Oakland, where Betty's father, Arthur Allen Whitaker, was born. Betty's paternal grandfather, Elijah Aaron Whitaker, combined entrepreneurial talent with a restless streak. Launching out from Whitaker, Pennsylvania, at twenty, E. A. was soon buying mining claims in Montana and later accumulated thousands of heads of cattle. By the age of forty-five, he had made his fortune. He sold his ranching interests and returned to Pennsylvania, where in 1884 he took Margaret Rudisill—twenty-three years his junior—as wife. By 1890 they had moved to Oakland, where they bought a commanding hilltop house at 2303 Seventeenth Avenue, formerly owned by a prominent San Franciscan, who had moved across the Bay to protect wife and children from "exposure to prostitution, saloons and communicable diseases." By the turn of the century, the Whitakers had four young sons. Elijah Aaron was sixty-three when his wife became pregnant yet again, this time with twins Arthur and Richard, born in 1901. Elijah Aaron lived on into his eighties, dying in 1925 as the owner of "half of downtown Oakland" and leaving his sons a great fortune. Margaret Whitaker, listed in her own right in *Who's Who Among the Women of California*, moved to the luxurious Hotel Oakland until her own death nine years later.

Betty's mother, Olive Irene Metzler, also came from a prosperous family. Olive Irene was just three when the Metzlers made the move to California in 1907. "My maternal grandparents had an enormous ranch in Wayne County, Ohio," explains Betty, but they heard of "an area in California that had the most fertile soil and the most wonderful growing conditions. My grandfather Abraham Lincoln Metzler sold his big spread, packed up, and came to Los Gatos." Nestled in "the Valley of Heart's Delight," Los Gatos was known as a gem of the foothills, rich in fruit, flowers, and blossom and blessed with a sunny, equitable climate.

Unlike Pescadero, Los Gatos was firmly on the railroad map: the Southern Pacific offered service to Oakland and Santa Cruz, and there was a streetcar line to Saratoga and San Jose. By the late 1920s, it was a small but thriving agricultural center, surrounded by orchards and vineyards, attracting artists, painters, writers, musicians, and other bohemian types. Violinist Yehudi Menuhin's family moved there, and novelist John Steinbeck built his first marital home a mile to the west. Here, he wrote *The Grapes of Wrath* in the late 1930s, while complaining that the area was increasingly populated (the 1940 population was still only thirty-five hundred and 99 percent white).

Olive Metzler, known as Irene, was the youngest of four children. "My grandparents brought two sons and two daughters and all the worldly goods they felt they couldn't live without." Graduating high school, Irene, a naive eighteen-year-old, entered what turned out to be a disastrous marriage to Arthur Allen Whitaker, twenty-two, in 1923. The couple had met at a community event in Los Gatos, where Arthur's wealthy parents owned a summer place, "a big spread" on Foxworthy Avenue. Irene was awed by its English-style furnishings and gorgeous carpets. Betty explains, "It was not the norm for the countryside, for a ranching area."

In 1925 Arthur and his brothers were left rich men when their father died. This proved Arthur's undoing. Betty, not born at the time, later learned the whole story. "Things went to heck and gone. He started buying cars and running with women. Those young men didn't know how to handle money, and they were all given millions." The crux came as the young couple was supposed to head east to take a liner across the Atlantic and tour Europe. Instead, "Mom was pregnant; she had morning sickness. That straw broke the camel's back. It made my father disgusted, and things went from bad to worse." Betty Irene, their first and only child, arrived on January 9, 1928. The arguments between the couple grew fierce, culminating in an incident in which Arthur left his wife and young child on the roadside and drove off alone in his car. "There was a newspaper article about my father pushing us out of the car. My mom said, 'Hey, this is it. I don't need this.' She filed for divorce right away."

Divorce in that era was rare, scandalous, and disgraceful. Struggling but resolute, Irene, with baby Betty, temporarily found refuge with friends. For the long term they moved back to the ranch in Los Gatos, with her parents, Abraham and Mary Metzler. Betty's psychological defense was to deny memories of the divorce, while allowing its lessons about the dangers of flaunted wealth and the virtues of a rock-solid man to sink deep into her psyche. "My mother told me they had divorced, but for years I put that out of my mind. She said once, 'Your father was so mean to me. He deceived me. It hurt me so badly that I can't even talk about it.'" In due course, Betty would marry Gordon, a man of quiet conviction, and agree with him on the need to shun ostentation, to guard their children from wealth's dangers, and to have her devote herself wholeheartedly to the well-being of the family.

Arthur himself moved to remote, rural Humboldt County, two hundred miles north of Oakland. With his divorce from Betty's mother finalized, he quickly entered a brief marriage with a twenty-two-year-old Kentucky-born woman. As he fought his inner demons, he surely preferred to be away from the eyes of his brothers, mother, and first wife. His second marriage soon failed, and for a period he appears to have been homeless. Admitted to the hospital in the spring of 1933, he was listed as a single man. He died in Eureka, California, on May 29, of pulmonary tuberculosis complicated by four years of acute alcoholism.

Arthur was quietly buried in the Whitaker plot at Mount View Cemetery in Oakland. At least he did make a will, providing modestly for his daughter, Betty. "I'm sure that my father felt guilty. He set up a trust fund for me before he died. My paternal grandmother, Margaret, kept in touch with my mom, and every month we had a payment. Back then most people didn't get provided for like that. Until I was eighteen, it took care of me." Betty rarely saw Grandmother Whitaker. "She was living way up in Oakland. It took a long while to get there. From Los Gatos, it was a windy, slow road, and the cars were not speedy."

Betty found a lesson about the perils of wealth in the downfall of her father, but in Grandmother Margaret, "a grand lady, looking like the dowager queen," she sensed the possibilities that wealth, properly handled, could bring. "She had come from a background of comfort and would send me wonderful things. She went to Hawaii on one of those Pan-Am clippers and gave me a little silk kimono. She brought me a wonderful doll from Europe, with a bisque face and real hair; a gorgeous thing. That was the most expensive thing I owned as a girl." When Margaret died in October 1934, she set for Betty a benchmark quite beyond anything Gordon's family could envisage, a bequest of two thousand dollars expressly to "put the six-year-old Betty through college." Otherwise, Betty and her mother knew no more of the Whitaker family.

Life on the Ranch

As a young man, Abraham Metzler had studied farming and animal husbandry in Ohio. In Los Gatos he and his wife had twelve fruitful acres, near the intersection of Camden Avenue and Union Avenue. They also owned and worked a nearby vineyard. Their sons, Russell and William, attended local schools, as did daughters Ruth and Olive Irene. A further daughter, Ada Marie, was born in 1911. Betty's maternal grandparents were educated, comfortably off, devout Protestants. Her grandfather was a gentle and kindhearted man who was always "fun to be with," and he had a passion for politics. A staunch Republican, he had held office in Ohio. Mary Metzler was also politically minded, but in the opposite direction. "She was interested in the New Deal. She liked President Roosevelt," says Betty. "When I was a little girl, we had a radio that sat on a table, and we'd push up around it and listen to the 'Fireside Chats.' My grandparents would cancel out each other's vote. When they were older, they both said, 'I'm not going to dress to go to the polls. You vote me away anyway.'" In a manner that Betty would later emulate with Gordon, Mary Metzler "ran the show within the home."

Irene's older siblings all went to college, married, and moved away. Both her brothers were Stanford students and stayed on for graduate work at the fledgling institution. Russell loved history and would later teach citizenship classes to aliens, "to make sure that they were knowledgeable about our country." William earned his PhD, became a well-known sociologist, and combined teaching and research with government work. Betty recalls, "When I was small, he was in the Southwest for a while, studying or working there with the Indians." Irene's older sister, Ruth, graduated from UC Berkeley and married fellow student Carl Schiller, who became a builder. They brought up their three children in San Anselmo, Marin County.

Irene was the deviant, marrying in haste while young, instead of heading to college, and then rapidly returning to the ranch to repent at leisure. Of necessity, she became practically oriented. As her parents aged, she took control of every ranch detail, from hiring workers and contracting out irrigation to pruning and picking crops. "It was full-time work." Even so, the 1940 US Census listed Irene as having no occupation. Her father, Abraham, disabled from strokes, was described as the ranch "operator," signaling that "it was unusual for a young woman to be running a ranch. It's a man's job: it certainly was then. She took a lot of flak when she went down with the crops to San Jose: 'Here comes the widow.' She had to be strong willed because she had so much on her shoulders."

Irene was a woman of small stature, but tough. Betty today thinks of her as "the Barbara Stanwyck of the West Coast." (In the late 1930s, Stanwyck was a star in the fast-developing movie industry, playing strong female characters. In real life she was orphaned at a young age, divorced like Irene, and owned ranches in California.) Describing her mother's personality and habits, Betty prefigures her own role in relation to Gordon many years later: "She had quick reactions to things: not always correct, but she meant well. You cannot be wishy-washy when you have so much responsibility." Watching her mother at work, Betty learned much about human nature. For example, it was crucial for Irene to monitor closely the transient workers she hired at the ranch.

> You have to trust what they're going to do on your property. You're paying by the hour, and they're going to take advantage. That was a nightmare. With the migrant workers at harvest, often it was a whole family—a swarm

> of people, six or eight at a time. The kids would crop pick if they were old enough. Other times we had a lot of little ones making messes and terrorizing the rabbit hutches. Gypsies would come through and stay: they would steal everything they could get their hands on. The Mexican workers would sometimes take tools that were sitting out. We had a dog, and the dog disappeared.

Despite these unwanted incursions, Irene Whitaker "was always good to anybody and everybody." She was a religious person, "not a religious nut, but she believed in the Golden Rule; 'Treat people like you want to be treated.'" If someone came to the house hungry, they were fed. During the Depression Irene and her mother would make meals for tramps, but soon wised up that the little stacks of rocks left outside the ranch meant that this was a good home to tap: "We'd get rid of those." The ranch suffered financially during the 1930s, the Metzlers even coming close to losing it. "My family had to put mortgages on the property. They were afraid the bank was going to foreclose. They had another eighteen acres below the main ranch, and they almost lost that, too." Betty's income from her father's trust fund came to the rescue. "The payments pulled the property through, to keep it in their hands."

Irene had married and become pregnant soon after leaving high school. Her younger sister Ada followed an eerily similar pattern. Ada's husband was abusive, too. Having seen Irene act decisively to exit a bad marriage, Ada did the same, divorcing her husband and moving back to join Irene and Betty on the parental ranch. "Her husband played in a band. He drank, and he'd come home and beat her. She said, 'Hey, I'm outta here.'" Ada's sons, George and Richard, born three years apart, were a little younger than Betty, who quickly became close to these two delightful playmates. "For a while, there were seven of us on the ranch. They were like brothers to me. We've kept in touch all our lives."

At holiday time other Metzler family members descended on the ranch. Uncle Russell, Aunt Sadie, and the cousins from Fresno would join Aunt Ruth and Uncle Carl and their children. Uncle William and his wife, Julia, came, but had no children, only pets. "We had huge meals. There was the big people's table and the kids' table." Like the Moores, the Metzlers were a large, extended family, settled throughout the area. From this network flowed an atmosphere of security and constancy.

The Metzler ranch grew nuts, principally apricots, and plums destined to become prunes. Additionally, the family sold grapes from its vineyard to a local winery. "The ranch was not huge," recalls Gordon, who first visited it in the late 1940s. "Except for the harvest, you could do it with just the family." The two-story ranch house was in the Victorian Gothic style. To Betty,

Betty Whitaker (*far left*) with (*left to right*) her mother; her aunt Ada; and cousins George and Richard.

SOURCE: KEN MOORE.

it offered security and an aesthetic she would always remember. "The rooms were large and comfortable. The parlor had an organ and a piano, as well as a big davenport and rockers. The house had wonderful stained-glass windows at the landing. There were all these banisters and all this woodwork." Los Gatos itself was predominantly "ranches and very narrow roads."

Betty's earliest memories are intensely evocative of the colors, smells, light, and sounds of settled agricultural life in the Valley of Heart's Delight, today all but obliterated in its transformation into Silicon Valley. "The area was completely tree filled, big locust trees, great flowering trees. I loved the smell of them in the spring. You knew every car that went up and down the road. You knew when the vegetable man was going to come along, not that we needed a vegetable man. We were self-sufficient in that respect. There was also a man who would call out, 'Rags, bottles, and sacks.' Our parrot picked it up and would yell, 'Rags! Bottles! Sacks!'" Irene "was busy running around—doing this, doing that, managing the ranch. She was so full of business that she was never there for me. She said, 'I never went to school enough. I can't help you.'" Men might make the world, but women would do whatever the situation demanded to keep the show on the road.

Like Gordon, Betty spent much of her childhood outdoors, in close engagement with the natural world. She tended to the ranch's hens, goats, ducks, chickens, rabbits, horse, and cow and collected manure. She also raised vegetables, growing corn "tall as an elephant's eye" for a contest at school. She laughs in recalling how Babe, the horse, demolished her prize crop. "Aunt Ada was trying to save my corn. She was out there with Babe, pulling. Well, Babe sat down. How are you going to pull a horse when she's sat down? It was like something out of a silly movie. I never was able to put my wonderful corn in the contest, but I was really proud of what I had grown."

"Betty learned a lot about life, from her grandmother," says Gordon. What Betty imbibed would stay with her all her life: a confident, pragmatic approach and a commonsense mind-set that fitted perfectly with Gordon's own view on life. Her grandparents and mother inculcated Betty, from a young age, with the need for constant practical effort. "There was laundry day, baking day, slaughtering day, canning day. Every day was something." Grandmother Mary taught Betty to cook and to bake. "She'd say: 'All right. You feel this dough. That's the way it's supposed to be.'" In early summer Betty's job was to cut apricots. Laborers would bring the fruit to the cutting shed, and she would sit and "do trays" to be put in the sulfur house. Later, she helped can peaches. As the prunes and nuts came in, there was more work to be done.

> We had a dry shed. We'd dump the prunes after they'd been out in the sun for a period of time and hope the rodents didn't find a way in. If there was any problem, we couldn't sell the fruit to Sunsweet Growers. The nut crops came in after the prunes. We had to hull them. My grandfather had the blackest hands in the West. In the wintertime, when it was raining outside, we all sat and cracked walnuts for the candy store down in San Jose. The halves made good money; the broken ones didn't.

Throughout Betty's years of grade school and junior high, summer break was simply a season for hard work on the ranch. Unsurprisingly, she became a no-nonsense "can-do" person—a trait that would serve her well as the wife of a busy, preoccupied, career-oriented technologist with ever-ramifying obligations.

Growing Up

If Irene Whitaker could spare her daughter, Betty, little time on a daily basis, she was determined that in the bigger picture, her only child would benefit from a good education. The ranch was in the catchment area for a small elementary school, with "one teacher for everyone," but sisters Irene and Ada shrewdly rented a small property whose address would entitle their children to placement at the larger Campbell Union Grammar.

Until she was in high school, Betty had no female peers. Like Gordon, she thrived on solitude. "The next residence was acreages away. There wasn't anyone out there in the ranch area," she explains. "I didn't feel I was missing anything because I was kept busy." Her bond with her grandmother was strengthened when Betty suffered a bout of rheumatic fever, "going in and out of consciousness, my ears rumbling, thinking some horrible creature was coming through the window at me. My grandmother slept with me, putting wet towels on my head. It was terrible." Afterward, Betty developed a heart murmur and would tire easily, yet was expected to continue her farm chores.

Piano lessons were a staple item. "To practice, practice, practice. That was my explicit job!" Shirley Temple, born three months after Betty, had achieved fame in *Bright Eyes,* in 1934, and Betty now began to attend music and dance classes in San Jose. "We were all pushed to the nth degree. Every girl was going to be another Shirley." Betty played piano for the Olmstead studio orchestra, learned to play the xylophone, and was even encouraged to play the accordion. Despite all the chores and commitments, she still managed to spend time at a little desk in a corner at home, reveling in reading, writing, and studying alone. "There's something about having your own little private corner," she recalls. Unlike Gordon's family, Betty's loved books. Classics such as *Peter Rabbit, Alice in Wonderland,* and *Sherlock Holmes* were staple fare, together with Kay Thompson's *Eloise* books. A more unlikely inspiration was Longfellow's epic poem *Evangeline.* "Going to Nova Scotia was one of my dreams."

Betty was unusually practical and independent, yet valued intellect and the life of the mind. These characteristics helped link her to Gordon. Later, the couple's rapport developed through their shared love of fishing and outdoor life. "When I met Gordon, I found out he was trout fishing off his back porch. That sounded great. Soon we went trout fishing together." Fishing had been one of her first loves as a child: "I learned to fish with a handover line. My aunt and my mom used to drive the three of us cousins—George, Rick, and me—to the pier in Santa Cruz. We would fish and fish. We learned how to do pole fishing and get trout, in the streams over by Almaden Mine (now Quicksilver County Park)."

The little "hump radio" in the corner of the breakfast room was a vital feature of the family's daily life, particularly as America entered World War II. "We listened to all the speeches that Winston Churchill gave. They'd say, 'Churchill's going to be on,' and we'd all clamor to it." Signaling the growing empire of the vacuum tube in the mid-1940s, Aunt Ruth and Uncle Carl brought one of the very early black-and-white television sets to the Los Gatos ranch. "My grandparents never went out, and it brought the world to them. It made everything alive. When Reverend Billy Graham started out, they were right there with him." The Metzlers had brought a sturdy midwestern Protestantism with them to California, and it became part of Betty's life.

Campbell High School was close to the ranch, but once again Irene Whitaker had grander plans. She used another alternative address, this time to place her daughter at Los Gatos High, a school renowned for its scholastic standards. It was fashionable, too. Graduates included three future movie actresses: Olivia de Havilland, Joan Fontaine, and Audrey Long. In the fall of 1941, Betty became a pupil. For her, these were happy days. "My high school years were the best." Though she failed to shine at sports and was "last to be chosen for everything," she developed a quirky talent for archery. Soon she discovered her real passion: journalism. She joined the school newspaper, the *Wildcat,* and became both reporter and advertising manager. "I would run around town and collect fees from merchants and put the ads in. I was very businesslike." In addition, she noticed "how Los Gatos High School was managed by a clique. "This is not fair. If you're not from a wealthy family, you'll never get in the clique." Gordon, too, would become impatient of the status born of privilege, not earned through achievement.

Betty began to branch out socially, developing a close friendship with schoolmate Barbara Kinney. As her aunt Ada had remarried and moved out with her two sons, Betty was now the only young person on the ranch and took on still more chores and duties. "At high school, I had a home economics class. The teacher said, 'How the heck do you know all this stuff?' I said, 'Because I learned from my grandmother.' I wanted to live in foreign countries and started to learn Spanish so that I could work in an embassy somewhere. That idea came directly from my uncle William Metzler. He was of the 'Go out and see the world' opinion."

In the spring of 1945, as she reached the end of her high school career, Betty won a coveted spot on the *Wildcat* reporting the United Nations' inaugural sessions in San Francisco. My mother said, 'Now, you know you must sit and be a lady.' I had to wear my suit. We all had hats and gloves. It was very proper." In her role as reporter, Betty talked to such luminaries as Nelson Rockefeller (later vice president of the United States) and William Averell Harriman, who had negotiated with Stalin and was the US ambassador to the Soviet Union. "At a luncheon, I was across the table from the governor of our state, thinking, "Oh, my goodness, pinch me. I'm not sure I'm here."

Betty had outstanding academic ability. Unlike Gordon, she was invited to membership in the California Scholarship Federation, winning a financial award that, together with her grandmother Whitaker's bequest, she put toward her studies. She also won a five-hundred-dollar scholarship for an essay she wrote. In addition, she won two consecutive scholarships for international study at Mills College, Oakland, in the summers of 1944 and 1945. She attended both programs along with her good friend Barbara Kinney and remembered being "very sociable." Mills was a college for women, but men attended the program that second summer. "It seemed weird hearing men's voices in the hall, so I kept my door locked. The United Service Organizations [USO] set up dancing there. The dancing instructor said, 'You're a natural.' He'd swing me around and say, 'Not many people get the beat of the tango.' Sailors would come and dance, then have to leave. It was very monitored."

Through the UN event, Betty tasted the excitement of an international career. Through the Mills College programs, she glimpsed the possibilities of study. At home, she learned to be self-sufficient: "Eight years is a long time when you're young. As a child, I ferreted out things I needed to do for myself. That made me a strong person, made me realize I could do many things. All I had to do was set my mind to it." Together, these experiences inspired Betty to head away from home for college. She and Barbara plotted a move to study in Claremont, in Southern California, but suffered the sort of parental veto to which young women were prone in that era. "My mother said, 'I don't wish to have you that far South. I would never see you.' Barbara's mother and grandmother did not wish her to go either. We got zeroed out. Barbara and I were disappointed. We wanted to see the world, yet we were only seventeen."

Like Gordon, Betty was loyal to her family. Conforming to the mores of the time, she settled for San Jose State, enrolling there in the fall of 1945 and commuting from Los Gatos on the bus, followed by a ten-minute walk to campus; tuition cost fifteen dollars a quarter. Few young men had yet returned from the war, and women were very much an established presence in colleges and careers. "It was about equal until 1946, when the veterans were coming back by the hundreds," she remembers. "That diluted the whole situation. My classes became terrible. Women were being pushed away, or pushed around. Quonset huts were set up, and we were put out in these huts."

Betty opted for the social sciences, following the example of her uncle, but the influx of GIs made that difficult. "I switched to journalism, because my niche was blown apart. I wasn't going anywhere with sociology. I thought I'd better give up that idea; it was easier than fighting." She could stand up for herself, but she knew how to pick her battles. Journalism might be second best, but it was fun, and it suited her lively, intelligent, inquiring nature. "You weren't focused like a horse with blinders: you had a wide range, the whole gamut from music to art to architecture."

Photography, useful for her work on the San Jose student newspaper, the *Spartan Daily*, became a part of her repertoire: "I had photography labs and learned how to carry this huge Hasselblad camera on a tripod, walking the streets of San Jose."

Betty Whitaker was attractive, intelligent, well schooled, and well balanced. As she met and danced with Gordon Moore at Asilomar, she was ready for anything. Yet in the wake of World War II, the options for a young American woman were curtailed. For most, as for Betty, "anything" would entail marriage and a career in the home.

The Chemistry of Romance

The harsh experiences of the Great Depression and renewed world war were replaced by domesticity as the postwar era's central value. Love and marriage, the baby boom, the suburban home, the breadwinner, and the housewife and mother: these themes swept the nation with hurricane force in America's greatest age of prosperity, power, and progress, the mid-1940s through the mid-1960s. Within these themes, Gordon and Betty would find each other and make one of the great marriages of their age, for, while this book is a biography of Gordon Moore, his life and profound contribution to our world are fully understandable only within the context of Betty's crucial contributions to their partnership.

In the first weeks of the fall 1947 term, Gordon Moore and Betty Whitaker met only occasionally. Even so, powerful human chemistry was at work. "At San Jose State, Gordon and I found each other, but not right away. I was doing my photography for the paper in the darkroom, and Gordon was always in the science building. If I hadn't met him at Asilomar, I would probably never have run into him at all."

The couple arranged a fishing date. "It was the one outdoor thing Betty liked to do," recalls Gordon. "That was fine with me." On weekends they began to dance together at proms and to attend local parades and ball games. "We couldn't go out during the week; we were too busy," says Betty. "We had early-morning labs that started at 7:30 a.m. I had to get in from Los Gatos. It was murder. We didn't have much time." The romance matured only slowly. There was no obligation on either partner unless they were "going steady," and Betty herself had plenty of other options. "A lot of guys were coming back from the war. They wanted to marry. Most of the time I said no to dates. One guy in particular was after me. He was ten years older. He told me, 'We have to get married,' but I said, 'No, I'm sorry. I'm finishing school.' This was before I was serious with Gordon."

An impartial observer would have seen Betty Whitaker as a much better bet than Gordon Moore, destined for higher things. She came from a family familiar with wealth and education. Her uncles included a widely traveled, respected sociologist (William Metzler) and a successful commercial builder (Carl Schiller). She had won prizes and scholarships. She was outgoing, attractive, decisive, and intuitive. Betty was of an altogether broader gauge, better read and possessed of obvious social skills. Gordon, in contrast, was a "good average" student, used to television rather than books, and the unremarkable, quietly competent product of a deep-rooted but distinctly backwoods culture.

Seven decades later the observer has the benefit of hindsight. Gordon Moore's deep and focused skills were uniquely matched to the masculine technological world beginning to emerge around him at the end of the war. He would become the right person, in the right place, at the right time. Meanwhile, Betty's intuitive, decisive can-do approach would prove a highly effective foil to her husband's quiet, rational, deliberative, and understated approach to life's promises, challenges, and decisions.

By early 1948 it was clear to both Gordon and Betty that their relationship was not going to peter out. "In comparison to other men, Gordon seemed pretty sincere," observes Betty. Committing to "going steady," he gave her a birthday present that January: a garnet birthstone ring. In the spring she traveled up with him to Redwood City to meet his parents, Walter and Mira. Although smaller, Gordon's home on Westgate Street struck her as being similar to her own in Los Gatos. "It had modest furnishings and was very comfortable. I remember being at their house, having dinner. I helped Gordon's mom make a salad, and she said, 'Oh, you make good salads.' From then on, I was the salad maker! Gordon's mother seemed to be very open. Like any mother, she was protective of her son. She was looking at me, thinking, 'Here comes another one. How will this go?' We became very good friends."

Betty also got along well with Gordon's father, who "smoked cigars incessantly." With his love of horses and rodeos, Walter was "more of a farmer type" than Gordon. Betty quickly took his measure: "You always knew that Walter was in command. He was the one that ran on time, as does Gordon. Like him, Gordon lives by his watch. When out at a social occasion, Gordon's father would look down at his watch and say, 'It's time we got a humming.' That meant, if you were riding with them, 'Get your gear.'"

As Gordon and Betty were finding each other, Gordon was also realizing that his ambition reached beyond San Jose State. A far more intense scientific world beckoned in the form of Berkeley, fifty miles away. There, his commitment to Betty and to chemistry would deepen, with enduring consequences for his own life and for the wider world.

3

CHEMICAL APPRENTICE

BERKELEY AND THE BIG GAME

Cloyne Court

Gordon's ambitions were growing, his horizons widening. Early in 1948, halfway through his sophomore year, he and Bob Naughten reapplied to UC Berkeley. "I sent a transcript and a couple of letters of recommendation by professors at San Jose. The next thing I knew, I was accepted. I don't feel I suffered by going to San Jose State for my first two years." That summer Gordon Moore took a break from the cement plant to work for the highway department, hoeing weeds and cleaning up roadsides. Then he packed his bags, bid his home farewell, and moved across the Bay to Berkeley, to start his junior year.

California was on the verge of celebrating its first century. Its economy was booming. "The lights went on all at once, in a blaze, and they never have dimmed," wrote journalist Carey McWilliams. "One cannot properly place California in the American scheme of things. The gold rush is still on, and everything remains topsy-turvy. There is still too much commotion—too much noise and movement and turmoil." The port of San Francisco remained the busiest on the Pacific Coast, with a "crowd of ships riding at anchor in the bay." Writer Lewis Lapham, who grew up in the area in the 1940s, describes people dressed like characters from a mystery by Raymond Chandler. "The men wore hats and double-breasted suits; the women wore fur and high-heeled shoes. Together they danced to the music of Cole Porter and the Andrew Sisters." The computer was as yet unknown, "a giant robot confined to the realm of science fiction." Had San Francisco gossip columnist Herb Caen been asked to describe the meaning of the word *silicon,* "he most likely would have said something about an insect repellant or a Chinese tailor who had figured a new way to make silk shirts."

If San Francisco was a bustling, noisy city that in its enjoyment of postwar liberties was slow to grasp the possibilities of a more technological age,

Chemistry buildings of the University of California, Berkeley, 1940s.

SOURCE: BANCROFT LIBRARY, UNIVERSITY OF CALIFORNIA, BERKELEY.

Berkeley, a dozen miles across the Bay, was in the vanguard of the new era. "Cal" was a heady place in 1948, at the forefront of academic physics and chemistry and bidding for national and global leadership. The center of academic science had long been in the East, but in the years after World War I administrators at Caltech and at the University of California decided to foster research and began backing up their intentions in their budgets. At Berkeley Ernest Lawrence, a full professor by the age of twenty-nine, worked on ways to accelerate subatomic particles, leading to the first cyclotron, the renowned Radiation Laboratory, and the pathway to the atom bomb. The brilliant and influential dean of the College of Chemistry, Gilbert Lewis, built the College into a powerhouse, especially in physical chemistry, as electronic instruments began to transform the understanding of matter and its chemical properties.

By the late 1940s, Cal was on a roll. Two of Gordon Moore's chemistry professors, William Giauque and Glenn Seaborg, would receive Nobel Prizes within three years of his arrival. Seaborg had been a key player in the

Manhattan Project and would be at the center of subsequent thermonuclear studies and their Cold War implications. Using the cyclotron, he isolated and identified the chemical element berkelium. Seaborg would become the principal or codiscoverer of ten elements, including plutonium, einsteinium, nobelium, and element 106, eventually named "seaborgium" in his honor. Breakthrough discoveries were the stock in trade, with the university's roster including (as alumni, faculty, or students) a dozen others who were or would become Nobel laureates: Melvin Calvin, Joseph Erlanger, Lawrence Klein, Willis Lamb, Willard Libby, Douglass North, Thomas Schelling, Wendell Stanley, Otto Stern, Henry Taube, Harold Urey, and Selman Waksman.

Gordon Moore and Bob Naughten began humbly enough. They rented a shared room in a dilapidated house on Ridge Road, a block from campus. Naughten paints a colorful picture: "We had the craziest landlady. She made it clear that we were not to use Lifebuoy soap, because she didn't like the color it left in the showers! The place had about three rooms. It was really old; the mattresses were like padded foxholes." Aged nineteen, and lacking culinary skills, the pair quickly signed up for meals at nearby Cloyne Court, an all-male hotel turned student-housing cooperative. Built in 1904 and named for the region in Ireland where George Berkeley was once bishop, Cloyne Court had thirty-two suites, connected to public areas through paired private stairways. "The co-op really took good care of us. The bill was $26 a month for food."

Gordon remembers the situation as very economical: between them he and Naughten paid $30 a month rent, and tuition was free to California residents. A student who minimized his expenses could live for an academic year on $550, an amount Gordon could earn during a summer. Cloyne Court was entirely student run. Its male residents organized dances, inviting the young women who lived at nearby all-female halls. Residents, and "outsiders" who came in for meals, were required to do weekly chores to reduce running costs. Frugal with his time, Gordon quickly decided that "if I signed up for pot washing, at the bottom of the pecking order, I could have any time I wanted. For the better jobs, you couldn't choose good times. With pot washing, I could write my own ticket and schedule my two-hour stint to include lunch—I washed pots for an hour and a half, and I got to eat."

As when he encountered the chemistry set at age eleven and saw how easily he could request the ingredients and make explosives, Gordon took advantage of California's massive investments in public education and its uniquely low-cost system. It was a fortunate coincidence that he arrived at Berkeley as it rose to the heights of scientific eminence, a coincidence he would exploit fully.

Nobel Science

Gordon was still very much the small-town boy, an innocent abroad. As for other wet-behind-the-ears newcomers, the first challenges were logistical. Alongside arranging meals and work shifts, his principal difficulty was "to figure out how to get through registration." With so many veterans, the university was bursting at the seams, and he had to "finagle" classes. "It was very complicated getting the subjects you wanted, when you wanted them. There were a couple of professors you were told to avoid, but that was all the guidance I had." Courses were large, with fifty students minimum. Chemistry and physics classes "were too big to know who anyone was. I never kept any associations from that time."

Gordon had disposed of introductory requirements while at San Jose State. Now he could narrow his focus to chemistry, math, and physics. He also took a course in chemical engineering, covering material and energy balances as well as unit operations. While this gave him "an exposure that I would otherwise have completely missed," it only confirmed his decision to become a chemist. "I had a preconceived idea of where I was going, and engineering was not part of it." He remained self-contained, a nondescript face in a sea of students. He neither sought nor attracted a mentor. On occasion his education was painful:

> I had a physics class every term for two years. The one I remember was a horrible course: Introduction to Modern Physics. We'd make our way into the classroom, and the professor would have already filled the entire blackboard with small print, his lecture for the day. He'd just read down through it. I worked my tail off in that class and couldn't figure out why I didn't do better. I found out too late that he'd been giving the same test for years. Everybody else had copies; I was trying to learn the stuff! I don't have a very fond memory of that class.

Chemistry was increasingly a mathematical science, especially the subdiscipline of physical chemistry. Physical chemists sought to unpack phenomena in terms of molecules, atoms, and even subatomic particles, like electrons and nuclei. New instruments powered by electronics were indispensable, translating material realities into electronic signals. Math was at the fore. This suited Gordon well.

Math was memorable for other reasons. His teacher Pauline Sperry would be dismissed along with thirty other faculty members for refusing to sign the period's notorious "loyalty oath," denying membership in the Communist Party or in organizations that advocated overthrowing the federal government. At twenty Gordon Moore was already skeptical of "political" matters. Like literature, these subjects resisted the certainties of math and measurement. Tellingly, he remembers Professor Sperry as "emotional," someone who would "rant and rave" about her cause. "I learned some modern algebra, but absorbed more politics than math." Two decades later he would reprovingly declare: "We are really the revolutionaries in the world today—not the kids with the long hair and beards who were wrecking the schools a few years ago."

Despite dramatic left-wing activity, much of it associated with organized labor, the Republicans retained overall control of California politics in this era, particularly in San Mateo County and especially in rural ocean-side areas such as Pescadero. With little sign of any inward reflection, Gordon adopted the political identity of his family: California Republican. In the late 1940s and early 1950s, that implied a pragmatic orientation. With rapid, relentless growth in population, the state was contending with a host of problems, and Republicans responded with an approach that was low in ideology. When it came to taxation, the growth of government, and state requirements for businesses and individuals, there was an emphasis on getting things done as economically as possible. This approach entirely suited the frugal, business-minded Moore family. So too did the stress on freedom, self-determination, and entrepreneurship, all values that resonated with a family steeped in pioneer heritage. Governor Earl Warren—an accomplishment-minded Republican—offered a more plausible ideal than did Professor Sperry. Gordon instinctively liked his perspective.

Politics was of little interest compared to Gordon's animating passion for chemistry. As postwar optimism shifted to discourse on hydrogen bombs, the arms race, the Cold War, and the military imperatives behind the funding of research, he paid scant attention to the broader scene, focused as he was on the lab and the library. In this fresh milieu of world-leading chemistry, Gordon experienced a personal awakening. As his ambition grew by leaps and bounds, these were formative, significant years. Before, it had been sufficient to identify strongly as a chemist. Now, he was determined to become a first-rate chemist, someone who could work alongside and grow into a top researcher.

His calm exterior concealed an inward thrill at close proximity to major figures: "There was a bunch of well-known people around Berkeley. Starting with Nobel laureates, I had Glenn Seaborg and William Giauque. Knowing that Seaborg was going to be teaching nuclear chemistry was one of the reasons I took the course." Another well-known teacher was Wendell Latimer, who on the first day of class asked Gordon and other students to memorize the periodic table and tested them on it until they had it perfectly. "He gave a good overview of why the periodic table is set up the way it is. We heard about things like rare earths that we were not inclined to learn much about anyplace else. He must have talked about silicon along the way, but not very extensively. We learned how oxidation potentials varied systematically as we went down the periodic chart. That gave me a useful perspective."

Gordon took an organic chemistry course taught by another Nobel laureate-to-be, Melvin Calvin, noted for important research on photosynthesis in plants. "He was a good lecturer. I always enjoyed the lab work. Organic synthesis was fun." Donald Noyce, a newly arrived instructor, oversaw Gordon's organic chemistry lab. Eight years later Gordon would join Noyce's younger brother Robert at William Shockley's lab, and as their partnership blossomed, Gordon Moore and Bob Noyce together would launch two of the world's leading firms in electronics.

One chemistry class stood out because it first challenged and then confirmed Gordon's awakening ambition. It was a graduate-level thermodynamics course with William Giauque, another Berkeley star. Undergraduate Gordon wangled his way into this graduate offering: "Most of my chemistry professors won the Nobel Prize after they had experienced me! But Giauque won the Nobel Prize while I was there, and I wanted to see what he was doing. On the midterm exam I got a zero. No points at all! I'll never forget getting a zero on an exam. That really grabbed my attention."

Gordon wanted to play in the top academic leagues, but could he live up to his own ambition? "I worked on it a little bit and ended up with the second-highest grade on the final." Still an undergraduate, he bested all but one in a group of the nation's premier graduate students in chemistry. Roommate Bob Naughten noticed the change. "He became very, very studious. I was amazed by the superb academic work he did." Gordon was learning that if he focused with unwavering effort on a chemical task, there was little that lay beyond his talents.

Rockets and Rose Bowls

According to the university's catalog, students at Berkeley could enjoy mellow days picnicking in Marin County at Point Reyes; overnight camping trips to the Lick Observatory, Yosemite, and Lake Pillsbury; Sunday-evening dinners in San Francisco; and bull sessions with friends. Gordon had little opportunity for such leisure, given his "heavy bunch of classes" and pot washing at the co-op. The occasional spare hour found him taking in a ball game instead. "I went to the football games as a spectator rather diligently. In those days, there was a separate men's rooting section; no women allowed."

On weekends he and Naughten drove back across the Bay, not coincidentally avoiding demonstrations against the California loyalty oath (the largest student protests ever witnessed in the United States until that time).

Naughten recounts other boisterous events, such as a parade in 1948. "When a student float went by the reviewing stand, the students pumped water on all the dignitaries. The city really came down on Cal. We didn't participate." Instead, they would head south to be with their families. "In some ways, our lives were still centered there," reflects Naughten. "On Sundays we would play pinochle with my dad, then drive back to Berkeley."

Gordon's love of explosions matured into an appreciation of rocket science. In that first year at Berkeley, he belonged to the Rocket Society and designed a rocket engine, which he kept on his desk in Cloyne Court. On a visit home, he and Naughten set the rocket off: "It went completely out of sight. We don't know how high it went. It came down through a two-story house and exploded in the living room! We were at Gordon's home in Westgate Street. In came his father, who was livid about 'screwing around with rockets.' Nobody was hurt, but he knew the only two people who could have done that."

From Redwood City, Gordon frequently drove south in his Model A Ford to Los Gatos to see Betty Whitaker, now in her senior year at San Jose State. Sometimes he would head out with her and Bob Naughten on a jaunt. Naughten remembers that "once we had an accident driving back. The car ahead of us stopped suddenly, and we didn't. The guy was drinking a chocolate milkshake on the freeway. All of a sudden what looked like blood was all over his windshield. We said, 'Oh, God, we've killed somebody!' Gordon's father was the deputy sheriff of the county, and here we were being hauled off, the two of us, with this young lady. We were in the police station for quite some time before anybody confirmed that his dad was the chief deputy."

During the New Year's vacation at the start of 1949, Gordon and Naughten ventured to Southern California, to Pasadena, to watch the Berkeley football team—the Golden Bears—play Northwestern's Wildcats in the Rose Bowl. "Betty didn't go to the football games," explains Gordon. "Those were different times." Naughten adds, "Gordon and I went by train. Two lovely coeds came up and wanted to know if we would like to play bridge. We could play pinochle but didn't know how to play bridge. I said, 'Oh, God. Of all the skills we needed.'"

Betty in Berkeley

In 1949 Betty graduated from San Jose State with a degree in journalism. The job market was saturated with war veterans. With the renewed emphasis on marriage and domesticity, no employer was interested in her qualifications. She remembers how, for anyone female, it was an "awful time" to be looking for work. "I never, ever intended to stay on the ranch.

I went to all the major newspapers across the whole Bay Area, including the *San Francisco Chronicle*. They didn't want to hire a woman." The only offer was from a local paper on the peninsula, to organize its small ads and write obituaries.

Betty settled for a clerical position in San Francisco, at the US Rubber Company. There she worked "in the back, typing, in an area where the legal stuff was coming through. With my earliest paycheck, the first thing I did was to go out and buy a refrigerator for the ranch. It was the one thing I wanted to do for my mom." Betty lived in Berkeley with her favorite uncle, William Metzler, and his wife, Julia, who were childless. She could also be conveniently near Gordon, while commuting to the city on the F train, over the Bay Bridge. Earlier, Uncle William had nurtured her political interest, but he was "much too liberal. I never discussed politics with him, because you don't want to have that edge to things. I have never been involved in overt political activity. Neither has Gordon."

Among other obligations, William Metzler was a consultant for the US Department of Agriculture and traveled frequently. On one trip, involved in a serious road accident, he was hospitalized. "My aunt had to fly to the East Coast and be with him. I stayed in Berkeley, ran the house, and took care of all the pets, because she was never without cats and dogs—that was their family." Metzler was working on a history of the Southwest Indians. Gordon, accustomed to a hard-science approach—measuring, recording, performing analyses, and formulating conclusions—found Metzler's labors hard to fathom. "He was writing that book for years. It was an historical and sociological study of labor problems, social problems, crop failures—everything. You name it. He kept changing and rewriting it."

Gordon and Betty saw little of each other during the week. "Gordon had to wash pots, so we weren't able to date much," says Betty. On weekends they would drive home together to see their families. Gordon often stayed for dinner on the Los Gatos ranch. "My family liked him. One night, while I was preparing the dinner, I heard my grandfather say, without Gordon hearing, 'You know, it's time for these kids to get married.'" Indeed, as 1950 dawned, their relationship entered a fresh phase. This time Gordon took Betty with him to the Rose Bowl. The Golden Bears were playing the Ohio State Buckeyes. Betty stayed overnight with her elderly great-aunt, Mary Metzler's sister Cora. It was Gordon and Betty's first trip together. On a later occasion they would take his mother, Mira—a willing chaperone—with them to Lake Tahoe, stopping in Reno. "I had never been, and wanted to see what it was like," remembers Betty. Gordon, at this stage, had little desire to travel. He observes, "By the time I finished grad school, I had never been east of the far reaches of Southern California. I used to say, 'Never east of Reno,' but I find that Los Angeles is east of Reno."

Elite Aspirations

An important influence on Gordon as he entered his senior year was George Jura, an assistant professor, recruited in 1946 to help broaden Berkeley's physical chemistry program. He did this with the help of another new faculty member, Don Gwinn, whom Gordon later identified as the teacher "I knew best." Together, Jura and Gwinn introduced novel experiments, altering the undergraduate laboratory course by putting an emphasis on independent thinking rather than the ability to follow directions. Important to Gordon's growth was that students had to assemble their own apparatus. As a senior Gordon took Jura's class on surface chemistry and quickly became his star pupil.

Thanks to Jura, Gordon began to enter a whole new world: original research, with its triumphs and tribulations. He loved it. It was a matter of thought and trial and error, of experimenting to find answers, not simply looking them up in a book. "Jura's view was that the literature was almost all wrong: 'Find something and explain why it's wrong.'"

He began with a simple task:

> Looking at a complex—something with four halogen atoms. The chlorine version had been done: "Why don't you see what happens if you substitute bromine?" I started out on the wrong foot, using nitric acid, which is what I had available. It screwed up all my experiments. I had to go back and use hydrobromic acid. I spent one long, hectic afternoon collecting my data and getting it right. I was doing a lot of titration. I determined the chemical composition of the compound successfully. It had the same number of bromine atoms as the chlorine version had of chlorine. I wrote a paper. He was very happy with it. I remember him saying it was the best paper he'd ever had.

Jura was interested in why unusual colors appeared when certain chemical ions—electrically charged atoms and molecules—were absorbed into silica gel. He asked Gordon to figure out which ions did it and which didn't. "I spent time trying every metal I could think of as the cation [positively charged ion]. Eventually, I discovered the anion [negatively charged ion] was the difference." The electronic nature of the materials—whether they had a surplus or deficit of electrons, the carriers of electrical charge—determined their behavior. Establishing "what was going on" provided enormous satisfaction. "Jura pointed me in a direction. I did the experiments and collected the data. I was a volunteer—just learning. I was happy to do it."

Jura also initiated Gordon into another skill: glassblowing. In later years, as he built innovative equipment for semiconductor research, this would prove immensely useful. "I started by watching him, then trial and error. If you wanted to build something complicated, by putting together a series of stopcocks and the like, you could teach yourself. I enjoyed that. I did a lot of glassblowing in my next years." Gordon by now was fully focused on physical chemistry, a field that was mathematically rigorous, whose preeminent researchers were American, and that was growing in theoretical and practical importance.

Molecular structure was beginning to be understood, in its role as a determinant of chemical properties and behaviors. Chemists were employing vacuum tube–based instruments to probe molecules, using radiation—from visible to infrared light and from radio waves to microwaves—to generate electronic signals through which to "read" structure. Gordon himself used one of these instruments, a Beckman spectrophotometer made in California, as he moved closer and closer to independent research. "I really liked it."

With Betty living nearby, he increasingly stayed on campus on weekends. "Study came to Gordon easily," remembers Betty. Life remained busy—"I don't remember having a lot of time to waste"—but he was "pretty happy. I was certainly getting the education I wanted." With science enrollments expanding and university departments growing, the next step for any talented student was a PhD degree. Gordon knew he wanted to be on the frontier: "I needed an advanced degree." Remaining at Berkeley was not a realistic option. "They give you a strong push, unless there is a compelling reason for staying."

Gordon discussed with his roommate, Bob Naughten, the merits of various graduate schools. Already comfortable among the emerging elite of American (and world) science, Gordon wanted somewhere at least equal—and perhaps even superior to—Berkeley. He made the California Institute of Technology, in Pasadena, Southern California, his first choice. He also began to consider heading east and applied to both Princeton and Chicago. "Princeton had an excellent reputation, and it was in a different part of the country. I thought it might be interesting to see."

At twenty-one Gordon was poised to venture well beyond the familiar arena in which he had grown up. He had discovered and begun to display a clear intellectual talent. He had developed special skills and found fresh passions that would bring him into alignment with the dawning age of high technology. Though he was as yet little aware of commercial realities, he had realized the importance of focus and hard work. Old habits, however, died hard, especially for a conflict avoider. "I'm a terrible procrastinator. I waited until the last minute to apply to graduate schools. There was some concern from my adviser that I might be too late." Fortunately, Princeton offered him a place without requiring him to visit: "If Caltech hadn't come through, I would have gone there." As it was, he belatedly received an acceptance telegram from Caltech. It meant he could stay in California.

By now, Gordon was establishing a quiet reputation as someone to watch. "He knew people were looking to him, because he was able to get out of Berkeley and immediately go to Caltech—a highly regarded school," Betty observes. Bob Naughten, top of the class at Sequoia High, was not so fortunate: "Gordon was really kind. Even though he had been accepted at Caltech, he was waiting for my acceptance before we had a party about it. That didn't occur. I applied to the West Coast med schools—Stanford and Cal—but the overwhelming number of veterans applying meant, 'No.'" The two friends graduated from Berkeley on June 15, 1950. After that their paths quickly diverged. The Korean War started ten days later, and Naughten, a reserve hospital corpsman, was called up immediately. Gordon went back for a final summer at the Pacific Portland Cement Company. This time he had a much more responsible position, in quality control. It was his first opportunity to observe how chemical expertise could directly influence commercial results.

> The silos were full of mud and limestone. You had to mix them so they'd fire properly. You threw the slurry down into a big kiln. You had to get the right percentage of calcium carbonate, or the cement would set in ten minutes instead of two hours. For the people driving the trucks, that could be a big problem.
>
> This was basic analytical stuff. I enjoyed it. I was telling people what to do, mixing all the reagents, and defining procedures. There was concern about air pollution, so I set up carbon monoxide analysis, took samples, and analyzed for residual oxygen. That was a big deal for me.

TWO'S COMPANY

Life Choices

Accepted at Caltech, Gordon faced a major life choice. Would he ask Betty Whitaker to come with him? In 1950 such an invitation was tantamount to a marriage proposal.

Unsurprisingly, Gordon hesitated. His deep-rooted pattern of avoidant behavior loaded the decision with tension, uncertainty, and internal conflict. Gordon had been raised in a world of clear divisions between the sexes. He knew well the role of a "Moore male," while his parents had also modeled the benefits of a steady marital partnership, and in Pescadero his aunt Louise had provided a vivid example of female ability to handle social relations. When he could postpone no longer, he chose Betty and marriage.

If belated, the decision made late that July would remain firm and unwavering for the rest of his life. "I was about to run off to graduate school. Betty was concerned because Pasadena was a long ways away. So she encouraged me, and we were married."

Betty's version asks, "Why didn't we get engaged sooner?" and correctly answers:

> He wasn't ready to ask me! And I wasn't waiting for him to propose. I was focused on getting my degree and going out to work. I landed at US Rubber. After a year, my job ended. I went back to the ranch, started interviewing in San Jose, and quickly found a position working for the city government. Then Gordon said, "Hmm. Wait a minute." He was going away in the fall and had to think about whether he wanted to take me with him. All of a sudden, he decided he wasn't going to go without me. He said, "How about getting married?" He never asked me about whether he should go to Caltech. It was, "I'm heading there. Do you want to go on the ride?"

Gordon's proposal was "matter-of-fact, not from any movie. I would have been very embarrassed with something like that." Having agreed to marry, Betty called her new employer to resign. "That ended wherever I was going to go in that career." In the 1950s, for better or worse, a man's decision on employment took unquestioned priority. Marriage occurred early, and children were the obvious and natural next step. Neither Betty nor Gordon wished to question the prevailing social order. Instead, the challenge was to make a success within it. As Betty recalls, "I was ready. I was really gung ho then. I wasn't afraid of anything."

Gordon was due in Pasadena on Monday, September 11, 1950. The wedding was set for the Saturday two days before, which made the month of August a hectic time for Betty: "It was a record short engagement." Her family was thrilled and excited about the forthcoming nuptials; an engagement party took place in Pescadero. Betty recalls, "Gordon's father told him he was crazy to get married at twenty-one." Walter Harold Moore "always enjoyed Betty," according to Gordon, but feared the marriage would suffer from the couple's youth, compared to his and Mira's timing. Moreover, remembering his own experience, he saw the oncoming Korean War as a threat. Yet Walter showed his fondness for Betty by giving her a precious diamond out of his deputy sheriff's badge for her wedding ring. "It was perfect, a wonderful color. I never did have an engagement ring."

Gordon's sequence of behavior—procrastination, a low-key proposal, no engagement ring, ducking the wedding preparations, and being unavailable in the run-up—displays in extreme form certain classic male behaviors and speaks volumes of his suppressed anxieties and need for control. He makes no apology for his avoidant actions, explaining, "I planned to work right up until I got married. I had some things to get finished. I was even going to work on my wedding day, but I needed time to get ready for the move." Perhaps heeding his father's words, he was unconsciously lessening the significance of the event, and hence the psychological risk should the marriage fail.

Within four weeks, showing her natural decisiveness and flair for organization, Betty assembled all the ingredients for the wedding: church, gown ("the dress with the train—the works"), fittings, and invitations, "all printed and out." Gordon's comfort zone was in the lab. He remained at Pacific Portland Cement until the day before the ceremony, sending Betty the message that he was in charge. Yet Betty understood the subtext and could on some level sense his vulnerability. She too was accustomed to solitude and autonomy and knew the profit and the pain of being alone. However casually their desire was expressed, both Gordon and Betty believed they had found a fit.

The couple was married in the Advent Christian Church on Homestead Road, Santa Clara, a small building distinguished by a steeply pitched roof and pointed arch windows, with one central aisle. The wedding took the form of a candle-lit service at eight o'clock in the evening, similar to Walt's marriage ceremony in Half Moon Bay, two years before. "I loved the thought of it," says Betty. "A nighttime wedding was so different from the norm, particularly in September. We had candelabras set up all across the front, all along the side, and big tapers. It was already dark by eight. It really was beautiful."

Betty's big disappointment was that her uncle and mentor, William Metzler, by now living in Washington, DC, was unable to give her away. Even so, their extended families made an impressive crowd (as they did at the family's centennial celebration picnic, two weeks later, an event necessarily missed by the newlyweds). "The church was full. All of Gordon's relatives, all of mine, eighty or a hundred wedding guests, as many as we could fit in." Uncle Russell stood in for Betty's long-deceased father. Later that evening Gordon and Betty held a short reception at the big downtown clubhouse of the Los Gatos History Club.

At last, the newly minted "Mr. and Mrs. Moore" climbed into a loaded-up car (a 1935 Dodge, the gift of Uncle William who had, on moving east, left it behind), said farewell to family and friends, and drove away toward Pasadena and the South. They stopped at a motel for their wedding night. The next day they resumed the journey, with Betty wearing a new off-white travel suit and hat. "It was hot, and the car windows were open. My hat went flying out. Gordon stopped and went back and found the hat. I kept it for years."

In an old car, with a new hat, alone together, and barely out of their teens, Gordon and Betty Moore—heeding the call of scientific research and

Gordon and Betty cut their wedding cake, 1950.
SOURCE: KEN MOORE.

Moore family centennial picnic, 1950.

SOURCE: SAN MATEO COUNTY HISTORY MUSEUM.

a brighter future—were setting out together. Neither knew where this move would take them, but both knew it would be away from the rural lives of their childhoods. And Gordon was not one to encourage introspection. "Our honeymoon," he observes, "was a day of driving from Northern California. I had to be in Pasadena on Monday to start taking tests."

A Fresh Milieu

Gordon and Betty had been to Southern California together once before, to the Rose Bowl. Now they were heading to Caltech, a far smaller and more focused school than Berkeley. Their immediate challenge was to make a success of marriage and graduate school, with all that they entailed. Gordon's initial aim was to find his way at Caltech and progress in chemistry as an experimentalist. His choice of graduate school was congruent with a wider high-tech migration to Los Angeles County, where enormous growth in the electronics and defense industries was matched by a population explosion: from 2.8 million residents in 1940 to 4.2 million in 1950. Pasadena and the Los Angeles Basin were playing a leading role on the new frontiers of electronics and war—both hot and cold.

Los Angeles County stretches from the ocean-side communities of Malibu, Santa Monica, El Segundo, and Long Beach eastward across a flat basin through Hollywood and central Los Angeles and onward to Pasadena and Claremont at the foot of the San Gabriel Mountains. The county had become home to the R&D and manufacturing operations of giant aviation and aerospace firms such as Hughes, Northrup, North American, Douglas, Convair, and Lockheed. The region, with its clear blue skies and proximity to the Pacific Ocean, was also dotted with major US Air Force bases.

Through World War II, and increasingly thereafter, aerospace technology had become inseparable from electronics. Vacuum tube circuits filled the communications gear, instrumentation, radar systems, and weapons aboard aircraft and missiles. Vacuum tubes also enabled the massive ground equipment that guided, controlled, and monitored those systems, and tubes suffused the R&D facilities behind these creations. For example, Northrup, an aerospace giant, was a major customer for IBM's tabulators and calculators and inspired IBM's Card Programmed Calculator, a protocomputer humming with relay switches and vacuum tubes, which crunched the numbers needed for missile design and testing.

When the newlyweds drove into the county, Northrup had recently taken delivery of the very first commercially produced electronic digital computer: the BINAC—Binary Automatic Computer—made by the Eckert-Mauchly Computer Corporation. This company was a Philadelphia spin-off from the pioneering ENIAC electronic computer project, at the

University of Pennsylvania, that had been funded by the US Army during World War II. ENIAC was slow, cumbersome, unreliable, and truly massive. It had almost eighteen thousand vacuum tubes and occupied more than a thousand square feet (see photograph on page 125).

BINAC aimed to be an improvement; it was the first commercial entrant in the hoped-for computer industry. Northrup could never get BINAC to work properly, but progress elsewhere in the county meant that by the end of 1950, the SWAC—Standards Western Automatic Computer (a homegrown creation of the federal government's National Bureau of Standards)—became the first working electronic computer on the West Coast, going online in Los Angeles. SWAC was also massive and clunky. Thousands of "tube" switches reduced mathematical calculations and information processing tasks to "digital" operations defined by the digits zero and one, by "on" and "off" switches. The speed at which tubes could switch became the speed of computation. As Gordon later observed, these were computers "in glass rooms tended by monks who knew how to do the proper incantations."

Digital electronics used the elementary "If X, then Y" format of formal logic to create a language in which a chosen logical reality became a specific digital series of zeroes and ones. Sequences of this kind were themselves ideally suited to being represented by the switching actions ("on" or "off") of vacuum tubes and, later, transistors. Amazingly and counterintuitively, it turns out that digital electronic codes and communications can thus be made to model, transmit, and replicate many human realities—from thoughts to poetry, plays, speech, and scenes, indeed the whole repertoire today communicated by computers, cell phones, cloud, Web, TV, and their endless adjuncts.

Caltech and Pasadena were very much part of the early developments that bridged the worlds of electronics and armaments. World War II had aided the fortunes of Pasadena, setting it firmly if cautiously on the pathway toward high technology as it evolved into a center for research on, and light manufacture of, electronic instruments. Caltech's serene campus was nestled between the historic downtown and a broad residential swath. Seven miles away lay the Jet Propulsion Laboratory, the army's guided-missile lab, an organization with which Gordon Moore would in later years be deeply involved. Caltech ran JPL for the government, working with aerospace contractors in the county and beyond.

In 1950 JPL was deep into the development of the United States' first guided missile with a nuclear weapon: the Corporal. Vacuum tube electronics were essential to its guidance systems. With the Cold War intensifying, reliable means to destroy Moscow and other priority targets with atomic (and soon hydrogen) bombs were a priority. The Corporal could climb into space, its trajectory tracked by radar and adjusted from the ground, before falling back to earth. When Gordon arrived on the Caltech campus, it had just set a record, reaching 244 miles, the highest ever for a missile.

Closer to the Caltech campus, and to Gordon's immediate interests, were two Pasadena firms—Beckman Instruments and the Consolidated Engineering Corporation—shaped by war, though also using electronics in other very different ways in 1950. Beckman Instruments, in South Pasadena, was the earliest spin-off from Caltech's own chemistry laboratories. Arnold O. Beckman, who would later enable William Shockley's move to Palo Alto, had earned his chemistry PhD at Caltech in the late 1920s and joined the faculty. From an East Pasadena garage, he also worked on his own inventions and consulted for other firms. Using vacuum tube electronics, he created a pH meter, an instrument that allowed for the quick and highly accurate determination of acidity, a fundamental chemical measurement. Beckman sold his first pH meter in 1935, for use in monitoring the acidity of lemon juice by-products for the area's citrus industry. The pH meter was a smash hit, and Beckman soon left Caltech to pursue his growing commercial activities. His vacuum tube–enabled devices found multiple uses in World War II and were invaluable to production of fuels, explosives, synthetic materials, and medicines, leading to rapid postwar growth in the commercial instrumentation industry.

The pH meter was both a market success and the herald of a dawning "instrumentation revolution" in chemical research. Chemists of many stripes quickly adapted the novel electronic devices, with infrared spectrophotometers especially popular. A graduate student or technician could suddenly accomplish in a week work that earlier would have taken an entire career. These instruments dramatically expanded the reach and ambition of researchers.

By 1950 the instrumentation revolution was at a rolling boil. Beckman now produced a range of devices that used vacuum tube electronics to measure matter's physical properties. The Consolidated Engineering Corporation was another Pasadena instrumentation firm with close ties to Caltech. It too manufactured electronic instruments for chemistry and was the leading maker of mass spectrometers, machines that broke samples into highly energized fragments, filtered through electromagnetic fields.

Another vacuum tube–based technology—black-and-white television—was seeing rapid growth. *The Jack Benny Show,* made in nearby Hollywood, debuted that year. Thanks to Aunt Ruth and her get-ahead husband, Carl, Betty's grandparents in Los Gatos had been among the first in the United States to experience the delights of this new medium; from here on, other American consumers would adopt television at an astonishing rate. In 1947 100,000 television sets were produced in the United States; three years later

7 million were being sold, and by 1955 more than half of all households had a set. A pattern rapidly emerged in which the average set was switched on for more than four hours each day.

With the screen and receiving circuitry both dependent on vacuum tubes, televisions were fully electronic creations. Growth in their production and sales was symbiotic with the growth in broadcasting itself. Hollywood fast became as great a center for the production of television shows as it was for films. The new reality of digital, electronic information was beginning to demonstrate its hypnotic effect on human minds and motivations. Systems and devices built with vacuum tubes—electronics—were on the rise. In the military realm, electronics were used not only to communicate, but also to generate fresh information and to deploy that information for many different kinds of control. The same trinity of digital information, electronic communication, and remote control spread to government offices and commercial workplaces.

Telephone networks increased, and electronic machines (tabulators, desk calculators, and the like) became essential tools. In factories digital electronics enabled novel automated machinery. In laboratories complex instruments for measurement, calculation, and information handling became ever more common. Electronics in the home became ubiquitous. More than 90 percent of dwellings now had a radio, more than had a refrigerator. Television purchases grew. In the more than 60 percent of houses with a telephone, usage doubled, though minuscule by later standards. A five-minute call from New York to San Francisco cost around thirty-five dollars in today's currency.

IMPACT OF ELECTRONICS, 1940–1950

	US households with:				
Year	**Telephone**	**Radio**	**Refriger-ator**	**TV**	**Total US electronics sales**
1940	37%	73%	60%	0%	$340,000,000
1950	62%	91%	85%	9%	$2,705,000,000

Year	**Vacuum tubes produced**	**Value**
1940	108,000,000	$27,600,000
1950	382,000,000	$250,000,000

Sources: Electronics Industries Association Electronic Yearbooks; US Census Bureau.

Pasadena was close to both Hollywood and defense industries and Caltech's advanced scientific and engineering research. It was a milieu rich in opportunity for newlyweds Gordon and Betty, who themselves were in the midst of profound changes in their own lives. In Pasadena they would establish the basic rhythms of their long years together. At the same time, much about this area and the wider world was quite foreign to them—especially to Gordon—and their focus remained narrow, even provincial, within this frame of emerging opportunity.

As a matter of practical reality, the couple's first nights in their new hometown were spent in a motel, looking at rental listings in the local newspaper. "We saw some pretty ratty places," says Betty. They settled on half of a duplex, on Catalina Avenue, less than a mile from Caltech. Betty remembers, "We had a strange landlady. She said, 'You can make a big pot of spaghetti at the beginning of the week, and then you don't have to cook.' She was snoopy. When we were away, she was in our apartment, looking at my wedding gifts. She used my brand-new Mixmaster and tried to mix something so heavy that it burned out the motor; after that it howled every time I used it."

In 1950 Pasadena itself had not fully shaken off the aura of a genteel town, with a benign climate and a low-key economy. A 1939 study by Dr. Edward Thorndike of Columbia University had named it the best US city in which to live, having more radios, telephones, bathtubs, and dentists in proportion to population than any other city surveyed. The Arroyo Seco Parkway offered a fast route to Los Angeles, and the establishment of Bullock's, the award-winning "department store of tomorrow," initiated an exclusive shopping area on South Lake Avenue. Even so, Pasadena had its share of quiet desperation. One observer in the late 1940s wrote, "The rich and retired live in seclusion so complete and so silent that in some of the residential hotels one scarcely hears anything but the ticking of the clock or the hardening of one's arteries."

Pasadena was very different from the Bay Area. Soon, Betty was homesick. "It was hard to leave Los Gatos." Familiar scenes and extended family were her bedrock, and she was troubled to think that her aging grandmother Mary—with whom she had a special bond—might die while she was away. Keeping busy was the remedy. While Gordon tackled his entry tests at Caltech, Betty reviewed the help-wanted advertisements. She had dismissed the prospect of becoming a journalist, but as the wife of a PhD student, she expected to support her husband through graduate school, seeing this as a necessary step on the pathway to a financially secure future. An agency soon placed her with the Consolidated

Engineering Corporation, in public relations, helping to prepare advertising for the firm's electronic devices. Betty's jobs included answering phones and typing materials to be printed. "We all worked very hard, nine-hour days. We were running slick magazines to promote the leak detectors that the corporation made. The brochures for trade shows had to be ready to go with the salesmen."

The company had adapted its technology to make hypersensitive leak detectors for vacuum systems, used, among other things, in the production of uranium for atomic bombs. Detectors sold for several thousand dollars each. Gordon, accustomed to working at the cement plant and on his own laboratory experiments, was struck by how technology could be adapted to such needs. Naively, he was also stunned by the financial opportunities: "I was flabbergasted that anybody would pay that much for a leak detector. I couldn't imagine spending thousands of dollars on one. I did it with a little spark coil that went on the outside of the glass. How the world was changing! That was my reaction."

Gordon and Betty moved twice more, remaining in their third rental, the downstairs of a house, for the rest of their time in Pasadena. It was "nice," but not quite what Betty had hoped for. "I would have loved to have been on the other side of town, more towards San Marino." Gordon matter-of-factly explains, "There's a reason we didn't go over to San Marino: the high rent."

Tracking Down Badger

Compared to Berkeley, Caltech was small, but Gordon experienced the chemistry department as "pretty big." He was one of twenty in his entering class, in what was still an all-male institution. "The first woman graduate student entered as I left, in August 1953."

Caltech calls its departments "divisions," and the chemistry and chemical engineering division required incoming graduate students to take a battery of tests. Married on Saturday and arriving on Sunday, Gordon presented himself on Monday morning for the round of rigorous, consequential exams. With characteristic modesty, he offers, "They wanted to see if you had organic, inorganic, and physical chemistry. If you weren't up to snuff, you had to take a remedial course. I was the only one from another school who didn't have to take one of the remedial courses, so Berkeley did pretty well."

The tests were part and parcel of the matchmaking between the faculty and the fresh arrivals. Professors were looking for talented, dedicated students to further faculty research plans. Students sought mentors who might enhance their career prospects. It was a time to decide on mutual investments. Some students arrived with clear-cut plans to work with particular professors: Linus Pauling, whose star in the chemical community was burning bright, was especially popular. Gordon was not yet worldly wise enough to understand the game. "I went down to Pasadena with no real idea of whom I would work with. I started interviewing the various professors to see who was interested and had something interesting to me." Jack Kirkwood, a prominent physical chemist and theorist, looked at the test scores, saw Gordon had performed well in physical chemistry, and invited him on board. Gordon declined. "I wasn't a theoretician by inclination. I wanted something much more experimental."

Instead, he was attracted to the work of Richard McLean Badger. Caltech had been home to Badger for more than three decades, first as an undergraduate, then as a graduate student, and finally as a professor. Now in his midfifties, he had established a reputation as an experimentalist. His area of particular expertise was the study of molecules—their structures and bonds—using infrared spectrophotometers. His recent studies tied him to the instrumentation revolution and to the growing interest in infrared. Gordon had already used a spectrometer at Berkeley. Hence, even though "I'd never heard of him" before arriving at Caltech, it was only a short step for Gordon to decide that "Badger was doing things that looked fun," pursuing novel facets of infrared spectroscopy and exploring its power as a technique.

Though well known and well connected in the expert community, Badger was not a star like Pauling. He supervised only a small group, two graduate students and a postdoc. "He was getting along in his career, and that was all he wanted to worry about." Gordon clicked with him on several levels. Badger, the same age as Gordon's father, was low-key and reserved. Like Gordon, he enjoyed the California outdoors and was an avid camper. "He seemed somebody I could relate to easily." His style of work was also appealing: the hands-on crafting of inventive experimental equipment and its use in precise, exacting measurements. Here, practical, technical problem solving could be coupled with fresh scientific understanding. Badger's postdoc in this period, Oliver Wulf, described him as a "very careful investigator" who showed "meticulous care in his scientific work." For Wulf, the essence of Badger's style was his innovative use of instrumentation, opening up fresh lines of research. As Gordon recalls, "Spectroscopy was being used by other people as a tool, but for Badger spectroscopy itself was the main focus."

The chemistry faculty of Caltech, 1952. Linus Pauling sits at far right. Richard Badger stands behind Pauling, to the left.
SOURCE: COURTESY OF THE ARCHIVES, CALIFORNIA INSTITUTE OF TECHNOLOGY.

Professor Badger in his basement spectrometry lab.
SOURCE: COURTESY OF THE ARCHIVES, CALIFORNIA INSTITUTE OF TECHNOLOGY.

Badger was happy to have Gordon join his small group. As a low-key, reserved, but enthusiastic experimentalist with excellent entrance tests, Gordon was a good fit. Badger's laboratory was in the subbasement of the Crellin Laboratory, itself a major expansion of the division's resources, completed in 1938, and thus the newest of Caltech's chemistry labs. A brochure from the period describes how "the organic chemists occupy the second and third floors and the auxiliary rooms on the roof. Conveniently close, occupying the first floor, basement, and sub-basement are the physical chemists with their appliances for the study of molecular structure through photochemistry, magneto-chemistry, spectroscopy, and x-ray and electron diffraction." Put simply, it was lab benches upstairs and instruments downstairs. Infrared is closely associated with heat (much of the heat radiated by an object takes the form of infrared light). As Badger's instruments were extremely sensitive, his lab was purposely in the subbasement. As Gordon explains, "You need a very stable environment. If you left your desk, you had to leave a lightbulb on to make up for the heat loss in the room."

Badger's spectrophotometers used vacuum tube electronics, translating infrared into electrical signals that, with care and precision, could be recorded. Other Caltech professors in the subbasement were also using electronic instruments. Don Yost, who had likewise joined the Caltech faculty after earning his PhD there, was pioneering the microwave spectrometer. Yost had been a contender for Gordon's interest, but since his work centered on the instrument rather than on the chemistry, Moore chose Badger. "I talked to Yost, but I wasn't really an engineer. I was interested in the science."

Gordon also considered joining the much younger Verner Schomaker, who similarly took a PhD in chemistry at Caltech and then joined its faculty and also worked with delicate instrumentation in the Crellin subbasement. Badger, however, proved "very nice to work for." Gordon became ensconced in the lab, happily spending long hours underground. "In the morning, I'd go down and—except for classes—spend my whole day there." As is often the case, disorderly desks and workspaces went hand in hand with disciplined work and organized minds. "Badger lived in an office that was absolutely buried under piles of paper. I remember one of the other professors saying that he wondered if anybody who worked for Badger ever regained his neatness again. His office was a mess; his lab was a mess. I took to that very readily. You ought to see my office. It's pretty close."

Badger himself was hands-off. He would come around every once in a while:

> I didn't see him an awful lot. If I wanted to track him down, he was available, but I could go for weeks without seeing him. It wouldn't bother him a bit. He'd suggest I modify a piece of equipment a certain way, and I'd go off and do it. He didn't interfere with what I was doing. If I was going to take on a fresh project, I'd discuss it with him, and he'd decide if it would fit or not. He had to squeeze it into his ONR [Office of Naval Research] contract.

The funding from ONR was unremarkable. The US military, mindful of the role of science-based innovation in World War II (such as the atom bomb and synthetic rubber), supported an extensive array of basic research at elite academic institutions such as Caltech.

Luminous Personalities

Gordon's initial support came from the ONR contract and from his assignment to a two-year teaching assistantship under Linus Pauling. Pauling gave the general lectures in freshman chemistry and encouraged first-year graduate students to sit in. "He talked about whatever he was interested in that week. He was a lot of fun, but he didn't teach the freshman the chemistry they needed; that was left to us," says Gordon, who took to teaching after some initial jitters. "I started out not doing a very good job. I was overimpressed by the students at Caltech. The first session I had, I told them everything I knew in an hour! Once I slowed down to a reasonable rate, I enjoyed teaching and did a pretty good job at it. Undergraduates from other sections were soon sitting in on my classes before the exam, because they thought they'd learn more."

For Moore, the teaching experience confirmed his developing vision of himself as an experimentalist and academic. A university career equaling that of Badger, Yost, or Schomaker looked appealing and possible. His orientation placed him at the junction of physics and chemistry; now his concentration on infrared spectrophotometry intensified his engagement with both disciplines. "The kind of spectroscopy I was doing was done in physics departments as often as in chemistry departments. I was almost a physicist in that respect, more inclined to go to a physics seminar than an organic chemistry seminar." Still, Gordon attended the chemistry faculty's weekly seminar as well as "probably half" of the physics graduate seminars. He took three graduate courses from Pauling, including one in quantum mechanics and another on the nature of the chemical bond, in which Pauling "essentially went through his famous book."

While Gordon found Pauling "a pretty good teacher," something about the man unsettled him. As chairman of the Chemistry and Chemical Engineering Division, Pauling was one of the biggest personalities on campus. He was well on his way to establishing his reputation as one of the greatest chemists of the twentieth century and would win the Nobel Prize in Chemistry in 1954; being prominent on campus and off for his political activities, he also won a second Nobel Prize, in 1962, for peace. Gordon Moore was unnerved less by Pauling's accomplishments than by his personality, which communicated both charisma and caprice.

Quiet, reserved, and methodical, Gordon could not read Pauling or even—in Pauling's presence—retain his own composure. "He looked like a clown, after a clown had put on his makeup. I found him very intimidating. Even though I had several classes from him, he could ask me my name in a manner such that when I was asked, I wouldn't know the answer." Gordon's discomfort was so great that he resorted to finagling the exam committee for his PhD. Because of his extensive course work with Pauling, the latter was a logical choice. Gordon, however, took care to arrange it so that the exam occurred "while Pauling was out of the country, so he wouldn't be on my committee." It was a neat trick. After nearly five years in college, Moore was slowly shedding his youthful, backwoods naïveté. And he was practicing his technique of conflict resolution by avoidance.

Richard Feynman was another luminous personality, a famously charismatic, extroverted, and talented physicist hailed as one of the leading theorists of his generation and an exceptional teacher. Gordon arrived at Caltech around the same time as Feynman and was soon eagerly attending his seminars. "Listening to Feynman was a treat. It was physics taught by a Brooklyn taxi driver. He had the ability to make you think you understood everything he told you. I'd sit through one of his seminars and think, 'Boy, this is marvelous. Now I understand.' I'd try to explain it to somebody and had no idea how you got from *here* to *there*. Feynman would pull the wool over your eyes, and you didn't realize what was happening. It was a lot of fun."

His new milieu further energized Gordon. Diving deeper into physical chemistry, instrumentation, experimentation, and physics, he surfaced with renewed ambition and a clearer vision of himself as a future professor. He began to enjoy his teaching responsibilities and became more confident. While learning from stars such as Feynman and Pauling, he also had the benefit of opportunities for close collaboration with the much more stable Badger. Gordon's trajectory remained steady, marked by occasional "awakenings." Ten years earlier his first encounter with his neighbor's chemistry set "really got my attention." This interest led to explosives and the idea of becoming a chemist. At Berkeley his "zero" in Giauque's graduate course demanded he rise to the occasion. Now Caltech, too, was demanding his mastery of diverse scientific issues. Here, he recalls, "I found a lot of things interesting."

Life Outside the Lab

In keeping with the times, and with his own proclivities, Gordon assumed complete responsibility for the couple's budget and spending decisions. "In the early days, we didn't discuss money," says Betty. "He just took over: 'I'm the bean counter.' I went, 'Okay, good.' It seemed a natural order." While no stranger to frugal living from her childhood on the Los Gatos ranch, Betty found Gordon to be especially careful. Both shared an aversion to debt, and both desired to put away as much as they could. Neither could tolerate unnecessary or frivolous spending. They shared an understanding that their way of living would be built on hard work and good management. Even so, Betty found Gordon more intense than she was. After six decades of marriage, Betty observes that her husband, having amassed billions of dollars, "is still very frugal. If it's wearable, he'll wear it. I'm going, 'Honey, the collar has been out of style for ten years. Hello! Throw that shirt away.' He'll say, 'I have a drawer full of those shirts.' I'll say, 'But you don't need to wear them. Give them away.'"

Married, Gordon embarked upon a discipline that he would continue for years. He used an ordinary green speckled composition notebook to keep a detailed ledger of income and expenditure, together with notes about hypothetical and actual investments. "We didn't have a heck of a lot of money. We started accounting for where it came from and where it went." This record keeping reflected both his care for the couple's financial well-being and the propensity for meticulous data collection and quantitative analysis that also characterized his experimental work. "For a while there, we tried to keep it to the penny, not always successfully, but pretty close." The ledger would doubtless have met with the approval of his maternal grandfather, the prudent shopkeeper Josiah Williamson.

Gordon recorded not only the prices of movies but also the movies seen, such as *The Red Badge of Courage* in 1951. One entry records an expense of seventy-five cents for "laundry done by Betty." "We must have joked about something, and he gave me seventy-five cents for it," she says. "I did all

4

Date	Item	Income	218 35
10-18-50	Misc.		18
	Bread		22
10-19-50	Misc		17
	Southern Calif Gas Co. (Check #5)		1 75
	City of Pasadena (Check #6)		2 24
	Mrs. Whitaker (for repairs) (Check #7)		10 00
10-20-50	Misc		17
	Betty's paycheck to 10-15	68 40	
	Entertainment (Broken Arrow & The Furies)		1 10
	Parking		25
10-21-50	Brake fluid		50
	Groceries		8 06
	New Check book (Taken from account)		1 00
10-22-50	Church		50
10-23-50	Community Chest (Pasadena)		1 00
	Groceries		1 91
	Misc		10
	Laundry (Self service)		1 20
10-24-50	Groceries		15
	Misc		17
	Star-News collections		1 50
10-25-50	Brake fluid		50
	Groceries		22
			25[illegible] 24

5

Date	Item	Income	251.24 / ~~250 24~~
10-25-50	Lunch for Betty & Misc		62
10-26-50	Misc		17
10-27-50	Misc.		08
	Entertainment (Mr. 880 & Three Secrets)		1 90
10-28-50	Shampoo		62
~~10~~	Brake Master Cylinder Overhaul (Check #8)		9 45
	Groceries		9 19
10-29-50	Church		50
	Gas & Windshield Wiper Blades		3 97
	Lunch (San Bernadino)		1 50
10-30-50	Misc.		44
10-31-50	Misc		32
	Gord's Pay to 10-31	105 46	
		372 21	280.00
	~~Final~~:	Final	Original
	Money in B of A	298 23	231 30
	Money on hand 1. cash (silver rosy ration)	9 12	89 47
	2. check	105 46	0 00
		412 81	320.77
	Amount total income		372.21
			~~69[illegible].98~~
			692.98
	Total Expenses		280.00
			412.98
	Net deficit: $0.17		

Gordon's ledger of his personal finances, starting in 1950. *SOURCE:* GORDON MOORE.

of his laundry, every shirt." The ledger also records donations to the Advent Christian Church in Pasadena, where they were members. This was the small beginning of their lifetime commitment to philanthropy. Caltech itself would, in due course, become one of their main beneficiaries, and UC Berkeley would be far from neglected.

The ledger shows that across their Pasadena years, Gordon and Betty stayed ahead of their expenses. Not just that: by recording, measuring, and analyzing, Gordon had begun to look at his own life as an experiment—even as a business. He used the language of business in the home ledger, distinguishing "reserves" from "working capital" and "revenue" from "cash on hand." Small amounts mattered: from the nickel found in the pocket of Betty's coat and the quarter found in a Caltech hallway to the dime spent on a "pencil (red)." Years later, son Ken commented, "My wife says, 'Don't you guys ever talk about people or feelings?' I tell her, 'Nope.' There's no doubt about it: the Moore family is a business-structured family. We operate like we're a little corporation."

On their arrival at Caltech, Betty had immediately taken a job, but it soon became clear she would not survive long at the Consolidated Engineering Corporation. "Things were not going the way they should. My overseer was working me to the bone, acting like an old mother hen, ratting on me and doing all kinds of awful stuff." Showing her unwillingness to tolerate nonsense, and her willingness to confront problems head-on (a quality that, throughout their marriage, has offset Gordon's conflict avoidance), in early 1951, "I went to my typewriter and wrote my letter of resignation. I knew it was time to move on."

A headhunter quickly placed Betty at the Ford Foundation. The foundation had been created in 1936, but it was only when Henry Ford died some eleven years later that it began to prepare to become the largest philanthropic organization in the world. Betty was hired as it began to gear up. Two days after leaving her former position, she started at Turk House, a mansion with large grounds and a swimming pool located conveniently near San Marino, purchased to provide "quiet, pleasant, and scholarly" offices for Ford's president and a handful of senior executives (a style the Gordon and Betty Moore Foundation would emulate in its own start-up a half century later, in San Francisco's Presidio). "The Ford Foundation was very small. I fell into the position not even knowing it was there." It was Betty's first taste of organized philanthropy. Her job was project control: tracking letters, requests, and initiatives and screening proposals, "trying to decide to whom they ought to go to and which ones went in the crackpot file," as Gordon puts it.

In addition to Betty's income, Gordon had a stipend of around a hundred dollars a month and tuition remittance as a teaching assistant. They also received fifty dollars a month from Gordon's parents. With this help, along with remaining funds from Betty's inheritance, the couple managed their quiet life, while also accruing savings of two thousand dollars. Their most expensive purchase was a silverware setting for their new household. "I think I saved a larger percentage of our income then than I ever have since!" Gordon says. For his last year at Caltech, he was awarded a prestigious DuPont Fellowship, freeing him from the need to teach and allowing him to focus fully on his experiments and writing. He also took the bait when Hollywood—more specifically, Paramount Pictures—came calling for technical help.

> It was the early 3-D movie deal, with polarized glasses. The Polaroid Corporation had all the patents. Paramount hired me to go through them and summarize, hoping to find ways around. They paid four dollars an hour, which was a lot. Also, their high-powered lights were burning out, so I ran the spectrum of the floodlights at various voltages and showed them how to change. I even started looking for space to set up a laboratory to work on Polaroid material, examining storefronts to find a suitable space. Then they decided not to do it. It was a good part-time job for a graduate student. I could drive down there, go in the yard, and talk to the guys. That was my movie career!

It was also a step up from the cement works and another demonstration of Gordon's comfort with applying his knowledge to practical ends.

Gordon and Betty were early risers and hard workers, often spending a full six-day week on the job. They lived too far for Gordon to walk to campus, so he drove Betty into town each morning. Betty had a driver's license, but the Dodge was old and as soon as she saw the narrow roads and the speeding traffic, she decided driving was not her cup of tea. In the evenings Betty would remain at the Ford Foundation in Pasadena until Gordon came by to pick her up. "I always had to wait for him. Even after I'd worked my hours, he was still lost in the lab. Gordon was driven! If he had a project running, it would last into the evening. I would go back with him and sit in his basement lab. I took my reading with me."

On Sundays they attended the local Advent Christian Church, in a congregation of several hundred. Gordon became a deacon. "We went fairly regularly," he says, as he conformed to this aspect of his wife's agenda. Betty explains:

> Gordon's father was Catholic, but he never pushed his children to be in Catholic school. His mother went to a Protestant church by herself and near the end of her life was very religious—let's say spiritual. On Sunday she would go to church early in the morning, and then they'd go to Pescadero in the afternoon.
>
> Through his mother's influence, Gordon defined himself as Protestant. He briefly went to the church in Santa Clara where we were married, and he also went to the church in Pasadena. I had not been baptized; neither had Gordon. After we married we were both baptized by total immersion in the large pool behind the parsonage.

In Los Gatos the church had been a big part of Betty's life. In Pasadena, where she was at a distance from her close-knit family, she found comfort in the familiarity of church involvement, particularly in music. She attended choir practice every Wednesday evening and arrived at church early on Sundays to dress in choir robes for the services. "I was working six days a week, so it really cut into our brief time together." Being active in organized religion was in keeping with the times and the tradition of their families. In Pasadena, as a young couple without children, they were involved in choir, deaconry, and baptism classes. If participating in the Advent Christian Church provided a valuable anchor and a sense of continuity, later these would seem less urgent. Once they had a young family of their own, it was not easy to make time in the same way. When they left Pasadena, they would barely fit organized religion into their lives again.

In their apartment, Gordon and Betty indulged a shared love of cats, taking in strays. "We had so many kitties there that we had to start farming them out," remembers Betty. They ran ads in papers and found homes for most of them, but took two back up to the Los Gatos ranch. One, "Little Hitler, was a character," Betty recalls. "He would attack us in bed and tangle with our toes. He was hit by a car and died, but Prunella lasted. My mother had her for years."

Apart from a calm home existence of cats, each other, and the church, their social antennae remained directed toward family in Northern California. Betty explains:

> We were home for all the holidays. We came up north at least three or four other times a year—whenever we could get a break. We went to stay with Gordon's parents. His mother was family oriented. Holidays meant a great deal to her. When we'd come on trips, my mother had a big spread for everybody, and his mother had a big spread, sometimes on the same day.

Betty Moore in Pasadena, with Lucifer.
SOURCE: GORDON MOORE.

> Gordon and I would overeat so badly. We went away saying, "Do we really need to do this?" The answer was "Yes," because we had to keep them both happy.

Friends, Acquaintances, Losses

In that era the majority of Caltech graduate students were unmarried. Accordingly, Gordon found himself cut out of many of the more lively social goings-on and "didn't fraternize much with other people in the department." If they went to a rare evening chemistry event, Gordon "would be yakking in the corner with the guys about science," while the women were off in another area. "It was quiet and very staid." The women had their own semiformal group, the Chem Wives, which met regularly. Betty recalls a party given by Richard Badger. "There was a funny professor who would sit and read the dictionary in the corner. I thought, 'What is going on with this man?'" "That was Oliver Wulf," Gordon explains. "He was a very unusual personality. He spent most of the evening reading the dictionary. Betty was impressed."

They did occasionally socialize with two other young couples, Roger and Carolyn Newman and Tom and Addy Danforth. "Roger was working with Badger on the spectroscopy of crystals. He was very efficient in the way he did things. He would spend a month in the library finding a problem that he could do in a week. About every six weeks he turned out a paper. He selected his projects very carefully. Betty and I probably had more to do with the Newmans—Roger and his wife—than almost anybody else. We met together reasonably often for dinner. We had tacos at their place or ours."

The friendships were not especially close. After leaving Caltech, "Newman bounced around," working in academia and then going into industry, where he took research jobs in aerospace labs. "Our paths crossed two or three times, but I never saw very much of him." Betty's closest friend was Addy Danforth, head of the printing division at the Consolidated Engineering Corporation. "Addy and her husband, Tom, were our first real 'couple friends,'" she recalls. They took several weekend trips together into the desert, visiting Death Valley, where they slept out in Uncle William's old Dodge car or stayed in a cheap motel.

Betty took care to maintain her older friendships. One was with her cousin George Mattos, son of Aunt Ada, who lived with Betty in childhood. An All-American pole-vaulter, he was now one of the world's top ten in the sport, competing for the United States at the Olympic Games in Helsinki (1952) and Melbourne (1956). "At the ranch George was always taking the props from the trees and practicing pole-vaulting in the backyard over boxes and ladders," recalls Betty. In preparation for his first Olympics, "he came to visit us in Pasadena, with this pole stuck across the living room and dining room of our little apartment. He stayed for weeks, for the Trials in Los Angeles, and we fed him. He qualified. I guess it was all that good farm cooking I did for him." Betty's friendship with George was lifelong. In 2012 she visited him in his last illness.

Much less robust was Gordon's relationship with his undergraduate friend and roommate, Bob Naughten. On leave from the Korean War, Naughten visited them in Pasadena, but the friendship faded to occasional contact, years or even decades apart. "Gordon didn't want to keep up with Bob. You can't push somebody to want to have people come back to your house," observes Betty. Close friendships that involved meaningful interpersonal commitments were not a priority for Gordon. He remained absorbed by, and satisfied in, his scientific research. That work was "fun" and "interesting." The lab was the place where he felt confident. Challenges at other levels of life—such as adjusting to the need for intimacy in married life—were less easy. Betty discovered early on that Gordon, as she puts it, "does not like human emotions." When she—and he—suffered two early personal losses, his abilities to express those emotions, and make himself available to his wife, were put to the test.

In January 1953 the young couple conceived their first child. Betty was anxious to have a family "before I was too old. I was hoping Gordon wasn't going to be a perennial student. If you didn't have your family by your early thirties, everybody thought you were not going to have healthy children. Doctors were saying, 'You have to have your babies before thirty.' I thought, 'I'd better get busy.'" During this first pregnancy, Betty turned to her mother for support through letters and phone calls.

One morning in late April, Betty received a call with the news that her grandfather Abraham Metzler had died of a stroke. Quickly, they drove up to the funeral in Los Gatos in their old Dodge. On the way back, in the car, Betty experienced severe pains and, later that night, miscarried. She was heartbroken. Tellingly, she does not recall talking to Gordon about her emotional state or describe his reaction to the news. "I was terrible. I was hurting, because it was my first. I had already told so many people that I was pregnant. That was the worst part. I went right back to work, and everybody was supportive, but it was hard."

Gordon is laconic. "I certainly remember the event," he comments. However, he lacked the language to contend with the profound emotional consequences of the miscarriage. Instead, he turned to his familiar defenses: practical, measurable activity and intellectual analysis of more tangible problems. Betty eventually came to view this lack as a prominent Moore trait. Long aware that "Gordon never gets very emotional," late in life she reflected that "Gordon doesn't even like to go to movies, because he doesn't like human emotions. When Gordon and I were dating, the only films he would go to were things like *The Red Badge of Courage* or science fiction."

A pragmatist streak ran through both Gordon's and Betty's families. "Everybody in the family is practical, wanting to understand how things work," observes Ken Moore, their eldest son. "My grandmother on my mom's side had a very strong social interest, but she was also very practical. 'You have to work,' she would always say. My dad's dad was very hands-on, and mechanical. He built things. I look at all my cousins, and they're the same way. We don't have any dreamers." Ken and Steve, the second son of Gordon and Betty, would share this inclination. "We're similar in being logical, analytical, and numbers and fact based. If you have a continuum between emotional-based and logical-based people, we're hard-core logical.

The typical comment is that we all listen. We listen, and we like to fix problems. We like to analyze things."

Ken can recall only two occasions on which he witnessed his father express real anger: once visibly, once subtly.

> The first was when I was sixteen or so and having one of those teenage angst periods. I came down to the breakfast table, grousing about a bunch of clothes that I didn't like. My exact quote was, "I'm not going to wear these crappy pants and shitty shoes." I don't think my dad had ever heard me use the second word. He blew up. He stormed toward me, "Rrrrrrr!"—the whole bit, as if he was going to slug me. It's the only time I've ever pushed enough buttons to hit launch sequence. It was an impressive display, coming from somebody who is as levelheaded as they come.

The other incident occurred when Ken and Gordon were out fishing. They and another fisherman were throwing chum into the water to attract fish. "Nobody was catching anything. The other guy decided he was going to start shouting at our boat. It was really stupid." Gordon was angry, but held it in. "I can tell when my dad gets mad. His lip will be a little bit tighter and his voice will slightly change tenor, but he's not one to do something in a situation like that." Younger son Steve concurs: "If he became angry, he would just not be communicative. He'd leave. He wouldn't really let it out. He would disappear or be quiet."

While Betty was frustrated by Gordon's inability to engage emotionally and his lack of empathy, she and his sons came to recognize that feelings did lie beneath. On rare occasions they could surface. Ken says:

> Dad is an intellectual debater. You're not going to get a rise out of him. If he's arguing, he argues purely from an intellectual standpoint. He rarely raises his voice. Traditional PhDs tend to be very bright, focused individuals. They don't consider how other people think, how other people feel. Their approach does not lend itself to that. On the other hand, when his younger brother, Fran, was about to die of cancer, my dad was choked up. He was definitely sad when his own dad passed on. He does have emotions.

If Gordon rarely showed anger or empathy, he was equally slow—according to Steve—to display jubilation. Once again, this was in keeping with Moore family habits: "When something really great happened, he'd be in a good mood, but there were no major celebrations. My mother would smile a lot, but my father doesn't show emotion in that direction either. Other people hoot and holler when they're happy; that's not us. We don't show much emotion overall, especially my father."

Research and Publication

Badger set Gordon to work. His first task was rounding out the experiments of one of Badger's prior students, Llewellyn Jones. The goal was to collect infrared spectrophotometer data on the nitrous acid molecule (HNO_2) in order to draw conclusions about its structure. Part of the challenge was that nitrous acid was a weak acid, prone to decomposing into other molecules. Gordon was familiar with its transformation into nitric acid, a key ingredient in nitroglycerine and TNT. The other part of the challenge was to get the most meticulous measurements, using the best instruments. It was only with precise data that Gordon, along with Badger and Jones, would be able to perform the complex mathematical analysis, transforming data into conclusions about molecular structure.

Infrared light, like any other form of electromagnetic radiation, is defined by its wavelength. Spectrophotometers can detect how much of a particular wavelength a material absorbs. These absorptions constitute the characteristic spectrum of the material, its "fingerprint." Each of the three spectrophotometers in Badger's lab was suited to its own range of wavelengths, and each was the product of a Californian instrumentation virtuoso. There was a Beckman Model IR-2, one of the very first commercial spectrophotometers. A similar but more powerful custom-built instrument had been a gift to Badger from Robert Brattain, a top infrared spectrometrist at the oil giant Shell in the Bay Area, whose brother would share a Nobel Prize with William Shockley. The third instrument, the most powerful in terms of sensitivity, was Badger's own creation. It included a vacuum enclosure that removed the residual gases that could cause error.

With these instruments Gordon could survey the entire infrared spectra of nitrous acid and explore segments of it in detail. "Badger had half a dozen problems to get me started on," he recalls. "It was an easy time to find a thesis topic, if you look at it that way." Badger's research contracts revolved around compounds of nitrogen, such as nitrous acid. "It's an interesting molecule because of cis- and transconfigurations [different three-dimensional arrangements of the molecule]. Jones and Badger had observed a phenomenon on a couple of absorption bands that they thought was an interaction pushing the energy levels apart in the same molecule. Badger wanted me to look at that."

The nitrous acid experiment was the first in a series of investigations of simple nitrogen-containing molecules that Badger encouraged Gordon to undertake. Gordon had both firsthand knowledge of and a visceral interest in nitrogenous molecules, that is, potential explosives. It was a happy fit with, though far distant from, the realities of Badger's US Navy sponsors. Nitrogen is central to conventional explosives, and, with the Korean War in full swing, US Navy ships and aircraft were hurling untold tons of nitrogenous explosives at enemy positions.

In the subbasement in Pasadena, to obtain data about the structure of these simple molecules, Gordon needed to push his instruments to their limits. He would have to adapt them with gear built with his own hands, while creating complex setups that could produce, purify, and handle volatile molecules. He modified Badger's instrument, the most powerful in the lab, with an improved infrared detector, containing what was, in essence, a thin layer of lead and sulfur mixture. "Most of what I did was on a homemade grating instrument with pretty high resolution, using our lead-sulfide cell. You had to have an 'in' with a guy who knew how to make lead-sulfide cells."

The lead-sulfide detector enlarged the instrument's sensitivity to the "fingerprints" by which spectral data signaled molecular structure. Through Gordon's PhD research, "we were able to start seeing some molecular features." The detector that Gordon integrated into Badger's instrument also gave Gordon his first real exposure to semiconductors, materials that—as the name suggests—are neither good at conducting electricity nor good insulators from electricity, but instead "semiconduct."

Semiconductors are useful in devices that switch, amplify, and convert electrical energy because their conductivity can be altered either by the deliberate addition of impurities or through the application of electric fields or light. Lead sulfide, in particular, changes its ability to conduct electricity when exposed to infrared light, producing an electrical signal in response to that light's presence. In his first practical work with a semiconductor, Gordon modified the vacuum tube amplifier in the instrument that handled the signal from the detector. The modified electronics provided him with a sensitive detector that revealed the desired infrared molecular spectra. He also heavily modified the other custom spectrophotometer, given to Badger by Robert Brattain, and in doing so he designed and built an automatic electronic control system.

Gordon also built an intricate system of glass tubes for handling the gas molecules. "I did a lot more glassblowing and a little electronics, which I knew nothing about as I started and a little bit about when I finished." Spending most every nonclassroom moment in the lab—handling the instruments, wiring vacuum tubes, blowing glass, and recording data in his laboratory notebooks—Gordon was running as fast as possible toward completion of his PhD. The pace of his experimental work reflected his ambition and his absorption in research.

Only Badger fully appreciated the rapid strides his student was making. As always, Gordon was low-key and unassuming. He did nothing to promote himself or cultivate power connections among faculty, fellow students, and visitors. As a "grind, not an operator," he was entirely focused on the task:

"I spent most of my time in the subbasement, churning out data." This was where happiness lay. He was tackling fresh problems and generating publications. He desired to achieve the recognized goal of scientific work—original publication—as quickly and efficiently as possible. He drove himself hard.

Just over a year after pulling into Pasadena, Gordon published his first scientific paper, in December 1951, when still only twenty-two. The paper, coauthored with Llewellyn Jones and Richard Badger, was titled "The Infrared Spectrum and the Structure of Gaseous Nitrous Acid" and appeared in the *Journal of Chemical Physics*. It presented Gordon's experimental work on the HNO_2 molecule. "There was some competitive pressure. Somebody else was looking at the spectrum of nitrous acid," he recalls. He next pursued other molecules with nitrogen in them. "You pick one, and you get some results that look interesting; if you do, you keep it up." He began to look at nitrogen dioxide, NO_2: "A simple molecule, but its structure wasn't known. The shape of the molecule was completely up in the air."

Saliently, but of little interest to Moore, the molecule was an important element in the smog that increasingly beset Los Angeles and Pasadena. "We had an instrument that let us resolve the infrared spectrum of NO_2. It had aspects that weren't easy to explain initially. Soon I was looking for other molecules, ones with no more than four atoms that nobody had done the spectrum of!" One possibility was the chloromines NH_2Cl and $NHCl_2$. "It was publishable journeyman work that I was reasonably happy with, the kind of experiment I enjoyed. I could build the apparatus and make the molecules and occasionally blow up a test tube! The chloromines decompose into NCl_3. If you leave them overnight and wash the test tube out next day, it explodes." Other publications quickly followed, demonstrating Moore's use of cutting-edge infrared measurements to draw conclusions about the structure of various nitrogenous molecules.

In the world's leading publication for the field, the *Journal of the American Chemical Society*, he published "The Infrared Spectra and the Structure of the Chloramines and Nitrogen Trichloride" in 1952, with Badger listed as the second author. In early 1953 Gordon submitted his first solo article, "The Spectrum of Nitrogen Dioxide in the 1.4–3.4 Mu Region and the Vibrational and Rotational Constants of the NO_2 Molecule," published in the *Journal of the Optical Society of America*, an important outlet for spectrometrists.

By the early summer of 1953, still shy of completing a third year at Caltech, Gordon decided that his rapid pace of experimentation, combined with a good publication record, entitled him to his PhD. "My thesis consisted of investigations of molecular structure. I had looked at enough molecules. I asked if I could finish." Badger agreed. Gordon set to writing up his results. Finishing a Caltech chemistry PhD in fewer than three years was a noteworthy achievement. It reflected Moore's ambition and focus. He was also a married man, with a family of his own very much on the agenda. Later in life, Gordon Moore discussed his rapid completion of a PhD with Caltech's provost. "Were you married when you were a grad student? That seems to be the single thing that gets people through in a hurry." It was not only Gordon who was ready to move on; Betty, too, was "happy when it all wound up at Caltech." She received a certificate then in its heyday, a "PhT," for "Putting Husband Through," bestowed on her by the wife of Caltech's president.

Gordon Moore was largely self-made, having reached where he was through his own hard work and by taking advantage of California's public education system and very modest parental help. He had traveled far beyond any Moore in his immediate family. Now, to make good on his investment in education, he needed to start earning a living as a chemist and experimentalist.

Once the work was written up, Betty typed out a professional-looking dissertation that Gordon submitted in the summer of 1953. Its core was his published and "in press" papers. A lengthy appendix detailed some of the complex mathematics he used to tease out molecular structure from the infrared data and described the electronic system he built to control the instrument from Brattain. By doing this, Gordon was demonstrating his expertise in experimental innovation and infrared interpretation.

In keeping with an idiosyncratic practice of Caltech's Chemistry and Chemical Engineering Division, his thesis closed with a section called "Propositions." In it the dissertation author could briefly memorialize observations, conclusions, and speculations that he thought significant, whether connected to the thesis or not. The section took the form of a numbered series of paragraphs. Gordon penned ten. Around half were statements about the spectra of the molecules he had studied. One recorded his belief that certain lines in the spectrum of nitric oxide were of a previously unobserved series. In his last statement, Proposition 10, Moore—possibly with a twinkle in his eye—connected this culmination of his chemical education to its origins in his boyhood endeavors. "To alleviate the shortage of chemists," he wrote, especially physical chemists like himself, "a laboratory course in explosives and pyrotechnics should be included in the high school curriculum."

10. To alleviate the shortage of chemists, especially physical chemists, a laboratory course in explosives and pyrotechnics should be included in the high school curriculum.

Gordon's PhD "proposition" on explosives.

SOURCE: GORDON MOORE.

Gordon now considered himself a bona fide chemist and experimentalist. His expertise was the result of a significant investment, one that should reap dividends in the form of a well-paid and interesting job. His commitment to infrared spectrometry and instrumentation put him at the center of major developments in research. It was time to look for an academic position. After three years of intensive effort, he professed himself "tired of going to school."

Professor Moore?

As the young couple began pondering life after Caltech, Betty learned that the Ford Foundation was moving to New York, and she would be required to help close operations in Pasadena. "The timing was perfect," says Gordon. "They gave her a big severance package—big by their standards—and she would have had to quit anyhow." Betty remembers, "There was a lot to do, buttoning up and winding down."

At Caltech Gordon was driven and intensely focused, spending every day in the subbasement running experiments. He gave no talks, polished no useful contacts, but simply worked hard to achieve results in an area he enjoyed. Now his mind turned to the obvious next step: finding a job and continuing with his science. He had a healthy appreciation of his own abilities and of his suitability for an academic life. He could count on receiving Badger's support and help. And it was fully acceptable for a research-oriented chemistry professor to be outwardly quiet and self-contained, even conflict avoidant and lacking savvy.

Moore envisaged a job at a leading school. He was looking for a combination of three things: a prestigious institution, a group with a reputation in infrared spectrometry, and a pleasant locale. Having invested in what Caltech offered, he wished to move to a center that would make use of his education, a post in which he "wouldn't have to start over again." If an elite institution with the requisite reputation, such as Princeton, had offered him a job, his life would have started down the expected path. "I would have been a professor," he says.

Unhappily, as he looked around in 1953, he uncovered few options to match his vision. The GI boom had passed, and the "baby boomers" were still in their cradles; faculty positions were rare. The only obvious opening was at the University of Oklahoma. Its group had a solid standing in infrared spectrometry, and Gordon "might have applied," but the position failed to match his other requirements. "It wasn't a top school. Oklahoma was not my ideal place to live." He didn't even visit; the opportunity simply did not measure up to his ambitions. Beyond the Oklahoma possibility, there wasn't anything that fitted scientifically. In the face of this dearth, Badger advised him, "Take a look at industry—I think you'll like it."

Gordon Moore, 1950s.

SOURCE: GORDON MOORE.

Gordon was still not very self-aware: "I have no idea what led him to think I might find industry interesting. I had been doing academic research and imagined I would get an academic job. There were no obvious industrial groups in infrared, so although that was the area where my skills would be most attractive for an employer, I figured I'd have to do something else if I started looking in industry." Badger, through his long experience and connections, knew otherwise. Infrared techniques were increasingly in use in manufacturing. With Gordon's practical, experimental flair, there was potential for a good fit.

Firmly rooted in California, "going along pretty well," Gordon and Betty wanted to stay in their home state, "if something were available." Badger sent him 150 miles north to the Naval Air Weapons Station in China Lake. The station had begun as a US Navy–Caltech collaboration in World War II and was now developing the Sidewinder air-to-air missile, which used infrared detectors to find its target. Badger was a consultant. "He asked me to go out to interview. The job was relevant to my research." Gordon took Betty along, driving in the old Dodge. She was dismayed. "China Lake is like falling off the edge of the earth, out in no-man's-land in the desert, run by the government. As we were driving out the gate, I said, 'No way, José. I'm not going to live like that." Gordon was similarly unimpressed. "There was nothing around there, nowhere to go. It wasn't an outstanding group to work with." The desert missile lab failed to meet at least two of his three criteria.

Another possibility was back in the Bay Area, in Richmond, working at Standard Oil's sprawling refinery. The refinery was one of the oldest and largest in the West, producing a huge array of petroleum products. Infrared and other electronic instruments were vital to its operation. Betty recalls how they traveled up to that "smelly place out on the other side of the Bay, an industrial facility. What a hole! It was too far to commute from the peninsula; we would have had to live there. You don't spend that amount of time at college to live in a situation like that." Again, it was not a good return on investment. Showing how high he would set the bar after his Caltech experience, Gordon—despite acknowledging that "it was California"—dismissed the opportunity as "not very interesting."

As his search widened, Gordon interviewed with the R&D organizations of two chemical industry giants, DuPont and Dow. They both had top research reputations in chemistry, and it was common for their staffs to stay in close touch with academia. Having held a DuPont Fellowship, Gordon was under some obligation to interview with the company. He flew east to DuPont's fabled Experimental Station in Delaware. "The basic research prestige of that operation was high, and things coming out of there—like nylon and Teflon—were significant." Infrared was vital to the development of fresh products. The R&D center met all three of Gordon's criteria. However, during the visit, Gordon was unable to convey what his PhD work was about, instead getting stuck on one minor aspect: "They asked only about the photochemistry, a two-week experiment. They took that as my whole thesis and thought it pretty weak. I answered their questions, but I didn't say, 'That's only a little bit of what I did.'" The downside of Gordon's devotion to the experimental grindstone in the Crellin subbasement was apparent. Having avoided an active role in informal scientific discussion, he did not know how to promote himself and his work; he did not present well.

While in the East, he also interviewed at DuPont's extensive production site in Deepwater, New Jersey, a facility not wholly unlike Pacific Portland Cement. To Gordon, its work seemed trivial, and he saw his interviewer, who extolled the virtues of industrial projects, as lacking in sophistication. "He cited a purification process that used activated charcoal: the chemist initially used whatever supply happened to be available but, by optimizing it, could change the yield dramatically. He thought that was a major contribution. Working at that level didn't sound very exciting."

The Experimental Station was different. Had this arm of DuPont made an offer, "I would have gone—but they didn't." He crossed DuPont off his list. Gordon made a more promising start with the Dow Chemical Company, which had a California-based research lab at Pittsburg (its large manufacturing site thirty miles northeast of Berkeley) and wanted to expand. He met with a Dow representative at Caltech and then traveled up to Pittsburg "to see what they were doing." Soon it emerged that there would be a possibility for Gordon to work at Dow's headquarters in Midland, Michigan, then return to Pittsburg in a research management role. "That sounded pretty attractive."

Dow and other companies, following the vogue of that era, used industrial psychologists to screen recruits and evaluate their management capability. Gordon was duly sent to a psychologist in Los Angeles for a battery of tests. The report was negative; he showed little managerial aptitude: "good technically, a good research scientist, but he'll never be a manager." His hopes were quashed. "They were willing to offer me a job in Midland, but it no longer had this management possibility associated with it. After those tests, they would consider me only for a technical job." A move to small-town Michigan, no hope for managerial advancement, and no prospect of an early return to California: Gordon decided to pass on Dow, too.

Finally, another opportunity arose, more intriguing if somewhat oblique. Chemist Gordon Teal arrived at Caltech to recruit. Teal was just months into his post as director of research at Texas Instruments (TI) in Dallas, where he would help to build the firm's innovative capacity. Formerly, he had been with the leading industrial research laboratory of the era, the Bell Telephone Laboratories, where his expertise in materials had proven invaluable to one of the most exciting innovations in electronics: the "transistor," a compact alternative to bulky and frail vacuum tubes. In 1953 the transistor was just beginning to make its mark, as a creature of military electronics. Seeing little connection to his infrared expertise and "not anxious to consider Texas," Gordon let what was—in hindsight—the most promising of his opportunities pass him by. Teal instead hired another young physical chemist finishing his PhD at Caltech, Morton Jones.

Heading East

Badger came to the rescue again. Through his connections in the infrared community, he brought Gordon into contact with the Applied Physics Laboratory, where a colleague worked in an important infrared spectrometry group in a new basic research center. APL was operated for the US Navy by Johns Hopkins University (in an arrangement similar to the way Caltech operated the Jet Propulsion Laboratory for the US Army) and was located in Silver Spring, Maryland, just outside Washington, DC. In 1947 APL opened a small basic research center, which included a group devoted to advanced infrared and molecular spectrometry. Gordon flew East for a visit, traveling alone, as he had to DuPont. He quickly saw that this was "a research organization buried in a large government-contracting lab, doing work similar to what we were doing" in Badger's basement.

As he learned more about the APL infrared group, comprising a mixture of physicists and physical chemists, Gordon decided that he had found the best available combination to meet his criteria. APL was a thriving, prestigious institution, offering high-quality experimental activity in his specific field of expertise, in an attractive locale. The research program fitted well with his previous work and offered opportunities for the equipment building he so enjoyed. "It was a fairly wide-open research center, doing things that people were interested in. I thought this would be an opportunity to use what I had learned." To his delight, he was offered a position as the sixth PhD in what was known as the "flame spectroscopy" group. Gordon saw that the operation was well funded, with "great" instrumentation and access to "significantly better infrared detectors" than at Caltech. It was August 1953, and he was ready to move on. "I decided to give it a shot."

Betty had never been to the East Coast, or even to the Midwest. While Gordon took seriously her opinions about where she did not want to live, he alone made the decisions about career and location. In his search up to this point, Betty's distaste for locales such as China Lake and Richmond had matched his own lack of enthusiasm for the jobs offered. Considering a move East, Betty was more excited. She had heard about Washington, DC, from her uncle William. Yet her interest was mingled with worry, for a move to the East Coast would take her much farther from her beloved family. Even though pioneer days were a century past, long-distance travel remained a serious venture. "It was very hard for me to go East that September," she remembers. "My mother was still caring for my grandmother at the ranch. She never said anything, but my grandfather had died, and I had lost a baby. I was afraid I would never see my grandmother alive again."

Gordon and Betty decided to drive to Maryland. The first challenge was transportation. Their car, the 1935 Dodge, was almost twenty years old, and their mechanic could scarcely believe they dared even drive it between Pasadena and the Bay Area, let alone across the continent. In the closing of the Ford Foundation, they found their solution. Not only did Betty get her severance package, but one of its executives, Chester C. Davis, sold them his car at a discount. "We bought his 1950 Buick. That was a big step forward." Excitement outweighed doubt. Betty remembers thinking, "Instead of going west, we're going east. My attitude was, 'We are heading out, gung ho, and that's it!'" Gordon had done well on his tests and his PhD oral exam before leaving. Betty looked at her husband and saw he was determined to "find a place in the world." Importantly, they were in it together. "I was at his side."

4

SCIENCE, SHOCKLEY, AND SILICON

COLD WAR

Missiles, Bombs, and Electronics

With belongings crammed into the secondhand Buick, Gordon and Betty Moore set off across the country. They took their time, weaving their way through national parks and staying in simple lodges and motels, a route reflecting their preference for the great outdoors. Eventually, they arrived in Washington, DC, at rush hour, reeling around DuPont Circle four times before snagging the correct exit. For their first few nights in the East, they stayed with Betty's uncle William and aunt Julia.

The young couple rented an unfurnished apartment in Silver Spring, Maryland, on the northwest edge of the District. From there, Gordon had but a short drive to the APL Research Center in Maryland. Betty's severance payment from the Ford Foundation enabled them to buy Scandinavian-style furniture and a secondhand piano for her use. They quickly settled into a routine. Gordon was gone most days at the lab, as at Caltech. Previously an apprentice in chemistry under Badger, he was now a full-fledged, if junior, practitioner within a well-funded research center of fifty-plus staff.

Developments in aerospace and nuclear technology, highly dependent on the vacuum tube, were transforming both offensive and defensive warfare. Since the Applied Physics Laboratory's formation in the early 1940s, it had been a player in this transformation. As the Cold War thickened and congealed, doctrines of air dominance, strategic bombing, and "total war" were becoming ever more complex and recondite. Aeronautics morphed into aerospace, as captured German V2 missiles were upgraded and enabled to shoot to the stratosphere, before (theoretically) falling onto distant cities and military formations. The APL research center began primarily as a home for the high-altitude experimentation of distinguished physicist

James Van Allen, who took the lead in working with captured V2 rockets. The V2 had a range of around two hundred miles but potential for traveling many times farther, a potential fulfilled in the intercontinental ballistic missile (ICBM) of the 1960s and subsequent decades.

Nuclear weapons were creatures of electronics. Calculating machines helped design them, electronic circuitry was central to their mechanisms, and electronic instruments and equipment facilitated the enormous technical and industrial effort to produce their deadly cargo. A nuclear explosion offered the means for instant destruction of a city. First bombers and then missiles were the delivery systems for nuclear weapons. In the late 1940s US military funds poured into aerospace and nuclear technologies. Throughout the 1950s a growing flood of military dollars would support the majority of the country's R&D effort and the bulk of university-based research. Gordon was already a beneficiary of these military dollars, since his career at Caltech had been partly underwritten by the Office of Naval Research, through its grant to Badger.

The US Navy began funding the design of guided missiles to be launched from ships. Code-named "Bumblebee," this work encompassed all aspects of design, from warheads to jet engines and from guidance systems to telemetry (electronic communication of data from the missile). The task of spectrometrists Robert Herman and Shirleigh Silverman, prominent analysts of molecules through infrared wavelengths employed at the research center, was to extend the realm of missile knowledge through their "flame spectrometry" group, by deploying infrared to study the molecules present in rocket flames. Studying the fundamentals of infrared spectrophotometry was seen as a legitimate part of this work.

By the time Moore joined in 1953, the research center had amassed an impressive array of instruments. He again had access to a high-resolution infrared spectrophotometer, similar to the one made by Badger. What was new was the existence of skilled technicians, "instrumentation people." While free to use his own hands to make or modify equipment, Gordon welcomed this, his first experience with trained assistants. The work was comfortingly similar to his previous experiments. He was asked to focus on a simple but tricky-to-handle molecule, consisting of an atom of hydrogen bound to an atom of chlorine. This molecule, hydrogen chloride, is a major by-product when rocket fuels are burned, hence its relevance to APL's work on guided missiles.

Using the spectrophotometer, Gordon shone infrared light at samples of hydrogen chloride that absorbed particular wavelengths, thereby creating "absorption lines." A lead-sulfide detector such as he had already used registered the lines' occurrence. So sensitive was this instrument that Gordon and his colleagues could analyze the fine structure of these absorption lines, not just their intensity, but also their width and shape. Deep in the details, he could search for the effects of the molecule's vibration—the shaking of its chemical bonds—and its rotation, clues to, if at a great remove from, the actual firing of a rocket engine.

The research required delicate and subtle work. Since temperature and motion were closely linked, heat had to be carefully controlled, with samples contained in special furnaces. Gordon spent many hours blowing glass, winding metal filaments, and designing gas supplies to create a furnace with the ultraprecise high temperature required. His group also needed a container, a "cell," to hold the hydrogen chloride. The cell had to withstand the corrosive high-temperature gas while retaining transparency to infrared light and to be of an exact size. Here was territory in which he was comfortable and happy: building equipment, running experiments, recording data with meticulous care, and making painstaking analyses, worthy and tangible contributions to the research program. And, without knowing it, he was developing a subtle appreciation of the power of advanced instrumentation, miniaturization of chemical processes, and exacting performance standards, all essential to his as yet unsuspected vocation.

Meanwhile, at Home in Maryland . . .

On the surface, 1954 was a year of settling down. Later, Gordon and Betty would reflect that it was the quiet before the storm. As at Caltech, Gordon quickly lost himself in the practical and intellectual challenges of his work. His routine was repetitive: up early, off to the lab, back home for dinner. "He never had time to read the paper in the morning. He was up and blasting out," recalls Betty. He took the Buick for the short drive to work. Betty had not liked driving in Pasadena and liked the prospect even less in this new area. Even though there were no shops within walking distance, she stayed close to their ground-floor one-bedroom garden apartment, "somewhat isolated" but busy setting up a household. She had no plans to find a job, being preoccupied with the desire to start a family of her own. That spring, to her relief, she became pregnant again.

Gordon's small portions of free time were spent with Betty. On occasion they would socialize with neighbors and Caltech contacts or visit Betty's uncle and aunt in nearby Alexandria, Virginia. Several people from APL, including his boss, lived in the apartment complex, but the young couple mostly led a quiet life. Gordon maintained his meticulous personal ledger, detailing their finances and his earnings. With Gordon working five and a half days a week, as was still customary (the half day being Saturday), they traveled little beyond the metro area. Instead, from their apartment windows, they watched the novel phenomena of thunderstorms and snowfalls. They also marveled at the strained race relations of mid-1950s Maryland, a southern state completely unlike California. "In Palo Alto, nobody gave a hoot where they were placed in a restaurant," recalls Betty. "In the East, you could see the difference in where people were seated. That made a real impression on me."

In their move from the San Francisco Peninsula to Pasadena, Betty's church affiliation had been one constant and the local congregation an important focus. In an unfamiliar eastern world, and without an Advent Christian church close at hand, they dropped out of regular churchgoing. On the advice of her doctor, Betty guarded her health and took care not to exert herself. Her only exercise was in brief walks around the apartment complex: "I couldn't even go out to see the fall leaves. I had to be terribly careful." Instead, she relied on the telephone to keep in touch with family and to get things done. "I would call stores and have them deliver baby goods, such as the crib and everything for the nursery." During the day she spent time with a pregnant neighbor two doors down, but her closest companion was the kitten she had found freezing under the apartment stoop and took in, against the rules. The cat eventually gave birth to its own four kittens, who offered diversion and company.

The days shortened toward Christmas, and on December 7 Betty went into labor. By evening the couple was hurrying the seemingly very long miles to the George Washington University Hospital downtown. There, Betty labored, while Gordon waited nearby. Her physician became concerned that the baby was much too large for her small frame. He ordered a cesarean section (at the time a far from routine operation), and in the early hours of the next morning a healthy son, weighing more than eight pounds, was born. Kenneth Moore was the biggest baby in the nursery, robust and energetic. Unlike other newborns, he frequently moved about in his bassinet. His mother recalls seeing this as a premonition of a spirited youth to come: "Oh, boy."

The cesarean's aftereffects tempered Betty's joy at becoming a parent: catgut surgical sutures caused an infection that put an end to breast-feeding and made everything "goofy." She stayed in the hospital ten days. Once back home, with no extended family to support her and little help from her aunt and uncle, she relied on Gordon to prepare Ken's bottles. "He was good at mixing formula, because he's a chemist. 'How many ounces of this?' He could sterilize bottles. It worked fine." Christmas—the first to be celebrated with their own child, alone on the East Coast—came as something of an anticlimax.

Computers and Transistors, Ahoy!

Gordon lived in two wholly separate worlds: an isolated home, previously quiet but now overtaken by the noisy cries of Ken, a colicky baby, and the intense laboratory. APL itself was entering a period of unsettling change, fueled by two recent developments in electronics: the digital computer and the transistor. Digital computers—"mainframes"—were beginning to prove effective in solving tough data-based problems in cryptology, bomb design, scientific calculations, and the like. The US Army and Navy were major sponsors of efforts to improve these huge, temperamental vacuum tube–intensive machines, as well as major customers for them. The SWAC and BINAC had been joined by AVIDAC, Atlas 1, and Whirlwind 1, among others, and a nascent industry was offering commercial machines to the military, to government, and to large corporations and their R&D organizations.

On November 4, 1952, the night of the US presidential election, the public had its first insight into the power of computers. UNIVAC 1, a machine destined for the Atomic Energy Commission, was borrowed by CBS Television and used returns early in the night to predict a landslide victory for Eisenhower (then rated at odds of one hundred to one against). The computer maker, convinced there was some mistake, initially withheld the result. The next day the success of the computer shared headlines with news of Eisenhower's victory. Within five years more than two hundred digital computers had been ordered and installed, most by the military and their aircraft and missile suppliers: APL ordered one of the largest and most powerful. In 1957 Katharine Hepburn would star as a librarian in the movie *Desk Set,* competing for her job with just such a computer (not ENIAC or UNIVAC, but EMERAC). Data and the possibilities of digital data manipulation were beginning to impinge on the popular imagination, if only in the purest fantasy.

Side by side with these early computers came a new device—the transistor—that would first rival, then eventually replace, the vacuum tube and enable electronics to take center stage in the story of world revolutions. Created at Bell Labs in New Jersey in December 1947, the transistor was an electronic component with the same functions as the vacuum tube, but a very different physical incarnation. Both could amplify, and switch on and off, the signals they received. Tubes were built around the flow of electrons from a glowing wire that passed through a vacuum inside a hot, fragile glass bulb, while transistors depended on electrons moving through a *solid,* a semiconductor, at much lower temperatures.

Unlike a vacuum tube, the flow of current in a transistor has everything to do with the chemical nature of its different regions and nothing

ENIAC, the breakthrough electronic digital computer, 1940s.
SOURCE: US ARMY/WIKIMEDIA COMMONS.

to do with high heat or a vacuum. Semiconductors may be compounds, like the lead sulfide in Gordon's infrared detectors, or chemical elements, such as germanium or silicon. Through the addition of specific chemicals—called "dopants"—a semiconductor's conductivity can be varied almost at will. The key to the transistor is the creation of chemically distinct regions in a single piece of semiconductor material. These "doped" regions have their own electrical behaviors and in appropriate combinations can produce both a switch and an amplifier. Transistors are not only solid but also much smaller than the smallest vacuum tube.

The transistor was greeted by outside commentators as an "interesting little device." Those in the know hailed it as a significant breakthrough in telephone technology. Compared with the fragile vacuum tube, the transistor was not only smaller, but less power hungry, and in theory more reliable. Its proliferation by the trillions and quintillions would eventually revolutionize daily reality, yet at the moment of its invention, this was a

SELECTED EARLY US DIGITAL COMPUTERS, 1947–1953

Name	Location	Began operation	Floor area (square feet)	Number of vacuum tubes
ENIAC	Army Aberdeen Proving Ground	1947	1,800	18,000
BINAC	Northrop Aircraft	1949	20	700
ATLAS I	National Security Agency	1950	110	2,700
AVIDAC	Argonne National Laboratory	1950	500	2,800
SEAC	National Bureau of Standards	1950	150	1,300
Whirlwind 1	MIT for US Navy and Air Force	1950	3,300	6,800
Harvard Mark III	Naval Proving Ground	1951	1,600	5,000
UNIVAC 1	Six installments; government and military	1951	1,250	5,600
EDVAC	Army Aberdeen Proving Ground	1952	1,200	3,600
IAS	Institute for Advanced Study	1952	100	2,300
ILLIAC	Military; University of Illinois	1952	25	2,800
MANIAC	Los Alamos Scientific Laboratory	1952	20	2,500
ORDVAC	Military; Aberdeen Proving Ground	1952	800	2,700
SWAC	National Bureau of Standards, Los Angeles	1952	75	2,300
ATLAS II	National Security Agency	1953	300	4,500
ERA 1103	Multiple installments; government	1953	300	4,500
IBM 701	Eighteen installments; government, military, and industrial	1953	500	400

Source: US Navy, "A Survey of Automatic Digital Computers," 1953.

future quite hidden. The transistor would become the essential everyday brick in building the digital revolution and would be closely tied to Moore's future. What was its genesis?

Leaders at Bell Labs in New Jersey wanted to bring the switching speed of the vacuum tube to the connecting of telephone calls. Telephone exchanges employed complex mechanical gear, far slower to "switch" than vacuum tubes. The charm was that these mechanically controlled switches consumed less power and were much more robust. Bell Labs' speculative yet focused researchers looked to combine benefits, while somehow bypassing the vacuum tube itself: to replace electromechanical switches with a smaller, less power-hungry, faster, more reliable solid switch. They turned to semiconductors for inspiration.

Before the 1920s crystal radio sets had been popular among amateur enthusiasts, using a fine wire—a "cat's whisker"—attached to a carefully identified point on the surface of a semiconductor composed of crystalline mineral. This mechanism could change an alternating current radio signal to a direct current, allowing it to be converted to sound. From 1920 on, however, vacuum tubes could perform the same function more predictably, and cat's-whisker devices fell out of favor as vacuum tube–based radios spread across the land. Bell Labs' researchers revisited these semiconductor materials, seeking to create a "solid-state" device that could amplify, transmit, and switch like a vacuum tube. By the end of 1945, William Shockley was leading a team of scientists in pursuit of this idea.

Two physicists on the team—John Bardeen and Walter Brattain—created just such a device around Christmas 1947 by placing two electrically charged wires very close together on a piece of doped germanium crystal. The key to the success of this "point-contact" transistor was the electrochemical alteration that instantly occurred in the semiconductor crystal, changing it from a conductor of electricity to a nonconductor, or vice versa. Its smallness compared with a tube, and its reliability caught the attention not simply of telephone engineers but also of military communities. Like the vacuum tube, the germanium transistor could amplify, transmit oscillations, and switch, yet it used less power and promised greater reliability.

The military underwrote research on, and limited production of, point-contact transistors. When it first went into production, the germanium transistor was roughly the size of the eraser atop a pencil. Then, in July 1951, Bell Labs announced that William Shockley had conceived—and its labs had built—a faster, more reliable device: the "junction" transistor. Its promise was much greater. Rather than relying on wires pressed onto the surface of germanium, the junction transistor used boundaries (junctions) between three distinct regions of chemically changed material within the crystal. Particular elements were "doped" into a region to alter its electrical character; the resulting chemically complex solid was the device. Junction transistors were better performing and more reliable than their point-contact predecessors. By 1954 Bell Labs was engaged in a big push to transistorize the US Army's nuclear-armed Nike missiles. At APL an effort was under way to do the same for the US Navy's competing Talos, a ship-launched missile. The year also saw a far different use of germanium transistors: their first commercial employment in hearing aids, rendering these cumbersome devices slightly less clunky.

Electronics firms—most clustered on the East Coast—began to develop germanium junction transistors. Many (such as RCA, Raytheon, and Philco) were already producers of vacuum tubes. Others, like Texas Instruments, were fresh to the field. Although the transistor, its manufacture, and its applications had all been widely patented, this posed no barrier to entry. Mindful of ongoing antitrust negotiations with the federal government, AT&T promoted inexpensive licensing of its transistor patents, while its Bell Labs arm held symposia for licensee companies to transfer to them the latest in transistor design, manufacturing, and use.

As the military supplemented its use of vacuum tubes with transistors, changes in computer and transistor technologies began to feed off each other. A generation of slightly smaller, more compact computers housing transistors appeared for control functions in airplanes and on ships and to crunch data in cryptology centers. Most used germanium, but in the fledgling semiconductor industry and its predominant customer, the US military, interest began turning to a new alternative, the silicon transistor. Silicon transistors were much better able to withstand the high temperatures ubiquitous in jet aircraft, missiles, and big computers. Texas Instruments scored a major coup in bringing to market a silicon junction transistor in 1954. Initial orders followed from the military. Even so, the transistor remained a niche technology, in the shadow of the vacuum tube. In 1955 the latter reached its all-time peak production of well over 1 million tubes a day.

COMPARISON OF TUBES AND TRANSISTORS, 1954–1957

Year	Units of tubes	Units of transistors	Value of tubes	Value of transistors
1954	385,089,000	1,320,000	$275,999,000	$6,100,000
1955	479,802,000	3,690,000	$358,110,000	$12,300,000
1956	464,186,000	12,820,000	$374,186,000	$37,400,000
1957	456,424,000	28,700,000	$384,402,000	$69,700,000

Source: Electronics Industries Association Electronic Yearbooks and Market Data Books.

US PRODUCTION OF VACUUM TUBES, 1930–1957

Year	Units	Value
1930	40,213,000	$37,930,000
1935	75,962,000	$26,565,000
1940	108,476,000	$27,610,000
1945	139,478,000	$68,500,000
1950	382,961,000	$250,000,000
1955	479,802,000	$358,110,000
1956	464,186,000	$374,186,000
1957	456,424,000	$384,402,000

Source: Electronics Industries Association Electronic Yearbooks and Market Data Books.

Gordon Moore's first encounter with the transistor came toward the end of 1954, when he was twenty-five. Taking a break from his experiments one fall evening, and leaving his heavily pregnant wife at home, he traveled downtown along with several of his colleagues to the prestigious Cosmos Club, long a meeting place for scientific and political elites. The lure was a lecture, under the auspices of the Philosophical Society of Washington. Its title was "Transistor Physics." The lecturer was the world's foremost expert on the topic, William Shockley. Moore was impressed both by the technology and by the man promoting it. Two decades older than Moore, Shockley exuded self-confidence: "He was a real showman, a very engaging speaker." At the end of his talk, Shockley tossed handfuls of transistors—the size of peanuts—into the audience, with a flourish. Moore perceived him as potentially "an exciting guy to work for."

In March 1955 it was Moore's own turn to take the stage. The APL group had assembled its most recent results into short papers for presentation at the annual meeting of the American Physical Society, the primary professional society for physicists, in nearby Baltimore at the Lord Baltimore Hotel and on the Johns Hopkins University campus. Around a thousand attendees were registered for the multiplicity of offerings. Moore would present the second of his group's two talks. This was Gordon's first formal presentation of research to a national assembly of his scientific peers, a necessary rite of passage. He had so far avoided or evaded public performance, preferring time alone in the lab. Beyond writing up his dissertation in notably short order, he had done little to promote himself and had displayed no obvious desire for glory.

However, the time had come for Gordon to demonstrate his bona fides as a professional scientist: that meant not only publishing papers but also delivering them. Faced with the unavoidable, he became "ridiculously" nervous about his talk. Despite the speech's brevity and straightforward nature, he suffered an upset stomach for days before, while lacking the vocabulary or inclination to discuss his nerves with Betty. His paper was tucked obscurely into one of the concurrent sessions, of interest only to others with equally specialist credentials. Unsurprisingly, he spoke about his work with the hydrogen chloride molecule, discussing the relationship between the shape and width of spectral absorption lines and the electronic structure of the molecule. He raced through his ten-minute presentation, shaking as he spoke. Despite everything, he survived; it was an instructive experience and an important step in Gordon's overcoming his fear and realizing that he could indeed speak in public: "I was so darn nervous, but I learned."

At the conference the buzz was certainly not about arcane infrared results, but rather about the transistor. A banquet at the meeting honored John Bardeen and Walter Brattain, the two physicists from Shockley's group who had fashioned the first transistor. In contrast, Gordon's contribution to the meeting was brief, quiet, and all but unnoticed, yet it marked a milestone in his nascent career.

Looking for Change

On the home front Betty provided unquestioning, staunch support, following the standard script for a young scientist's wife. Naturally outgoing, she formed a close friendship with neighbor Charlotte Bruce, also married to a scientist and with a son born shortly before Ken. Ken Moore was a difficult baby, crying for long hours. The two women would take their baby carriages outside on the grass, while conspiring to keep Betty's kittens a secret from the apartment's manager. The truth came out when Gordon and Betty moved to a larger two-bedroom apartment in the family area of the complex. "Our cats liked to go on the bed in the master bedroom, which meant I couldn't hide them underneath the crib in the baby's room," recalls Betty. "I explained to the manager, 'This cat found us.' He said, 'Not to worry. I have four of my own!'"

In 1955, pacing the small apartment at night, trying to soothe his son and give his wife some relief, Gordon had time to think. The disruption of his nights was matched by his mounting disquiet at, and with, his work. His research was solid, but the Applied Physics Laboratory was not Caltech; nor, in his eyes, did the Washington area measure up to California. In addition, he was now a father and missing his wider family. It was still less than two years after his arrival on the East Coast, but he believed the situation ripe for reevaluation.

The research center had had bouts of instability in the past. Its character as a small, curiosity-driven venture was at odds with the military agenda of APL as a whole. This caused tension and fracture. James Van Allen, the doyen of high-altitude research, had already left abruptly to take up an academic post, believing the laboratory's leadership was trying to exert too much control over his work. Robert Herman and Shirleigh Silverman, the leaders of the flame-spectrometry group, also began to chafe and look elsewhere for opportunities. Late in 1955 Silverman quit for a management position at the Office of Naval Research itself. Herman also left, later becoming a group leader in General Motors' research lab.

Managerial conflicts heightened, along with Gordon's unease. The unease was not just situational but existential. He began to doubt that his work was worthwhile. Infrared spectrometry, as part of chemistry's instrumentation revolution, had a tangible value in a broad array of activities, such as the manufacture of medicines and synthetic fabrics, like nylon. Yet Moore's work was connected only remotely to any practical output. He was on the research frontier, pushing the science forward by teasing out fleeting signals of molecular behavior. Although these insights might one day have implications for manufacturing and production, that day was far away. If he continued on this path, he could foresee only more of the same: recondite experimentation, abstruse interpretation, and specialist publication.

To assuage his disquiet, Gordon turned to his preferred introspective mode of understanding. For this, he by now routinely resorted to a ledger or laboratory notebook. In it he would choose a "hard" quantitative measure to generate results for analysis. Soon, he found his metric: it was economic, as it would be so often in the future. What was the cost of the published articles that he and his group were delivering, their "product"? Who read these words? How did the cost measure up against the benefit? "I literally sat and calculated the cost per word of our published articles," he recalls. "Maybe a handful of people were reading them. I wondered, 'Is the government getting its money's worth?' That's when I became nervous. Looking forward, I didn't see the activity as viable." Measurement and analysis failed to calm his concerns—just the opposite.

Major disagreement over who was in charge of the lab was also turning into all but open warfare. "We weren't shielded at all. There was a running battle about what was going to happen. People were beginning to look outside for jobs. I decided I'd better start searching, too." Gordon's perspective had shifted. He had yearned for an academic job at a prestigious institution, at a moment when opportunities were scarce. Now his analysis implicitly called into question the worth of even academic research. The image of himself as a chemistry professor, tucked into a specialist corner of the lab, was no longer compelling. "I thought I ought to get closer to something practical." Through such a move, he could amortize the expense of his work not just through words but through deeds and tangible output. Despite his lack of interest in engineering while in school, "I guess I had an engineering inclination someplace that was beginning to come out." Badger, his Caltech adviser, had sensed Moore's practical, commonsense approach long before Moore had an inkling of it himself. Belatedly, he now focused on seeking opportunities in industrial and military labs.

One of the first possibilities to emerge was at a US nuclear weapons R&D center, the Lawrence Livermore Laboratories, inland from Moore's undergraduate stomping grounds. Named for Berkeley scientist Ernest Lawrence, it had opened three years before and was recruiting talented physicists, chemists, and engineers. The organization certainly met Gordon's primary criterion. As he later joked, "What they were doing was practical: making bombs." A job at Lawrence Livermore would enable a return to the Bay Area. For Betty, housebound with her baby son, the lure of moving back to family was especially strong. Gordon, too, desired a speedy return to California. He applied. Lawrence Livermore was interested and flew him out. Betty came too, even though pregnant for a third time. They combined the trip with visits to their families, in nearby Redwood City and Los Gatos. Their demanding travel experience—including flying through a major thunderstorm—was followed by another miscarriage for Betty.

Gordon had his own disappointment. Lawrence Livermore offered him a job, analyzing the spectra of nuclear explosions, but the work looked less interesting intellectually than he had hoped. He was comfortable working on nuclear weapons, but believed there was little future in nuclear bombs; as a career move, it hardly made sense. "It sounded like a one-shot deal, if you'll excuse the expression! What do you do next? It was a big government lab, and the stuff they were doing didn't really intrigue me." To Moore, the prospect seemed a rabbit hole. It would lead only to another uninviting specialist career, rather than being a good investment of his time. He turned the offer down. However, the lab was impressed by him and kept his résumé on file.

Looking for leads, Gordon contacted Roger Newman, his former colleague in Badger's Caltech laboratory, who was by now at General Electric's central research organization in Schenectady, New York. GE was a major military contractor, government supplier, and commercial manufacturer. Its laboratory had a reputation for technology leadership and a wide reach; its R&D ranged from vacuum tubes to atomic energy to aircraft engines. Newman invited Gordon to come up and visit.

One particular venture at GE snagged Gordon's interest and was, in retrospect, a clue to the direction of his ambition: a program to make artificial diamonds. This project involved high temperatures, high pressures, and large pieces of manufacturing equipment. Here was a novel effort in science-based manufacturing that struck him as "a new, different territory," deploying a difficult technology within a first-rate organization. Schenectady was not San Francisco, but an interesting job with obvious practical utilities was, on the balance sheet of his reasoning, more important than getting home quickly. As with Princeton's infrared group and the DuPont Experiment Station, "if they had made me an offer, I would have taken it."

The company did make Moore an offer, but not in artificial diamonds. Instead, it was to join an R&D group for atomic energy. GE's atomic power division had its headquarters in San Jose, California, and its interest in atomic energy (the Cold War's civilian twin to thermonuclear weapons) was part of a gathering wave of commercial interest. The Vallecitos atomic laboratory, near Livermore, was a major facility, centering on a small nuclear reactor used for a host of research projects. The hope was to establish a full-fledged power station, generating electricity. Indeed, by 1957 the firm's Vallecitos reactor would be the first privately owned plant to deliver significant quantities of electricity to a public grid. Gordon temporized; procrastination was a well-practiced response.

Christmas came and went, and they were still on the East Coast. Then another opportunity hove into view, one that would utilize his investment in infrared expertise, bring him closer to practical matters, and allow him to return to the Bay Area. A position was open at the Shell Development Company's laboratory in Emeryville, adjacent to Berkeley and beside the Bay Bridge, connecting to San Francisco and the peninsula. The Emeryville lab was the American R&D center for Shell Oil and boasted a storied history in infrared spectrometry. Physicist Robert Brattain—brother of Bell Labs' Walter and the donor of an instrument to Richard Badger—had developed important infrared spectrophotometers in his career there. The lab invited Gordon to come for an initial discussion. He quickly went west again and liked what he saw. He returned to Silver Spring, waiting for an offer.

He decided to turn down General Electric. The Cold War focus on nuclear bombs and atomic power stations did not excite him; it seemed too limiting. It was not enough simply to find a job: the cost in a narrowed career path must not be greater than the benefit of an immediate return home. To close the only fully open door was tough, as both he and Betty were anxious to head back west. "We could not have stayed much longer," she says. "It was a different culture. Returning to family was the major part of our desire to get back to California. We were hoping something would open up." Then came William Shockley's call. Through it, Gordon Moore would learn the truth of Louis Pasteur's dictum that "chance favors the prepared mind."

"Hello, This Is Shockley"

In February 1956, when Shockley telephoned, Moore knew the voice and the name right away, thanks to Shockley's public lecture at the Cosmos Club. William Bradford Shockley was one of the world's leading solid-state physicists. As the call made clear, Shockley had by now left Bell Labs in

New Jersey to start his own venture in California, under the aegis of Arnold Beckman. Gordon had long been aware of Beckman Instruments as a leading player in chemistry's instrumentation revolution and had often used their products in his research. Shockley explained his intent to hire the best and brightest young PhDs to join him in the little-known agricultural town of Mountain View, immediately to the south of Palo Alto on the San Francisco Peninsula.

Palo Alto was "nowhere land," many hundreds of miles from any transistor maker and thousands of miles from the main East Coast manufacturers and researchers. However, Shockley had spent childhood years in Palo Alto, and he yearned—and had received Arnold Beckman's okay—to set up shop close to his mother, who still lived there. No young PhD in physical chemistry from the major leagues had ever heard of Mountain View, no one, that is, except Gordon. Mountain View's fruit orchards lay barely fifteen miles from his parents' home in Redwood City and a similar distance from the Metzler ranch in Los Gatos.

Bell Labs and the Pentagon viewed Shockley as an extremely creative and competitive physicist, valuing his broad-ranging insight. During wartime he had pioneered "operations research," applying scientific analysis to military activities such as strategic bombing and submarine hunting. With a track record that was second to none, he had the highest-possible security clearance, including access to nuclear secrets. Yet his own robust self-regard made him intensely competitive with fellow physicists, including, disastrously, those who worked for him at Bell Labs. Now launching out on his own, Shockley aimed to perfect and mass-produce the most advanced form of silicon junction transistor, still barely in development at Bell Labs. It promised faster switching and greater reliability. Shockley was focused on military markets, where performance always trumped mere price. Making transistors was fundamentally a chemical and materials challenge, and he knew that at Bell Labs chemists and metallurgists were mainly responsible for this latest breakthrough.

Intent on assembling his own small team, Shockley sought a talented experimental chemist; indeed, he had already advertised in *Chemical and Engineering News*. Shockley next explored which, if any, interesting chemists might have applied for California jobs. He perused the classified personnel files at Lawrence Livermore Laboratories. There he found Gordon's name. "Shockley had had chemists in his group at Bell Labs, and they did useful things, so he thought he needed a chemist," says Moore. "That was the role I was going to fill: to do useful things."

Shockley's interest was sparked by Moore's earlier track record of high promise at Berkeley and Caltech. Also intriguing was that Gordon had turned down an offer from the prestigious Lawrence Livermore Laboratories.

Shockley was fascinated by virtuosity and felt that even at the rarefied Bell Labs, mediocrity was disagreeably dominant. He believed real contributions were made by that rare handful of individuals who possessed elevated "mental temperature": an unusually strong, genius-level capacity for creative output. In examining Gordon's file, Shockley saw not simply the technical skill sought by recruiters at Lawrence Livermore, but also the self-confidence to spurn their offer. Might this be the chemist he sought?

Chemists and Metallurgists Needed for Semiconductor Research and Development[1]

BY WILLIAM SHOCKLEY[2]

RECEIVED JANUARY 15, 1956

Semiconductor chemistry and metallurgy are expanding fields. Examples of the creation of new science in the field of solid solutions in silicon and germanium are easily found[3] and this science is being effectively applied to transistor technology.[4] There will be many opportunities for professional advancement in this field as semiconductor electronics production will grow rapidly[5, 6, 7] with transistor production increasing by a factor of 100 to 1,000 in the next 5 to 10 years.[8]

New semiconductor devices and manufacturing methods will be developed by the Shockley Semiconductor Laboratory of Beckman Instruments, Inc.[9] The Laboratory will also afford opportunities for basic research and publication to its scientific personnel[10] and will be located on the Stanford University Estate a short drive south from San Francisco. One or two chemists and metallurgists are needed now to participate in the formative phase of the activity. Experience in semiconductor work is desired, but it is less important than outstanding technical ability and personal maturity.[11] All inquiries will be promptly answered and held in confidence if so desired.

William Shockley's advertisement for chemists, 1956.

SOURCE: STANFORD UNIVERSITY LIBRARIES, SPECIAL COLLECTIONS AND UNIVERSITY ARCHIVES.

There was no obvious connection between Gordon's expertise and the chemistry of silicon transistors. Moore had already absorbed the idea that to be hired to undertake "practical" work might entail writing off his investment in infrared. Hence, he was open to Shockley's approach, which impressed him on several levels. First, it would be a chance to do experimental chemistry alongside one of the world's top scientific minds. Second, the aim of the work was directly useful: to create an advanced transistor, a task that would take him to the forefront of an intriguing recent development in technology. Third, Shockley and his sponsor, Beckman, were both Caltech alums, like Gordon, and finally, but not least important, Mountain View was home, or near enough. Here was a *real* chance for Gordon and Betty to return to the environs—family, geographical, and cultural—that they loved.

Shockley asked Gordon if he was interested. Would he fly out to talk it over? Without hesitation, Moore agreed. On short notice he would again go west. He hung up and shared the incredible conversation with Betty. Her instinctive reaction was "Shockley is a very serious proposition." Moore did not know it at the time, nor did Bill Shockley, but the two men also shared a distant ancestor, John Alden of the *Mayflower,* whose eighth-generation descendants included Shockley himself as well as Gordon's maternal grandfather, Josiah Williamson. The common link lay in Marshfield, Massachusetts, Josiah's hometown. Shockley's aristocratic *Mayflower* lineage was no secret. Moore, altogether a quieter character, lived in simple ignorance, knowing little of his own family history.

Gordon quickly booked a cross-country flight. In 1956 air travel from the East to the West Coast was still something of an adventure. The journey was extremely costly, took more than eleven hours, and involved at least one stop. Nonetheless, in short order Moore found himself back in familiar territory, driving up San Antonio Road along the boundary between Mountain View and Palo Alto. While the area was still largely agricultural, the move of Lockheed's division (responsible for developing the US Navy's submarine-launched nuclear missile) into facilities in Palo Alto to the north was already triggering subtle changes, as suburban development encroached upon the orchards and fields.

Beckman Instruments' Shockley Semiconductor Laboratory, Mountain View.

SOURCE: CHEMICAL HERITAGE FOUNDATION.

The building to which he went was an unprepossessing sight, a modified Quonset hut, one of the prefabricated metal and timber buildings made in great numbers by the US military during World War II and afterward sold as surplus. This example had a low-slung addition on the front, providing additional office space with ample windows. It was an unlikely setting for a venture hyped as leapfrogging to the forefront of high technology. Inside, 391 San Antonio Road was very much a work in progress, with Shockley in the early stages of converting the space into offices and labs. He already had a handful of very young scientists on board, most of them fresh from their PhD work. He had also recruited a small group of production engineers from Western Electric, the Bell telephone system's fabled manufacturing arm.

No one had a higher opinion of Shockley's abilities than Bill Shockley himself. He knew that Moore, a reserved and hard-to-read young man, had turned down prestigious opportunities. On both sides the stakes were high. Gordon Moore would need to be persuaded of the worth of the project—to be wooed—before he signed up. And Shockley would need to be sure of Moore's talent. Chemistry, after all, was the essence of making transistors. The group gave a brief tour of what Gordon remembers as "the beginnings of the facility that they were trying to put together" and then quickly moved to a discussion about what was planned.

Gordon did not yet appreciate how complete a departure for Bill Shockley this makeshift, small-scale, entrepreneurial venture was. Shockley was used to working in large organizations, occupying top research posts, and leading free-ranging groups. At Bell Labs his competitive management style had cast a shadow over his otherwise stellar reputation. The leaders there wanted to have him doing innovative work, but also to keep him safely within bounds. However, Shockley himself had become restless, as he watched a whole industry begin to take shape on the basis of his junction transistor, while profiting him little. The early 1950s had also been a more than usually turbulent time in his turbulent personal life. In 1953 Shockley's wife, Jean, was diagnosed with uterine cancer. While she was convalescing, he announced that he was withdrawing from the marriage, abandoning her and their three children.

Shockley took leave from Bell Labs (while maintaining his position there) and left the East Coast to become a visiting professor at Caltech. Once in Southern California, he opened discussions with technology firms about a transistor venture. Finishing 1954 with several tentative conversations, he still had no clear route to his goal of total control of technological direction and a substantial equity stake. He now switched to a top scientific job in Washington, DC, with the Pentagon's high-level Weapons Systems Evaluation Group, producing strategic reports on Cold War questions such as how to prosecute a thermonuclear war or when to use chemical and biological weapons.

As 1955 progressed, Shockley read with growing excitement a cluster of reports from Bell Labs that indicated a huge opportunity. Chemists there had found a fresh way to make the doped layers of a silicon transistor: "diffusion." Diffusion promised transistors with faster switching and also the ability to make many transistors in one batch. That summer Shockley took the plunge, giving up his by now nominal position at Bell Labs and exiting his Pentagon job. For a month, he worked with financier Laurance Rockefeller and his family investment company, Rockefeller Brothers, in the hope of assembling a deal for a diffusion-based transistor project. Arnold Beckman was another prospect, one who seemed more promising.

That February Beckman had invited Shockley—along with vacuum tube pioneer Lee de Forest—to be the guest of honor at a gala organized by the Los Angeles Chamber of Commerce. Beckman was ten years Shockley's senior and three decades older than Gordon Moore. Already in the 1920s he had taken time out to work at Bell Labs, where he encountered the emerging world of vacuum tube electronics. In 1935, four years before Bill Hewlett and Dave Packard set up operations in Palo Alto, Beckman rented a garage in Pasadena, four hundred miles south, for his own venture. By 1955 he had made automation—an electronics-linked technological concept that was beginning to set business and scientific circles abuzz—the central organizing concept of his well-established company. He believed that the fully automated factory was imminent and that it would revolutionize industrial production, using signals coming into computers from electronic sensors.

Shockley was equally taken with automation, especially with his own concept of "automatic trainable robots," which would fill factories and homes and transform American manufacturing. He spun together detailed designs for electronic detector "eyes" for these robots and took them to Georges Doriot, a Harvard Business School professor who was the leader of a novel company that made speculative investments in high-tech start-ups coming out of MIT. With Doriot's help, Shockley filed for a patent on an electronic eye.

At the gala in Los Angeles, Beckman and Shockley quickly uncovered their mutual interest in automation and began an intellectual love affair, based on their shared enthusiasms and their many similarities: Caltech, high-level security clearances, competitiveness, belief in virtuosity, and inventive drive. Shockley sent Beckman his robotic-eye patent to incorporate into Beckman's automation efforts. Beckman reported back that while his engineers could find no immediate use for the patent, he hoped they would continue their association.

In August 1955 Shockley approached again. Would Beckman be interested in partnering to create a company to mass-produce diffusion-based silicon transistors? This time, wooed by Shockley's arsenal of arguments, Arnold Beckman could not resist. To him, it all made sense. Mass production of electronic goods for defense purposes already provided a large proportion of his profits, so he was au fait with growing US military needs and emerging opportunity. Moreover, silicon transistors could be ideal for automation, a perfect fit for sensors, controls, and computers in his improved automatic chemical factory of the future. It was not a question of whether this future would arrive, but rather a question of who would best capitalize upon "first mover" advantage. Few outside Bell Labs yet understood what was possible.

The official name for Shockley's organization was the "Shockley Semiconductor Laboratory of Beckman Instruments, Incorporated." Beckman believed that Bill Shockley—with his track record of moving from one blockbuster innovation to the next—was the man to create the next big thing in electronics. Beckman wanted Shockley's ideas, but most of all he wanted Shockley himself. A firm believer in the power of individual genius, Beckman ran his company on a hub-and-spoke model, with himself at the center of a circle of talented individuals, each running large swaths of the organization and reporting in. Shockley agreed with the need for uniquely gifted individuals; was he not such a virtuoso himself? Both men were fundamentally antimanagement. Enduring progress in high technology could not, in their view, be adequately nourished by organizations with prescribed processes and procedures, like Bell Labs. True success came from finding genius and letting it run.

The two shared huge ambitions. Beckman was on the board of Caltech and had flatly declined a buyout offer from giant IBM. Shockley, in turn, envisaged a significant venture under Beckman's sponsorship, through which he would reap major financial reward. In August and September 1955, the pair hashed out a plan for their domestic partnership. Shockley would work for Beckman Instruments as an employee for two years, the "initial project" being the mass production of diffusion-based silicon transistors. There was also ample scope for flexibility: "We propose to engage promptly and vigorously in activities related to semi-conductor research, patenting of inventions, licensing, production, and other developments in the field of semi-conductors." Beckman promised a handsome salary, stock options in Beckman Instruments, and funds required for the venture, estimated at three hundred thousand dollars for the first year. After two years the venture might become a wholly independent corporation.

Shockley would have no responsibility for business functions such as accounting, sales, and marketing, which would be more than three hundred miles away at Beckman Instruments' headquarters in Fullerton, near Los Angeles. He would instead enjoy unfettered control over research. Shockley made clear that Arnold Beckman himself was the reason he was working with Beckman Instruments, and he reserved the right to terminate the agreement should Beckman leave. If after two years they decided to launch a freestanding company, Beckman would take 51 percent, Shockley the rest.

The decision to locate the venture in or near Palo Alto was a concession to Shockley. As well as family and childhood connections there, he was well acquainted with Frederick Terman, Stanford University's powerful dean of engineering, who dreamed of seeding a semiconductor electronics community in the region. Also, Beckman had recently purchased two small high-tech Bay Area firms. He agreed that the laboratory could join this cluster. Shockley set the seal on his reinvented life by traveling to Columbus, Ohio, in November 1955 to marry Emmy Lanning, a psychiatric nurse three years his junior, whom he had met the previous year.

Silicon, Chemistry, and Bell Labs

As Gordon Moore would discover and Beckman already knew, Shockley possessed a huge advantage when it came to making transistors: his inside track at Bell Labs. He remained close to colleagues there, in particular Jack Morton and Morgan Sparks, both key personnel. Morton, a hard-driving electrical engineer, was by now the undisputed leader of development activities on semiconductor devices. Sparks was the physical chemist who had fashioned the first of Shockley's junction transistors. Together, they supplied him with all the latest thinking and gave word of advances as they broke.

By mid-1955 US producers were churning out almost a half-million transistors a month. Nearly all—99 percent—were junction transistors created by chemically treating germanium crystal. The military was the prime consumer, but commercial uses were growing. Texas Instruments had recruited chemist Gordon Teal from Bell Labs and was well on its way to being a top producer of transistors for military use. In October 1954 it had already announced a civilian first, the TR-1 transistor radio.

The TR-1 was robust, small, lightweight, portable, and quite unlike the familiar radio set characterized by vacuum tubes, a stationary location, and finicky reception from an outdoor aerial. Within a year one hundred thousand transistor radios had been sold, for more than five hundred dollars each in today's money. Soon, portable and automobile radios would be the primary users of germanium transistors in consumer electronics, with sales in the multimillions and prices dropping. Thanks to Teal, 1954 also saw Texas Instruments offer limited quantities of a junction transistor made from silicon. These new transistors were more rugged and heat resistant

US PRODUCTION OF TRANSISTORS, 1954–1957

Year	Units of germanium transistors	Units of silicon transistors	Total number of units	Ratio of silicon to germanium units	Total value
1954	1,300,000	20,000	1,320,000	1.54%	$6,100,000
1955	3,600,000	90,000	3,690,000	2.50%	$12,300,000
1956	12,400,000	420,000	12,820,000	3.39%	$37,400,000
1957	27,700,000	1,000,000	28,700,000	3.61%	$69,700,000

Source: Electronics Industries Association Electronic Yearbooks and Market Data Books.

than their germanium counterparts, perfect for military use in missiles and aircraft. Their major drawback was that they did not switch nearly as quickly. By 1955 ninety thousand a year were being manufactured, most for military use. Transistor—including silicon—technology was changing rapidly.

Bill Shockley was in a hurry to get going. As Gordon Moore toured 391 San Antonio Road, Shockley let him in on a secret of the Bell Labs reports. Researchers there, he explained, had very recently come up with a fresh set of chemical techniques to transform silicon crystal, forming doped layers with the needed junctions. These techniques depended on mastering diffusion as a means of doping. In theory this would allow greatly improved manufacture and performance, giving him an ability to seize the initiative from Texas Instruments. Prototypes had already been made at Bell Labs. The diffusion technology developed there—and revealed to Shockley—would allow nothing less than the chemical "printing" of silicon transistors.

Shockley glossed over the uncharted complexities of undertaking unfamiliar microchemistry on a commercial scale. As with traditional printing, in which chemical and mechanical processes first create paper and then print on it with special inks, the making of silicon transistors would now begin with chemico-mechanical processes to create silicon wafers—the "paper"—and then to "print" on them, using diffusion as the printing process and dopant chemicals as the printing ink. Techniques had already been developed whereby a "crystal puller" machine could draw out a near-perfect silicon crystal from a pool of the molten element. Once this long, cylindrical single crystal had cooled, wafers (roughly the size of a nickel, but far thinner) could be cut from it as circular slices. Wafer quality profoundly affected transistor function; purity was vital to success.

Paradoxically, Bell Labs' first junction transistors had been produced by deliberately upsetting the purity of semiconductor crystals. By adding dopant chemicals to the melt during the crystal-pulling process, researchers had created distinct chemical layers and the junctions between them, the essence of the junction transistor. Now, a better technique had been developed. If undoped wafers were instead placed in a high-temperature furnace with a *gas* as the dopant, its atoms diffused into the wafer in much the same way as a dollop of cream diffuses into a cup of hot coffee. In theory, *diffusion* could become a large-scale batch process in which many wafers were treated at once in the same furnace, even as many transistors were made on each wafer.

Bell Labs' chemists then hit upon another advance, oxidation, a chemical reaction in which materials incorporate oxygen, as with the formation of iron oxide (rust) on an iron surface. If water vapor was admitted to the furnace during a diffusion operation, a layer of quartz (silicon dioxide) formed on the surface of the wafer. This oxide layer proved to have many virtues. It was highly durable, serving as a protective "mask" and electrical insulator that protected the wafer from damage or stray electrical charges during and after the diffusion. The same Bell Labs chemists soon found that the mask was impervious to subsequent attempts at diffusion, opening yet another possibility. If researchers etched windows through the mask, then they could diffuse further dopants into the crystal in specific areas. "Oxide masking" combined with just two diffusions could cover a whole wafer with thin sandwich layers. Thinness was the key to fast-switching transistors. Oxide growth, window etchings, careful diffusions: here was the crux of a chemical printing technology for making fast silicon transistors.

In March 1955 Morris Tanenbaum, a Bell Labs chemist who had worked in Shockley's group, made a laboratory example of a silicon transistor entirely by diffusion. If such techniques could be refined and standardized, they would provide the blueprint for a manufacturing technology. Anyone who brought this vision to fruition would undoubtedly make a fortune. There was a ready market. As the Korean War ended and Cold War tensions escalated further, the US military's need grew ever greater for advanced electronics in its missiles, aircraft, computers, proximity fuses, communications, and nuclear weapons. In some advanced uses, germanium transistors were already nudging out vacuum tubes. It seemed possible that a fast silicon transistor, one remarkably tolerant of extreme heat, might beat out germanium in crucial areas of military aerospace.

In the lab in Mountain View, it was clear that Bill Shockley intended his team to leap in front of Bell Labs and start running. He believed himself poised to dominate the military market. The Quonset hut might be uninspiring, but the vision was impressive. Moore, characteristically low-key, recalls: "Shockley told me he was going to make a transistor; more precisely, a diffused silicon transistor. He planned on selling them in some quantity.

He was a very convincing guy; he was the guru. I didn't know enough technically about the field to question anything. He said there was interesting technical work to be done, and the eventual product was useful. It wasn't a very hard sell."

The military market was key to Shockley's sales pitch. But for Gordon, getting in on a burgeoning area of research—silicon electronics—was the exciting thing. It opened up the possibility of a demanding, promising career. "I liked the sound of the project. It was a whole fresh field. That what I'd be working on had a practical application was part of the attraction." And, of course, a job at Shockley's lab would mean that Gordon and Betty would be back home. Even his early Pescadero haunts were only an hour's drive through the coastal mountains. The location was impossible to beat.

Gordon was also strongly attracted by the prospect of working with the very best. At Berkeley he had sought classes with Nobel laureates; at Caltech he had worked alongside Pauling and listened to Feynman. Shockley's accomplishments were proven and his mastery of his field unparalleled. The Quonset hut's other occupants were bright young PhD scientists from the very best schools, an "intriguing bunch, the kind of group I could fit into pretty well." In turn, he endeavored to make a good impression. In the advertisement in *Chemical and Engineering News,* Shockley had deemed experience in semiconductor work desirable, but "less important than outstanding technical ability and personal maturity." Moore came across as able, if a little reserved. Now Shockley needed to test his competence: "Shockley described how a chemist at Bell Labs had figured out that he could get rid of copper contamination by rinsing with cyanide, thus complexing the copper. Would I think of something like that? That was one of the particular things he asked. I said, 'Yeah, probably!'"

Shockley was far from finished. He gathered the staff for an impromptu seminar to discuss and question Moore's work at APL and Caltech. Primed by his previous public speaking and interview experiences, Gordon displayed none of the nervousness that had marked his conference debut or the stunned passivity that had botched his presentation at DuPont. Instead, he fielded a string of "good questions" from the group. In the end, "I knew more about the subject than Shockley, so I passed." The visit wrapped up with dinner, a mile from the lab at Rickey's, then Palo Alto's main haunt for business meals. Knowing that Gordon was making the long trek back east the next day, Shockley instructed him to appear in New York City by way of follow-up, for a battery of industrial psychology tests, like the ones Moore had taken for Dow.

Gordon arrived home in Maryland with the welcome news that the Mountain View opportunity looked as if it was becoming reality. At work, when he requested another day off, no one raised an eyebrow; most

EMPLOYEES
8/28/56

1. W. Shockley
2. L. Valdes
3. W. Happ
4. S. Horsley
5. V. Jones
6. R. Grunewald
7. T. Zinn
8. L. Bolender
9. A. Pretzer
10. S. Roberts
11. C. Himsworth
12. H. Breen
13. D. Knapic
14. R. Noyce
15. R. Wagner
16. J. Blank
17. G. Moore
18. D. Allison
19. G. Stout
20. J. Last
21. W. Gadbury
22. W. Pleibel
23. E. Kleiner
24. V. Grinich
25. G. Troyer
26. S. Lee
27. M. Asemissen
28. C. Sah
29. J. Hoerni
30. K. Jacobsen
31. Bill Stansbeary
32. D. Farwell

Shockley's staff in late August 1956.

SOURCE: STANFORD UNIVERSITY LIBRARIES, SPECIAL COLLECTIONS AND UNIVERSITY ARCHIVES.

colleagues already had one foot out the door. Quickly, he scheduled a visit to New York for the personality assessment. Gordon was dubious about its worth but played along: "They give you a picture of some somber-looking people; you can make up any story you'd like about it. It's hard to understand how they learn much about you." In fact, the test results were remarkably consistent with the conclusions drawn by Dow's industrial psychologists. Shockley's consultant reported that while Moore was bright and very capable technically, he would never be a good manager. This was just fine with Shockley, who had not the slightest desire to relinquish control of any aspect of his lab. He wanted not another manager but an obedient chemist, sufficiently talented to do "useful things." Gordon passed the test.

In the last week of February, a letter arrived to seal the deal: "Dear Dr. Moore, I should like to offer you a position in my organization at $750.00 per month. I shall be glad to have you come as soon as you can arrange your departure appropriately with the Applied Physics Laboratory, but not later than May 1, 1956. Since we are actively seeking chemical candidates, I should like your decision within two weeks." After details about moving expenses, the letter concluded, "I believe that you are well enough acquainted with the nature of my project through our discussion, so that no additional comment is called for here. I hope we shall enjoy a rewarding association for many years." Against both their hopes, Shockley and Moore would work together, increasingly at odds, for barely eighteen months, but in February 1956 Betty Moore needed no convincing. "When Shockley made the offer, it was a no-brainer to take it." Gordon accepted.

Subsequent events would bring into sharp contrast the differences in style of the two men. If Shockley's vision of the significance of silicon transistors was prescient, his ability to realize that vision was sadly deficient. His main advantage, the possession of inside information, would soon be worthless, since Bell Labs was about to disclose to the nascent industry the secrets of the diffused silicon transistor. To establish a lead in the field, team building and focused action were of the essence—and utterly beyond him.

The contrast with Moore is instructive. As he tuned into the realities of the silicon transistor, Gordon became every bit as much of a visionary as Shockley. The scale of the agenda that would grow in his mind over the next decade was truly revolutionary, yet Moore was eminently practical in every respect: a hunter, gymnast, sports player, and skilled experimenter. He was the patient fisherman, the record keeper who would use his notebook to measure, analyze, and decide. In due course he would master or find proxies to execute the skills needed to be a team builder, a successful and admired CEO of a major corporation, a salesman, and a spokesman for the agenda he developed. "Moore's Law" would come to epitomize his quiet ability to focus, execute, and deliver: to raise a human vision to the status of a technological certainty. In February 1956 all these matters lay in an unknown future, in a California forever at the edge of novelty.

SHOCKLEY SEMICONDUCTOR LABORATORY

Beckman, Shockley, and Affairs of the Heart

As Gordon and Betty packed up their belongings, Shockley's plans were featured in *Chemical and Engineering News*. A short article described "free-style research and development which, initially, would involve transistors and other semiconductor devices." The writer noted that Shockley "found it difficult to present a detailed program" and expressed surprise that Arnold Beckman was backing "so vaguely defined an enterprise." What Bill Shockley did have was "definite ideas" about the modus operandi: "We want unusually capable people. We have no fixed principles. Flexibility will be the keynote. The organization will grow at the rate at which capable people become available, rather than according to a fixed pattern."

Shockley's vision might have been appropriate for a high-level basic research group, whether academically or industrially based. In relation to the state of semiconductor technology, it was naive. Bell Labs had already demonstrated that the diffused silicon transistor could be made. Commercial production now required the development of manufacturing techniques, not basic research. Focused effort was needed to engineer the transistor into a more workable form, to refashion lab-fabrication methods into robust manufacturing processes, and then to mass-produce and market the product. Success depended on close liaison of development with manufacturing and sales. These latter organizations were hundreds of miles away in Fullerton, and not especially interested. Neither Beckman nor Shockley appears to have considered the issue.

Shockley's reflex was to focus back on research, on novelty and game-changing action, and on the group he had led at Bell Labs. At lunches in New York and through visits to California, he sought to interest the lab's top transistor researchers in his Mountain View start-up—courting, in particular, Morris Tanenbaum, who had been part of his group, and Morgan Sparks, who had kept him informed about events at Bell. Neither opted to join him. He did manage to hire Robert Noyce, a promising young MIT physicist who was working for Philadelphia-based Philco, a company making the fastest germanium transistors on the market. In this era the center of gravity of electronics was still in the East—around Boston and New York and in the mid-Atlantic belt from Bell Labs through Princeton to Philadelphia. The idea of relocating to Mountain View, California, was to most as puzzling as Bill Shockley himself. One former Bell Labs engineer, Leo Valdes, was lured from Pacific Semiconductors, a Los Angeles subsidiary of a nuclear missile contractor, together with three mechanical engineers from Western Electric—Julius Blank, Eugene Kleiner, and Dean Knapic—possessed of experience in automatic production methods for telephone exchange equipment. None had worked on transistors.

Having reluctantly abandoned his attempts to lure experienced Bell Labs researchers, Shockley turned his attention instead to young PhD physicists, chemists, metallurgists, and electrical engineers out of MIT, Caltech, Berkeley, and Stanford. He was looking for people of high talent, suitable as staff to a scientific virtuoso like himself. This was an error born of hubris. A further mistake came as Shockley rewrote the scope of his ambitions and efforts. Not only would he be the first to commercialize the Bell Labs diffused silicon transistor, but he would launch another device as well. He would deliver not one but two initial products.

Shockley intercepted Victor Jones, a fresh physics PhD from Berkeley on his way to take up a post at Bell Labs, and recruited him to Mountain View. He tasked Jones with mastering all the reports on an as yet entirely theoretical device: the four-layer diode, another intriguing item that had been conceived by Shockley himself when still at Bell Labs. Pursuit of this holy grail—a long-dreamed-of, reliable solid-state replacement for the electromechanical switches of the world's telephone exchanges and possibly a key component in digital computers—made perfect sense for the free-ranging Bell Labs, but not for Shockley's commercial agenda. The focus on the main task—to mass-produce diffused silicon transistors—was already lost.

In the first feverish months, the consequences of these mistakes had yet to emerge. Shockley himself remained busy recruiting staff and talking up both diffused silicon transistors and four-layer diodes to potential military customers. His hires, meanwhile, were struggling with the fundamentals of chemical printing: crystal growing, diffusion, and oxide masking. As the young recruits trickled into the modified Quonset hut, more mistakes followed. Shockley decreed that his laboratory would take on an ambitious subsidiary project: to build the world's most sophisticated crystal puller, to produce silicon of the highest purity and crystalline perfection. "He thought he saw how to melt a puddle enclosed in a big piece of solid silicon and grow the crystal out of it. The solid silicon itself was the crucible," remembers Moore. The project was huge and expensive. As *Chemical and Engineering News* broke the story of his venture, Shockley admitted privately to Beckman that he was spending at a much stiffer clip than agreed upon in the original deal. And, as yet, there was little to show.

Enter Gordon

While Betty and Ken traveled west by air, Gordon drove in the Buick. A moving van, paid for by Beckman Instruments, hauled the couple's piano and possessions. Gordon's cross-country trek in April 1956 was no straight shot. He received "a little education on the way out west" through several meetings set up by Shockley. At Bell Labs he held lengthy discussions with Shockley's former colleagues, including Tanenbaum. At the University of Illinois, he talked in depth with Paul Handler, an expert in the surface chemistry of semiconductors. "Shockley understood that was an important area, so he was trying to bring me up to speed. I spent two or three days at each place. I was completely green. I had a lot I could learn."

For a brief time, while they searched for their own accommodations, Gordon and Betty occupied his former bedroom in Walter and Mira Moore's modest home in Redwood City. Betty noticed, after two years in the East, how much the area was changing. "All the cities had started to run together. El Camino was not the sleepy little street it used to be." Gordon, buoyed by his prospects, was keen to buy a house and establish roots; he and Betty had saved enough for a down payment, and, coming home, they wanted to purchase their first property. His careful accounting in the ledger was paying off. Such were his hopes for Shockley Semiconductor that he was even persuaded by Betty to look for a home in the quiet but pricey settlement of Los Altos. Both Gordon and Betty valued privacy, and neither wanted to live in liberal "Pinko Alto."

The house they agreed upon was on Alford Avenue, with open-beam ceilings and three bedrooms. It cost around fifteen thousand dollars. "We had to start pretty low because we'd only saved five thousand dollars," recalls Betty, but that was still "a great amount of money." Relishing home ownership, Betty ordered up-to-date appliances and eventually a whole new kitchen. She also acquired her own car. Home alone with Ken, she looked out for nearby "latchkey" children. "I knew that I would never let our kids go through moments like that. Even when [my sons] were older, I was always there."

Arriving in California, Gordon reverted to familiar routines. He left the house early and put in long hours at Shockley—including weekends—as he had done at Berkeley, Caltech, and APL. Bob Noyce had arrived at the lab on a Friday, Gordon the following Monday, and the two found themselves compatible. Noyce's outgoing personality, like Betty's, complemented Gordon's quiet reserve. There was much to do. Gordon, as employee number 17, found start-up realities challenging. "It was comparable to alchemists' laboratories—a few benches and furnaces you could buy out of chemical equipment catalogs." His understanding of transistors and their production, if basic, was fast developing, matching the state of the Mountain View lab itself. He studied the standard text on transistor physics, Shockley's *Electrons and Holes in Semiconductors, with Applications to Transistor Electronics.* "I don't know if I began reading before I arrived at Shockley's lab or afterwards, but by the time he was handing out his book to employees, I already had one. Since Shockley was an international celebrity, he gave me one in Polish, which I still have. Autographed by Shockley in Polish!"

Gordon was a steady presence in the lab. "It was an open space, but we had separate benches. Nobody else could work on my bench. It was always such a mess." Using diffusion to make transistors was to use chemistry, and Gordon's insight, perspective, and understanding would be crucial. He knew that having the right equipment was key to moving ahead. Before the lab could truly establish its processes, it would need, among other things, better furnaces and reliable gas-supply systems.

> I was a good glassblower, so I fashioned gas-handling systems and a system that had several five-gallon jugs of distilled water siphoning from one to another so we didn't have to fill it every day—very simple things to get us going. You didn't have to know a lot to do something useful.
>
> It fit my inclination, to build apparatus. I could see things I could get up to speed on pretty fast. At APL I had to make furnaces with long zones of flat temperatures. Diffusion is done in a tube furnace. At Shockley they were using a furnace with a temperature profile that was parabolic rather than flat.

Here was a challenge of the kind that Moore enjoyed: adapting equipment for more demanding use. "A temperature profile that was parabolic" is a physical chemist's way of saying that the temperature inside the furnaces was far from uniform. This would not do for making transistors. Gordon set to work with gusto: "I started building furnaces that were more appropriate. You had to mix gases in order to get the atmosphere you wanted. I'd always enjoyed blowing glass. I built my so-called glass jungles and stuck them on the front end of the furnaces. I was in a position to start doing R&D on diffusion processes."

Bob Noyce, Sheldon Roberts, and Jay Last were key colleagues. Roberts, described by Last as a "young fogy," was a PhD metallurgist from MIT, familiar with "fancy metallograph microscopes and that sort of thing." Jay Last, a physicist and fresh MIT PhD, tackled the issue of meticulously cleaning, polishing, and flattening the lab's limited supply of silicon wafers, each one the size of a nickel. Only Noyce and Leo Valdes had any semiconductor experience. Morning seminars became a regular occurrence: "We were all in a learning mode. Even though Noyce's experience was with germanium, he could help us learn. He said we had a tremendous advantage, not knowing how easy germanium was to work with compared to silicon. In the seminars, one guy would take the lead, try to learn something beforehand, and then teach the rest of us. We were helping one another along, wherever we were going."

Gordon found Sheldon Roberts "a good resource when I was looking for defects in my silicon" but no soul mate. "Personality-wise, we didn't mesh especially well." Roberts was friendly with Last and with Jean Hoerni, a brilliant Swiss physicist with two PhDs, whom Moore helped Shockley to recruit from a research post at Caltech. Hoerni and Last became close friends. "They spent a lot of time climbing mountains together." Shockley, too, was a keen climber, having scaled Mont Blanc with his daughter, but he did not join these trips. Gordon formed a strong working bond with Jay Last. For the most part, this remained a workplace friendship. "Jay was fresh out of school," remembers Moore. "He had a really interesting sense of humor, and I always had a kick out of him."

Through Shockley's book and lab conversations, Moore grasped that the very essence of the transistor lay in its chemically distinct regions and the interfaces, or junctions, between them. The fundamental task was to prepare appropriately pure chemicals and to create separate regions and their junctions through precise chemical diffusion. Unless Gordon, and the lab, could figure out how this could be done reproducibly, the game was lost. Diffusion was a way to introduce the wanted chemicals—dopants—into silicon. Minute, exact amounts of dopants must be diffused into the right place at the right moment. Even the smallest traces of unwanted materials (contaminants) could wreak havoc, like spies and fifth columnists in an army. Shockley and his team had quizzed Moore on an aspect of this subject in their interview in February. "They'd discovered at Bell Labs that tiny amounts of residual copper made the transistor degrade dramatically. Copper wasn't good, but neither was iron or gold or several other elements in tiny trace amounts."

Specific chemicals were needed to make "P-type" or "N-type" layers in silicon and the "PN" junctions between them. Phosphorus, arsenic, boron, and gallium were all possibilities. Other elements—copper or gold, say—prevented the transistor from functioning. Precise details of the wafers, the diffusions, and the contaminants meant everything. Moore sought to make junctions with the expected electrical properties, and, if not, to understand why not. The task was unfamiliar "microchemistry" in which minute traces of any unwanted chemical spelled disaster. As Gordon applied his considerable skills, "I started learning 'in the school of hard knocks.' It's hard to realize how unclean an environment we were working in compared with what's actually required to get good electrical characteristics.'"

Failure was the order of the day, followed by fresh experimentation. Successes often came accidentally: "The times when I achieved good electrical characteristics were often when my system had gone awry. My colleagues then worked out the dynamics of silicon junctions from my accidents." In transforming diffusion—however haphazardly—into a viable manufacturing process for making silicon transistors, Moore was proving his considerable chemical skills. Years later, he would reflect on how in the early semiconductor industry, "physics and chemistry problems abounded. Celebrating Halloween as an industry holiday was suggested," since there was so much witchcraft in the manufacturing process.

Complications Galore

Shockley had declared that he would operate without any set principles. He was true to his word. Increasingly, he pushed the lab in multiple directions: some employees worked with him on scientific papers on transistor physics, others explored how to make four-layer diodes, still others worked on the crystal puller, and some were instructed to follow up, fruitlessly, on an idea for a novel form of device based on heat effects. Lacking experience to question this diverse activity, his young colleagues were caught up in the novelty and excitement. Shockley was supremely confident in his own decisions and exuded his belief to others. Multiple ideas and experiments meant more progress.

To Gordon, initially it all made sense. The pace might be hectic, but Shockley was very inventive. It was a research laboratory. "We were randomly doing things that were interesting, and learning. We certainly weren't moving in a linear fashion toward diffused transistors. Although we were trying to make them, there were plenty of sidetracks along the way." Then matters became really interesting. In November 1956 Shockley won the Nobel Prize in Physics for his part in creating the transistor. He shared the Nobel with John Bardeen and Walter Brattain, who five years earlier had found Shockley's relentless competitiveness intolerable and had demanded that Bell Labs' management move them out of his sphere.

To celebrate the prize, Shockley treated his Mountain View staff to a champagne breakfast at nearby Rickey's. Spirits soared as congratulations poured in; Nobel Prizes were rarely bestowed upon individuals in for-profit enterprises. In December Shockley left for Sweden for the prize ceremonies. As 1957 opened he traveled extensively in Europe and the United States. Enjoying the applause and adulation, his visits to the California lab became brief and sporadic. His restless mind was now occupied more with the promise of the four-layer diode—representing uncharted territory—than

Shockley and his lab celebrate the announcement of his Nobel Prize.

SOURCE: INTEL CORPORATION.

with winning the race to bring the diffused silicon transistor to a waiting market.

Alongside the letters of congratulation for Shockley's Nobel, there landed on Arnold Beckman's desk in Fullerton a more ominous communication, an anonymous letter addressed to Beckman personally and dated December 8, 1956, two days before the Nobel ceremony in Sweden. It was signed, "Senior Members of the Technical Staff, Shockley Semiconductor Laboratory." The letter claimed that matters in Mountain View were reaching a breaking point. The advanced crystal puller had suffered catastrophic breakdowns and had yet to produce silicon reliably. It was an embarrassing failure. In front of other staff, Shockley had fired Leo Valdes, then deeply engaged in the project. "The anonymity of this letter is dictated by the circumstances. We are scared stiff and afraid to talk." The letter, most likely authored by Valdes and possibly backed by others, went on to say that the laboratory had lost its "best man; the only one who had the courage to stand up for his own ideas. Reason for the firing—he refused to be psychoanalyzed by Shockley's wife [Emmy Lanning, the psychiatric nurse]." A torrent of distress filled out the missive: "Please help us immediately. Our days are numbered. We do not know who will be fired next. Please send an administrator to run the laboratories and prevent further injustices and excessive spending." Beckman simply filed the letter.

Shockley contributed further to the confusion by complaining to Beckman about another key staff member involved in the ill-fated crystal puller. In an early description of the lab in the annual report of Beckman Instruments, Dean Knapic had featured prominently. Now Shockley pronounced him a pathological liar and declared him responsible for the venture's failure to meet expectations. This time Beckman did act. To placate Shockley, he asked consulting psychologists to interview Knapic. They reported back that he was no liar, simply "a very typical European refugee, surrounding himself with a cloak of magical thinking which tends to give an air of unreality." It is unclear if this nonsense reassured Beckman, yet there was no avoiding the fact that in Mountain View something was amiss.

Despite Shockley's shifting research directions and long absences, Moore believed that he and his colleagues were making progress toward the diffused silicon transistor. There were several missteps—one involving the accidental evaporation of a few thousand dollars' worth of platinum heating filaments, a mistake that the frugal Moore never forgot—yet he managed to create high-quality diffusion furnaces, controlled with the accuracy necessary for making junctions. He also developed the procedures for the diffusions. By March Gordon had succeeded in fabricating a whole series of silicon diodes, simple semiconductor devices made from two layers of doped silicon. Each diode contained a single PN junction, a definite step toward the diffused silicon transistor, and perfect for testing equipment and procedures. Shockley and Noyce, along with Chitang Sah—a young Stanford PhD in electrical engineering who had recently joined the lab—used the diodes to coauthor a major paper on the physics of a diffused junction. With this understanding of the fundamentals and his own growing mastery of the necessary experimental techniques, Gordon was confident that the lab could soon achieve a diffused silicon transistor.

Meanwhile, Shockley was shedding the remaining vestiges of his interest in, and commitment to, this goal. Thanks to his inside track at Bell Labs, he kept abreast of their pioneer focus on the four-layer diode, a task involving numerous highly specialized scientists and engineers. That it was not the raison d'être of Mountain View mattered to Shockley far less than his strong connections to—and rivalry with—Jack Morton, the head of the Bell Labs' program. Shockley's desire to realize the device increased by leaps and bounds. "He felt it would be a revolutionary product," remembers

Harry Sello, a PhD chemist, who that summer joined the still small group in the lab. "The idea was true. The trouble with the four-layer diode was that it was years ahead of its time in processing complexity."

Already eighteen months into his venture, Shockley began feeling the heat. Arnold Beckman had agreed to fund the lab for two years; negotiations would begin soon about whether Shockley Semiconductor should become an independent company. There was little to show: large cost overruns and no real product. In an effort to demonstrate the value of his work, impress Beckman, and increase his chances of retaining his independence, Shockley decided to push his young PhDs harder to create the four-layer diode he lusted after. He yanked several key staff from the transistor effort and dropped them into a "crash program" on the diode.

Because Shockley was highly impatient, those on the crash program became subject to withering criticism. Hoerni, Last, and Roberts, friends outside the lab, had never encountered such derision in their short but successful careers. It rankled, and they murmured. Moore, too, was unhappy, as was Bob Noyce, but for other reasons. These two men were now the technical leaders of the laboratory. Moore's quiet determination and experimental skills had led to considerable craft knowledge of diffusion (the technology at the very heart of Shockley Semiconductor), while Noyce's charisma and extensive transistor physics background made him the laboratory's device expert, second only to Shockley himself. Moore and Noyce each supervised their own small team. The emphasis on the four-layer diode disrupted their progress in the race to bring the diffused silicon transistor to market.

The laboratory was close to realizing the transistor, and the military and aerospace market was large and growing. Yet it was clear to Gordon that the silicon transistor was no longer Shockley's main concern. He discussed his fears with Noyce. They decided to pen a joint memo, urging Shockley to maintain focus on the transistor rather than becoming diverted by the diode. Shockley was unmoved. Moore's dissent went underground.

At Beckman headquarters in Fullerton, others were looking askance. Managers Robert Erickson and Max Liston favored known products, such as the transistor. In a meeting with Shockley, they expressed their concern about the laboratory's costs and skepticism about market prospects for the diode. Shockley responded with a barrage of information, filling a blackboard with a statistical argument that suggested a large market within a few years. Erickson said he was not convinced. Shockley, unshaken, retorted, "Show me my mistake." The crash program continued apace. Moore talked with others about the neglect of the transistor. By common consensus, dealing with Shockley was becoming intolerable. As Charles Townes, inventor of the maser and Shockley's colleague at Bell

Labs, would later comment to Moore, "Bill was so smart he understood everything—except people."

Out of the Loop

At home on Alford Avenue, Betty Moore was delighted to be near her wider family, but in daily life felt out of the loop. Her husband, conducting some of the most critical experiments of his life, was away for twelve-hour days and frequently on weekends. "I'd be lucky to see him by 7:00 p.m." Betty was also less than impressed by the Quonset-hut setup. "It looked like an old garage." Bill Shockley, in contrast, was highly intimidating. Betty took a brief walk with him during a technical conference in Monterey. "We were beside this great estuary. There were some swans—white and black. Shockley said to me, 'Do you like swans?' I said, 'I like all birds.' He gave me this strange look. He was feeling me out, and I startled him by giving an opinion. I started to worry. He just walked on. He didn't have much to say to wives. He was on another level."

Betty was an intelligent, creative woman. In the mode of the times, she was alone at home with an active toddler, keeping house. What most frustrated her was Gordon's inability to share his day. At least when he was a student, she had gone to sit in the lab to keep him company.

> Gordon was up and down, but when he came home from work, I never had a blow-by-blow of what was going on. He was very quiet about work. He has always been very secretive about how he feels. I had to pick up bits and snatches. Somebody could have had quintuplets, and I wouldn't have known. I took Gordon to task, but the situation didn't change. That was not his cup of tea. It's the attitude of "Why do you need to know?" It was how his dad treated his mom. She didn't know a lot of things that were going on. There was this absolute wall.

After more than six years of marriage, Betty had discovered other ways to sense her husband's mood: "I could tell whether he had had a good day. If he was contented, he was jovial, 'ready for dinner.' He wanted to sit down and have a good meal and then look at the paper or the news. Sometimes things were very silent; it was depression. I knew not to ask for money then. On those days, I didn't tell him that we were supposed to go to a party or that someone wanted us over to eat."

Betty made up for her quiet home life through family camaraderie, especially on the weekends: excursions to see her mother and grandmother, picnics and barbecues, and fishing trips with Gordon's father, catching lingcod off the rocks while Grandmother Mira helped little Ken build sand castles on the beach. Following the celebration of Shockley's Nobel Prize, she also began to develop relationships with the wives of Gordon's colleagues.

> That event made everybody from the lab get together. We all began to drift in and out of one another's lives. We'd play bridge. Everyone except Jay Last was married, with small children. Once, we had dinner at Jean Hoerni's. They made bouillabaisse. It took forever to put all these fishes in this enormous dish.
>
> I liked Hoerni's wife, Anna Marie. Gene Kleiner's wife, Rose, would get us all together. She and I were great friends, and Sheldon Roberts' wife, Pat, and I were close, too. We came to know Bob Noyce, but I was not best of friends with his wife, Betty. She loved the East Coast and never accepted that we were human on the West Coast.

Other wives kept Betty up-to-date on what was happening at the lab, and she gleaned news of missed invitations and important developments. "I would say to Gordon, 'Why don't you tell me things? It puts me in a foolish position. People start talking, and I don't have a clue what it's about.'" Outside the lab the couple's social life developed a little through trips with the Noyces and the Robertses. "We went up to Crater Lake National Park in Oregon. Our kids were all dusty and dirty. Gordon loved the fact that we could go hiking and learn about the area. I preferred more amenities. I was just not settled, falling into some old, dirty sleeping bag!"

In 1957, as matters at Shockley Semiconductor went downhill, Betty's family received a shock of a different kind: the Los Gatos ranch, home to Irene and Grandmother Mary, was condemned. "It was the last undeveloped acres in Los Gatos. They wanted to take the land for an elementary school. My mother and my grandmother had to be out within six months." After fifty years of occupation by the Metzler family, the ranch was demolished. The only consolation was the retention of a few big oak trees and a promise that the school would be named after Betty's maternal grandfather, Abraham Lincoln Metzler. Betty became fully occupied in resettling her mother and grandmother, finding a house in nearby Saratoga that suited their needs. "This place had a great spreading oak tree my grandmother could see from her bedroom. Before we could move in, we had to strip everything and start over. The kitchen was chartreuse green! We organized it and moved them. Suddenly, they were in a small place. It was very hard on my grandmother."

When Mary Metzler suffered a fall, Betty bought a wheelchair, beginning her lifelong engagement with the realities of nursing. Gordon engineered a ramp from the back door so that Mary could still go out on the patio. At home in Los Altos, Betty suffered yet another miscarriage and struggled to keep up with her son, Ken, now two, "a holy terror, into all kinds of experiments." The novelty of daytime television programs was not enough. Only a nursery school, she concluded, could offer the required stimulation. "Ken needed activities. He needed something other than his mother and the TV."

With time to herself, Betty became increasingly aware of Gordon's difficulties. Shockley had begun "doing strange things and questioning people," she remembers. "As a group, we became aware there were problems. Someone would say, 'Now what?' Shockley had gone off on a fresh tangent. People were disgruntled with what was happening."

Things Fall Apart

Facing technical setbacks and business pressures, Shockley resorted to paranoia. He declared that the laboratory was harboring a saboteur, trying to injure staff. The allegation arose from a simple incident involving his secretary, who cut her hand on something sharp sticking out of a door. He announced that all staff would take a lie-detector test to get to the bottom of the matter. Quickly, Sheldon Roberts used his metallurgical microscope to examine the offending object: it was nothing more than a pushpin with a broken-off top. Shockley's outburst alarmed his already concerned researchers.

Matters reached a head when Arnold Beckman visited Mountain View in May 1957. Beckman Instruments' fiscal year had just ended, and the results were sobering. He had doubled spending on R&D, largely in support of Shockley Semiconductor. Profits had evaporated. Beckman spoke urgently to the group about the need to rein in costs and come up with a product that, through sales, would quickly generate revenue and profit. As exhortations go, Beckman's was respectful and straightforward. Moore and his colleagues decided he was urging them to focus on the four-layer diode. Shockley heard something quite different and exploded. In an aggressive, disrespectful manner, he asserted that if Beckman did not like what he was doing, he would leave with his researchers and find a fresh backer. Beckman ducked the challenge; the meeting broke up awkwardly, with Shockley's threat hanging in the air.

Over lunch shortly afterward, Moore, Noyce, Hoerni, Kleiner, Last, Roberts, and electrical engineer Victor Grinich expressed their mutual outrage with Shockley's blithe assertion that they would jump ship with him. Grinich, a Stanford PhD with experience in computers, declared that the time had come to do something. They had to call Beckman and tell him the truth. Who would make the call? There was a lot of hemming and hawing.

Then Gordon declared, "I'll do it." The rest of the group was more than happy to have Moore take on the uncomfortable job of spokesman.

Why Gordon? It was not in his nature to crave the limelight. "I don't think I enthusiastically volunteered." At the same time, he was trusted by his colleagues, a trust grounded in the very reasons that made him willing to make the uncomfortable call. He had measured and analyzed the realities, and he knew the lab's work was foundering. "I was willing to do it," he remembers. Gordon was calm and steady, believed wholeheartedly in the power of scientific and technical truths, and was unafraid of reporting them. Just as his colleagues knew he would record with precision the results of his latest experiments, they knew he would report to Arnold Beckman the realities of the situation.

After the lunch broke up, Gordon placed the call to Fullerton, with members of the group gathered around him. Jay Last recalls a "quavering" in Moore's voice as he asked to speak with Beckman, but—anticlimactically—the boss was not in. Moore was instructed to call again later. This he did, at Jean Hoerni's house, after working hours. Again, the dissidents gathered around. This time Arnold Beckman *was* available. A short conversation took place between the two Caltech physical chemistry PhDs. Moore told Beckman that Shockley's claim—that he would take everyone with him—had no ground in reality. "He'd have to go almost by himself." "Things aren't going all right up there, are they?" Beckman asked. "No, really a lot of problems," said Moore matter-of-factly. Beckman quickly made arrangements to fly up and meet Moore for dinner, with his colleagues but without Shockley.

Meanwhile, Shockley attempted to heal the damage to his own cause by sending a letter to Beckman's right-hand man, Jack Bishop. He explained that he planned to sell four-layer diodes to government laboratories and hoped soon to be producing "many hundreds per month." With this reply he indicated, however obliquely, that he had heard the exhortations about costs and revenue. Beckman himself flew to the Bay Area two or three times for dinner meetings with the dissidents and listened to their analysis of the root problem: Bill Shockley's devastating, capricious management. Progress was being thwarted by the boss's shifting priorities and by his damaging competitiveness and outbursts. Beckman asked for their best recommendation. The group suggested moving Shockley into an advisory or consulting capacity, perhaps in connection with an academic post. Beckman seemed genuinely interested. Gordon—encouraged, but still naive on management realities—believed their attempt to "neutralize" Shockley would solve the problem.

Shockley, intensely competitive and paranoid rather than naive, started to lay the groundwork for an imagined break. He drafted a fresh business proposal, indicating that he would bring in another investor. Beckman would then face a choice: to be bought out or to partner with Shockley and the fresh investor. It is unclear whether he sent the proposal, but its very existence suggests he was serious about a split. The two-year anniversary deadline for a decision on the laboratory was looming. Then Shockley received wind of the dissidents' activities. Leo Valdes told an acquaintance, Louis Ridenour, a prominent physicist, of the unhappiness at the laboratory and revealed the private meetings with Beckman. Ridenour thought Shockley should be in possession of the information and called him, just as he was putting the final touches to his proposal for Beckman.

On Saturday, June 1, Shockley had the tip-off confirmed by Beckman himself. Beckman invited Bill and Emmy Shockley to dinner in the Bay Area and laid his cards on the table. He had met with the dissidents; they had told him they would not leave with Shockley if he were to find a fresh backer. Their belief was that a management change was necessary and that Shockley should move to an advisory role. The news stunned Shockley. He said very little, but on Monday morning he promptly began to interrogate his staff. His plan was to call researchers into his office one by one and ask about the dissident group. This surprise confrontation—challenging individuals who had no idea their discussions had been revealed and diminishing their power by catching each one off-guard—was a typical Shockley tactic.

First he called in a recent recruit, Chitang Sah, who said honestly that he had no knowledge of the goings-on. Next, he moved to the heart of things, summoning Gordon Moore. Shockley asked if he had knowledge of the secret meetings with Beckman. Gordon, with his habit of straightforwardness, told his boss that he might as well stop. Yes, he, Gordon, had been involved. Further, nearly everyone on the senior staff agreed with the dissidents, and several had participated. Moore, self-contained and by nature ducking conflict that might arouse difficult emotions, did not find these truths easy to deliver. There was no triumphalism or satisfaction in his words. More than a half century later, he can recall vividly that Shockley stood up and walked out with his head down, chin to chest. Gordon avers, "It was not one of my most pleasant moments."

Shockley called off his serial interrogation. Later in the day, he regrouped and decided to gauge Bob Noyce's thoughts on the matter. He phoned Jack Morton at Bell Labs—still a close and trusted friend—and asked him to call Noyce. Morton did and rang back with the response. In the best case, he said, Noyce was sure that Hoerni, Last, and Roberts would leave. While Noyce was confident that he and Shockley could work out their issues, he was disappointed about the technical situation, that is, Shockley's decision to disregard his joint memo with Moore, urging focus on the diffused silicon transistor.

First Moore; now Noyce. Shockley faced dispiriting echoes of his situation at Bell Labs in 1951, as his original transistor team fractured. When he returned home that night, his wife noted that he behaved uncharacteristically, lying immediately on the couch, ashen. Within two days, as his sense of technical competence and intellectual superiority reasserted itself, he began to recover. To prefer the transistor over the four-layer diode was to misunderstand the competitive potential of the latter. And the idea that he should step aside and take on an advisory role was ridiculous. He would come up with a countersuggestion. Shockley jotted down notes about how he might bring in a manager between himself and the rest of the laboratory.

While Shockley temporized, he turned again to Bell Labs for support in deterring Arnold Beckman from the idea of making him a consultant. He asked Morton and Sparks to telephone Beckman and persuade him it was necessary that he, Shockley, lead the lab. They quickly obliged. Sparks called back to report. In their call Beckman had led with the suggestion of Shockley becoming a professor at a nearby university, consulting with the laboratory. Morton and Sparks told him this would be a disaster. Indeed, they testified that Shockley's choice of the four-layer diode was a good one. While the performance of the lab had been "creditable, not outstanding," they averred that without Shockley as its leader, it stood very little chance.

Knowingly or not, Morton and Sparks struck a chord. Their arguments confirmed Beckman's belief in the virtuosi approach. His own "rules for success" included the maxim "Hire the best people, and then get out of their way." With Shockley, he had followed just this rule. By linking the success of the laboratory to Shockley's genius, Morton and Sparks chimed in with Beckman's own beliefs. They assured him that Shockley understood the serious error he had made in selecting Hoerni, Last, and Roberts for the diode crash program. Beckman ended the call by saying he was happy and relieved. Moore and his fellow dissidents had no idea that the tide had so quickly turned against them.

On Saturday, June 8, Beckman flew up to the Bay Area again and met Shockley privately to agree on terms. A manager would sit between him and the rest of the laboratory, but Shockley would retain technological and strategic direction. They also agreed that Shockley would not fire any dissidents for at least six months, that they would reach an agreement about the future of the laboratory by September, and that Beckman would reestablish Shockley's authority. Beckman made short work of this last task. He immediately called a meeting of all laboratory staff, with Shockley present. This time he was decisive. Beckman informed the group that Shockley was the director. There would be no change in leadership. He also spoke of hiring a manager, while making it clear that no one would be fired. Moore and his colleagues hardly heard him. They were, to a man, surprised and dismayed. Beckman had let them down, utterly. He was on Shockley's side, not theirs.

Last and Hoerni, "flabbergasted" by Beckman's U-turn, left the meeting and drove directly to Yosemite. On an arduous climb the next day, they contemplated what to do. Moore, for his part, belatedly realized that it was "hard to push a Nobel Prize–winning scientist aside" and understood that there was no going back. The news meant only one thing: he and the other dissidents had burned their bridges "pretty badly." In trying unsuccessfully to get Beckman to fix the problems with Shockley, they were now hopelessly compromised with both. How could the pair forget what Gordon and the others had said and done? To Moore, it was clear that he and his group had only one option: they would have to quit.

WHERE THERE'S A WILL . . .

Realization Dawns

As far as Beckman and Shockley were concerned, the situation was now in hand. Several of Beckman Instruments' senior staff from the Fullerton headquarters visited as candidates for the manager position. Tellingly, each rejected this "opportunity." In early July, Maurice Hanafin, cofounder and general manager of Spinco, a Bay Area start-up acquired by Beckman, was given the job. Spinco had pioneered ultracentrifuges for the dawning world of molecular biology and had moved into the fledgling Stanford Industrial Park in Palo Alto (its facility would one day be an early home of Facebook). Under Hanafin's management, the work of the start-up was refocused. The four-layer diode effort moved in with Spinco, while the original transistor project, including Gordon's work on diffusion, stayed in the Quonset hut in Mountain View.

These changes reassured Bob Noyce, at least for the moment. Some months earlier, Moore had seen Noyce become immensely annoyed after sharing a particular novel idea with Shockley. "Shockley immediately called his friends at Bell Labs to see if it was right or not." Noyce nevertheless swallowed his pride and determined to stand behind Shockley. Others, including Gordon, believed nothing would help. With Shockley reaffirmed as leader, the focus on the four-layer diode was unchallengeable; the diffused silicon transistor would be left in the dust. Gordon, whose faith in the latter device remained constant, organized an evening meeting at his Alford Avenue home, inviting the nine or ten colleagues with serious engagement in the underground conversations. Six showed up: Blank, Grinich, Hoerni, Kleiner, Last, and Roberts. As they talked, the assembled group came to realize two things. First, it was clear that they possessed between them all the expertise needed to make the diffused silicon transistor, expertise in chemical, metallurgical, mechanical, and electrical matters and in production engineering. Second, it was plain that all—young scientists and engineers of considerable talent and ambition—desired to finish what they had started and produce the device.

Each man was confident that he could, if necessary, find another job. In the small, growing transistor industry of 1957, demand for people with appropriate education and experience far outstripped supply (a condition that amounted to negative unemployment). What was novel in their conversation was the thought that, instead of leaving one by one, they could leave together. Maybe some other firm would hire the group, allowing them to continue in the race to manufacture the diffused silicon transistor and conquer the lucrative military market. In Gordon Moore's living room that night, a new reality crystallized. The seven would keep up appearances at Shockley, while searching for a pathway to leave en masse.

It was a heady evening, yet Gordon maintained his habitual calm. As Betty recalls, "Our house on Alford Avenue was quite small, with an open plan. There was no privacy. What you heard is what you knew. Afterwards, I said to Gordon, 'Tell me what is happening.' I don't remember any great excitement from Gordon. He was, necessarily, in a confrontational role, but I don't think going against Beckman was a big issue for him."

The next step for the group was to consider not what they knew about transistors, but whom they knew who might help. The answer was Gene Kleiner's father and his financial contacts. Gene wrote a three-page letter with the assistance of his wife, Rose. It went to a New York investment bank with which his father had dealings, Hayden Stone and Company. The letter requested help in finding a corporation to hire the group of seven, a corporation wanting to enter the silicon transistor business and able to supply "enlightened administration and financial support" (as Shockley had expected of Beckman Instruments). In return, the group would deliver the diffused silicon transistor within a year.

While Kleiner's proposal was similar to Shockley's to Beckman, it had important differences. One was the existence of a team having experience with the device in question; this was not a bunch of wet-behind-the-ears recruits. Another was that there was no mention of dependence on an individual star performer or of starting a new company. Rather, the dissidents were searching for an existing company willing to take them on as employees. They were all financial innocents. Start-ups, spin-offs, and stock options were not in their experience or vocabulary. Silicon Valley was not yet born, and the concept of breaking away as a group and launching a new company, today a well-worn trope, was all but unknown.

also became board chairman of IBM. Fairchild expressed interest and asked Rock to explore the matter with his managers at Fairchild Camera.

Three-quarters of Fairchild Camera sales were to the military. Already supplying radiation detection, reconnaissance, satellite, and missile equipment, Fairchild Camera was in a position to install diffused silicon transistors in updated versions of its products, to give higher performance. It would also be able to sell transistors directly to customers in the aerospace and computing industries. A final sweetener was that its Los Angeles–based subsidiary, Fairchild Controls, competed directly with a division of Beckman Instruments in precision potentiometers for the aerospace market, and diffused silicon transistors would give it the edge. That Beckman Instruments had invested in Shockley's team gave the proposition further credence.

Sherman Fairchild had previously charged Fairchild Camera vice president Richard Hodgson—a Stanford engineering graduate who had worked on vacuum tube electronics for the Atomic Energy Commission—to watch for opportunities. Talking with Rock, Hodgson saw a strong connection between the firm's interests and what the "California group" could offer. At the start of September, he took a flight to San Francisco along with Coyle and Rock to meet the eight dissidents. Fairchild would later describe, in a letter to his shareholders, just how well the interests of Fairchild Camera fitted with the California group's talents. The transistor industry was growing in importance, he wrote, but while Fairchild was keenly interested in entering the field, his company was unable to do so "without very large capital investment, and the prospect of years of research before production might be feasible." The challenge was to find "qualified technical personnel," who were in "extremely short supply. The answer came when a group of leading semiconductor scientists appeared with a proposal that the company finance and administer them as the nucleus of a specialized semiconductor business."

Hodgson took very little convincing that the opportunity made sense. He felt confident that the group could produce diffused silicon transistors, that a market existed, and that they could agree on terms. He arranged for the experienced and charismatic Noyce, together with the venture-capital instigator, Kleiner, to come east immediately and meet Sherman Fairchild himself. During the visit Noyce's charm and Kleiner's gentlemanly bearing played well. Interest solidified. Hodgson returned to San Francisco for final negotiations with the California group and Hayden Stone.

They agreed on a contract that would run for eight years, creating a Fairchild Semiconductor Corporation, with 1,325 ownership shares. The eight dissidents, as cofounders, would pay $500 each, and each would receive 100 shares. For orchestrating the deal, Hayden Stone would receive

225 shares. The remaining 300 would be reserved for future hires. Fairchild Camera would provide up to $1.4 million across the first eighteen months. Beyond that first period, the agreement covered various scenarios. Fairchild retained an option to pay $3 million to purchase all shares—to buy out the start-up—until the point that the firm generated $300,000 in profits for three years running. After such a three-year streak, and up to seven years in, $5 million would be required to buy the company out. If the start-up was successful but Fairchild did not want to exercise its option, the participants could repay Fairchild's advances and continue independently. For the financially minded Gordon Moore, the agreement made sense. Whatever happened, he would make a good return on time invested with Shockley and exploit his status as an expert in silicon diffusion. He was keen to sign up.

The Traitorous Eight

As the eight dissidents were finalizing the deal, Shockley returned from the East Coast. Concern about the group was not at the forefront of his mind. Instead, he once again huddled with Maurice Hanafin on how they might convince Beckman to turn Shockley Semiconductor into a freestanding company. Beckman and Shockley had originally agreed to reach a decision by September 3, but that day had come and gone. The existing agreement envisioned the lab carrying on inside Beckman Instruments, something neither Hanafin nor Shockley desired. By now, both Bill Shockley and Gordon Moore's group of dissidents were concerned far more with outside negotiations than with anything going on in their workplace.

In the third week of September, the agreement between the California group and Fairchild's Dick Hodgson was ready for signature. Decades later, Hodgson recalled that Sherman Fairchild had directed him to contact Arnold Beckman before signing and claimed to have received Beckman's assurance that he could accept a deal with this group of his employees. There is, however, no record of such contact or of Beckman taking any step based upon it. This time, neither he nor Shockley had an inside informant. Indeed, Beckman's actions reveal that he was wholly taken aback by what transpired.

Gordon and the rest of the group believed it appropriate to tender letters of resignation before signing their deal with Fairchild Camera. On Wednesday, September 18, they handed in letters to Hanafin, who quickly informed Shockley that the most important members of his senior staff had just resigned en masse. Shockley never spoke to any of the dissidents again. Beckman immediately booked a flight to San Francisco. On Thursday, September 19, as the dissidents took on the status of company cofounders by signing the contract that created the Fairchild Semiconductor Corporation,

Beckman met with Hanafin, Shockley, and Shockley's wife, Emmy, in Palo Alto. He was furious about what he took to be a collective act of betrayal. Both he and Shockley were galled and hurt, reportedly giving the breakaway group a label memorable for its venom and resonance: "the Traitorous Eight," a name the group would never entirely shake.

Beckman and the Fairchild Semiconductor cofounders faced each other the next day in the conference room of the Spinco building. Beckman—a former US marine, now in his late fifties—was tall, imposing, wealthy, established, accomplished, and incandescent. He laid into the group. They were "disloyal," their leaving "an act of conspiracy." He labeled them as "young and emotional, running away when the going has gotten tough" and "costing him a million bucks down the drain." They had "stabbed him in the back," and the rest of the electronics industry would see them as "shameful." He threatened that to avoid legal complications, they should get out of semiconductors altogether. His lambasting of Gordon and the others came from personal feelings of betrayal and anger, while as a businessman he saw that it was critical to dissuade these young men from becoming his competitors. Neither side could appreciate at the time what, in retrospect, was a moment of profound historical significance.

In October 1957 Gordon Moore had been on the San Francisco Peninsula just shy of eighteen months. He and Betty were back where they belonged, in the Valley of Heart's Delight. Under Shockley, Gordon had completed an intense apprenticeship and become an expert in diffusion. Betty had established their household while caring for their young son, Ken, an energetic, bright, and demanding boy. The year from October 1956 had been a roller coaster: from the high of Shockley's Nobel Prize to the lows of capricious management, staff distress, and fragmentation. Betty had seen the maturing of Gordon's remorseless work ethic and how his calm self-possession and fortitude had helped to expose and resolve his group's troubles. His approach had been consistently quiet, practical, and low-key.

Compared with Moore, Noyce had been indecisive, moving into, out of, and finally back into the dissident group. Gordon was not concerned that the start-up might fail. He did not expect it to, but he was confident he could "get a job someplace. We weren't going to starve to death." Betty shared his outlook: "I had faith in what they were doing. They were all young and very talented, and they meshed well. Why not go for it? I know how to roll with the punches. Gordon had the same attitude."

Bill Shockley had a harder row to hoe. He saw the departure of the dissidents as a temporary, if serious, disruption and believed he could achieve a quick fix through European recruits, who would be less likely to cause trouble than Americans. In 1958 he finally succeeded in getting the four-layer diode into pilot production, but while government laboratories and military contractors ordered evaluation samples for possible use in missile fuses and detonators, no commercially viable diodes were ever produced. In April 1960, after nearly five years of investment in his fellow virtuoso, Arnold Beckman had had enough. He sold the faltering enterprise to the Clevite Corporation, a transistor producer, for a mere $1 million.

Shockley gave up his dream of entrepreneurial greatness and accepted a professorship at Stanford. There, he began an engagement with eugenics that led to his increasing marginalization and isolation. Convinced of the importance of virtuosi, he saw individual capabilities as fixed, determined by biological heredity. He went on to promote the idea that African Americans were mentally inferior to whites and made various proposals in eugenics, such as discouraging certain groups from reproducing and promoting "genius sperm banks." When he died in 1989, he had lost contact with his three children, who learned of his death from the newspapers. Ten years later *Time* named him one of the most important people of the twentieth century. Fittingly, Gordon Moore penned the entry. He was as evenhandedly factual about Shockley's accomplishments as he was about his shortcomings.

5

LAUNCH

START-UP

Inventing Fairchild Semiconductor

At Fairchild Semiconductor Corporation, Gordon Moore and his seven cofounders were raring to go, to vindicate their decision. Pending the lease of a work space, and to avoid the distractions of wives and children, they met at Jay Last's apartment. Last did not even have enough chairs for everyone to sit down. Shockley Semiconductor had given the group an unparalleled apprenticeship and a strong mutual regard. Gordon had learned much there, but "it was a tremendous advantage, to be able to throw it all away and start with a blank sheet. At Fairchild Semiconductor, we were focused on getting into the transistor business, not on research or interesting sidetracks." While Beckman had initially promised Shockley three hundred thousand dollars to fund a year's work, Moore and his cofounders could draw down almost five times that amount, over eighteen months.

Shockley's eccentricities and overreach had driven the dissidents toward one another, but what made them a cohesive group was their shared desire to manufacture the diffused silicon transistor. Fairchild Semiconductor was a partnership, not an enterprise subject to a single unstable genius. The Traitorous Eight were no longer working for a scientific guru but for themselves. Where Shockley had been a micromanager, keeping individuals isolated, Moore and his colleagues agreed on the need to work together, to harness egos to a common goal. As when Gordon Moore's forebears ventured west, self-selection also played its part: those with the courage and will had made the leap from Shockley Semiconductor and now were determined to press on and reap the reward.

If the summer and early fall of 1957 had been a tumultuous time, things were about to become even more exciting. In the very week of Fairchild Semiconductor's launch, the Soviets, without any fanfare (the risk of failure being so high), blasted *Sputnik* into space. The world was stunned.

Gordon Moore in the early days of Fairchild Semiconductor.

SOURCE: WAYNE MILLER, MAGNUM PHOTOS.

Using its most advanced rocket, the Soviet Union had placed the first artificial satellite ever into an orbit 150 miles above the earth's surface. "A brilliant and deafening detonation of smoke and flame illuminated the Soviet Union's rocket test site near Tyuratam, Kazakhstan, as the 32 nozzles announced the rise of the Russian R-7 missile. 295 seconds and 142 miles later, the last of the R-7's engines shut down for good. Then, in one final act that signaled the dawn of the space age, a pushrod connected to a bulkhead of the R-7 was activated, shoving a 183-pound beach ball–sized aluminum sphere into the cold, harsh blackness of space. Sputnik had arrived."

Sputnik orbited the earth every ninety-eight minutes to awed, then hysterical, Western response. A successful satellite implied the Soviets' ability to send aloft a nuclear bomb. From its start, the space age was linked to the looming prospect of thermonuclear war. The United States, ostensibly focused on the "International Geophysical Year" and still developing its own *Vanguard* Earth-orbiting satellite, was upstaged. In November the Russians followed up with *Sputnik II,* carrying the dog Laika.

Sputnik created "a boom within a boom," highly advantageous to Fairchild Semiconductor Corporation. Silicon transistors promised greater robustness, miniaturization, and lower power usage, all desperately needed for missiles, satellites, and much else. The United States' first response to *Sputnik*—quickly labeled "the Flopnik"—exploded on launch, on December 6, 1957. Congress soon thereafter created the National Aeronautics and Space Administration (NASA), a leading source of lucrative development contracts. By 1961 President Kennedy was standing before Congress, promising to put a man on the moon.

World War II had seen tremendous expansion of the military's direct presence in California via naval, army, and air bases, and hundreds of thousands of servicemen had moved through the state en route to the Pacific theater. The role of the military was reinforced through contracts for shipbuilding, aircraft, electronics, munitions, food, and supplies. To these were added industrial and university R&D and then nuclear weapons laboratories such as Lawrence Livermore, as the Cold War intensified and the Korean conflict of the early 1950s escalated. By 1957 California was well established as a major force in aerospace, with nearly one-third of federal purchases within the state being connected to this domain, sustaining some four hundred thousand jobs. With its defining aim of manufacturing a reliable silicon transistor, Fairchild Semiconductor fitted easily into this milieu. From its outset, it proposed to become a supplier of military electronics, a player attuned to the California contract-garrison state. Gordon Moore and his colleagues were untroubled that their work would enable the creation of sophisticated weaponry of mass destruction. It was sufficient that the nation's Cold War interests demanded advanced transistors. Fairchild Semiconductor was of its times.

Following the announcement of *Sputnik,* the cofounders were buoyant. Free from Shockley's capricious management, and with the backing of Fairchild Camera to create the product they had long envisaged, all was set fair. The world's attention was focused on a frontier to which their expertise was highly relevant. The sense of opening possibilities roused Gordon to engagement as never before: "Charging off on the new venture had everybody enthused."

Quickly, the group leased a work site. Despite the euphoria, Moore was conservative in his estimate of requirements. "We saw a building that Varian Associates was vacating in San Carlos, but it was 40,000 square feet. We couldn't imagine needing a facility that large. We weren't thinking of a company of more than a hundred people." Instead, the answer lay a mile up San Antonio Road from Shockley's Quonset hut, "out in the boondocks" at 844 Charleston Road, in a modest, empty shell of 14,400 square feet, "a tilt-up, an R&D space—two-story in front and a one-story open bay in the back. It looked about the right size."

The building had no equipment, electricity, water, or other services. The eight set to work, creating a style for the many spinouts and start-ups that would follow in subsequent decades. Straightforward functionality was the theme, not flash or polish. "We were doing everything cheaply," recalls Moore. His conscious frugality, encouraged by many hours in the family store, took charge. "We put in all the utilities and benches. Instead of lab furniture, which seemed expensive as heck, we bought aqua-blue kitchen cabinets and Formica to put on top of them. It looked very nice, at a quarter of the cost."

At least they now had somewhere to sit down. Their former boss's prideful attempt to create the world's most complex silicon crystal puller had ended in major failure. The group, guarding against hubris, agreed that their tools and equipment needed only to be good enough for the job, no more. Parent company Fairchild Camera sent "financial guys" out from the East Coast to "help us set up the books." Otherwise, the cofounders were the workers, slowly supplemented by new hires, including several fugitives from Shockley Semiconductor. It took a while to equip the facility. In the first days, the environment was primitive. The group met every Monday morning to "figure where we were at," worked while the sun shone, chased the occasional jackrabbit out of the building, and used the bathroom of a gas station down the street.

Bell Labs had by now disclosed its chemical printing approaches to the silicon transistor, and the scramble was under way at several companies to translate the invention into a reliable manufactured product.

The Fairchild Semiconductor cofounders.
Left to right: Moore, Roberts, Kleiner, Noyce, Grinich, Blank, Hoerni, Last.
SOURCE: WAYNE MILLER, MAGNUM PHOTOS.

Unlike its competitors, Fairchild Semiconductor was an unencumbered start-up, without distracting commitments to vacuum tubes, germanium transistors, or even other silicon technologies. Its simple, all-consuming aim was to be the first to manufacture a diffused silicon transistor, taking advantage of the Bell Labs breakthrough. From October 1957 into the following February, Moore and his cofounders worked as a self-organizing team, aiming to develop chemical printing technology into a working production process.

"We knew the bits and pieces of the manufacturing technology we had to develop," recalls Moore. "We split those up among the group." In hindsight it was a remarkable period, says his colleague and friend Jay Last. "About the only thing we could buy were microscopes. Everything else we had to make. It was a cooperative effort. We had no real boss. Ten months after we went into this empty building, we had a commercial product. That showed we were cooperating, each of us depending on the rest to do their part." As the group's chemist and diffusion expert, Gordon was at the

The setup of the Fairchild Semiconductor facility, in process.
Note kitchen cabinets and benches for diffusion.
SOURCE: CHRISTOPHE LÉCUYER, FAIRCHILD.

center of the enterprise. Freed from the ambiguities, changing signals, and missteps of Shockley's laboratory, he could focus on the chemical tasks that, once completed, would enable them to achieve their goal. In the most literal sense, he was in his element.

Making the Mesa

Moore and his cofounders began to refer to their target, the diffused silicon transistor, as a "mesa" transistor. The name signaled its profile. When looked at from the side, the tiny device resembled one of the flat-topped mesa formations dotting the landscape of the American Southwest. Like all junction transistors, the mesa had three layers of doped silicon: the emitter, base, and collector. The emitter and collector were doped in the same fashion, while the base between them was doped in opposing fashion. Two categories of transistor resulted: one with a P-type base between N-type emitter and collector (an NPN transistor), the other with an N-type base between P-type emitter and collector (a PNP transistor). The two categories offered slightly different possibilities in electronic circuits.

Making a mesa transistor began with a "crystal puller" (far less complex than Shockley's) that was used to slowly draw a long, cylindrical strand, less than an inch in diameter, from a melt of liquid silicon at more than 2,500 degrees Fahrenheit, creating a single near-perfect crystal. Depending on the dopant added to the melt, the crystal was either N- or P-type silicon. A saw was used to cut the ingot into thin disks, like slicing a salami. The resulting "wafers" were cleaned and polished flat. They were now ready for mesa transistors to be chemically printed into them.

A set of wafers—N- or P-type silicon—were put in a diffusion furnace. With high heat, a supply of dopant gas, oxygen, and precise timing, a diffused layer of the opposite silicon type formed on the wafer's top surface. The process also capped the wafers with a layer of silicon dioxide. The now adjacent N and P layers would form the bases and collectors of the transistors. A second diffusion was needed to make the emitters, the third required layer. First, many windows needed to be cut through the oxide now covering the wafer. These windows would allow diffusion into the wafer, creating the emitters, while also defining the location for each of the many transistors forming on the wafer.

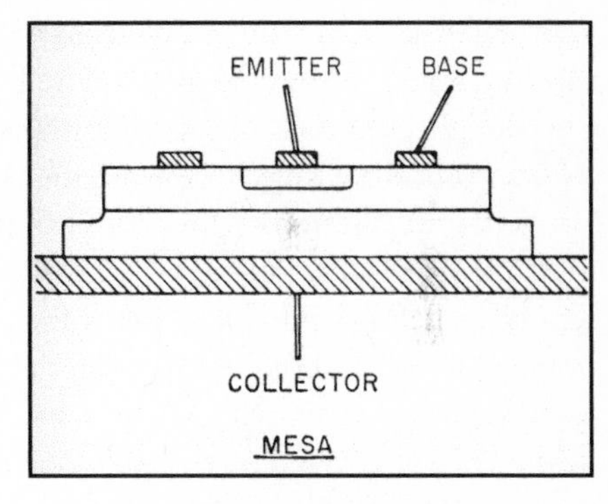

Mesa transistor in profile.

SOURCE: FAIRCHILD.

Photoetching, or "photolithography," was used to create the windows. In photolithography the wafer is coated with a light-sensitive chemical, called a "photoresist." A specially patterned stencil, a "mask," is placed over the wafer and the whole exposed to bright light. The pattern of the mask is the pattern of the desired windows. Where the light reaches through the windows to the photoresist, chemical changes occur. The coating now has some areas that are resistant and some

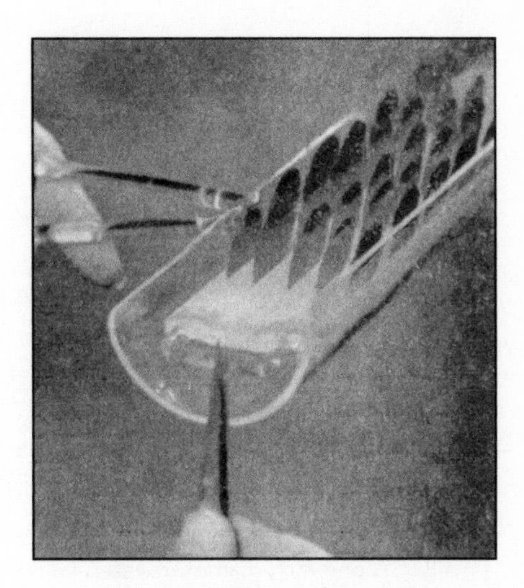

One-inch wafers prepared for insertion into a diffusion furnace.

SOURCE: MCGRAW-HILL.

that are vulnerable to being etched away in a strong acid. When the wafers are dipped into an etching bath, the vulnerable parts of the photoresist, and the oxide layer below, dissolve away. The wafer's oxide layer now has multiple windows. After stripping off the remaining photoresist, these windowed wafers are put back into a furnace, and the emitters are diffused into the wafers.

The oxide layer, "dirty" from the trapped dopants within it, is then removed in another acid dip. All of the required layers for the mesa transistor are now in place: each wafer has a collector and a base layer, with many distinct emitter layers dotted like islands across the base layer. The next step is "metallization," forming electrical contacts to the transistors by metal layers evaporated onto the top and bottom of the wafers. These layers enable wires to be attached to the transistors. The bottom layer is the collector contact. On top, photolithography is used to pattern the metal into an array of dots with surrounding rings. These are the emitter and base contacts.

With metallization the chemical printing of all of the elements for multiple transistors on each wafer was complete. The individual transistors were defined in a next step called "mesa-ing." Black wax was dotted over the wafer, covering each emitter area and a bit of the surrounding base layer, and the wafer again dipped in an acid bath. The acid removed the base layer not covered by the wax, down to the bottom collector area. Now the wafer looked like a plain covered with multiple mesas. To separate out

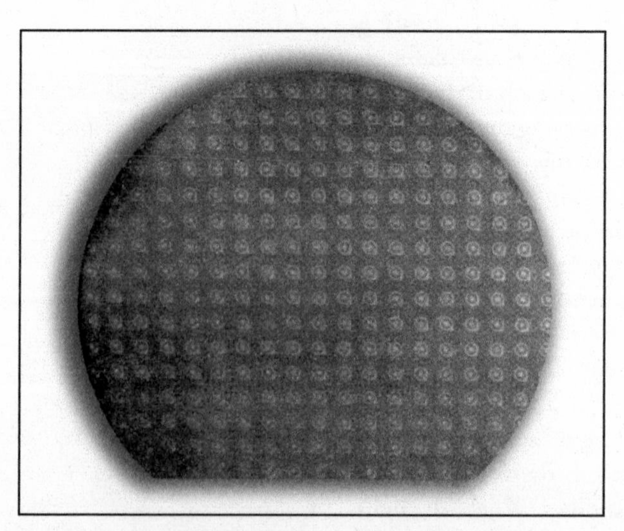

A one-inch-diameter wafer, with chemically printed transistors.
Each dot and ring on the wafer form the metal contacts for a single transistor.
There are approximately two hundred transistors printed on this wafer.
SOURCE: MCGRAW-HILL.

the individual transistors, the wafer was simply sawed apart into small squares called "dice." Each die was a chip of silicon wafer with all of the elements of a diffused silicon transistor: emitter, base, collector, and metal contacts to each.

An early Fairchild transistor. The metal-can package (about the size of a pencil's eraser) is cut away to reveal the silicon die. Wires attach to the metal contacts on the silicon die.

SOURCE: FAIRCHILD.

This excruciatingly intricate sequence of batch chemical processes yielded a set of dozens of mesa transistor dice. To make these into finished transistors required just two more steps: package and test. A worker, usually a woman, mounted the transistor die into a protective package—a small metal can with three wire leads. The contacts were wired to the leads and the can's lid welded shut. Once packaged, the mesa transistors were put into electronic testing machines that determined if they actually worked.

The ability to produce transistors by the batch was an immediate advantage of the chemical printing process; previously, transistors had been made individually, using alloying techniques. Batch production favored miniaturization: the smaller the transistors, the more on a wafer and the lower the cost—if they functioned. "The wafers were minuscule by today's standards," recalls Jay Last, "but there were a lot of transistors on each of them. We had to develop all the technology; figuring out how to put metal interconnections on the transistors, how to put them in packages, how to put leads on the package, how to test them—how to build a high-reliability device." Most important of all was the complex chemical ballet by which the different diffused layers of each transistor were endowed with exactly the right properties.

One marker of the novelty of the group's approach was that no commercial company yet produced diffusion furnaces for the electronics market. Diffusion was an experimental procedure, not a familiar manufacturing operation. Moore had to build from scratch the apparatus he required and discover the combinations of time, energy, and reagents that would produce successful results. The furnaces were long, narrow cylinders, operating at extremely high temperatures. They required elements that could both produce and withstand the necessary environment and precision systems to maintain exact temperatures for minutes, even hours. Fortunately, Moore had done work like this before. At Shockley's lab he had made his costly mistake with platinum. This time, he used special material from Sweden, durable at high temperatures: "I built some good furnaces. I wasn't an expert at designing control systems, but I could read the catalogs well enough to figure out how to make them work."

The most critical task—overseen by Gordon alone—was to establish a reliable, production-worthy diffusion process. This was immediately obvious to any visitor to 844 Charleston Road, where most of the working space was soon devoted to diffusion furnaces, fueled by a "glass jungle" of supply lines, blown by Gordon himself, lines that also piped in dopants and other required chemicals. Moore laid out the diffusion area with six tables, two furnaces mounted on each. "Jean Hoerni thought if we staggered the tables, we could get a seventh one in there, in case we had to expand." The arrangement worked well during the development period; thereafter, as production ramped up, "we were piling furnaces two high." Gordon enjoyed using his college-learned skill in glassblowing to create gas-handling systems. "The only one who knew how to blow glass, he was making all the jungles for the diffusion," recalls Jay Last.

Moore's responsibility soon had an added dimension. His dedicated work ethic, calm air, and track record proved so reassuring to his cofounders that they took it for granted he would succeed in making reliable furnaces. They agreed that once he had established these furnaces, he should immediately use them to develop diffusion procedures for the NPN mesa. Even at this stage, the eight realized not only that the processes for the NPN and the PNP versions of the mesa would differ, but that the NPN version would be easier and was thus closer to production. NPN diffusion needed only development; PNP required further research.

Hoerni, the firm's theoretician and expert in the physics of diffusion, took on the PNP mesa. Unlike the other cofounders, he was not "very useful in the original setup of the laboratory." Instead, he spent time ruminating on diffusion's inner workings, while others tackled the remaining bits and pieces of the manufacturing technology. Gene Kleiner and Julius (Julie) Blank started a machine shop adjacent to the building, making tools and equipment. Kleiner was the equipment manufacturer and also took on administrative tasks, while Blank worked on infrastructure: electrical service, plumbing, and the like. Sheldon Roberts, with assistance from Kleiner and Blank, built crystal pullers to supply three-quarter-inch silicon wafers. Vic Grinich, "the one who really knew what transistors were, and what they were used for," developed procedures for testing the mesas. Others such as Dave Allison—an early hire who, as a Shockley Semiconductor employee, had backed out of attending the dissident gathering that summer evening at Moore's house—also made key contributions.

Bob Noyce and Jay Last teamed up for the key task of photolithography, or photoetching. "We were applying it on a scale that hadn't been done previously," says Moore. For this particular step in making transistors, a "step-and-repeat camera" was needed to make matched sets of multipattern masks. The masks would define the windows for diffusion and the metal contacts for the transistors. The camera itself was necessarily a complex optical device. It had to be steady enough to miniaturize patterns faithfully and have lenses that matched sufficiently closely to align the patterns on the masks. Finding no manufacturer, Noyce and Last—alert to the dangers implicit in Shockley's crystal-puller catastrophe—decided to build their own.

Another tricky area was using the photoresist. This delicate operation was set off by itself in a room with filtered air and yellow light because, Gordon notes, "the stuff was UV sensitive." Bell Labs had photoresist problems, finding it, Jay Last recalls, "impossible to work with." The problem was not putting the resist on but getting it off without destroying the underlying layers. Moore remembers a plethora of challenges: "It doesn't scrub off very easily. Remember the old chemists' cleaning solution, chromic acid? We used that to get it off; then we had to get the chromium off! Eventually, we worked with the big chemical suppliers to make what became electronic-grade chemicals. We had to specify all the acids and solvents and demand unprecedentedly low levels of heavy metals, to avoid contamination."

As an exact and exacting experimentalist, with unrivaled practice skills and a broad-based education, Gordon reveled in the challenges, the long hours of absorbing chemical puzzles, and the sense of a project with a defined, practical goal. Quite unwittingly, he was also slowly stumbling on the exacting standards necessary for chemical manufacturing on the microscale, in terms of the purity of reagents and of the workplace itself.

Where's the Boss?

The building at 844 Charleston Road had no reserved parking spaces. Bob Noyce's beat-up old 1948 Ford caused general embarrassment. "One of the guys put a note on the window one day, 'Would you please park in the back? You're giving our company a bad name, parked in front!'" In contrast, when Fairchild Camera's president, John Carter, came to visit, his driver waited outside in Carter's limousine: "We couldn't imagine having somebody sitting around like that all day." The early months were "very egalitarian." Each cofounder had an equal stake and knew his place in the emerging mosaic. To establish the technology, they needed no advice, nor did they have any obligation to explain their work, since manufacture of the mesa transistor was the very basis for Fairchild Camera's investment. At Shockley Semiconductor, Moore and his colleagues had taken fitful steps toward a chemical printing technology and a manufacturing process. Now they ran full speed, implementing their shared vision in rapid practical steps. Weekly meetings detailed "problems rather than progress. We took advantage of what we had learned at Shockley. No secret projects!"

The cofounders quickly brought in reinforcements. Murray Siegel, an electronic technician recruited to work for Grinich on transistor testing, became employee 9. Sam Fok, a PhD chemical engineer from Shockley, was involved with the step-and-repeat camera. Technical conferences, and other transistor makers, proved good recruiting grounds. Other hires occurred through serendipity. Lars Lunn, a Swede, became employee 29 after visiting the facility on spec:

> I arrived at noon and met two guys on their way out. They asked what I wanted. I told them I was looking for a job: "Okay, come along and have lunch." It was Bob Noyce and Jay Last. We went to a restaurant on El Camino, and talked about Europe, skiing and life in general. On return to Charleston Road, they said, "You're hired!" This was the first and only time I had the feeling of being evaluated solely on my behavior. Of course, after three months, they could have said, "Okay, you're fired!"

The eight were dogged in their commitment to finish what they had started. They worked well together, yet were increasingly convinced of their need for an overall boss, a general manager. Bill Shockley had given short shrift to manufacturing and the business side of things and failed to get a product to market. Gordon Moore and his colleagues would not make the same mistake. Novices, in a start-up together more by accident than design, they comprehended their own lack of insight into the realities of business. "We didn't want anybody to tell us what to do technically; we could handle that ourselves," says Moore. "We knew, vaguely, that to run the business, a lot of other things had to be done. We needed somebody who could give us leadership. We still weren't quite entrepreneurs. So collectively, we set out to hire a manager."

Displaying their heightened ambition, they placed national advertisements: "Fairchild Semiconductor Corporation has opening for Vice President and General Manager," announced the December 8, 1957, *New York Times*. "The man we are seeking must have wide administrative experience in the semiconductor or electronic components field, and broad industry relationships to help establish the company as a strong competitor. He will assume full responsibility for the management and expansion of this growing company." The masculine pronouns reflect the reality of the day. While women were prized for their manual dexterity as assemblers and line workers in vacuum tube and transistor factories, executives and engineers were exclusively male. When Bob Norman, who joined Fairchild Semiconductor as an applications engineer, had the audacity to promote a woman technician to "lead girl" in the early 1960s, she was immediately shunned by the rest of the lab and the promotion rescinded.

To find a top manager, a display of chutzpah was essential. In the transistor business, negative unemployment was a fact of life. The specifics of one's experience mattered less than whether one *had* experience. Earlier in the year, the *New York Times* had covered a national meeting of electrical engineers with the headline "Job Bidding Is Brisk," adding, "Inducements offered for changing jobs are financial, or a more desirable climate, or a more promising future—one concern offers a $550 reward for an electronics engineer; 'your man or you.'" This "job-hopping" climate had paved the way for the cofounders themselves to jump ship and now enabled them to lure colleagues to their start-up.

Response to the advertisement was immediate and strong. Moore quips that they received "a whole bunch of résumés, mostly from salesmen. Every salesman thought he could run a company!" Sales engineers were not what was wanted for a boss, but they did find a badly needed sales manager among these résumés: Thomas Bay. "If you asked which of the salesmen looked right, it was Tom Bay: a big, lanky guy, easy talking, bright, and with strong opinions." Like most salesmen in the transistor industry, as well as most customers, Bay was an electrical engineer.

For general manager, Moore and his colleagues chose Ewart Baldwin, known as Ed. He was nearly a decade older than most of them, having served in World War II as a paratrooper immediately on graduating from college. Baldwin earned a PhD in solid-state physics in 1950 at the Carnegie Institute of Technology (today's Carnegie Mellon University) and then worked in the semiconductor manufacturing operation of Hughes Aircraft, in Southern California. As Hughes became a significant player in the industry, he rose to head of product engineering.

Baldwin's experience caught the group's attention. Many of the components he was making at Hughes went into military and aerospace products, giving him valuable exposure to market realities. During discussions with Moore and the others, Baldwin showed his superior knowledge of the semiconductor business by advocating an engineering stage between prototype and production, for detailed process and product specifications and for rigorous testing with specialized instruments. As Moore explains, "He brought a lot of ideas, things that any business-school student would learn the first year. They were new to us, but sounded right. We decided he was the guy." Baldwin had a good position at Hughes, so hiring negotiations stretched on through December.

Transistors, Texas, and Virtual Reality

By 1957 radio was a domestic staple, and television was coming into widespread use: every home with a television set became the receiver of novel and mesmerizing images. Television was the harbinger of what would become a metastasizing reality over the next several decades: that the consciousness of individual human beings might, through new technologies, attain liberation in a simple and routine manner from the geographic and temporal limitations implicit in their physical incarnation. For thousands of years, sages, seers, poets, and storytellers had offered glimpses of this possibility (whether around a crowded campfire, within a religious congregation, or in plays at Shakespeare's Globe Theatre), yet ordinary individuals had little opportunity to enter mentally into a different time or place. Such experiences were rare and fleeting.

Late in the nineteenth century, the developments of mass literacy, magazines, and steam printing began to offer a fuller and livelier sense of this option (as in the possibility of becoming "lost in a book"), but it was not until the early years of consumer electronics—in the form of vacuum tube–powered radio sets and black-and-white televisions—that the possibility of "mind travel" came within routine reach of ordinary people. Today, thanks to Moore's Law, the electronic digital reality incarnated in the smartphone, the tablet, and social media is part of the fabric of life and consumes several hours of most days for a growing fraction of the more than 7 billion inhabitants of earth. At the same time, "virtual reality"—total mind travel via headset devices that offer full three-dimensional immersion while shutting out mere present and local actualities—is an emerging possibility.

In 1957 there were some 130 million radio receivers in use in the United States, with an additional 14 million rolling off production lines annually. The majority employed vacuum tubes. Already, however, germanium transistors were beginning to drive a different, more robust, and (soon) cheaper class of radio receiver: the portable. Color television was also making its first entry into a scene dominated by now-familiar monochrome sets. Some 47 million TVs were in use, a number greater than that of the households in the nation. CBS debuted *Leave It to Beaver* the same week that *Sputnik* and Fairchild Semiconductor launched. The program joined *I Love Lucy, Gunsmoke, The Guiding Light,* and *Meet the Press* on the airwaves.

PERCENTAGE OF US HOUSEHOLDS WITH ELECTRONIC TECHNOLOGIES, 1940–2010

Year	Radio	TV	Cable TV	Personal computer	Internet access	Mobile phone	Smart-phone
1940	33	0	0	0	0	0	0
1950	91	9	0	0	0	0	0
1960	94	87	0	0	0	0	0
1970	99	95	7	0	0	0	0
1980	99	98	20	8[a]	0	0	0
1990	99	98	56	15[b]	<1	1	0
2000	99	98	68	51	41	36	0
2010	99	99	90	77	71	48	35

Sources: Electronics Industries Association Electronic Yearbooks and Market Data Books; US Census Bureau; US Federal Communications Commission; International Telecommunication Union.

[a]In 1984.

[b]In 1989.

Today, approaching sixty years later, cable, satellite, and the World Wide Web provide instant access to a myriad of TV channels and many millions of movies and videos. The average American adult reportedly spends eight hours a day abstracted from immediate physical reality through screens of one kind or another. This is one of the most obvious ways in which, thanks to the transistor, the electronic revolution has transformed and continues to change our individual lives.

The high point in the production of vacuum tubes for televisions and radios, and for use in military and industrial controls, instruments, computers, and communications, came in 1955. Nearly a half-billion tubes were manufactured, with a market value approaching $400 million. Transistor manufacture, almost nonexistent in 1950, was far smaller but growing steadily. Nearly 30 million transistors were manufactured in 1957. More than 95 percent were of germanium, using alloy methods. A mere million were silicon transistors. Because of their superior performance (to say nothing of the challenges of manufacture), the average price commanded by a transistor was $2.50, compared with only 83 cents for a tube. Though transistors were expensive, the business showed promise, and if small was growing rapidly; there were fortunes to be had.

The two dozen firms vying for these fortunes included many of the leading producers of vacuum tubes. Tube firms that had jumped into transistor making included General Electric, Raytheon, RCA, Sylvania,

Philco, and Westinghouse. These East Coast titans had built-in conflicts of interest, as their novel transistors were in direct competition with their own well-established tubes. Enjoying greater success were the newcomers, unburdened by previous investment in tubes. Numerous optimistic beginners bought the low-cost patent licenses that Bell Labs was under a government mandate to offer to all comers. Texas Instruments leaped to the front of the pack, securing a commanding 20 percent market share in transistors—mainly familiar germanium and some novel silicon—by 1957. Hayden Stone had helped finance the Boston-area start-up Transitron, which had a 12 percent market share. Hughes, the Los Angeles–based aerospace giant, also moved into military-oriented components and won 11 percent of the transistor market.

A further important shift was that mainframe computer makers announced their interest in transistorized computers. Advantages of size, power consumption, and reliability made the idea attractive not only for the military but also, increasingly, for civilian business use. Late in 1957 Texas Instruments forged a deal with International Business Machines to become its primary transistor supplier. The deal confirmed TI's transformation from a purveyor of instruments for oil exploration into the leading transistor producer.

TI's head, Pat Haggerty, an experienced electronics engineer, had earlier poached Gordon Teal from Bell Labs. It had been Teal who in 1953 had failed to snag Gordon Moore's interest, either in transistors or in the firm's Texas location. Teal developed the standard crystal pullers for germanium and silicon and established the value of pure single crystals. Convinced that silicon was the future, TI produced its own ultrapure silicon, its own crystal pullers, and its own supplies of high-quality wafers. It was a unique capability, and with it Haggerty and Teal pushed their firm out in front.

TI had surprised the industry in 1954 by being the first to bring a silicon transistor into production. In 1957 the company enjoyed close to $70 million in sales, about half from semiconductor devices. It was roughly twice the size and more than three times as profitable as Fairchild Camera, the backer of "the Traitorous Eight." As 1958 dawned, no firm had yet brought a diffused silicon transistor to the open market, that is, had succeeded in moving it from laboratory to factory. Bell Labs, having pioneered its creation, was a natural competitor, yet, to the delight of the Fairchild Semiconductor cofounders, operated under a grave restriction. The year before, AT&T had settled US government antitrust suits with a major consent decree, restricting it to transistors for its own communications and military project uses.

More immediately threatening were Texas Instruments and Gordon Teal. At Fairchild Semiconductor, as Moore raced to complete his furnaces and establish the diffusion process needed for NPN silicon mesa transistors, he found himself in direct competition with Teal. IBM's deal with TI was hardly welcome news for Gordon and his colleagues. David was facing Goliath. To an outside observer, Texas Instruments was the company to watch. In actuality, Fairchild Semiconductor was progressing at a far faster pace, in the period leading up to Christmas 1957. The firm had found Ed Baldwin, and, as Moore points out, this was also the period when "Jean Hoerni sat at his desk and made the seminal invention of the planar transistor." This invention would first transform the fortunes of Fairchild and, subsequently, the nature of the whole industry.

Hoerni's attention had alighted on the step in which the wafer's oxide layer was stripped off. Bell Labs, which in 1956 had broadcast its silicon diffusion techniques, taught that stripping was essential. Hoerni had the courage to ask the heretical question. Diffusion made oxide dirty, but did the oxide layer really need to be removed? Perhaps it was without any effect on the underlying device. Perhaps it might also physically protect the surface of the device, stabilize its electrical properties, and provide an insulating coating. Hoerni and Moore would soon begin making mesa transistors, and the PN junctions of these transistors would be exposed at the top surface and edges. If the oxide was retained, it could completely cover these, preventing contamination and electrical faults. On the mesa, the contacts to the emitter and base were made at the top, while the collector contact was at the bottom. In Hoerni's imagined flat, or "planar," transistor, all contacts would be at the top, improving performance, while simplifying production and hence reducing cost. As Moore recalls, "A lot of the electrical properties were determined by whatever crud you left where the junction came to the surface—an area of very high electric field, perhaps a million volts a centimeter over a small area, which attracted all the dust particles and crud. The conventional wisdom was that the oxide was awfully dirty, and you had to get rid of it. Hoerni thought, 'Why not leave it there?'"

Hoerni was so convinced of the promise of his planar process that he wrote an entry in his patent notebook on December 1, 1957, and had Bob Noyce read and sign it that same day. Quickly, the pair shared the idea with Gordon. All, including Hoerni himself, remained unsure if the idea could actually work.

As a group, they opted to put the idea aside and go with the silicon mesa, which they were confident *could* be made. Moore recalls that "paper invention was ahead of science and technology. We couldn't try Jean's idea right away, as it required four masking operations, and Bob Noyce had only bought three lenses." For the next eighteen months, Hoerni's "planar" heresy, a technological milestone, would lie dormant in his notebook.

The Call of the Valkyrie

As 1957 ended, Moore was busy on "good chemical problems," building furnaces and control systems and beginning to refine diffusion. Despite his and others' cost-cutting approach, they had already burned through nearly a fifth of Fairchild Camera's promised $1.3 million, the bulk spent on buildings, materials, and equipment. In addition to the eight cofounders, Fairchild Semiconductor now had twenty-two employees. The need to build relationships with potential customers was urgent. Fortunately, outside work came easily to Bob Noyce, whose personal charisma and competitiveness (classic ingredients for success in sales) would one day be the stuff of legend. With a PhD in physics from MIT and experience at Philco in fast germanium transistors and military projects, he shared a vocabulary with the firm's intended customers: engineers and researchers in military and aerospace computing. He was perfect for the job. On December 20 he made his way east. Noyce had been invited to discuss a match between the group's mesa transistor and IBM's work on an advanced aerospace computer. IBM and Texas Instruments might be developing a close business relationship, but IBM's board chairman, Sherman Fairchild, was also the backer of the Traitorous Eight.

IBM had a military products division in Owego, New York, keyed to Cold War competition with the Soviet Union. The division had secured a contract related to a proposed supersonic bomber carrying thermonuclear weapons, the B-70 Valkyrie. The engineers assigned to the project wanted a fast-switching silicon transistor; the problem was how to get it. Hearing of Fairchild Semiconductor's plans, they requested a visit. Quickly, they made Noyce privy to detailed information about Owego's needs and valuable intelligence on the competitive landscape. His hosts wanted low-power but fast transistors for the data-crunching "logic" of the computer, medium-power fast transistors as "core drivers" to hold this data in the computer's "memory," and high-power transistors for "servo drivers," the automatic controls that would allow the computer to guide the aircraft. General Electric and Texas Instruments, along with Transitron, Raytheon, Motorola, and Hughes, were all interested. Whoever offered a reliable product first would take the prize.

Suppliers would face a costly regime, putting their transistors through punishing physical and electrical tests. If IBM had to guarantee the US Air Force a computer whose reliability could ensure thermonuclear deterrence, it followed that suppliers must guarantee the reliability of their products. Though his company was not yet three months old, and though he had nothing whatsoever to show, Noyce was confident that Fairchild Semiconductor could create silicon mesa transistors with properties matching the need for a core driver; conceivably, it could even beat the competition to the punch. Fast-switching medium-power silicon transistors were, he perceived, "a vacant area at the moment." An opportunity existed at IBM; how many other corporations had a similar need?

Full of excitement, Noyce returned to Palo Alto. His colleagues required little persuasion to agree that a mesa aimed at IBM's core driver should be their first product. Typical transistors might bring a price of $2 or $3, but for the Valkyrie the metric was different. Military realities meant that (if reliability and speed met specifications) paying $100 for a transistor was a mere bagatelle. No doubt, prices would fall in time. What no one anywhere conceived, even remotely, was the extent of that tumble over the years. An ordinary, run-of-the-mill transistor cost, on average, $2.50. Today, the same money—in constant dollars—would purchase more than *30 billion* transistors. The silicon transistor has become the cheapest object ever manufactured. In 1958 such a reality would have seemed utterly fantastical, particularly to Gordon Moore as he struggled to figure out how to make his first mesa.

As an untested start-up, Fairchild Semiconductor had neither made nor sold a transistor. An order from IBM could change the game. Tom Bay joined the group in January to head marketing and sales and worked with Noyce to secure the deal. On February 8, as Ed Baldwin arrived at last from Hughes, they submitted their quote. Then Baldwin seized the baton. If the first months of the start-up had been expensive, he surprised his colleagues with the news that such expense was unproblematic, even insufficient. He argued that the mesa would enjoy a large market; military needs, mediated by firms such as Ramo-Wooldridge, Hughes, Litton Industries, and Fairchild Controls, would ensure that. In his opinion, Fairchild Semiconductor needed to get much bigger, fast, and to establish a separate manufacturing plant for its mesa.

Meanwhile, Noyce and Bay were doing the rounds in the Los Angeles area. Responses were similar to IBM's: speed, lower power consumption, and utmost reliability were needed, and needed right away. The news bolstered Baldwin's argument that Moore, Noyce, and their colleagues must rapidly expand, doubling staff for functions such as preproduction engineering and quality control, and start to construct a large factory. The idea was both heady and challenging. "We were all learning on the job," recalls Moore. "We could only grow so fast, and do a limited number of things." Baldwin understood that, sooner rather than later, the founders must bet the firm. Fairchild Semiconductor would succeed or fail on the success of its mesas. That meant the bulk of committed funds should be spent on staff and equipment, while Moore and his fellow founders figured out how actually to make the mesa. It was a high-stakes approach, but the cofounders did not balk. Gone were the days of working as a leaderless group of equals. Baldwin was in charge.

Moore and Hoerni, responsible for establishing sound diffusion processes, spent February and March 1958 running their experiments. Noyce concentrated on linking the inner work of R&D with the outer world of customer demand. It was a role he would inhabit and refine for the rest of his career. Behind the scenes, Baldwin negotiated aggressively about his equity stake in the start-up. The original deal to create Fairchild Semiconductor had reserved a block of shares for distribution to future key employees. Baldwin, well established at Hughes, had been offered a stake equal to that of each of the Traitorous Eight. However, he now went directly to Fairchild Camera's leadership to ask for an even larger equity stake: he was, after all, the boss. Less than six months in, the start-up was beginning to shift from a communal enterprise to one with much narrower definitions of self-interest.

While Baldwin negotiated, and Noyce and Bay communed with the IBM engineers in Owego, other private discussions took place. Sherman Fairchild (the largest individual IBM shareholder and chair of the executive committee of its board) met with IBM's famed leader, Thomas Watson Jr., and reassured him that Fairchild Semiconductor had the brainpower and resources to bring the mesa into production. An IBM order came through in early March. It was for a sample lot of one hundred mesa transistors at $150 a transistor (approaching $2,000, in today's dollars), with a delivery date of August 1, 1958, six months away.

Realizing the Product

In Fairchild Semiconductor's new organizational structure, the cofounders fitted into a hierarchy under Baldwin, with Moore and Noyce taking the lion's share of responsibility for R&D. Noyce—charismatic, ambitious, and experienced—was both director of research and development (overseeing the work of most of the other cofounders) and the company's public face, liaising with customers, competitors, and the leadership of Fairchild Camera. Gordon, still very much the nerd at heart, was satisfied to take the lead on device development, directing Hoerni and Last and assuming responsibility for producing the transistors for IBM. His background in chemistry, together with his experimental talent, diffusion experience, quiet determination, and measured evenhandedness, made him the obvious choice.

Moore soon worked out satisfactory diffusion processes, yet both his work on the NPN mesa and Hoerni's on the PNP were bedeviled by problems stemming from a later stage in the process: that of forming electrical contacts for the mesa's constituent layers of emitter, base, and collector.

These contacts formed the interface between the transistor and the circuit of which it was a part. Finding metals with the adhesive qualities to make robust connections was relatively easy, as was evaporating those metals to form the contacts. Far harder was the task of creating a connection of metal and doped silicon. Repeatedly, the chosen metal formed an unintended PN junction with the silicon, making the transistor unusable. Gordon knew he faced a major technical hurdle. Compounding the complexity, different metals were compatible with differently doped silicon. A metal might make a good contact for N-type silicon but form an unwanted PN junction and thus be a bad contact for P-type silicon. What was needed was a metal or alloy that could form a workable contact to both P- and N-type silicon. To Moore, the experimental chemist, the hunt required familiar skills.

Aluminum was one possibility. The element was lightweight, corrosion resistant, and abundant. Though it made good contacts with P-type silicon, it formed an unwanted PN junction when fused to the transistor's N-type regions. Moore made an alloy of aluminum and phosphorus in hopes that the phosphorus would prevent the formation of a PN junction, but the experiment was disappointing: "I was fiddling with all kinds of complicated alloys." In frustration, and needing fresh ideas, he turned to Noyce, his close colleague of two years, who urged him to persist with pure aluminum for both P- and N-type contacts. Using only one element would simplify fabrication and increase the yield of good mesas within each batch, lowering overall costs and leading directly to marketplace advantage.

On the face of it, Noyce's suggestion was ridiculous. Aluminum would not form a good contact with N-type silicon. "Noyce knew that better than any of us," says Moore. Yet Moore also understood that while many of Noyce's ideas were impractical, he sometimes generated brilliant insights. Maybe, Gordon mused, there was something in Noyce's all-aluminum idea. There was much to the behavior of materials beyond simple composition: their performances could be shaped dramatically by handling and processing. Perhaps, with appropriate witchcraft, he *could* make workable aluminum contacts to both P- and N-type silicon. In April 1958 he succeeded. Increasing the doping levels of the N-type regions and alloying the aluminum contacts at a very high temperature, 600 degrees Celsius, were what was needed. Moore's discovery was a major step, first for Fairchild Semiconductor and soon for the whole semiconductor industry. "We tried it. It became the standard of the industry for the next thirty-plus years." For those with eyes to see, the success demonstrated a powerful dynamic arising from Noyce's prolific enthusiasms coupled with Moore's insight, experimental abilities, and systematic, dogged approach. In the future they would together make multiple advances in the new territory of silicon electronics.

Success built faith. Gordon became bullish about his work. It would soon be time for Fairchild Semiconductor to throw its weight behind one version of the mesa or the other—Gordon's NPN or Hoerni's PNP—in order to meet IBM's deadline. To be first to market, they had to pick the right version, a version that worked and one that they could make quickly. Moore, as head of device development, would have the final say, and the presumption was in favor of NPN. However, in what by now was a characteristic mode, he wanted not only to be fair but to be seen as fair. The answer was to measure and analyze and then decide and invest.

For eight years Gordon Moore had maintained his personal ledger of daily expenses and calculated the return on educational spending against his salary. As a chemist he had been trained to measure precisely and analyze with exactitude. Now, in the biggest technical decision of his early career, he sought safety in his preferred method of understanding. In early May he drafted a memo to his fellow founders, defining the measures he would use. He would compare multiple batches of each mesa, looking at yields, failures, and electrical characteristics: all quantifiable. The version that scored highest would become their first product. One might argue with the merits of the measures, but in circulating the terms of his decision in advance, he made the process nonpartisan, transparent, and fair.

Hoerni's PNP was still beset by metallization problems. Thanks to the success of the all-aluminum approach and the relative ease of its diffusion processes, Moore's NPN was the clear winner. Hoerni, brilliant and intensely competitive, could not hide his disappointment and anger. "It was the correct technical decision, but he was pretty bent out of shape about it," says Jay Last, Hoerni's climbing companion. "Jean was very complicated. He could be the sweetest guy in the world, and the nastiest. Fortunately for Fairchild, he did his best work when he was very irritated. Jean was irritated a lot, so a lot of good work came out."

Gordon Moore was growing in the job. Quietly, he was cementing his leading role, based on demonstrated performance. "I had been responsible for the development of the transistor. Now I had the job of making those developments into processes that were well specified and capable of being transferred into manufacturing." At the instigation of Ed Baldwin, he formed a preproduction engineering group, which grew to "a pretty good size, because it had a big job to do." He also became head of engineering, which—because of the firm's helter-skelter growth—quickly became a formal organization to parallel Noyce's R&D. "Baldwin taught us that you have to set up the manufacturing operation separate from the development laboratory. You have to engineer and specify manufacturing processes, which is completely different from getting something to work once in the laboratory."

The IBM deadline required a hundred mesas by the start of August 1958. Not only that, but Fairchild Semiconductor also planned to announce the mesa at a major West Coast electronics trade show—Wescon—later that same month. Moore now bore the full existential weight of the enterprise. He, and he alone, must launch production successfully and meet the deadline. If he failed, the start-up would go down in flames.

Going for Broke

Fairchild Semiconductor was committed to its batch process, by which multiple transistors were formed simultaneously on a single wafer. This process was more flexible and less costly than the highly automated "assembly-line" approach favored by competitors. Thanks to his experience at Hughes, Baldwin—the old hand among the innocents—believed that, once the mesa was in production, they could improve the chemical printing technology, lowering costs while delivering increased product. Still without any actual sample, he argued that by the end of 1959, the firm could move from 100 transistors a month in the laboratory to 100,000 transistors a week in the factory. Such staggering thousandfold growth would allow them to lower the price from $150 to $5, while still earning handsome profits. The aim was to capitalize on market demand, exploiting the potential of the firm's batch technology. By ignoring the possibility of engineering problems that might scupper yields in high-volume manufacturing, they would leap ahead. Not only that, but by the end of 1960, he argued, the firm would be able to pay off the cost of the factory.

As boss and battle-scarred veteran, Baldwin convened a meeting to persuade Moore, Noyce, and the other cofounders—shareholders with votes—to commit to a factory of more than six hundred thousand square feet. He told them bluntly that a decision was necessary, to capture consumer demand by achieving cost advantages. Gordon Moore and his colleagues were receiving a rapid education in the business realities of the industry they, like Shockley, hoped to join. Moore was confident in his technical abilities, yet still unprejudiced by much business experience. Now, under Baldwin's tutelage, his horizons were widening. The cofounders agreed to the plan.

By June Baldwin had also prevailed on Fairchild Camera to make the substantial investment needed. In return, he committed to a target of $500,000 to $750,000 in revenue by the end of the year, from sales of up to 20,000 mesa transistors. The numbers both reflected and magnified industry-wide experience; transistor production had grown from well below 1 million in 1951 to almost 50 million in 1958. If Gordon harbored private doubts about Baldwin's breathtaking scheme, he kept quiet. He had options. With negative unemployment pervading the industry, he could always find another job if things went wrong. As Betty put it, they would not starve.

While talking expansion, actual tangible transistors were needed to meet IBM's order. The effort was so all-consuming that no one had time to orchestrate their delivery. Jay Last recalls that "we finally had the transistors ready to go, but what were we going to put them in? I went down to the grocery store. I thought that the Brillo box looked nicest, so I bought a box of Brillos. I grew up in Pittsburgh, as did Andy Warhol. It turns out that we both depended on a Brillo box to get started." The transistors were wrapped, put in the box, and dispatched to Owego. It was a no-frills solution, entirely appropriate to the frugal style in which the work had been done. Equally urgent was the need to get word out. Tom Bay ordered a print advertisement to coincide with Wescon, positioning the mesa as perfect for aerospace computing: able to withstand high temperatures and switch quickly. The copy boasted of "a uniquely experienced team of research scientists and production engineers whose objective was to bring the advanced solid-state diffusion process under close control." It added, succinctly: "They succeeded."

Bob Noyce, by now the recognized public face of Fairchild Semiconductor, was the obvious choice to announce the firm's success at the Los Angeles conference. He garnered "big interest" both from potential customers and from competitors. They had "scooped the industry," he told his colleagues. There was "no prospect of anybody getting in our way in the immediate future." That August the cofounders toasted their own success. They had set up a company, turned an empty building into a pilot production line, crafted a workable chemical manufacturing technology, and—particularly through Gordon Moore's focused effort—delivered the world's first fast-switching diffused silicon transistor. Sold in two versions, the "2N696" and "2N697," following an industry naming convention, proved immediately popular with military contractors. Fairchild Semiconductor, Bob Noyce crowed, had won. They had jumped ahead of Texas Instruments and Bell Labs, two of the largest, most advanced transistor makers. It had taken less than a year and less than $1 million. It was a remarkable performance.

The firm had sole command of the narrow military market it had chosen, aerospace computing. Its advertisement elicited more than four hundred inquiries. Interest at the show itself was equally gratifying; a different branch of IBM was there and ordered a sample of five hundred mesas. Bay's sales engineers expected suppliers such as Sperry-Rand, American Bosch Arma, Burroughs, and General Electric's military and aerospace operations to quickly fall in line. Bay himself believed the Owego engineers would ask for up to three thousand more transistors. He believed his staff could sell fifteen thousand "without any trouble"; the price would "hold up" at between $40 and $50. By the end of September, Fairchild Semiconductor already had orders worth $65,000 for its NPN transistors. This figure grew quickly to $500,000, and the payroll jumped to ninety employees. Moore recalls, "Everything was working fine: the development and preproduction engineering for our processes and first products was complete, we had a thick process-spec book that recorded all the detailed recipes, and we had interested customers." Success felt good.

By now, staff were jostling for space at 844 Charleston Road and overflowing into a hodgepodge of rented offices nearby. The majority were lower-paid local women, given the repetitive tasks of the intricate manufacturing sequence. "In those days you hired a woman who might not even speak English and taught her to align two things under a microscope, and she did that all day. You train a dozen of them at once. People were hired principally on their manual dexterity."

Making sufficient mesas to meet the orders surging in became an urgent challenge. "We had all the growth we could manage," recalls Moore, who had resorted to outsourcing production of manufacturing equipment like furnaces to his assistant Art Lasch. "We encouraged Art to go into business nights and evenings." Art finally made that a full-time job, setting up Electroglas and also selling to other transistor companies. "A lot of infrastructure companies developed like that during those early days," says Gordon. At Fairchild the minimum was done in-house. "A key direction was set by accident, that of a horizontal industry"—companies would focus narrowly on their special expertise, buying many needed items from other firms.

Moore's own work articulating and codifying production routines was far from complete, and so-called yield—a perpetual issue in silicon electronics—remained problematic. Yield was shorthand for the proportion of good transistors on a wafer as opposed to defective, unusable ones. Yield gave a measure of the efficiency of the production method. High yields equated to lower cost per transistor. Gordon's task was clear: drive up yields. As usual, he sought a measure through which to understand the issue. In this case, his metric was "final seals," the last operation in sequence, involving welding the top "can" of the transistor package onto the header on which rested the actual silicon mesa. With the final seal in place, the transistor could be tested and certified as customer ready.

Following the week of August 18–22, 1958—Wescon week—Moore reported that one thousand "final seals" had passed the test. The next week saw an increase of 50 percent, to fifteen hundred seals. Gordon calculated the maximum number that his equipment and workforce could generate, going forward. With multiple diffusion furnaces up and running, he aimed to process seventy of the three-quarter-inch wafers each day. From each wafer, some ten transistors made it to final seal, giving a production capacity of seven hundred transistors a day. That amounted to thirty-five hundred a week, more than sufficient to meet the year-end target. The pressure began to ease. Bay and his sales engineers could make good on their forecasts.

One mesa customer was American Bosch Arma, a New York–based military contractor. It had developed the guidance system for Atlas, the US military's first long-range missile with a thermonuclear warhead. Its order gave Fairchild Semiconductor an entrée into this prestigious and well-financed program. Another order came from the Autonetics division of North American Aviation in Los Angeles. Autonetics had the contract for the onboard guidance computer for Minuteman, the Air Force's latest nuclear missile. It requested a sample of NPN mesas for evaluation. This put the firm in the running for another lucrative application at the center of Cold War strategy.

Gordon and his cofounders had designed their mesas to serve in IBM's guidance computer for the Valkyrie supersonic strategic bomber. In the end, only two prototypes were ever developed. It mattered little, since Moore and his team's pioneering played straight into the broader clamor for increased reliability and greater miniaturization. "IBM was not going to buy an awful lot of transistors," explains Moore, "but the mesas succeeded in other applications right away. A key one was the Minuteman program. North American Aviation needed new levels of reliability and wanted a silicon transistor." Price was not a major consideration for the military, and sales for Gordon's NPN mesa—recognized as a good general-purpose transistor—were overwhelmingly military.

Prices high, sales exploding: through 1959, the miracle continued. "It had been close to a year before we had product, but once we did, the devices were pretty widely accepted," recalls Gordon. "Germanium transistors give out at about eighty degrees. The military wanted something that would operate at a couple of hundred degrees. They really needed silicon. Each of the services was pretty convinced: the Air Force, the Signal Corps, the Army, the Navy." What Gordon did not know at the time was how the career of *Vanguard* would develop. The US government hoped this vacuum tube–controlled rocket would put up a satellite to answer *Sputnik,* but of its eleven launches through September 1959, eight failed. Clearly, the day of tubes and germanium had passed. In contrast, for silicon transistors that could withstand the vibrations, gravitational forces, and searing heat associated with rockets and missiles, no price was too high.

Fairchild Semiconductor's decision to go for broke was rewarded. Demand took off, the way rockets were supposed to. The first silicon spin-off, in what was destined to become Silicon Valley, was a spectacular success. Moore's unconstrained attention to a novel technology was central to Fairchild's triumph over a multitude of established organizations. The new business model for success was simple: break away, focus, and execute. And Fairchild Semiconductor vividly conveyed how that success could lead on to fame and fortune on the electronic frontier.

THINKING BIG

An All-Silicon Future

Throughout 1958 Jean Hoerni struggled with the diffusion and metallization processes for the PNP mesa. Customers would soon demand a PNP to pair with the NPN. As Last explains, "When we had a matched NPN and PNP device, we had the world." In October, with the PNP mesa finally stable enough for preproduction studies, Hoerni was released to use his creative genius in other projects. In contrast to Hoerni, Moore had succeeded quickly in his practical tasks, establishing the manufacturing technology and making the NPN. He, too, was ready for fresh challenges. His engineering group took up the PNP, striving to increase yields. With growing confidence in the group's abilities, Gordon declared that the PNP mesa would soon be in steady, reproducible production.

Hoerni returned to questioning conventional wisdom. In his notebook lay sketches for the planar process, reconceptualizing "dirty" oxide as an asset in making transistors even better than the mesa. In a second contrarian leap, he began to consider another "detrimental" material: the chemical element gold, known as "deathnium" for its destructive impact on transistor performance. Could he use deathnium?

While successful, Fairchild's mesas frequently failed. To identify and solve problems was tricky, but critical. When IBM's engineers reported that some of the mesa transistors would not turn off, Hoerni and others traced the issue to too much "gettering." Gettering was a process in which an element, commonly nickel, was added to a transistor to draw out and neutralize impurities that harmed its performance. Hoerni turned the problem on its head. Reflecting on a recent Bell Labs report, he reasoned that gold had an effect like that of gettering, enabling electrons to flow, if not necessarily to the desired destination. Maybe the right amount of gold, in just the right place, might make a transistor switch faster. Late in 1958, in a series of experiments in which he diffused gold into specific areas of the collector region, Hoerni demonstrated the effect he suspected. If diffused silicon transistors could be made to switch still faster and compete directly with germanium transistors on the all-important metric of speed, this—combined with their other advantages—would give them the potential to take over the entire market. Gordon recalls, "Gold doping took us into a lot of high-speed applications, just as people were starting to build silicon transistor–based computers."

Moore grasped the enormous implications of Hoerni's work and quickly linked it to the lessons he was absorbing from Ed Baldwin. Gordon was now established as the leader and go-to person in device development. His view on what to make and how to make it carried great weight. His knowledge of chemical printing technology was unsurpassed. His work to bring the solid-state diffusion process under closer control and to put out a transistor with the greatest possible performance and reliability had led to spectacular success and concomitant rapid growth. Turning thirty, enjoying practical results, and shouldering responsibility, with the first mesas under his belt, he pondered, "Where next?" Reflecting on Hoerni's demonstration, among other things, Moore and cofounder Vic Grinich, leader of applications engineering, addressed "the best direction for the long term objectives of Fairchild Semiconductor" in a joint memorandum to Ed Baldwin in January 1959, even as the firm began to put Hoerni's theories into practice, introducing fast-switching transistors based on gold doping.

Titled "Device Planning," the memorandum foreshadowed the strategy that Gordon would pursue a decade later, at a different start-up. The memo argued for a philosophy of standard types: the transistor business was cutthroat, and the companies that survived would be those making "large quantities of standard types" rather than "special purpose transistors, for which there is not a broad market." Gordon came to believe that "there is no value in developing something that cannot be built in volume." Products were costly to design, so the best economics lay in long runs of well-designed products. The firm's NPN and PNP mesas should be the first of "a family of generally useful transistors," produced in volume and sold at highly competitive prices. If Baldwin agreed, a crucial question followed, one that—as production escalated through the thousands and millions to the billions and quintillions—would take on special resonance for Gordon Moore: transistors useful for what? To Moore and Grinich, the answer in 1959 was "switching and computers." Silicon transistors as miniaturized, rugged, fast on-off switches—in everything from telephone exchanges to the control systems of military equipment and the mainframe computers appearing in government and business settings—could, they argued, yield "potentially the largest applications."

The best germanium transistors were still much faster switching than Fairchild Semiconductor's mesas. Germanium could be handled easily and was fine at the relatively low temperatures that characterized applications in the commercial sector, yet Moore and Fairchild were committed to silicon.

Could silicon really compete in the commercial market? Thanks to the pioneering work of Hoerni, Moore was confident that it could. His growing ambition, fueled by success, is evident in the memo. Gordon had entered the transistor industry at a moment of paradigm shift—the superseding of vacuum tubes—and now proposed that silicon transistors could supersede germanium in the switching market. A few years later, he would develop truly imperial ambitions, declaring that a new form of "microcircuitry" built of silicon transistors would pervade all of electronics. Harnessing the data to execute his vision—following his trusted formula of "measure, analyze, invest"—he would make revolutionary change his overriding agenda.

Moore and Grinich proposed a family of no fewer than ten standard switching transistors, covering the range of current and power levels: from very low-power computer transistors, through the low- and medium-power regime of their NPN and PNP mesas, to higher-power transistors for automatic controls and other equipment. The family should be fully available by mid-1961. To stay in the lead, Fairchild Semiconductor also needed a tiny "small geometry" mesa: smaller meant faster. The market's fastest transistors, of germanium, were offered by Noyce's former employer Philco; the latter's latest transistor, initially developed by Noyce himself, was a favorite in advanced military computing. "Smaller, faster, better" went the mantra in the high-stakes world of defense and aerospace. Philco's germanium transistor was the competition to beat. "Anyone who wanted speed was using its devices," says Moore. Hoerni's experiments implied that Fairchild's diffused silicon transistors would match Philco's devices and—possessing superior reliability—could displace them altogether.

Particles and Possibilities

Fairchild Semiconductor now came up against an unexpected problem, one that threatened its existence. Customers and the firm's own engineers began reporting that some of its mesa transistors—having passed a batch of tests—were not only unreliable but degrading severely in routine performance. Gordon knew that the mesa "should withstand 100 volts or so, nice and cleanly, but some of them started displaying all kinds of peculiar electrical characteristics." Since military and aerospace systems had stringent requirements, unreliable transistors were not just a poor advertisement, but also a very serious problem.

Moore launched a program to track down the fault. A team of engineers and PhDs searched for the source of what were quickly labeled "unidentified flying objects" (UFOs). Gordon also assigned his friend Bob Robson, production foreman, to the problem. "Bob cut the can off one of these packaged transistors and looked at it under a microscope, as he put a current through it. A spot on the side of the junction glowed. He knocked the spot off and, all of a sudden, normal electrical characteristics returned." The problem lay with tiny pieces of dust and metal, present in the transistor package after the final seal. Attracted by high electric fields where PN junctions came to the surface of the transistor, these particles could short out the device. "All we had to do was get rid of the particles," remembers Moore. Since their provenance remained a mystery, this was easier said than done: "We were having 7:00 a.m. meetings every morning, trying to figure it out."

Welding the can to the header, in the final seal of the transistor, was intended as a protective step. Ironically, this step was the main culprit. Engineers moved to refine assembly and packaging and modify the welding. They also instituted a primitive "tap test" as a way to induce failure. "You took the transistor and banged on it with a pencil while you watched its electrical characteristics; if one of these particles jumped, you'd see a change." The "pencil-battering" process was automated, but the problem of failures continued. One thing it highlighted was a pressing need for still greater cleanliness in the manufacturing process. Another was the appeal of Hoerni's alternative planar process, which—because oxide covered the high-field junctions—would provide immunity to this problem. Planar ceased to languish: it was suddenly "a heck of a lot more attractive," says Moore.

Hoerni was keen to demonstrate the full viability of his idea. His notebook-based explorations of December 1957 had shown how the dirty oxide could shield the PN junctions. He had been prescient in envisaging that this would "protect the otherwise exposed junctions from contamination and electrical leakage from subsequent cleaning and handling." Cleaning up might provide a partial solution, but the planar process would eliminate the problem. The founders had already discussed the developments they wanted to patent. One was the "single metal contact," Gordon's all-aluminum metallization process, which would become an industry standard and would in 1963, upon issue of a patent ("method for fabricating transistors"), become an important addition to Fairchild Semiconductor's portfolio of tradable patent assets. Another was Hoerni's planar process.

Hoerni returned in earnest to his idea, making prototype "dirty" devices with the help of a technician. They left oxide in place over the area where the emitter-base PN junction came to the surface of the silicon. "Dirty" transistors were superior to regular mesas in both performance and stability. Moore was immediately very interested. On January 14, 1959, Hoerni asked one of the firm's secretaries to type up his December 1957 notebook entry as a formal patent disclosure. Forwarded to the firm's external patent lawyer, the disclosure served to anchor his invention in time. From then on, Moore and Hoerni pursued separate but complementary solutions to the problem of unreliability: while Hoerni focused on creating a planar

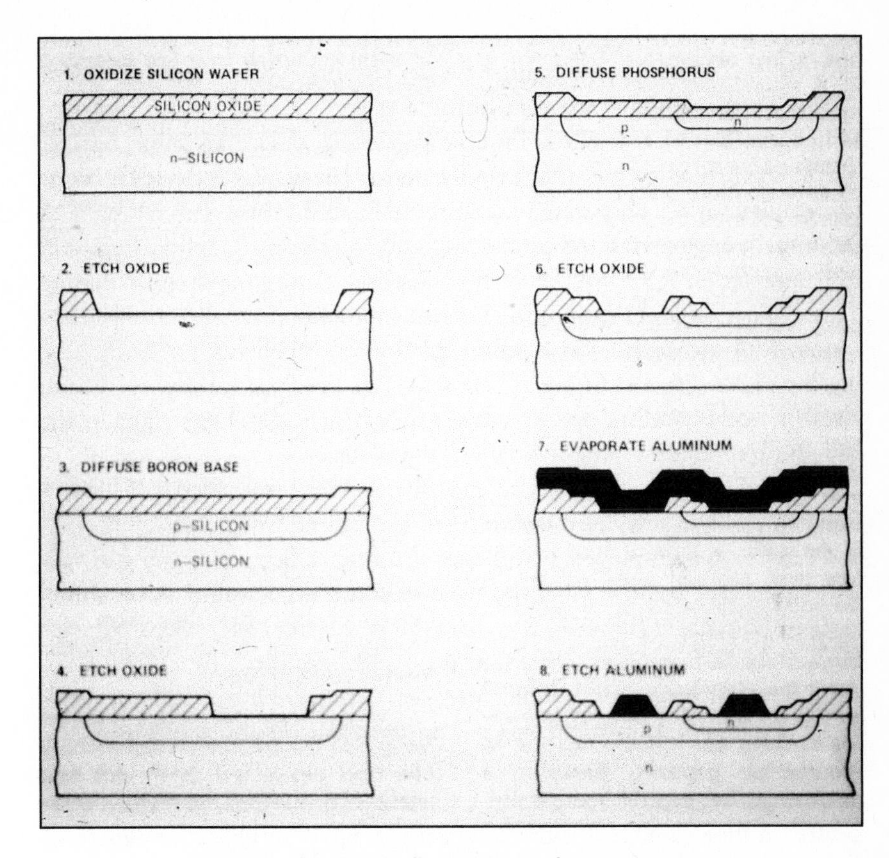

The steps for chemically printing a planar transistor.

SOURCE: GORDON MOORE.

transistor immune to UFOs, Moore concentrated on the more immediate task of cleaning up the production process.

The Minuteman program, in particular, required robust, reliable transistors. Moore recalls taking his team to meet Autonetics that spring: "During the preliminary banter before the presentation began, I discovered that they had omitted the percentage sign in the request for proposal. My carefully calculated plan missed their requirement by one hundred times! That was a short presentation." Moore quickly regrouped. If transistors for the Minuteman program required a daunting failure rate of one in ten thousand, production would have to be closely policed.

Production processes were one thing. The actual space in which to undertake production was something else. Following Fairchild Camera's agreement to fund a fresh factory, Julius Blank had canvassed sites, finding a suitable one on North Whisman Road, in Mountain View. This plant would be five times bigger than the firm's initial home and located fittingly close to the NASA Ames Research Center and the Moffet Field Naval Aviation Center. Coming onstream in the first part of 1959, it was designed to include "complete facilities for crystal growing, device assembly, testing, circuit research and applications engineering." The facility included sections for manufacturing, corporate functions (sales, accounting, purchasing, and personnel), engineering (preproduction and "sustaining"), applications and instrumentation (each with its own laboratory), a machine shop (including a dispensary, tool fabrication, stockroom, and maintenance), and mechanization. Murray Siegel, now a senior applications engineer, recalls discussing the plant with Vic Grinich. "He said, 'We need work benches.' I said, 'Well, I work standing up.' We designed a bench that came right to the belly button. That became standard in the industry."

Fairchild Semiconductor was growing at breakneck speed. Whisman Road was part of a wider move to provide customers with better products and services and employees with better conditions. Secure in its initial successes, the firm could even afford time to consider its image. A brochure, entitled *Welcome to Fairchild,* described the new factory as "pressurized with filtered air, and regulated in temperature and humidity," to ensure "clean conditions for transistor manufacture" and "contribute to employee comfort." While almost half the space was devoted to manufacturing operations, there was a generously sized lunchroom and "lunch patio." Further proof that the firm offered excellent working conditions were regular "musicasts" in the production room and "adequate parking space for 800 cars." The company—not yet two years from start-up—saw itself as having "one of the most functionally modern electronic component manufacturing plants in the nation."

Barely a week after Jean Hoerni typed up his notebook entry on the planar process, Bob Noyce wrote another notebook entry that would be just as important, if not more so, one that would secure his own lasting fame. In the entry, on January 23, 1959, Noyce imagined the *microchip*: an integrated circuit in which all the components of a complete circuit were made within a single piece of silicon. The launch of *Sputnik* had drawn Fairchild Semiconductor into military programs at the core of Cold War strategy, and large numbers of the firm's mesa transistors were being purchased for the missile and bomber programs central to America's offensive capabilities. The transistor reduced the size of switching and amplifying components, enabling smaller, more reliable electronics. Further shrinkage, or "micro-microminiaturization," was an obvious next step and a growing trend within military electronics. The military's needs were extraordinarily acute, even existential. Scientific, engineering, and business leaders also prized reliability.

By 1959 millions of dollars had been spent addressing the "tyranny of numbers," a concept articulated by Jack Morton of Bell Labs. His argument was simple—and brutal. Computers and other electronic systems required tens of thousands of interconnected components. Greater capability required a greater number of components and interconnections. Every addition increased the possibility for failures that destroyed reliability, required repairs, and made maintenance more costly. Limits of reliability and of repair set an upper bound on the size of electronic systems. This was the tyranny of numbers.

Military suppliers were also much concerned about the sheer size and weight of electronic components. According to Jay Last, "There resulted an enormous interest in microminiaturization for airborne devices; every company had a program." Among these programs was one to support research on using "modules"—very small ceramic plates of an agreed-upon standard size—with wiring and certain components printed on them and with individual transistors attached. Modules could be made for a range of standard functions and combined to make circuits. Some researchers speculated about whether the printed circuits could themselves include transistors. Perhaps all the components of a complete electronic circuit could be printed and interconnected in a single piece of semiconductor material?

On January 20, 1959, immediately following Hoerni's demonstration of his planar process, Gordon Moore, Bob Noyce, and their cofounders were directly introduced to the demand for highly miniaturized circuitry. They were visited by the noted engineer Edward Keonjian. Keonjian was heading American Bosch Arma's computer project for the Atlas intercontinental ballistic missile. As Arma had already ordered sample quantities of Moore's NPN mesa, this visit was high stakes. Keonjian filled them in on his activities. His team's computer would incorporate "microminiature circuits." Since Arma did not have the capability to make military-grade diffused silicon transistors to plug into its circuits, Keonjian proposed to ship his printed "plates" to Fairchild Semiconductor. They would make mesa transistors and put them into the printed circuits. Keonjian requested a quote for this work, "ASAP."

Bob Noyce, in addition to his unique charisma, possessed an intellect to rival those of Moore and Hoerni. Noyce had long chafed at the inherent inefficiency of making mesa transistors. His firm's chemical printing process leveraged the power of batch processing, with many transistors created simultaneously on a single silicon wafer, yet wafers were then laboriously cut into individual devices and each placed into its own package. Customers had to reconnect the transistors to make a circuit. Three days after Keonjian's visit, Noyce had a brainstorm: why not leave the transistors together in the wafer? If you could interconnect them there, a set could be detached as a single unit; it could be a reliably connected integrated circuit, or "microchip." The answer to the tyranny of numbers lay in the silicon microchip.

In his notebook Noyce argued that "it would be desirable to make multiple devices on a single piece of silicon." This was the easy part. He could use diffusion to make transistors and other similar components—diodes, resistors, and capacitors—for a circuit. He could even use diffusion to create extra PN junctions between components, preventing unwanted electrical interferences. The next trick, "interconnections between devices as part of the manufacturing process," was harder. Theoretically, the wiring could be done through an extension of the existing metallization process—making interconnections *between* components just as one formed contacts *for* components—but these interconnections would need isolating from the underlying devices in the silicon. Hoerni's demonstration of the planar process was fresh in Noyce's mind, and it provided a solution. The dirty oxide was an insulating layer over the whole wafer. Noyce could simply lay his interconnections on top of it.

It was a brilliant sequence of insights, grounded in manufacturing practice. Noyce showed how "planar technology could be used to do the interconnection, and additional junctions could achieve electrical isolation" (the latter being "at least as significant," in Moore's opinion, "in getting Fairchild's integrated circuits going"). Noyce was drawing from his own immersion in electronics technologies and his observations of the foibles and challenges of the fabrication line. He was acutely aware of commercial reality, too: the US military needed to overcome the tyranny of numbers and would pay handsomely for a credible solution. Yet just as Hoerni's planar idea had been set aside in the face of more immediate concerns, so now Noyce was forced to postpone development of his concept.

Keonjian's request necessarily took priority. And even more urgent was the problem of the mesa's unreliability. Weeks of effort to minimize "crud" through a cleanup of manufacturing operations had proved insufficient. Gordon remembers, "We cleaned as much as we could and decreased the incidence of failure in the transistors quite a bit, but we never got down to zero." With the Autonetics order—worth millions of dollars and potentially their largest contract—on the line, Noyce and Moore had to come up with a better answer, fast.

Defection and Reorganization

Unhappily, they suddenly had an even more urgent problem to deal with: a stunning defection. March 1959, an anxious time, became much worse: general manager Ed Baldwin quit abruptly to start his own firm. Given that Gordon Moore and his cofounders were in the middle of implementing Baldwin's aggressive bet-the-company strategy, this was a major blow. Further bad news followed: many of the engineers Baldwin had brought with him from Hughes—including Moore's lead engineers in testing and preproduction—were going, too. Just as Gordon and seven colleagues had quit Shockley eighteen months earlier, Baldwin's group was resigning en masse, a pattern that would come to characterize Silicon Valley. Such "Fairchildren" would, over the years and through successive defections, create dozens of silicon electronics firms tracing their origins back to Fairchild.

Behind the scenes, Baldwin had lobbied aggressively for the largest individual stake in Fairchild Semiconductor. When these negotiations stalled, he sealed a deal with Rheem Manufacturing, a large San Francisco company that in wartime had made aircraft parts and ordnance and in peacetime maintained a military products division. Knowing the profits that were to be had, and wanting to capitalize on his knowledge, Baldwin convinced the entrepreneurial leaders of Rheem that he could produce silicon mesas. Rheem Semiconductor aimed to compete head-on with Moore and his colleagues. Baldwin himself would gain a major equity stake (short-lived, as Raytheon bought the start-up in 1961). "He wanted his own company," says Gordon. "He didn't consider our enterprise to be his. He never bought in or received a share of Fairchild."

Baldwin's dozen defectors included instrumentation guru Bernie Elbinger and the head of preproduction engineering, Dave Weindorf. Both reported to Moore, who was "absolutely shocked" to hear the news; his business education was proceeding apace. "I was very friendly with Weindorf. Things seemed to be working fine, so the defection completely surprised us. They set out to make the transistor that we had developed. They even had our manual of how to make the first transistor. We had some patents and a lot of proprietary know-how. We sued them, but these suits are never especially successful. When an engineer moves company, you can't do anything. When we left Shockley, at least we were going to do something he had abandoned."

The defection was shocking, but it was also an endorsement. Baldwin believed in the silicon mesa enough to strike out on his own. Since it would be months before his firm could produce an actual transistor, the immediate challenge was organizational rather than economic. The Traitorous Eight met for lunch to discuss the situation: "The issue was, what are we going to do now? Baldwin hadn't been much of a hands-on manager. He pointed us in the right direction, but by the time he left we had recruited a bunch of people to handle various areas. Did we want to risk bringing in somebody else, or would we muddle along by ourselves? We decided to muddle along."

By the spring of 1959, Fairchild Semiconductor had more than two hundred employees, including some possessing wide electronics experience and a wealth of specialist knowledge. However, even to "muddle along" required a redistribution of power and responsibility. The group decided to retain Baldwin's organization of departments, heads, and labels. The natural candidates for leadership were Gordon Moore and Bob Noyce, whose abilities had, from the earliest days in Shockley's laboratory, set them apart. A decision came easily. Noyce, the idea man and "outside guy," would become vice president and general manager, taking over Baldwin's role. Moore, who possessed the greatest breadth and depth of knowledge about the firm's manufacturing technology, would extend his responsibilities and become head of R&D, while continuing with preproduction engineering and quality control. A majority of the cofounders would now report directly to Moore, the firm's undisputed leader in the fundamental technology. In 1956 he had been a novice. Three years later, as he entered his thirties, he understood the territory better than anyone in the world.

Men and Booze

While most of the firm's employees were female manual workers, its management and researchers were exclusively young men, typically in their twenties or early thirties. The Valley of Heart's Delight was now attracting newcomers from afar. Some were from the Los Angeles region, but many—as in Noyce's case—were from the East Coast and Midwest. Employees from Asia and Europe, such as Lars Lunn, were not uncommon. How properly to utilize the newcomers was the problem.

Lunn recalls that David James, a "long, skinny Englishman" with a PhD, called at Charleston Road one summer day dressed in a "white T-shirt, white shorts, white socks, and tennis shoes. This was when we all still dressed in shirts and tie, so it made quite an impression." James was on his way through the Americas by Jeep and asked to be given a job when he returned from his travels. "Nobody expected to see David again, but he turned up some months later to claim his job. He ran into trouble between Columbia and Ecuador and was forced to give the Jeep to the local military and get out of the country." Later, James would defect and form Signetics, which would in due course recruit Moore's son Ken.

Ed Baldwin had brought to the firm's operations a certain organizational discipline, but by the summer of 1959 the structure was again in flux. Charlie Sporck, a key figure at Fairchild Semiconductor in the 1960s, recalls that in August 1959,

> I answered an ad and came to an interview in New York City, in a hotel room. Two guys are sitting at this table, with all sorts of alcohol stacked up. It was around eleven o'clock in the morning. I hit it off with them; they got further bombed and gave me an offer. I was making seventy-two hundred dollars a year. They offered me thirteen thousand. I accepted on the spot.
>
> My wife was amazed to leave the East Coast, but we packed up the kids and drove to California. When I arrived, they didn't know me from Adam. They finally recognized that they had given me a job offer, so they put me in a room along with the other guy they'd hired for the same position of production manager.
>
> It was complete chaos. There was no understanding of how to manage a manufacturing organization. That's one of the reasons why it was so flexible—it didn't have any structure. For a guy who had been at General Electric for nine years, this was a hell of a shock. Having sold up back east, I had to stick it out, and was very fortunate that I did.

Alcohol was a major motif in American business culture in the 1950s. In the fledgling and still-to-be-labeled Silicon Valley, it eased the loneliness of fresh arrivals, aiding bonding and fraternity. Fairchild Semiconductor was no exception. The "three-martini lunch" was commonplace among male staff and management. Drinking and eating were the mortar that cemented customer relationships. The firm's sales force were young male engineers, and so were its customers.

Fairchild Semiconductor's salesmen were particularly flamboyant, reveling in aggressive technical arguments and refusing to take no for an answer. As one put it, "We were all young and full of piss and vinegar." Tom Bay—large, commanding, and impeccably dressed—assembled a bright, ambitious team, attracted by the firm's meteoric rise. Jerry Sanders, renowned for his Hollywood-style yellow and purple suits, who would go on to found Advanced Micro Devices, a rival to Moore's Intel, says that initially in 1961, he had no interest in leaving Motorola to join Fairchild Semiconductor but agreed to an interview, provided it could be over a weekend: "I went out there and was blown away by the caliber of the people. These were supersmart guys. The notion of playing around in California for the weekend vanished. Instead, it was 'How do I land a job with these guys?'"

It was common to stop for after-work drinks at Rickey's or Walker's Wagon Wheel in Mountain View. These bars became hubs for information exchange, deal making, boasting, and revelry. In the early 1960s Fairchild Semiconductor even appointed a San Francisco–based "entertainment" representative, Bill Herzog, whose job was to take out and entertain visitors from the Midwest and the East Coast. Over time, sales meetings began to be held in more exotic locations, such as Hawaii. In one infamous incident, in 1963, onboard a regular commercial flight, Fairchild sales engineers hijacked the plane's liquor, leading to the president of TWA requesting that the company's employees never fly on his airline again.

Gordon Moore was no teetotaler, but neither was he an especial devotee of the three-martini lunch or beer and cocktails at the Wagon Wheel. Unlike his colleagues, he was a member of a rare species in Bay Area electronics: a fifth-generation area resident, married to a locally rooted wife. A quiet revolutionary, his life was in his work, and his recreation was with his extended family and the traditional pursuits of Pescadero's outdoorsmen.

That recreation was fleeting, as 1959 proved to be another hectic and seminal year. With ambitions to conquer the switching market and devices to develop, Gordon had plenty to keep him busy. His life had a different tenor from his East Coast experience at APL, with its lack of practical products, or his time with Shockley, when arbitrary management and confusion over goals had led to increasing disquiet. Now he was driven and preoccupied to an entirely different degree. Betty, enjoying her modern house in Los Altos, was close to her mother and grandmother and other relatives. In addition to coping with four-year-old Ken, she discovered that she was expecting again, with the baby due in October. Over the previous six years, she had lost three unborn children to miscarriage. This baby was one she desperately hoped to keep.

Planar Takes Off

Hoerni at last created a prototype of his planar transistor, and it proved to be everything he, Gordon, and their colleagues hoped for. The planar transistor proved immune to the crud that bedeviled the mesa. Hoerni's innovation—not Gordon's cleanup—was the answer. Moore and Noyce put their full weight behind manufacturing a planar transistor that would work *with* the crud. While Baldwin had quit to race toward the mesa, Fairchild was leaving it behind; it would go planar. Autonetics and others were keenly interested in this development, which promised the reliability and performance they craved. Just as he had toiled to meet the IBM order by August 1958, Moore now focused on getting the planar transistor into production, to secure the Autonetics order. Then—as with *Sputnik*—came a shot from left field.

Texas Instruments announced it was producing a "Solid Circuit," invented by Jack Kilby in March 1959. Unlike Noyce's still-imaginary planar microchip, Kilby's device had been realized. Its very different design (with interconnecting wires crossing above its surface) was destined to make its production difficult, limited, and expensive; nevertheless, the Dallas giant had forged ahead of the little upstart Fairchild Semiconductor. In swift response, Noyce sat down with Fairchild's lawyers to draft a patent application on his own planar microchip, which featured interconnecting elements with aluminum laid atop dirty oxide. By May Hoerni's patent application on the planar process and the planar transistor was filed. Noyce's application was filed the following month, extending the manufacturing technology for planar transistors to microchips.

It was crucial, once more, to produce an actual device, as soon as possible. Jay Last took the lead in the effort to investigate whether Noyce's planar microchip could be made and could work. He began in late July. TI was already talking up Kilby's work at great length. Fairchild needed to keep pace through a "microcircuit" announcement of its own. With Wescon less than a month away, Last stood no chance of realizing either the planar transistor or a planar microcircuit in time. Instead, he went back to his "Brillo box" approach, making the most of what lay at hand. He took one of the standard metal "can" packages, mounting not one but *four* mesa dies inside. To make them into a circuit, he used point-to-point flying wires, as in Texas Instrument's Solid Circuit, to connect the transistors to one another. Last then drew thick pencil lines of graphite, to serve as resistors, on the header of the can: ever after he called them his "Ticonderoga resistors," after the brand of pencil. The device took just three days to make and was able to store only a single "bit" of digital information (a zero or a one), but it was a working microcircuit, and it was ready in time for Wescon. At Fairchild Semiconductor's booth, it was front and center.

Fairchild Semiconductor was now six hundred strong, if mostly women performing manual operations. *Leadwire,* the firm's internal publication, wrote a glowing account of Wescon, avoiding any mention of last-minute efforts. "Gordon E. Moore noted that the major advantage of our design is that it allows the use of conventional mounting and wiring techniques while still effecting considerable savings in size and weight. Plans are underway to expand the packaged circuit concept." The packaged circuit shared the Wescon stand with a "small geometry" NPN mesa—the shrunken, fast-switching, low-power transistor that was all set to compete with Philco's switches. The Philco germanium transistor had been "the best thing around," remembers Moore, but its production was "a one-at-a-time operation and heavily automated, which meant it became completely inflexible." Compared with Fairchild Semiconductor's batch processing of wafers, by chemical printing, "their economics didn't work. It was not a technology that could be extended." With its main competitor "stuck in a dead end," Fairchild Semiconductor could begin to demonstrate its prowess.

The market was already vindicating the founders' belief that diffused silicon transistors would enjoy immediate acceptance. With its planar successor in preproduction, Gordon Moore and Bob Noyce could—given the emerging space race and the escalating stakes in thermonuclear deterrence—see their way to even greater success. The company's "accelerated program of research, equipment design and pilot production" had paid off. Its leadership in R&D was matched by its volume production of "premium grade semiconductor products." These facts did not escape Sherman Fairchild. In the original agreement, he had reserved the option to buy the firm out, turning this freestanding enterprise into a division of Fairchild Camera and Instrument. Fairchild Camera's own stock had risen sharply and was routinely above two hundred dollars. As a result, far fewer shares would be needed to acquire Fairchild Semiconductor. Exercising the option at this early stage—and at a lesser cost—made sense. On October 16, 1959, Fairchild Camera, a large, well-established defense contractor based on Long Island, New York, bought the brash start-up in Mountain View, California, that had earned it such acclaim.

The purchase brought significant change. On the one hand, success and extraordinary growth remained a constant: the second full year set fresh records, with sales increasing tenfold. The head count was approaching twelve hundred, thanks to the expansion of manufacturing in Mountain View and the opening of an additional plant fifty miles north, in San Rafael. There were plans to double the size of the Mountain View facility and provide R&D with a new laboratory. Noyce remained general manager and Moore head of R&D, preproduction, and quality control, with other founders staying in place. Nevertheless, the business and employment practices of the parent had far-reaching consequences. All staff now became Fairchild Camera employees, and Noyce was required to answer to Long Island executives. Stock allocations and options, anathema to these executives, were off the table. As one division among several, Fairchild Semiconductor—an upstart, based in the West, remarkably profitable, and highly entrepreneurial—was decidedly an oddball.

By November 1959 the firm's planar NPN transistor was moving out of preproduction to manufacturing, with yields improving. The work of Jay Last's microcircuitry group was also looking promising. Across 1960 task number one in the R&D department was to create planar versions of all the firm's transistors. In March the division announced its planar NPN. Tom Bay and his salesmen quickly secured an initial $500,000 order from Autonetics for the Minuteman ICBM's guidance computer. The order required Fairchild Semiconductor to establish a major reliability program; thereafter, the firm won $8 million of Autonetics contracts, confirming to every other military electronics manufacturer the desirability of the planar transistor.

Sales of Fairchild Semiconductor tripled to $21 million. Employment expanded yet again, and Noyce won approval from the head office for the planned addition to the brand-new manufacturing complex in Mountain View. On the western edge of the Stanford Industrial Park, overlooking the serene Stanford foothills, the firm also broke ground for a custom-designed R&D laboratory, five times larger than the original building at 844 Charleston Road. Alongside these expansions, the division reached into the European electronics market, buying into a germanium transistor venture in Italy, Società Generale Semiconduttori (SGS), which forthwith began making silicon planar transistors for sale in Europe. At the end of 1960, Fairchild Camera's annual report announced that it had been "the year of the Planar Process." The recognition was timely, as Moore's R&D lab proved it could actually make a planar microchip. And the microchip would open the way to the revolutionary implications that Gordon Moore would encapsulate in "Moore's Law."

Gordon at Home

The purchase of Fairchild Semiconductor by its parent changed the fortunes of the founders. Each had paid around $5,000 in today's money, just two years before, for their share; each now received what would be close to $3 million today. With prudence, such a windfall could provide a lifelong income; success also guaranteed career employment within the electronics industry. Not long before, Gordon Moore, the low-key boy from Pescadero, was taking pains to balance his modest salary as a junior researcher in a government laboratory against his and Betty's spending. Now, at thirty, he had a modest fortune and was the technological leader of the most innovative organization on the advancing frontier of silicon electronics.

At work, throughout these two years, Gordon had been at the center of a hurtling stream of excitements and achievements. Just twelve days after the firm's sale, Betty gave birth to their second child, at Stanford Hospital, undergoing an elective cesarean section as Gordon waited in the next room. Their son Ken recalls his grandfather Walter Harold driving him to visit at the hospital, all the while "talking about the birth." Like Walter's wife, Mira, Betty had hoped for a girl but gave birth to a second son, Steven, whom she took delight in nursing. Complications surrounding the birth precluded her hopes of a third child.

For Betty, affluence somewhat compensated for Gordon's long absences at work:

> When did it come home to me that we were deep into a materially transforming set of experiences? When I realized we were able to pay off the house. I felt warm and fuzzy inside. I said, "Don't we want to have a party?" Gordon said, "No, I don't think so!"
>
> I've never been a materialistic person. I'm not one to run to funky fashion shows in Paris, or to spend thousands of dollars each month on clothing at the Stanford Shopping Center. I fished in the same clothes. I didn't have to have the latest outfit.

Gordon's responses to the wealth and the birth were equally low-key. Ed Baldwin's defection had nudged him into a more central position in the firm, the planar microchip was in development, a second son had arrived, and now he was affluent. These events brought no overt expression of emotion and no change in course. He had long been a man sure of his identity, grounded in solid associations on the peninsula. For Gordon, with his dyed-in-the-wool frugality, wealth was not a distraction; instead, it enabled him to journey without concern for status or financial need, to focus more deeply on what he wanted to do. Money was something unambiguous—you either had it, or you didn't—and something that could be measured and graphed. Gordon's chart of his salary over the years from 1953 to 1961, and on, had a reassuring exponential nature. However, the territory that absorbed his imagination was not money but the uncharted land of silicon electronics. He saw his role as that of both a pioneer and a settler, understanding what that land was, what it implied, and what it could yield. At thirty he was too young to entertain a midlife crisis; he would simply explore and follow where his analysis led him.

Outside the office the low-key rhythms of his life—upheld by Betty—continued to provide ballast, refreshment, and rest. Though he could stop by for an occasional beer with colleagues and tolerated the realities of alcohol, revelry, and after-hours socializing, his daily rhythm was more in keeping with that of a chemistry professor than the hard-charging engineers unwinding at the Wagon Wheel. As Ken recalls, "Mom made sure Dad would come home for dinner. A special meeting was okay, but not as a universal. Mom was pretty hard-core on that. 'You have a family. Your job is not to work to death.' At the dinner table, you'd talk about your day. For Dad, it was always the same: went to the office, went to meetings, and read papers. The most important message this sent me was that dads go to work every day, without fail."

Over dinner Gordon caught up with his wife and interacted with his sons. Yet there was still work to be done. Ken, who turned six in 1960, witnessed how his father would take his briefcase to and fro, arriving home with "tons of work." "Fairchild was a really young, aggressive company and

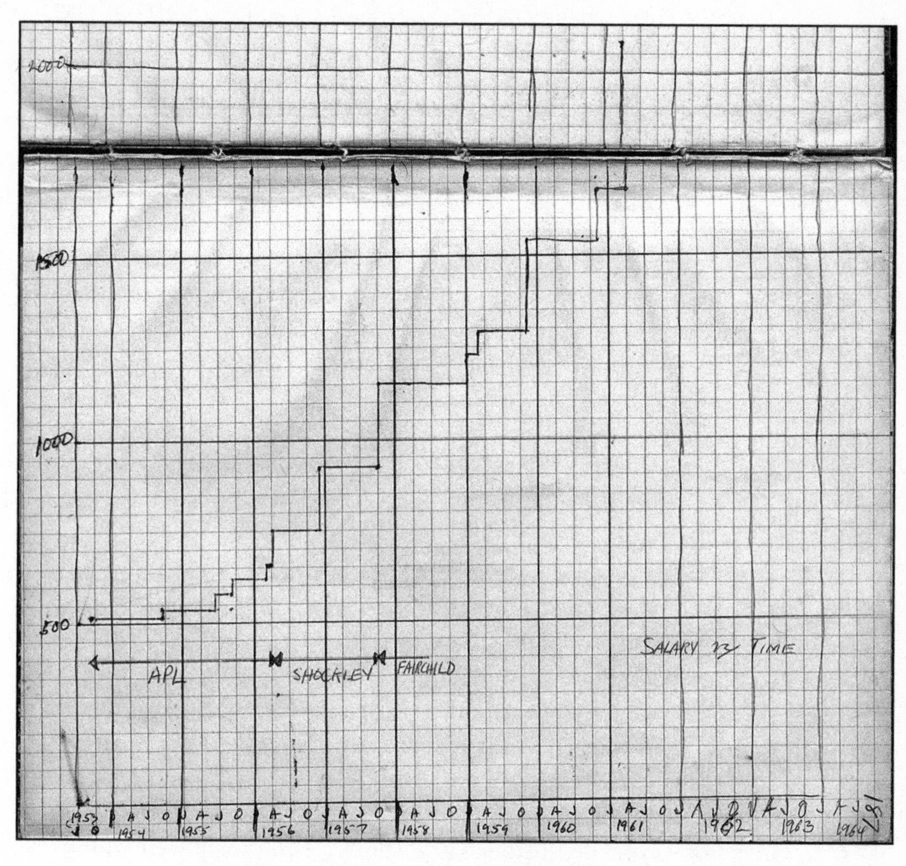

Gordon's charting of his monthly salary in his personal "ledger."

SOURCE: GORDON MOORE.

darn important. After dinner he'd process the daily mail. Then he'd go right back and work for a few more hours. He'd sit in our big family room. There was his chair and his pile of papers. He always says he's worked at least eleven hours a day, day in, day out, for his whole career; it's true. My father was always working."

On weekends, while Betty was with the baby, Gordon would necessarily make some time to play ball, walk, read, or go fishing with his first-born son. What enthralled Ken most was being taken to visit the Fairchild lab. "I remember going to the original building, on Charleston Road. It was pretty darn cool. They had all these oscilloscopes, microscopes, and technological doodads. He used to talk to me about technology. Once I started showing an interest in electronics, he was fostering anything relating to science."

If Gordon was comfortable talking about hard science, he avoided religion. On coming west, Betty had been keen to return to the Advent Christian Church in Santa Clara, where her mother, Irene Whitaker, still taught Sunday school. She even took the family on the church's summer camping trip, but her pregnancy with Steven, his birth, and a patch of ill health saw their attendance peter out. Betty then attempted to get involved in a local Congregational church, but found it too inflexible. "The church was so big, they had to have two services. I was told I couldn't go until the second session for the sermon, but they had our kids in the nursery at the wrong time. I thought, 'Well, this is not the army. Good-bye.' I didn't have any feeling toward anything that regimented." For Ken, Sundays were family days. His religious education came from his grandmother Irene. "My mom's mom was tolerant of people, even if they believed differently. She taught me the Lord's Prayer, but we were really sparse from that standpoint." Betty remembers, "My mom still had all of her little church-school booklets, and the children heard the historical Bible background and all the phrases."

Mary Metzler, Betty's grandmother, became bedridden following a fall at her new home in Saratoga and died in the spring of 1960, in her early nineties. Irene, who had passed up an opportunity to remarry, was now free to be with her own daughter's family and to make up for the time lost in Betty's youth. "I'd take my mother out for lunch. We'd go shopping. We did girlie things. Then as our resources grew, I was able to send my mom on trips with her sisters and her neighbor." With her mother also available to babysit, Betty began accompanying Gordon on business as the firm's reach and his obligations expanded through the 1960s: "I routinely went along, as he didn't like to take trips on his own." Ultimately, the delight in fresh places faded: "At first I enjoyed going, but Gordon would be caught up in business. I was often on my own and found that stressful. If it was just one person saying to Gordon, 'I want you to come,' there would be no social program, as there was for a conference. I would get to the hotel and have to figure out, 'Where am I going to go?' It was debilitating. After a while, I stopped going." Betty began to feel Gordon spent too much time traveling on business, but Ken was impressed. "Dad was taking trips because he was doing business; this was good."

If his mother was "hard-core" about Gordon coming home for dinner, she was also "strong" in ensuring her husband took vacation: a week as a couple and a week with their sons. As the boys grew older, the family would head into the Sierra Nevada, "to get Gordon out of his work routines," Betty recalls. Later, they bought a boat and "blocked out time" to go on fishing trips. Gordon and Betty mainly used weekends to visit extended family around the Bay Area. There were frequent trips to Los Gatos and Saratoga and to Redwood City and Pescadero, where Gordon would hunt deer and other game with his father and brothers. Ken, still young, would "play around" on his uncle Walt's farm.

Under Betty's watchful eye, family life was simple and unostentatious. Ken began to attend a preschool on San Antonio Road. Houses on his avenue were close together, with old prune trees dotting backyards and providing a reminder of the Valley of Heart's Delight. "All the kids would roam around and play. I recall helping mow the lawn, learning to ride my bikes, capturing bees in jars, and having fun with my friends. We had a lobster as a pet for a while." Thanksgivings and Christmases were family times. "Holidays would be at our house," recalls Ken, "Dad's mom and dad, my mom's mom, and the family. Mom's mom didn't like alcohol, so my parents had to hide all the booze every time she came over."

With the death of Betty's grandmother Mary, the couple decided to build a larger home. Betty had long loved architecture and home design and now had the ability to select her ideal location. She and Gordon scouted for land, considering a variety of options in the general area. "My mom was an obvious equal partner with my dad, if not more than equal," says Ken. "Many decisions were her decisions, like where we were going to live. There was no one side of those two, wearing the pants. It was and is situation dependent."

Soon a lot was found, on dead-end Jabil Lane in Los Altos Hills, an easy drive to Fairchild Semiconductor's planned laboratory and more rural than Los Altos itself, offering woods, valleys, and hill views. Superficially, the area resembled Pescadero, but in reality it was a center of burgeoning, affluent, suburban life, where neighbors would soon include a physician, an airline pilot, and a Stanford professor. The successes of semiconductor electronics, the spinouts and the start-ups, occasioned the first of numerous building booms. "All the houses on our street went in during a few years," explains second son Steve, who lived in the house for many years. "The owners of these houses were all professionals, and they all came in their early thirties. Everybody on the street had four kids, except my parents."

Betty hired a local architect and came up with "a really grand design"—an example of California modern, with an open floor plan of split levels and vaulted ceilings. The house was very much her creation. Ken describes a visit to see the as yet empty lot: "The Hills were still in the boonies. There was no Highway 280. The train ran down the middle of Los Altos, where the Foothill Expressway now is. For me, as a very young kid, it was a big adventure. Your parents are going to build a house. This is a big deal. Since I didn't have any school friends yet, it didn't bother me to move. It was fascinating, watching the house being built." Betty made many thoughtful design decisions, such as having the kitchen at the front with a view of the driveway: important for keeping track of two young sons. Gordon, for his part, had one stipulation: that his study should look out on a stand of California oak, a view that evoked memories of Pescadero and his youth.

In the main Betty coped alone, with Gordon off at Fairchild from morning until night. She ran "back and forth," dropping Ken at preschool and carrying Steve in a papoose, as she supervised the builders in her no-nonsense fashion.

> I would meet our contractors at the bottom of the driveway and check out the timber. If it wasn't A-1, I would say, "Sorry, back to the mill. I don't want it going up the hill to the house. I'm not having sapwood." The subs hated me, but it was in the contract and it was going to be right. I was a hard person to work for. I would watch every brick, every board. I like planning, I like style, and I'd been sketching ideas for years. A lot of my energy went into that house. Gordon and I were moving along. Life was as it should be. I started to meet with people from W&J Sloane's in San Francisco, because I wasn't going to move any of our old furniture into the new house.

Betty's work on the home's design was one signal of her assumption of responsibility for the family's private life. Her marriage was based on shared values, shared geography, and a clearly gendered approach to spheres of responsibility—very much of its times. Gordon, for his part, never doubted that his role was to be the breadwinner—providing as well as managing and controlling the finances. His sphere was the workplace, Betty's the home. There she could be "an irresistible force," according to Ken. In the designing, building, and furnishing of their home, Betty did "most of the pushing." She and Gordon lived in this Los Altos Hills house for four decades from 1961, even as their assets mushroomed, and son Steve would stay there after his parents moved out and begin his own family there.

Betty's priority was to keep the family close and secure. "My mom was always there for me," says Ken. "For technical questions, you went to Dad, but for everything else you went to Mom." The structure was straightforward: "Mom's job was shopping and cooking, and Dad's was to earn money." For a while, Betty shared a cleaner with a neighbor, but ultimately she preferred to be self-sufficient and marshal her husband and sons to the task. "Dad would vacuum the floors and rinse the dishes after dinner." More than a half century later, this latter reality still holds.

Betty also dealt with health matters and the care of elderly parents. Baby Steve's development was of particular concern, as 1961 went on. "He wasn't walking. I was carrying him on my hip. We didn't know what was wrong with him. He was eventually diagnosed as having an allergy. I had no idea that an allergy could make a child immobile for so long. With the right antihistamine, Steve was able to get up and walk, but that was not until he was about two."

Ken gained a place at the local elementary school. As time went on, he became the "school crossing guard" and was proud of being picked to run the flag up the flagpole in the morning. He loved school, joining the glee club, learning the clarinet, and eventually playing baseball. Back at home, there was potential to roam and, following in his father's footsteps, to make loud noises. "The homes had no fences; kids just roamed. There was less population, more open space, and not enough traffic to worry. You could go anywhere, do anything. I used to shoot my .22, and it would echo across the valley. A neighbor would say, 'Hey, what's going on there?' I'd say, 'Oh, just plinking.' I'd be on our front deck with what's known as a Big-Bang cannon. It generated acetylene gas and went 'Boom!' I'd sit out there for hours making these huge noises. It was fun."

Gordon and Betty made decisions in their respective domains, without much discussion. Gordon preferred to keep his own counsel about business and disliked discussing his work when home. Betty herself—accustomed to solitary life as a child, yet vivacious and energetic—wished her husband's absences were less frequent and at times had real difficulty in accepting the situation. Other people's lives intrigued her and helped to distract from her sense of isolation. Her close confidante was Rose Kleiner, the wife of Gordon's cofounder Eugene, but she also sought companionship from her sometime cleaning woman, who "came to be like family." She says:

> For a woman who is left to raise a family, the schedule of the house is not as important as the man's job. I had to make it work. Later on, Gordon was never there, even for the kids' birthdays. I thought, "There's always something more important than family to him." I would soothe the waters. Other families we knew had their own problems. Gordon never worried about household matters. He was always telling me that he was too busy and his time was too valuable. I'd reply, "All right. I'm handling it." I had to say: "This is my world and he has his." Gordon was always totally engrossed in his work. I know what I have—it's not too bad—but it has also been challenging because of the aloneness.

6

EMERGING REALITIES

SETTING A FRAME

More Defense for the Dollar

Gordon's world was Fairchild Semiconductor. In November 1959, as part of his new role within what was now a division of a large East Coast corporation, he sat at his desk to write a plan for his R&D lab in the year ahead. Its subject was as straightforward as its words would be revolutionary. He began with first principles, postulating four major responsibilities—principles from which he would not deviate in the next eight years. Gordon defined his tasks as supplying a flow of fresh devices to production, investigating how to extend established markets, improving the chemical printing technology, and enhancing the company's technical reputation, to attract top talent in a climate of negative unemployment. These simple, ambitious formulations would underlie his career, Moore's Law, and the electronic revolution of the next decades. Broad in their scope and implications, they capture how far Gordon had traveled in the years since leaving Berkeley.

His immediate priority was to fill out a family of transistors by making planar versions of the mesa and introducing additional complementary planar transistors—following through on his "philosophy of standard products." In addition, Gordon set Fairchild's sights firmly on "microcircuitry," that is, the planar microchip. In line with this aim, he first glimpsed the elements of a view that would eventually overturn received wisdom and change the game forever.

It was widely believed that small was expensive and that deliberately shrunken microcircuits (requiring greater technical skill for manufacturing) would always be costlier than circuits made by the wiring together of larger individual components. Although over a system's lifetime microcircuitry might lower total expense (thanks to economies in operation and repair), microcircuits were in and of themselves more expensive to produce. They were an exotic novelty whose cost made sense only within the context of exigent military needs.

Gordon now began to suspect that this received wisdom was wrong. Reality could be changed. If Fairchild Semiconductor succeeded in chemically printing planar microchips in a manner similar to planar transistors, it could disprove the "fact" that smaller always equaled costlier. The microchip would itself not only be smaller, but also both better *and* cheaper to make than an equivalent circuit of components wired together. Microchips could become cheaper than conventional electronics to *make* as well as to maintain. The implication was startling: the planar microchip could take over the whole of electronics, even as the whole of electronics fell in cost.

Moore articulated his insight, while planning ahead for heavy spending on the microchip; the novel realization was in his *why: why* would planar microchips conquer? "Micro-circuitry is an important trend," he began. "A most important point concerning our approach that has not been recognized by others is that the manufacturing cost for complete circuits," that is, planar microchips, "will be less eventually than it is for the conventional components necessary to duplicate the circuit's functions." A planar microchip would be cheaper to make than the *individual parts alone* for a conventional circuit, to say nothing of the cost of assembling them. And because "our transistor techniques are applicable to the production of microcircuits, sets of these components will be able to replace ninety percent of all circuitry" in digital computers.

As Jay Last's team worked to build a planar microchip, Moore saw that microchips could lead Fairchild, and the entire electronics industry, into a major revolution. The microcircuit could turn the world upside down. His written words in November 1959 mark a milestone in the history of technology, but until his lab succeeded in actually making microchips, it made no sense to broadcast his insight: first "I had to make these things work." Upon completing the R&D plan on November 5, he simply sent it to Noyce and to Dick Hodgson (the Fairchild executive to whom the division now reported) in Long Island.

Moore and his colleagues were acutely aware that Texas Instruments had announced Jack Kilby's "Solid Circuit" concept and was working toward the launch of its first commercial device. Fairchild quickly devised its own brand name—Micrologic—for planar integrated circuits (ICs). Both Texas Instruments and Fairchild marketers hoped their brands would become synonymous with the product, as with Xerox or Frigidaire. In February 1960, Bob Norman from the Micrologic team went to Philadelphia to the Solid State Circuits Conference, *the* place to present innovations in device production and use. The conference offered opportunities for young male attendees to gauge competitors' progress, understand frontier realities, and appraise possibilities. Four or five years before, the buzzword had been *diffusion;* now, all the talk was of *microcircuitry.* Norman, nicknamed "Mr. Logic" to Jay Last's "Mr. Micro," joined a panel of researchers from major players such as Texas Instruments, Westinghouse, General Electric, Motorola, MIT, Bell Labs, and the Stanford Research Institute.

Norman laid out Fairchild Semiconductor's plans to introduce a family of microcircuits for digital computers and other digital systems. In industry parlance then, as now, digital circuits are often called "digital logic," or "logic functions." The reason is the congruence between binary information—information turned into zeroes and ones—and the "If A, then B" formal logic of mathematicians and philosophers. Digital electronic systems process data by doing formal logic. Norman told the experts in Philadelphia that Fairchild was about to introduce "complete logic functions within a transistor package." He emphasized the potential of Micrologic circuits to reduce assembly cost and increase reliability, but did not yet publicize Gordon's revolutionary idea that the planar microchip might itself become the cheapest form of electronics, period.

By the end of May 1960, the Micrologic team had succeeded in producing planar microchips. However, the yield was shockingly poor, making routine manufacture impractical. In addition, they were horribly unreliable. Moore spent considerable time in his office pondering the issue—reading progress reports, technical publications, and the trade press and analyzing what he was learning. Last's team working on the integrated circuit had a variety of ideas.

Technological challenges like this were central to his view of his responsibilities in a division that was growing rapidly. The difficulty lay in balancing two requirements that were largely at odds: creativity and focus. He was coming to understand that novel ideas were very delicate and that "if you make somebody justify his idea before he tries anything, he usually gives up," whereas if "you give enough freedom, they go out and try an idea before they have to defend it to somebody." The creative work of innovation required "freethinking about novel ideas" and "a reasonable amount of flexibility" to try things out, even if the boss was skeptical. It was also essential to maintain focus and make good investment decisions, to "filter ideas and put enough effort in to ensure development into product."

Micrologic was a challenge through which Moore would learn how to be a somewhat more adequate manager, not simply an ace researcher. He slowly evolved a set of priorities—flexibility, care for delicate ideas, maintenance of focus, and selection of investment-worthy projects—that made his management style nonconfrontational and open, but not simply bottom up. While he expected his researchers to come up with ideas, he also analyzed these against his own understanding of the technology, its direction,

A Micrologic silicon microchip, with ant for scale.

SOURCE: FAIRCHILD.

and its economic consequences. That meant applying his preferred methodology ("measure, analyze, decide") to any idea. Those that came through would receive resources and manpower. A good idea alone was not enough: Gordon, the pragmatist, needed to be convinced. As Leslie Berlin, Noyce's biographer, puts it, "Where Noyce admired the inspiration, Moore foresaw the perspiration."

As sample quantities of planar microchips became available, Moore and Noyce were enthusiastic but cautious. Was the approach practical? Were the yields adequate? How reliable were the devices? Last's Micrologic group focused on finding the answers. Using a new diffusion process, they succeeded in printing microchips. "I've never forgotten the incredible experience of working with Jay and his people as we solved one problem after another to make Micrologic work," says Bob Norman, recalling issues such as "purple plague" and "tweezer scratches."

Soon, Norman was able to promote the team's success to an important gathering of "solid state" experts. He recalls being "in suspense because I could not present the paper unless we had working parts, and when we had, I did." Military contractors were deeply interested in the size, weight, power, reliability, and cost benefits offered by microcircuitry. Now Norman could publicly articulate Gordon's insight that planar microchips could become cheaper than conventional circuits, reducing the cost of manufacture as well as ultimate system cost. "Translated into English, the net result is more defense per dollar," he told the assembled experts.

By December 1960 Gordon Moore was fully confident that the effort by the Micrologic team would open the way to his cost breakthrough. Three months later, in March 1961, Fairchild Semiconductor announced the Micrologic family of microchips, successfully deploying diffused silicon transistors, by now everywhere seen as the reliable "bricks" of the electronic world, in the formation of complex microchips. These microchips then became the building blocks for digital systems and computers. "Noyce and Kilby are the coinventors of the integrated circuit," says Moore. "Jack made one by laboratory techniques. Bob showed the way to a practical product." Hoerni's contribution was also key, as "the technology we had at Fairchild became the path to the practical integrated circuit." Moore says, "I always measure it from the first planar transistor rather than from the first integrated circuit." The real electronic revolution had begun, but as yet not even Gordon could grasp the enormity of what was about to happen.

A Silicon Valley?

If Gordon Moore had a widening perspective on the microchip's potential to change the world, that of the firm's Mountain View office in Whisman Road was utterly blinkered. Its focus was on getting discrete transistors—basic "bricks," not exotic "building blocks"—to eager customers. Planar transistors were a hit; the sales organization had already achieved a massive profitable win in the guidance computer for the Minuteman missile. Helter-skelter growth, proliferating empires, and the earlier decision physically to separate manufacturing from the research labs were, in the words of one commentator, "the beginning of trouble." In 1960 the division had split not only physically but also psychologically. Fairchild's custom-built manufacturing plant and administrative offices, the "Iron Bucket" in Mountain View, were separate from Gordon Moore's organization, now occupying the whole building at 844 Charleston Road. "When Fairchild started to get successful," says Carver Mead, a Caltech professor who consulted for the firm, "they separated out the research labs. That never quite worked." Still greater separation was in store, with R&D's planned move to its larger site in Stanford Industrial Park, several miles away.

The Whisman Road complex was undoubtedly the mainland of the division, expanding steadily and absorbing workers, tools, and materials alongside top managers and major functions, as it ramped up production.

The manufacturing organization began to generate its own "R&D" activity, including production engineering, applications development, and device design. Frequently, this activity overlapped and competed with work done in Gordon's groups. And Moore was not the man to confront and challenge the change.

Moore's R&D represented the division's future. While his staff would extinguish frequent fires in routine device manufacturing, his primary responsibility was to keep in the vanguard of a fast-changing, highly competitive industry, through chemical printing and novel devices. Conversely, the work at Mountain View centered on immediate concerns. Charlie Sporck, the tall, hard-driving Cornell engineer hired in 1959 as product manager and now head of manufacturing, was introducing to Fairchild's booming plants, labor force, and products the discipline he had learned at General Electric. The name of the game was to maintain yields and volumes and build production. Sales engineers, too, dwelled in the present. They needed to book orders to meet targets and keep customers satisfied with actual deliveries. High-volume production ruled and required their urgent, continuous attention.

Gordon had foreseen that planar microchips could become better and cheaper than conventional circuits. This led to the strong, shared conviction within his R&D organization that it had achieved something of fundamental importance. The Whisman Road manufacturing and sales organizations did not agree. Their ambition was to saturate military electronics with silicon transistors and then to displace germanium transistors from commercial and industrial electronics. Microchips did not fit in. Would they be reliable? Affordable? Might they make their clients' engineers, who designed their own circuits, obsolete? It was a puzzling development, one about which customers and Fairchild salesmen were equally skeptical.

Late in 1960 this difference of opinion surfaced at a meeting that became heated when Tom Bay, the influential and physically commanding head of sales and marketing, lambasted Jay Last publicly for "wasting a million dollars" on Micrologic instead of supporting the immediate business: "Why the hell are you working on integrated circuits? Get these transistors straightened out!" By designing complete circuits, the firm was competing with its customers. Bay wanted to shut down the Micrologic program, a program within Moore's R&D organization. Such vehemence made Last, among others, feel that at Fairchild the microchip was overshadowed, even doomed. Moore was not at the meeting, and in any case his reflex was to avoid difficult encounters. "I didn't know Bay had gotten around me to do that," he recalls. "I was Jay's boss, but I was unaware of this."

In retrospect, Moore admits to "immature management." He preferred to direct by indirection, asking questions and then making investments. He hated loud, in-your-face confrontations and had learned to take pains to avoid being on either end. At Shockley Semiconductor, when summoned by Shockley himself, he had coped by straightforwardly telling the truth. Since no argument was involved, stating the facts had been sufficient. The situation with Last was far less easy to negotiate.

Avoidance and lack of communication might be Gordon's preferred modes, but in a competitive, growing organization, they carried a high price. Jay Last understood that it would be no good turning to Moore for direct support. Last could, however, vote with his feet. He concluded that Micrologic was only a "minor R&D thing" compared to the firm's "extreme success" with its transistors and diodes. Micrologic faced "benign neglect: the money was there, but nobody cared about it." In grammar and high school, and in job interviews and tests, Gordon's reticence in expressing himself had set him back. Now, despite being a superlative technological strategist, his failure to master certain management realities meant that he would be set back once more.

Moore believes that his failure to support Last was his "single biggest mistake at Fairchild." He modestly says that he too did not fully appreciate "how big the business for integrated circuits would be. We had barely scratched the surface of a technology that would be so important. It was just another product completed, leaving us looking around for a fresh device to make."

Gordon's mistake had disastrous consequences, leading to the disembowelment of his Micrologic group by not one but two sets of defections. The first was by far the most painful. In January 1961 Jay Last and Jean Hoerni quit, tearing apart the group of cofounders. For Last, Bay's public attack had been the last straw. At the urging of Arthur Rock and Bud Coyle, he agreed to meet with Henry Singleton, a talented electrical engineer who had cofounded Teledyne, to create electronic systems for military aerospace. Singleton, Rock, and Coyle proposed that Last and Hoerni launch an operation under the aegis of Teledyne, focusing on silicon microchips.

Last felt obliged to let Moore know the conversation was happening, but he downplayed it, saying, "Probably nothing will come of it." Something *did* come of it, and quickly. Singleton's technical brilliance and business ambitions impressed Last and Hoerni, while Singleton knew he had found the right players. "Singleton needed them, and I wanted to build them," explains Last. Never again would he have to justify his interest in microchips. Hoerni, similarly, was bubbling with dissatisfactions. Together they resigned to set up the Amelco division of Teledyne, in Mountain View. Last and Hoerni became vice presidents, with large equity stakes. Sheldon Roberts soon joined them. Eugene Kleiner also left Fairchild and began to consult for Teledyne. Other colleagues, including Lars Lunn, joined the exodus to a venture they perceived as more exciting.

Gordon was amazed and hurt and took the defection hard. His regard for Last had been strong, and he had made significant investments in man-hours, equipment, and materials for the Micrologic program. His mistake was in failing to understand the psychological realities with which two of Fairchild's cofounders, once his equals, grappled. Last and Hoerni felt they were now "two levels down" in the organization, and Last in particular was not fond of "somebody being my boss." The lure of a fresh venture, combining technological promise with opportunity to amass further personal wealth and renew a sense of ownership, proved compelling. Gordon belatedly got the message: "We'd started out as eight equals, and then Noyce and I became more equal than the others." In characteristic understatement, he concedes that this "had some influence on Last and Hoerni's decision to leave." The Amelco break was symbolic, splitting the Traitorous Eight exactly in half.

In the second defection, a couple of months later, another important group tore out of Moore's Micrologic program when microchip pioneer Lionel Kattner and transistor expert David Allison, along with engineers David James and Mark Weissenstern, left to create Signetics. Their goal was to produce microchips for mainframe computers in business use. New York investment bank Lehman Brothers assembled a consortium of investors to launch Signetics, which was headquartered in Sunnyvale, to the southeast of Mountain View. Both spin-offs, though not privy to Gordon's groundbreaking analysis of late 1959, believed that Fairchild Semiconductor, with its focus on transistors, was missing a trick. "I had a pretty good insight that we were really on to something phenomenal," says Kattner. "I left because I knew it was going to be an exciting time."

Launching out from Fairchild Semiconductor held potential for high returns and was appealingly low risk. Negative unemployment provided an ideal climate for entrepreneurial ferment. Switching jobs could be seen positively, as a source of needed experience, rather than as a sign of personal untrustworthiness. There was little downside in an attempt to create an enterprise of your own. If you failed, someone else would always hire you. Nor was company loyalty necessarily a virtue; it could be a millstone. The option to stay with the Micrologic group was low risk. However, thanks to policies set by Fairchild Semiconductor's parent company in Long Island, those whose work was opening a new frontier were not themselves in line for any stock options. Loyalty to Fairchild would actively exclude them from earning the kind of financial windfall that Moore and his cofounders had themselves enjoyed.

Spin-offs were beginning to be common, and copied, in the Valley of Heart's Delight. Many realities were coalescing. One was the "virtue of propinquity." Just as the dissidents from Shockley Semiconductor had moved less than two miles, the microchip spin-offs from Fairchild Semiconductor were siting themselves nearby. Company founders were most often young—family men, with children in school. Staying local made sense, especially given the area's attractive weather. Then there were the human contacts and the growing infrastructure of local suppliers. Real estate (orchards, vineyards, and grazing land) was cheap. A host of mutually reinforcing individual choices fed directly into the development of the region and gave momentum to its transformation. While the spin-off activity stung Moore, indirectly it enhanced his firm's standing by endorsing the area's emerging claim to be the world's premier center for silicon electronics.

If spin-offs and start-ups were becoming the order of the day, so too was the use of venture capital—a term coined by Benno Schmidt, an associate of the John Hay Whitney firm—to fuel the industry. Investors on the East Coast were directing their attention to high-risk, high-return efforts in defense-related technologies. Georges Doriot, John Hay Whitney, and Laurance Rockefeller dominated early investment in what Wall Street began to label "high technology." In 1958 the US federal government became a significant player through its Small Business Investment Corporation program, enabling financiers to leverage their own capital through government-backed loans. Some seven hundred SBIC firms, including many storied names, were established by the mid-1960s.

Another significant step came in 1961 with Arthur Rock's relocation to the West Coast. He was tired of endless red-eye flights, had spied fertile ground in the Bay Area, and decided to resign from his New York position at Hayden Stone. In San Francisco, with Tommy Davis, he created the region's first venture-capital partnership. The pair raised a $5 million fund, designed to run until 1968, to invest in high technology, especially defense electronics. Six of Fairchild Semiconductor's founders made significant investments in this fund, but Moore and Noyce, having asked Fairchild management for permission to participate, were told in no uncertain terms they could not. Since Rock had orchestrated the recent Teledyne-Amelco spin-off, this was hardly surprising, yet the investments of their peers marked the inception of a powerful dynamic on the peninsula, as the wealth derived from the first wave of high-tech development was reinvested to create subsequent waves. Gordon Moore comforted himself by investing privately with Rock's close friend Fayez Sarofim. (This relationship would continue for more than five decades, with Sarofim's investment advice proving highly profitable for Moore.) Later, he would also invest in Rock's second and subsequent partnerships.

As in the explosive economic expansions of California's past, a unique mix of ingredients fueled the early days of an embryonic "silicon valley." Venture capital, negative unemployment, benign climate, and alert academic instigators such as Stanford's Fred Terman: all favored risk, experimentation, and a life lived on the edge of novelty. And, as in the past, the generous ratio of land to population enabled rapid growth. The easy conversion of orchard, field, and pasturage into residential and industrial real estate in the Valley of Heart's Delight mitigated local price inflation, while the promise of entrepreneurial equity in a start-up mitigated wage inflation. As before, California remained extraordinarily well placed to nourish pioneers with ambition and stamina.

New Directions and Dynamics

Gordon Moore kept watch on outside developments. Every summer he attended the Solid State Device Research Conference (SSDRC)—where Bob Norman had laid out Fairchild's plans and where intense technical competition combined with young male revelry: "There were no old people. It was a new industry." One year he was part of a group who presented a set of limericks—witty, mocking, and off-color—at the concluding banquet on the Pennsylvania State University campus. Alcohol consumption fueled a contest about how far the industry's leading lights could urinate from the hotel's balcony. Unsurprisingly, the conference was not encouraged to return in subsequent years. Aside from such occasional excesses, Moore was always quietly gathering intelligence: "A tremendous amount of informal information was exchanged. People were sitting around the bar trying not to tell everything, but enough to show, 'Hey, we're pretty knowledgeable!' We didn't consider Bell Labs a commercial competitor, only a technical competitor. We were more tight-lipped with Texas Instruments."

In the summer of 1960, in the calm before events tore apart his Micrologic group, Moore listened to two Bell Labs presentations at the SSDRC having a direct bearing on Fairchild's technology. The first outlined work using a novel chemical process, epitaxy, to build a silicon transistor with superior switching speeds. Epitaxy was a chemical reaction that enabled the growth of layers of silicon atop a silicon wafer, perfectly matching the wafer's crystal structure. Such epitaxial layers could be doped as desired: P- or N-type, highly or lightly. Here was a fresh way to create varied layers of doped silicon, leading to better transistors and microchips.

Moore was already aware of epitaxy. His enthusiasm had been kindled in early 1960 by a visit from Merck, a leading pharmaceutical and fine chemicals firm in New Jersey, which had licensed patents to make ultrapure silicon, intending to provide the growing semiconductor industry with wafers. Merck had come upon silicon epitaxy accidentally and, seeing a business opportunity, had sent engineers to visit semiconductor firms and offer wafers with a variety of doped layers. In actuality, Merck was not ready to deliver its wafers commercially, despite several requests from Moore. Conservative with funds, he did not find the alternative (plowing time, money, and manpower into creating his own proprietary process) sufficiently compelling.

In contrast, at Bell Labs, Merck's visit spurred action: if Merck could do epitaxy, it could, too. Quickly patenting its own epitaxial transistor, the Bell Labs group presented it to the SSDRC in 1960, forcing Gordon's hand. He would have to develop and incorporate epitaxy into his planar manufacturing technology. In the event, since diffusion furnaces could be modified to provide the necessary tools, the barrier to entry was low, and his laboratory quickly established its own process. By February 1961 Fairchild's semiconductor division was providing sample quantities of a faster-switching epitaxial transistor to its customers. Epitaxy rapidly swept the entire industry.

The second Bell Labs presentation at the SSDRC debuted yet another novel silicon transistor, developed by Martin Atalla and Dawon Kahng. For several years, Atalla had been investigating the interface between silicon oxide layers and the underlying silicon, following a similar intellectual program to Moore's in looking at how oxide layers protected and electrically stabilized the silicon surface. The device he and Kahng described, a metal-oxide-silicon (MOS) field-effect transistor, was a radical departure from traditional transistors. The oxide layer atop the transistor was covered by a metal film, or "gate." Applying a charge to the gate created an electrical field—a "field effect." Like turning a faucet, this effect could start or stop current from flowing.

The appeal of the MOS transistor lay in its simplicity and the ease with which it could be made very small. Moore was intrigued, seeing implications for miniaturization and for manufacturing cost. By the summer of 1961, he had decided to invest in an exploration of MOS transistors. As with the microchip, he wanted his lab to extend its technology to novel possibilities. "We knew more about oxides than anybody at that time," he recalls, "so we started trying to make these things." Meanwhile, Fairchild's manufacturing organization belatedly adopted planar microchip technology, though yields remained worryingly low. Despite the challenge of spin-offs, the laboratory maintained its reputation as a global leader, with Gordon increasingly the champion of the planar microchip. One eloquent tribute to the technology's superiority came as Texas Instruments abandoned its own "Solid Circuit" designs in favor of planar microchips.

In 1962 the Fairchild R&D organization moved into the much larger home Gordon had designed for it at the western edge of the Stanford

Industrial Park, at 4001 Miranda Avenue. Thanks to his calm focus and unwearied assiduity, and experience ranging from the NPN mesa to the planar microchip, Moore had by now become the world's foremost expert, steeped in every aspect of the manufacturing technology and its use to make transistors, diodes, and microchips. No other living person had his scope of knowledge about, and access to, research on these technologies. The territory was his.

Gordon's team succeeded in adding epitaxy to Fairchild's manufacturing technology, late in 1961. This time Mountain View quickly took on the change, seeing dramatic alteration in its ability to deliver microchips to customers. Charlie Sporck reportedly boomed, "Good God Almighty! This epitaxy stuff is yielding fifty times what the old process did!" Meanwhile, Tom Bay quipped to Sporck that manufacturing had "ruined the market" for the Micrologic line. "You were so bad for so long," Bay explained, "that no one will believe we can actually make these things." Gordon was pleased with the change. "The epitaxial structure gave us a way to make these isolated regions for integrated circuits relatively easily. Instead of twenty-four hours, a two-hour diffusion was sufficient. That was a very important piece of technology that came along at the right time."

Success was vindicating Moore's approach. In running the laboratory, he worked by setting questions for which his colleagues would develop answers. This reflected the approach of Mervin Kelly, the renowned president of Bell Labs in the 1950s, who believed that finding good questions was much more difficult than generating good answers. Moore and Kelly were not unlike professors directing large academic laboratories. They formulated ideas, selected individuals or groups to pursue them, gave feedback or challenges, and facilitated independent curiosity-driven efforts that resulted in valuable results.

Moore's insight that the planar microchip would offer more defense for the dollar because more electronics for the dollar was slowly taking on resonance. When Bob Norman spoke at a conference of the US and Western Europe's top military and industry experts, in Oslo in the summer of 1961, he stressed that silicon planar microchips would "not only yield high orders of system miniaturization but also increasing reliability, while significantly reducing equipment cost." Smaller was better *and* cheaper. Smaller, better, cheaper would be the future.

Cost was one thing. Urgency of demand was another. Spaceflight by the United States and USSR had become a proxy demonstration of the power of missiles. In 1959 the United States had started Project Mercury with the hope of getting a US citizen into space before the Soviets. Instead, the USSR triumphed once again. In April 1961 Yuri Gagarin orbited Earth in a cramped capsule. Making a gambler's riposte, on May 25, 1961, President John F. Kennedy announced the Apollo program, with the goal of a US moon landing by 1970. Kennedy also thrust into consciousness the possibility of nuclear annihilation by missile, calling on US citizens and communities to build fallout shelters. (He had his own presidential shelters built in Palm Beach and Martha's Vineyard.) No one put it in these terms, but transistors in the form of microchips were now a central factor in what people perceived as the choice between ending and continuing human civilization.

Moore's microchips were selected for the onboard guidance computer for the Apollo program: quite literally, Micrologic would guide man to the moon and back. It was a major design win, with significant symbolic and market implications. By 1965 the Apollo computer was the largest single consumer of microchips, with $5 million of sales (200,000 chips) for Fairchild. The transistor was also beginning to find wider commercial uses. To cater to the expansion in television channels, the Federal Communications Commission mandated that all television sets receive both VHF and UHF transmissions. Silicon transistors (for which the military often paid upwards of $100 apiece) had the specifications to do the UHF job. Fairchild's star salesman, Jerry Sanders, presented the opportunity to Noyce. The market for UHF tuners required millions of transistors, but East Coast competitor RCA offered a traditional vacuum tube that, at $1.05, could do the job (albeit with less reliability, larger size, and greater power demand) at low cost. Noyce told an astonished Sanders to offer Fairchild's transistor to the market at the same price.

The move was bold and changed the game again. Moore and Noyce, having seen Ed Baldwin's prediction come to fruition, understood how the cost of manufacture could drop substantially over a production run; it was therefore appropriate to factor in confidence in future technology and lower costs into pricing. Sanders secured orders for 2 million transistors at $1.05 apiece. This was in striking contrast to the first deal Fairchild had struck five years earlier: to supply 100 transistors at $150 each. Cost would exceed price initially, but the deal would become profitable over the lifetime of a multiyear contract. The experience helped to cement an enduring strategy: to open fresh consumer markets by lowering prices. Silicon transistors became ubiquitous in TV sets as well as in nuclear missiles. Gordon observed that, as time went on, "the commercial market just completely dwarfed anything in the military."

With the departure in 1961 of Last, Hoerni, Roberts, and Kleiner, Gordon Moore and Bob Noyce grew closer, despite being in the same place at the same time only rarely. Moore spent his days in his R&D laboratory, while Noyce was based in the administrative and manufacturing complex in Mountain View. Together, they ran the division and enjoyed a solid, easy relationship based on complementary skills. Moore rated Bob not just for his technical judgment and ideas, but also for his personal charisma, "flair for corporate politics, and patience for dealing with nontechnical people." Noyce dazzled even the flamboyant Jerry Sanders and, as a leader, was "head and shoulders" above others in the industry. Gordon appreciated his creative style, in particular his bold solutions to problems and his success in dealing with customers.

Noyce, for his part, valued Moore's judgment, his measured and dogged approach, and his calm, self-possessed nature. He appreciated Gordon's insights into the technical and economic possibilities of silicon technology and his steady response to radical technological change. Moore was the inside man, directing investments to steer the division's future, while Noyce was the outside man, liaising with customers and senior management back east and devising strategies to open markets. Moore could entrust the outside world to Noyce, and the latter could rely on Moore for the very foundation of their enterprise: new technology. Fulfilled and respected in his sphere, Moore was untroubled by Noyce's obvious star quality, since this was the very thing that gave him space to operate. Far from competing, the two men were allies.

With their wives—both called Betty—they would get together for dinner on occasion. "We had more to talk about with them than with others," says Betty Moore. The Noyces moved to Loyola Drive, near Los Altos Country Club and close to the Moores' new home. "They moved to a very large house with orchards and this wonderful playland and water-sports area," recalls Betty. "I called it Disneyland North." Ken Moore was impressed by Bob and Betty's only son, Bill, who was older "and had all the neat toys," such as radio-controlled airplanes. "Plus, his dad had an oscilloscope, which to me was a big deal." Betty Moore was less impressed by Bob's wife, finding her impatient and uninterested in school events.

Whereas the Noyces' marriage was stormy, that of Gordon and Betty Moore was far calmer. Gordon remained too busy with work to be "around," as his own father had been. Still, he enjoyed doing things for and with his growing sons when he could. Noticing how Ken and little Steve would slide down the slope outside the Los Altos Hills house on cardboard boxes, he built them a soapbox-derby coaster. It had "an open frame with a simple steering wheel and pedal brakes," remembers Steve. "He found some Teflon plates, which was a big deal; getting these plates made the thing steer easily. We'd go down our driveway and down the street, time and time again." Seven-year-old Ken was already developing an interest in electronics and liked to fix things. He and a neighborhood friend "built everything out of cigar boxes. My grandfather smoked a lot, and we'd get his boxes. Electronics was always in the background. I wanted to know how things worked."

Moore's Law in Prospect

As his son became preoccupied at school with "the whys and hows" of science, Gordon was consumed by the "whys and the hows" of silicon electronics. He had perceived that the planar microchip could change the rules of the game. Although he discussed this insight internally at Fairchild, and Bob Norman used his ideas in a series of presentations, Moore had not yet made his argument in public. No other organization was trumpeting microcircuitry as the future. Microcircuits were rather seen as a solution for demanding military and high-end applications. In contrast, Moore was convinced that they could completely replace conventional circuits; buying microchips would be cheaper than buying individual components and wiring them into a circuit.

In 1962 Moore saw his insight beginning to be vindicated, through the dramatic improvements in the yields of planar microchips afforded by epitaxy. In response to an invitation from the McGraw-Hill *Yearbook of Science and Technology,* requesting a one-thousand-word article on "molecular electronics," he decided to put his conviction into print. His article asserted that planar microchips would vanquish the tyranny of numbers, by becoming the cheapest form of electronic circuit. He then embarked on a long chapter on the subject of integrated circuits, for inclusion in *Microelectronics,* a book edited by Ed Keonjian.

In this hundred-page essay, Moore would achieve the seminal intellectual breakthrough of his life, putting his earlier perceptions and insights on a formal footing, demonstrating once and for all that the microchip was the future. His arguments, formulas, and graphs offered a rigorous analysis of the technological and economic nature of the planar microchip, while his conclusions asserted that it would dramatically lower the cost of electronics, allowing novel systems, possessing previously unimaginable capabilities. In this, he accurately predicted the future.

The chapter's thesis about the "economics of integration"—that is, about how greater complexity of silicon microchips brings lower costs—would shape all of Gordon's subsequent efforts in the semiconductor industry. For the next four decades, his career would deliver on the implications of his breakthrough, and his work would affect every human being on the planet. The breakthrough itself lay in his examination of chemical printing technology. "The critical role played by the available manufacturing technology in determining the applicability of micro-circuitry to a given application cannot be overstated," he wrote. "It determines what can be done at all, as well as what can be done economically."

For a man such as Gordon Moore, so attuned to issues of measurement, it was not enough to know that, as chemical printing improved, microchips would be the cheapest choice for more and more electronic needs. He needed numbers to prove the case. He filled his chapter with equations comparing the costs of microchips to conventional circuits, then and in the future. Running the numbers, Gordon found that planar microchips already had a cost advantage. In a tenth of a square inch on a wafer, one could form a microchip with sixty-four transistors, or one could make twenty-five individual transistors. Since yields were the same, cost already favored the microchip. Factoring in packaging cost, the picture only improved further.

The numbers led Moore to a dramatic conclusion. For the data-crunching "logic" of computers, Fairchild's digital microcircuits were already the cheapest way to go. What of microchips with more transistors? Gordon's equations showed that they would be even cheaper, up to a point. Microchips "are favored more and more strongly, until one reaches the point where the yield falls below a production-worthy value, because of the complexity." At this juncture in the tiny, shrinking microworld, available techniques of chemical printing would break down, yields would fall, and the cost of the microchips would rise. The maximum possible complexity at any moment, where microchip electronics reached its cheapest level, would be determined by advances in the manufacturing art.

Moore was quick to observe that the maximum feasible complexity would increase over time. Manufacturing was an evolving art: as photolithography, diffusion, epitaxy, and metallization improved, the critical point would move toward ever-greater complexity: "As the technology advances, the size of circuit function that is practical to integrate will increase rapidly." Transistor size would decrease, microchip complexity would increase, and meanwhile costs would drop: a miracle! Gordon drew a triumphant conclusion: the planar microchip would allow computers of previously unimaginable size and power to become practical. At the same time, microchips would steadily provide cost and reliability improvements for existing computers. Eventually, the benefits of the planar microchip would expand beyond computers and "carry over into all of electronics," including consumer products. Moore correctly concluded that the economics of integration meant that the planar microchip would be the future of electronics.

Evidence was already available. In 1957, when the Traitorous Eight had established Fairchild Semiconductor, world production of transistor "bricks" was around a half million a week. Almost all were made of germanium. By 1963, when Gordon published his breakthrough analysis, transistor production had increased tenfold. Twenty percent were now made in silicon, and the planar microchip—built from silicon transistor bricks—was in production. The market for transistors totaled just over $300 million, while the integrated circuit market was already $20 million. The microchip was on the rise.

Gordon had made a trip to Caltech as early as 1960 to renew his connections to its faculty, hoping to recruit some of the best graduate students for his lab. At that time he had dropped in on Carver Mead, a young electrical engineering professor publishing novel work on semiconductor devices. Mead's students were the type that Moore wanted. Making a quick, silent decision on Carver's caliber, Gordon asked, "Could you use some transistors?" Mead, teaching a course that included a lab in the still-novel territory of transistors and transistor circuits, had access only to cheap devices that he found "god-awful." New military-grade silicon transistors would be a real coup; he answered, "Oh, that would be great!" Carver recalls what happened next:

> Gordon had one of those old-fashioned leather briefcases that opened up like a clamshell. He started rummaging around in there, next to a dirty shirt and a sock. Noticing me looking, he said, "I travel light," reached in, and pulled out a full-sized manila envelope. It was bulging. He set it down and said, "Those are 2N697s," then reached in and grabbed another envelope. "And these are 2N706s." I'd never seen so many transistors in my life. Gordon said, "These are cosmetic rejects. They'll be fine for what you want to do." I was floored.

Carver Mead was the surprised recipient of thousands of dollars' worth of the most advanced transistors on the planet, the original NPN mesa transistor (the 2N697) that Gordon Moore had shepherded into production and the latest miniaturized, gold-doped fast-switching version (the 2N706). Shortly after, Mead became a consultant to Fairchild, visiting Gordon in Palo Alto and lecturing to staff on his research, as well as meeting with individual teams to discuss their work.

From 1963 onward, Gordon Moore guided his colleagues according to his breakthrough analysis, toward refining the manufacturing technology for planar microchips. He knew that the vision he had outlined, and wished to realize and exploit, could be brought to fruition only through intentional, sustained action. His analysis was predicated on the cooperation of others: it required many actors, all in pursuit of the same goal. To push forward the manufacturing technology, engineers would have to design microchips that optimized cost, performance, and reliability, while system builders would have to form their products around those microchips. Gordon Moore's "economics of integration" required a social framework. Whole communities would have to direct their attention to realizing the promise of microchips within electronics. The microchips could not make themselves.

Moore now began the task of recruiting those outside of his immediate orbit to help realize his vision. To advance the revolution, he would learn to use his brilliance not just in improving the manufacturing technology,

Carver Mead teaching at Caltech, 1971.

SOURCE: COURTESY OF THE ARCHIVES, CALIFORNIA INSTITUTE OF TECHNOLOGY.

but also—increasingly—in advocating for the electronic future. For the rest of his career, he would push industry colleagues to practice the economics of integration. In time, this reality would become known as Moore's Law: a social law, dependent on many actors, and bringing transformational results.

The Changing Face of Electronics

The upward trajectory of transistor manufacturing continued with barely a wobble. In the second half of the 1950s, the number of computers in the United States rose from about 250 to around 5,500, one for every 31,000 citizens. Many of these systems were built using vacuum tubes, but transistors were in the ascendant. In 1959 IBM introduced the first of a line of transistorized mainframes: the 1401. It was smaller than IBM's massive tube-based computers, it performed as well or better, and was priced *six times lower*. Slowly, the public began to realize that major changes to the American way of life were somehow connected to this strange technology.

Thanks to transistorized mainframes, the number of computers in the United States increased to nearly 25,000 in 1965, or 1 computer for every 8,000 citizens. Close to half of these were IBM's 1400 series products. By 1975 there were 75,000 mainframes and smaller, cheaper "minicomputer" systems in use: 1 for every 3,000 US citizens. Years of intensive military support had resulted in transistorized computers achieving the economic velocity needed to burst into commercial applications. Digital computers were in wide use, not simply in Cold War military systems. Mainframes became familiar business machines, laboriously crunching data fed in on punched cards.

In 1957 the romantic comedy *Desk Set* had opened in theaters across the nation, while 1964 saw the appearance of a much darker satirical comedy, Stanley Kubrick's *Dr. Strangelove,* just as Fairchild's Micrologic microchips were finding their first uses in aerospace computing. The film is permeated by deep anxieties about the two superpowers and their complex schemes of mutually assured destruction. Computers run a strategic bombing system that allows a renegade US Air Force officer to provoke nuclear war. There is also an automated "doomsday machine," not unlike one eventually developed by the Soviets. At the end of the film, Dr. Strangelove (Peter Sellers) uses a computer to select American citizens to inhabit a deep cave, hoping to repopulate the planet, moments before the film closes with scenes of nuclear Armageddon (composed from clips of thermonuclear bomb tests).

During the 1950s transistors had entered American homes primarily inside miniaturized radio sets. By 1960 94 percent of households had radios (a growing proportion of which used transistors), and 78 percent had a landline telephone. For households of people with hearing loss, transistor-based hearing aids joined traditional tube models. However, the story of the 1950s was above all the story of television. At the start of the decade, 9 percent of households had a television set. In 1960 87 percent did. That rise was unprecedented. No other electronic technology, prior or since, has achieved as sudden and complete an adoption.

Unprecedented ability to transcend the limitations of geography and temporality came to characterize American homes. The choice to shift one's mind into electronic and away from physical reality involved unidirectional communication: a small number of senders and a vastly larger population of receivers of radio and television. Senders were predominantly corporations and, to a lesser degree, the government. From news to entertainment, the entire US population began to receive a common commercialized mass culture.

By 1960 major markets for transistors existed not only in the United States but also in Western Europe and Japan. That year Fairchild entered the European market through its joint venture SGS-Fairchild. Rivals such as Texas Instruments followed a similar pattern of creating foreign subsidiaries and opening factories in Western Europe. Sales into the Japanese market were more constrained, but US firms could use patent licensing to profit from Japanese interest. Bob Noyce and Fairchild's Dick Hodgson began to visit Japan, forming a strong relationship with the top management of electronics behemoth the Nippon Electric Company. In early 1963 Fairchild secured an innovative agreement: NEC would assist the firm in securing Japanese patents for Hoerni's planar process and then provide licenses to Japanese electronics industry. Fairchild would subsequently earn upward of $100 million in royalties from this deal. There was one unanticipated cost: Japanese industry became strong competition in the international marketplace for transistors and, eventually, microchips.

Noyce and Hodgson were champions of another form of globalization: offshoring, through which labor-intensive operations were transferred to countries or regions where wages were lower. When Fairchild's semiconductor division first offered the planar transistor in 1960, head count stood at thirteen hundred. Within seven years the division had expanded tenfold. By then offshoring was well established. Utilizing the new mode of travel offered by the jet aircraft—a mode itself dependent on the spread of transistors and electronics—Fairchild began to ship processed wafers from California to Hong Kong. There, they were diced, assembled into packaged transistors, and shipped back. Beginning in 1963 the Hong Kong plant assembled several million transistors a year, with workers receiving just $1 a day. An assembly plant was also opened in South Portland, Maine, which had an ample pool of people willing to work for a low wage.

Fairchild's success came in part from "using Asia and the big differential in labor rates with the United States" to drive costs down. "Fairchild saw that very, very early. Everybody else was slow. Many people starting companies didn't do that. And if they had, they would have been much more successful," says Jerry Sanders. By 1973, when Gordon proclaimed in an interview in *Forbes* that semiconductor makers were "really the revolutionaries," US firms employed more people in offshore assembly (eighty-nine thousand) than in domestic operations (eighty-five thousand). Gordon's work was set in a global context, with his Fairchild lab having a widening impact around the globe. The electronics revolution was becoming a worldwide revolution.

A less benign aspect of the revolution was the failure of the industry to keep pace with explosive growth in the use of certain chemicals. Semiconductor makers transformed their feedstocks into finished products by the controlled application of energy: chemical, electromagnetic, and mechanical. In the late 1950s fabrication of silicon transistors moved from small-scale production within the laboratory to large-volume manufacturing. With productive work came waste, by the late 1960s a matter of broad social concern. Two powerful chemical agents were key to photolithography:

a solvent, trichloroethylene (TCE), and an etchant, hydrofluoric acid (HF). "Trichlor" was familiar for its use in dry-cleaning, as an industrial solvent, and as an anesthesia drug. In the production of silicon devices, it stripped the photoresist from wafers. HF, a highly corrosive acid, could then dissolve oxide from the wafer in areas not covered by the photoresist. TCE and HF, working in tandem, were essential to the manufacture of transistors and microchips. Contaminated with the materials they had stripped or etched away, the chemicals became waste.

Fairchild and other companies disposed of this waste by pouring it down the drain. Those drains led to the municipal sewer systems of Mountain View, Palo Alto, Sunnyvale, and San Jose. HF attacked those systems, creating leaks. Soon TCE seeped through these leaks, forming large underground pools that flowed with the groundwater's hydrology. Over the decades large plumes of TCE (today understood to be a carcinogen) formed, particularly under Mountain View. With the Environmental Protection Agency (EPA) classifying many manufacturing locations as "superfund" sites from the 1990s onward, the area would eventually earn the dubious distinction of having the highest density of such sites in the nation. (Part of Google's headquarters sits today on the superfund site that was once Fairchild's manufacturing plant.)

Gordon Moore was blissfully unaware of these issues, as were his fellow scientists and engineers and the nation at large. In the Fairchild lab, "We had gallon jugs of hydrofluoric acid. The industry used it by the bucketful and dumped it down the drain." Jay Last recalls that Gordon did at least try to find out what might happen. "He went to some chemical engineers at Stanford or Berkeley and said, 'What are the health problems of using trichloroethylene? The answer was, 'If you use a big-enough tank of it and you fall in, you'll drown.' That was the existing view. Nobody knew these things."

Even so, it was troubling. The waste attacked plumbing. "We were having difficulty, with this corrosive mixture eating out drainpipes," explains Last. "One salesman said, 'I have the ideal material for drainpipes.' I poured it in. The corrosion never even slowed down; it went right through the pipe. That was the sort of practical problem we faced in scaling up." The issue literally went underground for many years. Gordon, who later invested billions of his own dollars into environmental stewardship, did not pursue the matter.

LEARNING TO LEAD

Vectoring a Lab

Gordon spent the bulk of his waking hours at the laboratory in Palo Alto. As the Fairchild Semiconductor Division continued to grow, he no longer spent much time at the bench, but was almost wholly engaged in directing the work of others. "We had a lot of technology that worked, but we had no idea why. We had to understand things better."

Commonly, Gordon carried with him one of the patent notebooks issued to his researchers, clothbound and olive drab with leather corners. Employees routinely recorded their activities, to back up patent applications. From 1957 into 1961, Gordon himself had done the same, documenting such efforts as his attempts to create the company's first transistor through the all-aluminum contact work. Now, he mainly used his notebooks to evaluate his colleagues' progress. Multiple entries each week reflected meetings with research groups, as he sought to guide Fairchild's R&D.

Gordon had two favorite terms: *vectoring* and *layout.* The first, describing his attempts to shape the efforts of groups and individuals, was drawn from mathematics, where a vector is an entity with a direction and a size: Moore's aim was to point researchers toward particular goals and ends, through a combination of advice, questions, decisions, and investments. He would encourage travel along certain paths and "vector in" on an endpoint. Moore believed strongly that his scientists and engineers required a reasonable amount of flexibility in how to move in the right direction. Years later he would comment, "I'm still a sucker for the bright, innocent young engineer with a delicate new idea and a 'best case' plan he's convinced he can pull off."

His second favorite term, *layout,* came from silicon manufacturing. Layout was a crucial stage in creating devices. Before making photolithography masks, engineers drew detailed schematics of a transistor or microchip and studied its layout for possible difficulties and errors. Gordon adopted this technique; before getting into activity at the bench, his workers would lay out the sequence and aim of the experiments to be performed, thereby generating a flow of questions for discussion. Occasionally, this required Gordon himself to make a hard decision on the fate of a project, program, or product, something he did not relish.

His was a very soft "command and control" approach. With areas he found interesting, Moore would invest in hiring suitable workers, giving them space and equipment and sinking funds into their pursuits. Employees could follow their instincts, yet Gordon's technological expertise was so highly regarded that his "guidance" usually had the effect of a direct order. Andy Grove, a brilliant PhD who joined the lab in 1963 and would become Gordon's right-hand man and a management guru in his own right, recalls that Gordon's inspiration had "a magical effect" on those who cared deeply about their work. "I would walk into Gordon's office and say, 'I want to do such and such.' He would tell me what was wrong with it, but he wouldn't tell me, 'Do this or do that.'" Although Gordon rarely gave specific directives, "I had such profound respect for his wisdom that he didn't need to."

Not all the laboratory's sections or projects are detailed in Moore's notebooks. Instead, he focused on efforts that were in keeping with his breakthrough insight about silicon microchips and the future of electronics. In keeping with his chemical expertise, he met most with teams working on oxides, metal depositions, and epitaxy, all territories that afforded good chemical challenges. He also met with manufacturing and marketing representatives on device development and devoted attention to the plans for individual transistors (still the mainstay of the business), but his priority was the development of future families of microchips. MOS microchips in particular were at the center of his attention.

Gordon remained acutely sensitive to demands on his own time and could (in direct contrast to his gregarious cofounder, and now boss, Bob Noyce) seem unapproachable. Within the lab he was often in his office or in a meeting room; moving between, he would walk with a downward gaze, notebook in hand, not breaking stride, and taking the staircase two steps at a time. While polite, he did not mean to be waylaid; he kept greetings to a minimum. Helen Bonfadini, previously a technician and now personal secretary to Moore, was a classic gatekeeper, employing an abrasive manner to shield him from interruptions. He did not want his time wasted or his thinking disturbed. Single-minded focus was becoming his hallmark.

"Layout" was equally important in catering to Gordon's need for private space. The fresh R&D laboratory in Palo Alto included an office suite off the building's entrance, physically separated from the rest of the lab by a complex of administrative offices. This suite was for the sole use of Moore and his associate director Vic Grinich. (Moore was always careful to assert that Grinich was the "associate" director, not the "assistant.") It was very isolated, recalls Andy Grove: "I tried to drop in a few times; it was painful to get there." When he knew Moore better, Grove learned how to work around his boss's need for private space. "I went to see Gordon just about every day. On my way out, nearing the exit, I would walk into his office, lean against the wall, and talk to him. Those were very useful conversations."

Between the offices of Moore and Grinich, and opening into both, was a small workroom with a powerful microscope and a variety of testing gear. The two men made use of this space, alone or together, to examine devices and processed wafers, combining their complementary expertise to puzzle over their findings. As boys of eleven, Gordon and Don Blum had "fooled around" with their chemistry set. Here, in this private lab, Gordon continued with the activities he loved best. He also guarded time alone in his office. "I have to sit down and think with a tablet and a pencil," he explains. "I needed time to read and think—private time. For me, it was very necessary."

While taking in the monthly progress reports from each section of his lab, Gordon also sought to stay abreast of countless memos from other parts of the division, in addition to a steady diet of reflection on the technical and trade literature. By the mid-1960s Fairchild's "corporate shtick" was well developed. One recruit found the whole memo culture unbelievable. "I remember two guys standing by their cubicles having a discussion. They talked; then they sat down, and each called a secretary in to take a memo to the other one, to explain what the meeting had been about. The memos would all be put in a file and also sent around. Everybody would read them."

Following the current literature helped Gordon keep close to the wider semiconductor community, and he supplemented this with reports from employees who had attended conferences. Moore had his own regular fixtures, including the Device Research Conference, gatherings that were abuzz with exchanges with industry and academic participants. Gordon also provided feedback to members of staff submitting their work to technical publications. Rarely, if ever, did he add his name as a coauthor. Had he treated their work in a way similar to that of many academic scientists running large labs, he would have been first author on hundreds of articles.

In this era his notebooks contain numerous to-do lists. He would exhort himself to read the lab's monthly reports, having himself in mind as their primary audience. If reading provided insight into it, Gordon exercised little active oversight over large swaths of his own laboratory. Much was left to Grinich and others. There were "significant limitations on the management crew," he concedes. "We were all going through on-the-job training. We were mining an extremely rich vein of technology, but the mining company was too small to handle it."

As Fairchild grew this lack of oversight led to problems. Moore's hands-off style might elicit great work from the best workers, but for those with less talent the opposite was true. Lack of internal discipline led to what Andy Grove describes as a casual "country club atmosphere" in which "people started work late and left early," with "very little expected in terms of tangible output." Some departments set their own priorities; one, in particular, became the butt of jokes for lack of productivity. Grove—who would go on to become a manager as feared as he was famous—became irritated. "You're dependent on someone else's collaboration to move your work along: his freedom renders yours useless," he explains. Gordon might actively navigate among research groups, notebook in hand, but those within a group often felt unconnected to other groups. A "silo" mentality developed. "He left it to chance that any product or knowledge would cross organizational boundaries. Gordon was a technical leader. He was either constitutionally unable, or simply unwilling, to do what a manager has to do."

The Last Entry in the Ledger

Moore might have been unwilling to engage fully with other people, but when it came to his own financial affairs, he let nothing slip by. Back in September 1950, he had opened a graph-ruled eight-by-ten-inch composition book and made it the ledger of his and Betty's personal finances. In his Caltech years, and through his time at the Applied Physics Laboratory, he recorded every bit of revenue, and each expense, down to the penny. A disciplined approach was good: they had little. Then came the offer from Bill Shockley and a roller-coaster eighteen months, then Fairchild Semiconductor. Following its buyout, he owned significant stock in Fairchild Camera itself. That stock quickly doubled and then continued to soar, fueled not least by Moore's own work with semiconductors. Even so, he remained frugal.

In 1961, some 140 pages into the ledger, he made what turned out to be his final entry. It described a changed reality. His salary as director was now more than $26,000 a year ($300,000 plus in today's dollars), an impressive sum for a man in his early thirties. His stock was already worth more than $400,000 (around $5 million in today's dollars), and he was adding to his assets, with holdings in Mead Johnson, Georgia Pacific, Olivetti, and Standard Financial. His focus on measuring, analyzing, and then making investment decisions fitted right into an American ethos of private capital, personal property, and financial markets; by 1966 his Fairchild stock had again doubled in value. He was comfortably rich, with resources greater than $10 million in today's money.

He and Betty lived within his annual salary, occasionally making big-ticket purchases outright. In 1961, despite his wealth, he was still driving an old Pontiac. Traveling fifty-plus miles on a regular basis to the new Fairchild plant in San Rafael, he realized the car had become a rattletrap. One day he stopped in at a dealership and told them that if they had a Porsche ready, he'd buy it on his way home that afternoon. Like Noyce, who owned a Mercury Cougar, Moore was helping to establish the stereotype of the young man who, in the throes of entrepreneurial success, splashes out on a high-performance sports car. Gordon, ever frugal, held on to the Porsche for many years; eventually, he gave it to son Steve, who still owns and drives it today.

Following the Porsche, the couple decided in 1965 to advance their favorite hobby: fishing. They bought a twenty-seven-foot Chris-Craft, which they kept at a marina off Highway 101 in South San Francisco, thirty miles north of home. Fishing was Gordon's main recreational passion. With Irene Whitaker watching grandsons Ken and Steve in Los Altos Hills, Gordon

Betty Moore in the 1960s.

SOURCE: KEN MOORE.

and Betty could take the boat out on weekends, sometimes with his brothers or with work colleagues such as Bob Graham. They would fish for striped bass, and if they went past the Golden Gate into the ocean, there were salmon to be had.

For family vacations the couple rented a cabin high in the Sierra Nevada near Pinecrest Lake, in Strawberry, just over three hours' drive from home. "We ended up there many years in a row," recalls son Ken. "We'd go trout fishing up and down the streams or out in a boat on the lake. I have wonderful memories of walking around the lake with my parents, fishing off the rocks, going to the dam and eating snow cones at the little concession stand. We'd hike. I loved it." These trips could turn into adventures. Once, armed with a government topological survey map, Gordon drove the family across streams and along dirt trails to find the Bennett juniper, the largest living western juniper tree.

Keen to provide their sons with exposure to the outdoors life they had enjoyed in their own childhoods, Gordon and Betty saw these vacations as an important part of family life. The couple also began saltwater fishing in more remote locations, first Baja California, then fishing camps in Mexico. The pleasure of these expeditions was twofold: the challenge of pursuing larger fish such as marlin and the opportunity for quiet time. Both Gordon and Betty enjoyed the escape from daily people-filled lives and the chance to explore undeveloped places teeming with diverse plants and animals. Both had grown up in this kind of environment, and each valued the solitary peace on offer.

The boat, the fishing vacations, the new cars: these were all things that Gordon could well afford with his annual salary, enabling agreeable variations on familiar themes. With assets enough to provide for the rest of their lives, Gordon and Betty's wealth served only to underline their rootedness. Wealth offered freedom of choice, to be indulged modestly. It was not something to be flaunted, but to be stewarded and prudently increased. It would not be spent but rather invested.

The Two Worlds of Fairchild Semiconductor

By 1963 the basic features of the chemical printing technology for making silicon electronics (diffusion, photolithography, planar devices, epitaxy, and thin-film depositions) were firmly in place at Fairchild and, increasingly, at other firms. Armed with this technology, scientists, engineers, and technicians could craft silicon diodes, transistors, and microchips and fabricate them in high volume. As Moore invested in the underlying technology, Fairchild learned to make better electronics ever more cheaply. This capability played out differently in the division's two primary organizations: Moore's R&D laboratory in Palo Alto and Charlie Sporck's manufacturing operation in Mountain View, an operation closely aligned and colocated with sales, marketing, and administration.

As time went on, Moore and Sporck created parallel organizations, each with considerable expertise in both manufacturing and device design, but with a fundamental divergence in agenda and priorities. The R&D laboratory's raison d'être was the future, while the manufacturing organization's was the present. As Gordon looked to processes and products on which Fairchild would succeed or fail in coming years, he oriented his laboratory to cutting-edge silicon microchips, convinced that if the required social investment were made, the microchip owned the future.

For the legion of engineers and operators in manufacturing, sales, and marketing, it was the opposite. Individual silicon transistors were the product to be made, sold, and promoted. The market for discrete silicon

TRANSISTOR AND MICROCHIP MARKETS, 1957–1967

Year	Value of discrete transistors	Value of microchips	Transistor units	Microchip units
1957	$70,000,000			
1958	$113,000,000			
1959	$222,000,000			
1960	$301,000,000			
1961	$300,000,000	$5,000,000		
1962	$291,000,000	$10,000,000		
1963	$305,000,000	$20,000,000		
1964	$336,000,000	$51,000,000	406,900,000	22,000,000
1965	$404,000,000	$94,000,000	608,100,000	9,500,000
1966	$476,000,000	$173,000,000	856,200,000	29,400,000
1967	$403,000,000	$273,000,000	759,800,000	68,600,000

Source: Electronics Industries Association Electronic Yearbooks and Market Data Books.

transistors was still exploding. In the transistor's battle to wrest control of electronics from the vacuum tube, 1964 was the crossover year: production of transistors surpassed that of tubes for the first time. Of the 400 million transistors made in 1964, more than 40 percent were silicon. In comparison, microchips, including Fairchild's, were still a niche market: 22 million were made, most for military and aerospace use.

For Sporck, better electronics meant individual silicon transistors that could knock out germanium transistors on performance and, most important, on price. The cost of putting transistors into metal "cans," and of the cans themselves, was more than the cost of processing silicon from the wafer through diffusion, epitaxy, photolithography, and metallization. While the manufacturing team enjoyed the benefits of improved yields and miniaturization (and thus lower processing costs), their main challenge was to reduce package and assembly costs. Improved assembly was the key to improving cost.

Moore was not deaf to current needs. Engineers in his laboratory continued to develop discrete silicon transistors to feed into sales, marketing, and manufacturing. They also helped troubleshoot the all too common collapse in manufacturing yields, when they "lost" the process and potential transistors turned into rejects. Silicon manufacturing technology was a complex assemblage of materials, equipment, processes, and operational recipes. Changes in any of these aspects, or even a seemingly natural drift, could cause derangement; the source needed to be rapidly tracked down and the system retuned (sometimes referred to by Moore as "witchcraft"). Gordon had himself done this detective work with Fairchild's first NPN and PNP mesa transistors, getting them into production and driving up yields. Now he made his staff available to do the same for Sporck's manufacturing.

This symbiosis worked well on occasion, as seen in an enormous order from Minnesota-based Control Data Corporation, booked by Fairchild in 1964. CDC was known as a producer of powerful mainframe computers and was developing a "supercomputer." This effort, led by the renowned designer Seymour Cray, needed very fast, very reliable transistors. Turning to Fairchild, Cray made a $500,000 investment to underwrite the creation of an appropriate device. Gordon's laboratory quickly developed a smaller, more reliable version of the planar transistor. This led to a $5 million order, the largest for a single product in the history of the industry. Each supercomputer required 600,000 transistors. Sporck's manufacturing operation ramped up its activity accordingly.

More common than symbiosis was a pervasive lack of interest. Andy Grove says, "For all practical purposes, everybody involved in R&D had given up on trying to transfer technology to the manufacturing world. We just did our stuff, wrote our paper; it was almost an academic environment. It was very rare for people from Mountain View to roam our halls or come to meetings. I didn't know any manufacturing people because I never saw them." Many of Gordon's own meetings involved attempts to hand off microchips to Sporck's organization, but these transfers often failed, not because Mountain View could not do what Palo Alto had done, but because they elected not to. Moore was unwilling to "go to the mat" for issues not very close to his heart. "As technical competence grew in the manufacturing organization, it became harder and harder to tell them what to do. They wanted to figure it out for themselves. People told me we ought to sit back and measure some fundamental things and let Whisman Road do development. However, we kept control of the development of really new stuff. I'd take a very strong position that I didn't want to give up development."

Irritating inefficiencies began to accumulate. As Jerry Sanders remembers, "Fairchild was a ragtag bunch of guys. They were bright, but they were very, very different. There were shouting contests." Tom Bay even grabbed him by the tie to challenge a point. Moore, so dedicated to nurturing delicate ideas and so uncomfortable with confrontation, found it difficult to maintain balance and "vector" his employees. One "pretty wild technologist" in manufacturing started to develop his own equipment to do epitaxial growth. Another, Lee Boysel, who would later set up the company Four Phase, "took off on a project all by himself" to design an advanced set of microchips in secret. "It really drove a big wedge between the two organizations," says Moore. Things became so bad that one R&D employee accused Sanders of "playing R&D like a piano." Sanders says, "All I was trying to do was get that great stuff, which was working fine, into production. That was a tough thing to do. It took somebody working hard."

Sporck's success with low-cost individual transistors and the collaborative success with high-performance transistors for CDC together enabled Fairchild to broaden its sales. "The military was key," says Sanders, "but Tom Bay, to his eternal credit, said 'We must get into the consumer market.'" Transistors had initially gone almost exclusively into military applications; Fairchild now began to aim squarely at commercial uses, from digital computers to television tuners.

Advances and Setbacks

Gordon Moore remained focused on advance. What paths were open? If he pushed for improvement in manufacturing technology—the ability to chemically print ever-smaller transistors on microchips at ever-greater densities—how far could he get? What might be the uses for such cheaper, better electronics? His immediate answer lay in mainframe computers. Even when "civilian," many of these mainframes went to corporations and laboratories engaged in Cold War efforts. Like all digital computers, mainframes had three functional parts: logic, main memory, and storage. Logic is digital circuitry—such as the central processing unit, or CPU, where information is processed through algebraic operations on the endless zeroes and ones of "binary" or "digital" data. Main memory holds the data that moves into and out of the CPU. Storage holds data longer term and provides it to the faster main memory.

Most computer logic used discrete transistors, with silicon a growing choice; a few computer systems, primarily for aerospace and the military, already used silicon *microchips* for logic. Main memory was not composed of semiconductors at all, but made of magnetic core arrays. Transistors could be used to read—and write information to—the magnetic core arrays. The arrays themselves were made from thousands of tiny doughnut-shaped magnetic beads, magnetized or not magnetized, to correspond to the zeros and ones of digital data. The third functional element of computers, storage, was also dominated by magnetic technologies—reels of magnetic tape and magnetic hard-disk drives the size of washing machines.

The immediate opportunity for greater production lay with logic. Moore saw that if he invested in improving the yields and complexity of Micrologic chips, they could become the cheapest and the best-performing option, and his firm could capture markets for digital logic. Yet the story was not as simple as that. Digital logic circuits came in a variety of forms defined by the components used to make the circuits and each having pros and cons. Engineers needed to employ creative flexibility in choosing the parts with which to work.

Fairchild's designers built their circuits solely from resistors and transistors; they were resistor-intensive, having "resistor-transistor logic" (RTL). Early in the microchip effort in Moore's laboratory, one engineer proposed a different form: "transistor-transistor logic" (TTL), which maximized the use of transistors. This suggestion complemented perfectly Moore's breakthrough realization about cost, complexity, and microchips. If transistor fabrication costs were going to fall, it made sense to use more of them; TTL microchips would have better performance, but could still compete on price. Moore focused to RTL, which was popular in military computing, and failed to seize the opportunity. The suggestion dropped away.

In 1964 the Micrologic spin-off Signetics emerged as a grave threat to Fairchild's business. It was using a middle way, "diode-transistor logic" (DTL), offering significant performance advantages yet with only a modestly more difficult production process. As DTL became popular, Signetics's circuits found an eager market. Sales surged, leaving Fairchild's RTL Micrologic chips trailing. Gordon belatedly awoke to the threat, and he commissioned a Fairchild family of DTL microchips to compete directly with Signetics.

Bob Noyce, without objection from Moore, ordered that Fairchild's DTL microchips be sold below cost and at half Signetics's price. Where once Noyce had slashed the price of planar transistors, now he did the same with these new chips. It was a doubly competitive move, expanding the market while massively undercutting Signetics's price, dealing the young spin-off a shattering blow. Fairchild could weather the near-term losses in its bid to capture the market; Signetics could not. "We almost sunk Signetics with that move," recalls an employee in marketing. Noyce and Moore showed their teeth in this and in other business decisions. If Signetics had been ruthless in its bid to unseat its former employer, Moore and Noyce took a no less brutal approach. Gordon might be mild of manner, but he was a determined competitor, one who could be highly aggressive in the impersonal domain of markets, technologies, and business competition.

Texas Instruments made another competitive switch, introducing a line of TTL microchips, the very transistor-intensive logic form imagined and then abandoned in Moore's laboratory. With access to massive resources, TI was able to price these high-performance microchips very competitively. No sooner had Moore and Noyce displaced Signetics than they found themselves pushed out by TI and playing catch-up. They belatedly countered with TTL products of their own, but not before their competitor had captured the lion's share of the microchip market. Why did Gordon

Moore allow himself to be overtaken? In the end, his failure came down to a narrowness of vision, a deep-rooted tendency to keep his head down. Moore was a physical chemist at heart; he could be determinedly competitive when roused to engagement, but in business matters his skills were still nascent and immature.

For Gordon, happiness lay in being immersed in developments that interested him; the corollary was that he easily missed less obvious trends. In this case, he failed to grasp the importance of TTL and to position his lab at the front of the pack. In so doing, he lost the chance to exploit the most commercially successful logic form of the time. It was an important lesson. Moore says that, in the early 1960s, "we didn't have any idea of the magnitude of the opportunity we were dealing with. We were still a bunch of guys in a laboratory, amazed that people actually wanted to buy our products."

The Promise and Perils of MOS

Among the developments that most interested Gordon was the metal-oxide-silicon transistor. In 1960 he had learned of Bell Labs' earliest investigations into this fresh breed. The MOS was itself a planar transistor, but could be made very small and required fewer masks than a standard transistor. On a microchip MOS transistors would be "self-isolating," avoiding the electrical issues that bedeviled other microchips, allowing more complex circuits to be packed into a given area of silicon. MOS microchips were attractive, and RCA, a strong East Coast player, was making a big bet on them. By late 1962 its labs succeeded in making a microchip incorporating MOS transistors. Gordon responded by putting his head of physics, Chitang (Tom) Sah, in charge of a "catch-up" project.

Now came another breakthrough: Frank Wanlass, fresh from his physics PhD at the University of Utah—a "phenomenally creative guy who once brought into the lab a list of something like forty-eight patentable inventions!"—joined Sah's group and quickly imagined another novel form of microchip, the "complementary MOS" microchip, or CMOS. Typically with chips, there was a trade-off between speed and power: the greater the speed, the greater the power needed; low power meant low performance. Wanlass's CMOS chip was configured so that its transistors drew power only when actually switching. As the majority of transistors were idle at any given moment, this was the key to a winning combination: low power consumption with high performance.

It was heady stuff, but Moore faced a number of puzzling challenges. MOS transistors were mercurial. Electrical performance was unstable, subject to "drift." Unless and until this was overcome, they could not be viable products. Gordon recruited additional chemical and physics talent and himself directed a "surfaces" group. No sooner had the group assembled than a deeper challenge arose. Key Micrologic staff announced their departure, this time to create an MOS microchip spin-off: General Microelectronics. Wanlass joined them. As Moore considered Wanlass and Andy Grove to be the two best guys in his lab, this was a major setback.

That summer Moore made his own (albeit temporary) departure from Fairchild, taking his very first journey overseas. His great-grandfather Alex Moore and grandfather Josiah Williamson were, by their midtwenties, well-seasoned travelers who had navigated long and perilous journeys into unknown territory. Gordon's pioneer journeys, in contrast, had been intellectual rather than geographical. Now he participated in a NATO-sponsored lecture tour of Europe, combining his voyages of the mind with a physical journey. Instigated by *Microelectronics* editor Edward Keonjian, the tour included events across France, Germany, Italy, and the United Kingdom, promoting the latest developments to the European technical community. Gordon and other American experts gave talks on "micropower electronics" to assembled groups of scientists and engineers.

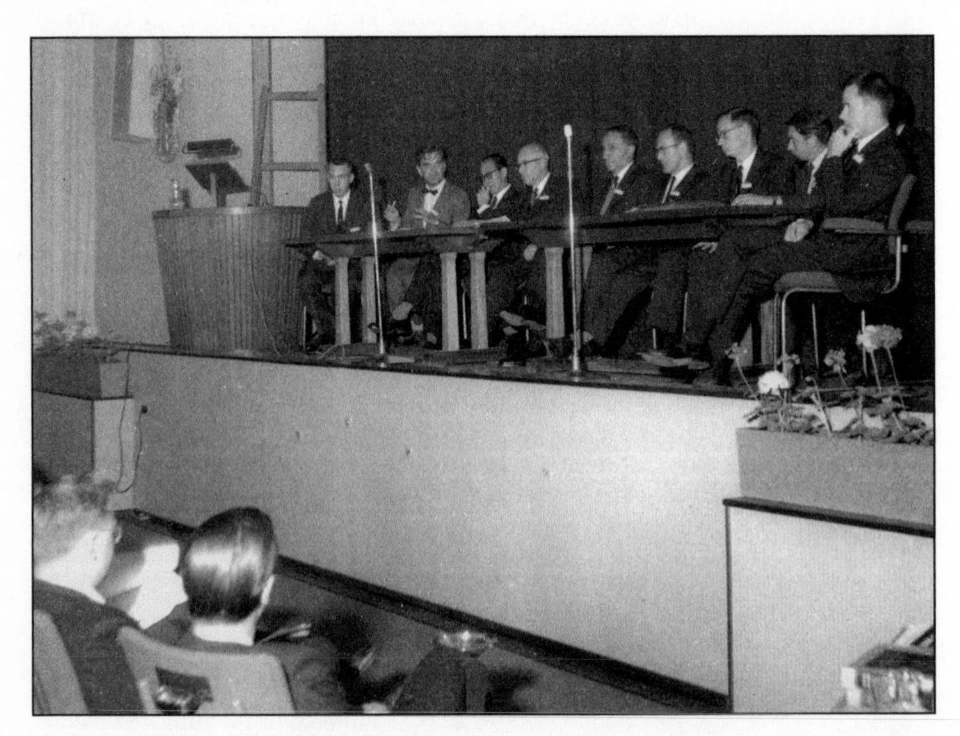

Gordon Moore discussing CMOS technology on his 1964 NATO-sponsored European lecture tour. Moore is seated fourth from the right.

SOURCE: GORDON MOORE.

Micropower electronics referred to highly miniaturized circuits that consumed very little power. NATO militaries recognized that this could mean vital new capabilities in satellites, missiles, and aircraft. Moore's lecture presented CMOS microchips, detailing how they drew power only when actually switching. He took pains to make clear that work was required to get the manufacturing technology "adequately developed" (a classic understatement, given the technical difficulties and the spin-offs from and fragmentation of his R&D team). Neither Gordon nor anyone else imagined that within two decades, the entire semiconductor industry worldwide would convert to this technology, with the overwhelming majority of silicon transistors produced being MOS transistors, resting within CMOS microchips.

What Gordon did foresee was an important future for CMOS microchips: they could crack open computer main memory. At the time, magnetic cores ruled in main memory. With CMOS, Gordon knew, "the power consumed by even a very large memory would be a fraction of a watt. Such an application will be practical when more experience has been accumulated, making integrated arrays of these devices." This anodyne statement signaled a huge opportunity, since most of the cost of computers was in main memory. Gordon was planning to pioneer a novel territory for the microchip.

Renewed Direction

Visions and lectures were one thing. Inconvenient reality was another. Gordon returned to Palo Alto to witness the prospects for CMOS at Fairchild deflate rapidly. The loss of colleagues to General Microelectronics had been an enormous setback; now Chitang Sah reduced his professional commitment by taking on a professorship at the University of Illinois. He would consult and provide occasional leadership, but Moore's twenty-man physics department was left without full-time management. "It was a peculiar period," recalls Gordon. "I didn't have a real replacement, so Tom would come out once a month, look into what people were doing, and go back." Moore's slowness in replacing Sah had serious repercussions. The CMOS effort all but collapsed.

Fortunately, there was a fresh force at work within Fairchild: Andrew Grove, who would one day become Moore's key enforcer, relieve him of direct management burdens, provide much-needed discipline, and enable Gordon to bring his vision for the microchip to fruition. Finishing a chemical engineering PhD at Berkeley, Grove set his sights on the semiconductor industry. Moore and Grove hit it off from their first encounter. Grove's academic supervisor had paved the way through a note to Moore in late 1962 that Grove was "a truly exceptional individual. Whoever hires him will be very lucky." As in most things, Gordon did not need to be told twice. Grove remembers that, during the interview, Gordon "absolutely got the significance of what I did in my PhD, was very interested, and had good questions. He was a smart guy—very personable, no airs, helping me see what I wanted to be." Moore's achievements, combined with his quiet intelligence and attention, were "a big selling factor." Grove might have gone to Bell Labs, which was courting him, "were it not for California and Gordon."

Grove's youth in Hungary had been troubled and deprived. Having fled during the brief, abortive 1956 Hungarian uprising, he had made his way through the City College of New York and Berkeley, a brilliant young émigré undeterred by his own strong accent and a degree of hearing loss. In his earliest days at Fairchild, facing many pressures, Grove needed a boss like Gordon: a paragon of accomplishment, but at the same time a reliable presence, quiet, engaged, and kind. In turn, Moore came to need Grove: a man whose considerable nervous energy was linked to high talent and an unlimited appetite for confrontation, in startling and welcome contrast to Moore's reflex of avoidance. Moore recognized Grove's abilities, admired his ambition, and felt they could work together: his "action orientation" complemented perfectly Gordon's "direction by indirection."

Moore knew that he needed to renew and expand his investment in MOS and tackle MOS drift once and for all. He set Grove to the task. "Andy Grove came in and almost immediately became the leader of a group that was looking at MOS devices and the surfaces around them. They developed a good set of experimental methods. Andy got right on the problem. He has a tremendous ability to simplify, and he trimmed that one down to things we could get our teeth into." Grove collaborated with two experienced semiconductor researchers, Bruce Deal and Ed Snow. While he bemoaned his new company's silo mentality, he also exemplified the power of Gordon's permissive approach to a highly talented staff:

> I don't remember a single department meeting. The three of us met by accident. We ran into each other in the cafeteria lunchroom. "Hi, I'm so-and-so." "I'm so-and-so." "What are you doing?" "I'm working on a capacitor." "Gee, that's funny. I'm working on the MOS capacitor, too." Claws come out. "What do you do?" "I'm trying to grow pure oxide." (A sigh of relief.) "I'm trying to analyze the theoretical capacity." (The other guy relaxes.) We were all working on parts of the same issue. If there was a master plan to pick the right three, it could not have been better designed. Without anybody telling us, we informally started to work together.

As Gordon watched, "Grove succeeded beyond my wildest dreams": his group grew to a dozen people. With cleaned-up processes, MOS transistors at last became commercially viable. "They pushed that area very fast and did an excellent job." Meanwhile, the semiabsent Sah took a portion of the credit. This caused Grove much aggravation, as Moore chose to detach himself from the unpleasantness: "Gordon always looked like he never heard us complaining." Belatedly, Moore did promote Grove to be a section head.

In March 1965 Gordon spoke to his peers on MOS microchips at the national Institute of Electrical and Electronic Engineers (IEEE) meeting in New York. Even though they were "sensitive to processing variables," MOS held great promise, as "the layout of structures is greatly simplified. Large digital functions made in this manner will be important." To drive home his point, Gordon quietly produced a photograph of an MOS microchip made recently by his engineers: a 64-bit random-access memory (RAM) that performed the same function as a magnetic memory array. "Random access" signified that any particular "bit" of memory could be accessed directly, rather than looked at in sequence. Gordon might be playing catch-up commercially, but his disclosure of this chip in 1965 reveals just how far his lab had come in two years. He now had not only logic but also main memory firmly in his sights.

THE MICROCHIP

From Proof to Persuasion

A half-million silicon *microchips* (as distinct from transistors) were sold in 1963, for nearly $20 million, almost all for military applications. Fairchild, which had scooped the industry with its first diffused silicon transistor and whose planar technology had triumphed, now lost ground to competitors such as TI, which won a contract for all the advanced microchips in the US Air Force's premier ICBM, the Minuteman II. Fairchild was in a trailing position. Its salesmen fought back as best they could. They had already chalked up several design wins for Micrologic chips in the US space program (including the huge contract for the Apollo guidance computer). In 1964 the NASA space probe *Mariner IV* also carried Micrologic chips. Eventually, this probe would return television images of the surface of Mars.

The microchip had come a long way, from an exciting laboratory development to an actual business. Gordon was armed with his own breakthrough analysis, yet faced customer skepticism. According to Jay Last, he became "frustrated that the world wasn't accepting integrated circuits."

However, news of the microchip's success in military aerospace was beginning to percolate throughout the electronics community. Industry leaders wanted to talk about microchips, and potential customers slowly became more receptive.

As widespread unrest and talk of social revolution gathered pace in the mid-1960s ("Never trust anyone over thirty"), those in the electronics industry were absorbing the reality that the microchip could change individual and corporate behaviors, communications and daily realities, and global relations. Moore, the director of the laboratory that had pioneered the planar microchip on which the entire international semiconductor industry was converging, put himself at the center of this conversation. His *Yearbook* article of 1962 and *Microelectronics* chapter of 1963 made his insight public, pointing out that, through a dramatic lowering of costs, silicon microchips could both pervade and expand the electronic world.

Gordon knew that his insight was contingent upon social acceptance. For it to become a reality rather than just an abstract concept, the industry would have to invest unprecedented sums in the development of manufacturing technology. The breakthrough he foresaw depended upon buy-in and cooperation, on legions of engineers dedicating their careers to technology, and on huge numbers of customers choosing microchips for their products. The semiconductor community as a whole would need to become convinced that this future would be not only achievable, but highly profitable. As the leaders of the US electronics industry echoed his arguments, adding their own analyses, predictions, and twists, Gordon's vision began to resonate.

In 1964 around 120 million silicon transistors, and 280 million germanium transistors, were sold. These transistors could be found in a range of locations, from domestic living rooms to missile silos, from satellites to factory floors. Most of the 10 million televisions made in the United States that year contained transistors. Transistors also filled more than 20 million radios sold and $400 million worth of home audio equipment: phonographs, tape recorders, and hi-fi systems. The transistor enabled the American public to enjoy the first Beatles albums, watch news reports of the Gulf of Tonkin incident, and listen to radio coverage of Martin Luther King Jr.'s Nobel Prize award.

US government and business spent $2 billion on computers, even as IBM announced its latest transistorized computer, the blockbuster System 360. Industry spent a further quarter of a billion dollars on electronic controls, and technical and scientific communities poured $360 million into electronic instruments. In addition, $1 billion was spent on electronic communications equipment for the expanding telephone network. During the escalation of the war in Vietnam in the late 1960s, a full 16 percent of all US Department of Defense spending went for electronics, some $8 billion.

Toward the end of 1964, Moore drafted a fresh talk, "The Evolving Technology of Semiconductor Integrated Circuits," in which he made the shift from proof to persuasion. In prior publications he had laid out his ideas; now it was time to convey his insights to the technical community as compellingly as he could and to convince others about the future of electronics. For Moore, who at school had been kept back for reasons of inarticulacy, this was a challenge. Preaching his message would take practice, hence a low-key start on home turf, with a talk on December 2 to a local section of the Electrochemical Society, on the San Francisco Peninsula. Matter-of-factly, almost negligibly, Moore used the *R* word. His talk's abstract simply said: "The evolution of integrated circuit technology will be reviewed and extrapolated into the future. An attempt will be made to indicate the extent of the revolution in electronics that will be precipitated as a result of these technological advances."

Moore's Law

In February 1965, Gordon found his opportunity to engage directly with the wider electronics community: a letter from Lewis Young, editor of the weekly trade journal *Electronics,* asking for an in-depth piece about the future of microcircuitry. *Electronics* was well established and widely read, with a mix of news reports, corporate announcements, and substantial articles in which industry researchers outlined their recent accomplishments. It covered developments both within the semiconductor industry and in electronics more broadly, giving technology and business perspectives.

Young was planning a thirty-fifth anniversary issue, including a series titled "The Experts Look at the Future." As the sole microchip expert in the issue, Gordon's words would reach sixty-five thousand subscribers. It was the moment that he had been waiting for. He made a giant asterisk mark with his pencil at the top of Young's invitation and underlined an exhortation to himself: "GO-GO." Answering Young, he admitted, "I find the opportunity to predict the future in this area irresistible and will, accordingly, be happy to prepare a contribution." Within a month he had drafted his manuscript: "The Future of Integrated Electronics."

The piece reiterated much of what Moore had already written, but sought to be more engaging. Gordon's confidence and comfort in his expert position shone through in his subtle use of dry humor and a clear, low-key style. His conscious attempt at warmth was designed to persuade readers both to buy into the future he foresaw and to help create it. Included, for the first time, were several explicit numerical predictions. He telegraphed the gist of his argument in a brief summary for the Fairchild lawyer who would review his draft: "The promise of integrated electronics is extrapolated into the wild blue yonder, to show that integrated electronics will pervade all of electronics in the future. A curve is shown to suggest that the most economical way to make electronic systems in ten years will be of the order of 65,000 components per integrated circuit."

The claim was nothing if not bold. Sixty-five thousand transistors per silicon microchip (up from sixty-four in 1965) would be a remarkable level of complexity. These microchips with sixty-five thousand transistors would represent the most economical way to make electronic products. Gordon's message was simple and stunning. Silicon microchips made better and cheaper electronics. Applications would widen throughout industry, technology, and society, and possibilities would emerge for computers to develop unprecedented capabilities.

In his opening paragraph, Moore set the tone: "The future of integrated electronics is the future of electronics itself." Since the actual future lay beyond his reach, he aimed not "to anticipate these extended applications, but rather to predict for the next ten years the development of the integrated electronics technology on which they will depend." Silicon microchips were now "an established technique." Nowhere was this truer than in military systems, where reliability, size, and weight requirements were "achievable only with integration," making silicon microchips mandatory. Beyond this, the use of microchips in mainframes was already surpassing conventional electronics in both cost and performance. Complex microchips of high quality would "make electronic techniques more generally available throughout all of society," enabling the smooth operation of "many functions that are done inadequately by other techniques or not at all." Existing technologies would be refashioned or replaced by electronics-based approaches, providing fresh technical, social, and economic functions.

The lower costs of systems, "from a readily available supply of low cost functional packages," would drive this expansion. He offered an impressive, visionary list of possibilities—"home computers," "automatic controls for automobiles," "portable communications equipment," and the "electronic wrist watch"—a list that today seems conservative but in 1965 was startling, exciting, and provocative. Silicon microchips provided a clear path to the realization of such futuristic, sci-fi possibilities. Here, indeed, was revolution.

Such change hardly seemed credible. IBM's System 360 mainframe started at $113,000 (more than $1 million in today's money). Fancier versions cost the equivalent of $7 or $8 million today. Less powerful minicomputers, like Digital Equipment Company's PDP-8, cost the equivalent of more than $150,000 today; even minicomputers were as expensive as houses. The implication of and evidence for Gordon Moore's argument—that microchips would bring "home computers" within reach of the ordinary buyer—was difficult to digest. He also made a more pragmatic point: in the near term, "the principal benefactors of the technology will be the makers of large systems." Mainframes would become available at much lower cost and with much more computing power.

Thanks to his experience in Shockley's Quonset hut, Moore had been an active participant in developing the core manufacturing technology. He had built diffusion furnaces and "glass jungles" from scratch, before handing the job on to a technician. Now chemical printing technology was fully in place and was robust. As far back as 1962, in a note stuffed into his olive patent notebook, he had written, "There are no major problems left in silicon device technology." For Moore, the technology was complete in the sense that he could grasp the fundamentals of its essential parts, but it was not limited by any immediate physical reality and was wide open to continuous development.

Uniquely among his peers, Moore predicated his vision on the idea that the manufacturing technology already had a trajectory of steady improvement. With intense effort and expensive investments, it could be remorselessly perfected to provide better yields of more complex microchips containing ever-smaller transistors. His philosophy of standard products for this future of complex microchips was central to his thinking, since such a future (with its ballooning costs for designing and developing ever more complex chips) was feasible only if large markets for high-demand, high-volume microchips were continuously developed. Meanwhile, he and his colleagues would improve each facet of the technology by concentrated hard work.

Photolithography could allow smaller patterns to be generated, with fewer yield-crushing defects. Better diffusion processes could improve chemical doping and reduce wafer damage. Epitaxy could produce better crystal layers, with fewer deformities. With novel materials and recipes to improve device stability and protection, oxidation could be refined. Contaminants could be more rigorously eliminated by cleansing of water, photoresists, acids, and gases. Larger, purer, and more perfect silicon wafers could be grown. Better metallization could provide more durable contacts and connections. Each aspect of the manufacturing technology could be enhanced to support steadily increasing miniaturization of transistors on microchips, to expand complexity and to improve yields. The power of Moore's insight—still true a half century after his 1965 article—is that the revolution in electronics depended on improving existing silicon technology, not altering its essential character.

The simplicity of Moore's belief allowed him to make the case for the future of microchips, electronics, and society. As chemical printing evolved, the economics of microchips would change. Over time, ever more complex chips would provide the cheapest electronics. To illustrate, he provided a plot:

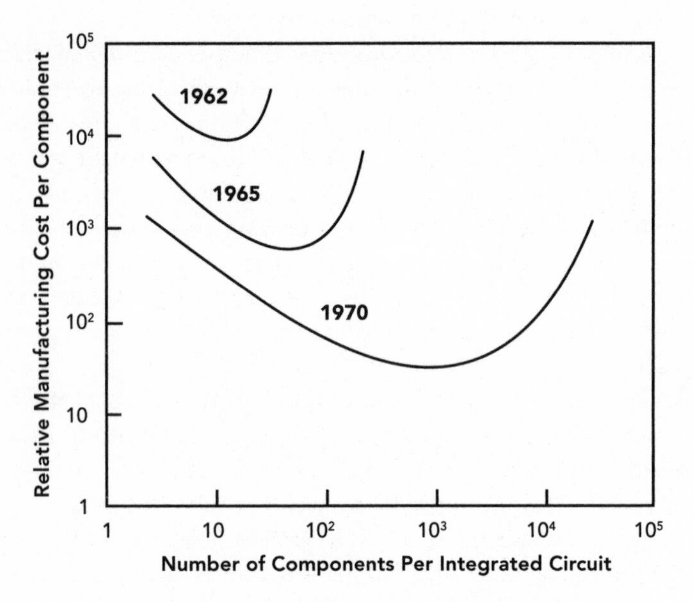

Gordon Moore's curves on this plot show that development of the chemical printing technology makes more complex microchips the cheapest form of electronics.

Source: Gordon Moore.

The vertical scale shows the cost for making a transistor on a chip, with each increment representing a tenfold difference in cost. The horizontal scale shows the complexity of the microchip, as measured by the transistors it contains, with each increment representing a tenfold increase in complexity. The relationships represented by the curves are not linear but exponential: small changes have great effects. The three swoops illustrate the relationship between cost and complexity in 1962, 1965, and (hypothetically) 1970. In each case cost falls to a minimum and then rises with further increases in complexity; each successive curve is lower on the cost and higher on the complexity scale. An astonishing change takes place: an 8-transistor chip, cheapest in 1962, changes to a 2,048-transistor chip, predicted to be cheapest in 1970.

Gordon combined the economics of integration with his philosophy of standard parts to assert that, in 1970, the transistor that could be made most cheaply would be on a microchip thirty times more complex than in 1965. The punch was in the tail: "The manufacturing cost per component can be expected to be at least an order of magnitude lower than it is at present." In other words, standard microchips could make electronics ten times cheaper in five years. Moore and the industry at large could deliver, steadily, exponentially increasing electronics for the dollar. Revolution, indeed!

A second plot answered the question: "What will be the complexity for minimum cost over time?" Gordon's answer, with a numerical prediction, strengthened the persuasiveness of his essay.

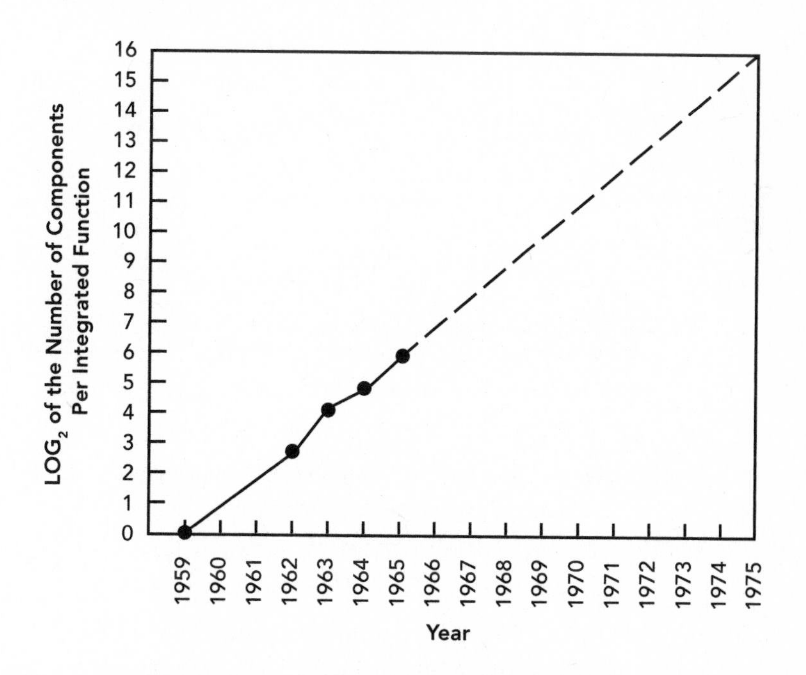

Gordon Moore's numerical prediction for the future of the silicon microchip: fierce competition leading to an annual doubling of complexity in order to minimize the cost of electronics.

Source: Gordon Moore.

This time the horizontal axis is linear, showing each year from 1959 to 1975, while the vertical axis is exponential; equal increments represent the doubling of transistors on a microchip. Because Fairchild and its rivals had focused on minimizing the cost of electronics by investing in technology, "the complexity for minimum component costs has increased at a rate of roughly two per year." Later this sentence would be seen as the first articulation of "Moore's Law," but the world's attention would routinely rest on the "what" (the doubling of complexity) rather than the "why" (the minimization of cost by investing in advances in chemical printing in order to gain competitive advantage).

On a linear plot, Gordon's graph would have been a hockey stick, typical of exponential growth.

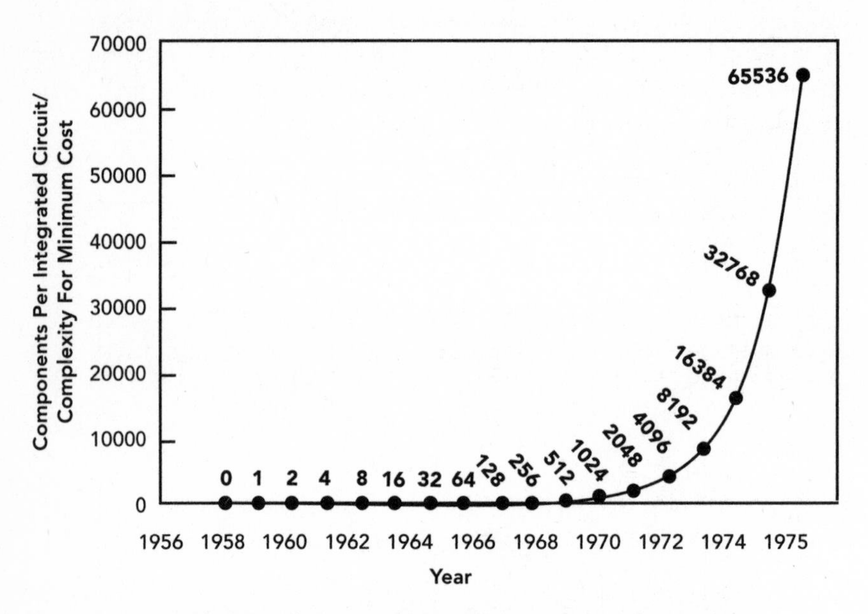

An alternate, linear rendering of Moore's plot of microchip complexity for minimum cost over time.

Source: Della Keyser.

Would the annual doubling trend continue beyond 1965? Moore's second graph shows his affirmative answer—in 1975 the cheapest electronics would offer more than sixty-five thousand transistors to a microchip. In words, he put it dryly: "Certainly this rate can be expected to continue, if not to increase. The longer-term extrapolation is nebulous, although there is no obvious reason for stopping the curve before it intersects the top of the graph." He closed with a quip as dry as the numbers themselves: "This curve was purposely plotted with a rather obscure unit as ordinate so that the logic of the extrapolation of the historical data might be appreciated without the confusion of the absolute numbers implied."

In the remainder of his manuscript, Moore deployed skeptical questions to examine the "reasonableness" of 1975's sixty-five-thousand-transistor chip. Could so large a circuit be made upon a single wafer? Yes, said Moore:

Picturing Moore's Law. Close-up photograph of silicon devices from 1959 to 1968. At lower left is a single planar transistor. At the upper right is a microchip made of hundreds of planar transistors.

SOURCE: GORDON MOORE.

there was plenty of room on an inch-diameter wafer to squeeze in sixty-five thousand transistors. The idea was just horrendously expensive in 1965; in another ten years, with improved chemical printing technology, it would be a different story. The crucial thing to remember was that "there is no fundamental reason why yields are limited, below one hundred percent. Nothing exists comparable to thermodynamic equilibrium considerations, which often limit yields in chemical reactions." To Moore, the physical chemist, perfect yield was simply a matter of massive investment. "Device yields can be raised as high as is economically justified. It is only necessary that the required engineering effort be committed."

Then came a question about the power consumption of a functioning sixty-five-thousand-transistor chip. "Is it possible to remove the heat generated?" The concern was prescient. Four decades later, heat would become one of the semiconductor industry's major worries. While Gordon foresaw this, he believed that the heat produced could be handled. Rather than glowing brightly like vacuum tubes, reengineered microchips would achieve an improved "power density," offering ever greater speeds for less power.

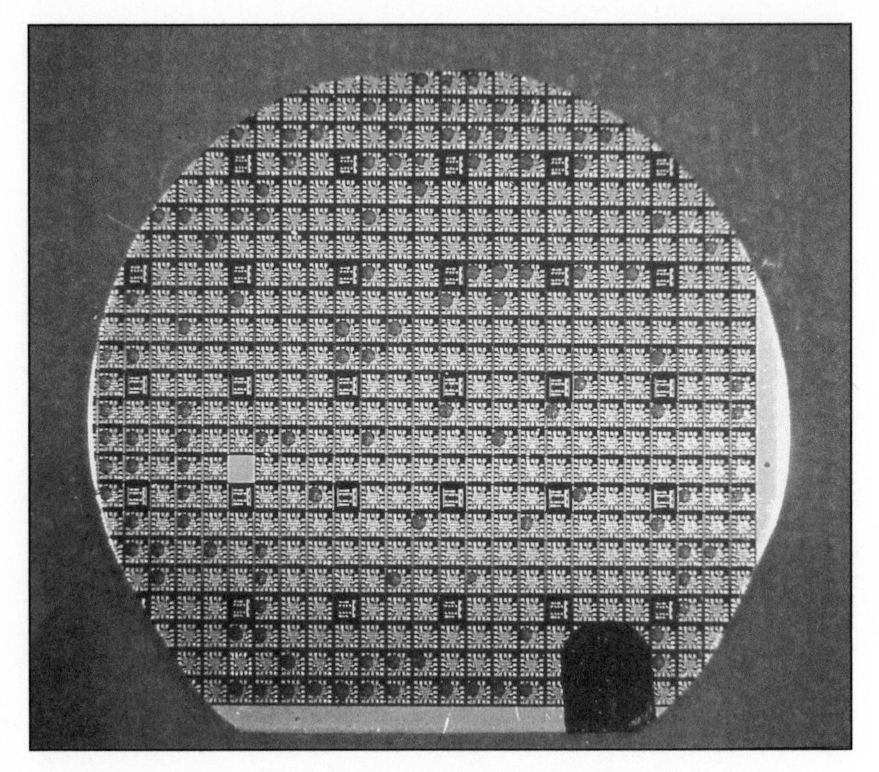

An inch-and-a-half-diameter wafer, with chemically printed microchips. Each patterned square holds the metal contacts and interconnections for a single microchip. There are approximately five hundred microchips printed on this wafer, each with dozens of transistors.

SOURCE: GORDON MOORE.

Gordon closed his discussion with the observation that the huge design costs for such a complex chip must be minimized, either by amortizing the engineering "over several identical items" or by evolving flexible engineering techniques, "so that no disproportionate expense need be borne by a particular array." This was his philosophy of standard products, honed in the earliest days of Fairchild Semiconductor. As with transistors, so for microchips: the best would be those achieving standard functions.

Moore submitted his manuscript to *Electronics,* where cuts and editing diminished his original clarity of exposition. Under a more awkward title, "Cramming More Components onto Integrated Circuits," the piece appeared in the anniversary issue of *Electronics* on April 19, 1965. Most of his key language made it into the published article, along with a cartoon indicating just how fantastic the idea of a home computer seemed, even to the editor of a main communication vehicle of the electronics community! There is no evidence that the article made a splash at the time. It may or may not have been widely read, but it was not especially cited or republished.

Gordon's published article in *Electronics* may have had little impact in the wider world in 1965, but it affected Gordon himself in a profound way. He had articulated his vision of a future that could be built. He now knew exactly where to go and took it upon himself to work to realize his vision and ensure that the prediction would come to pass. The understanding he had now fully achieved guided the rest of his career.

Through the later 1960s, Gordon expounded his analysis wherever he saw opportunity, in text and in oral presentations. He became as committed to persuasion as he was to developing the manufacturing technology itself: braving the limelight, trying to tame language (not his natural métier), developing rhetoric, and mastering presentation skills. Moore could diagram a sentence to perfection, but the art of persuasion depended on manipulating passions and emotions. It was not until the later 1970s that he began to make comments in his talks about the "sheer excitement" of his business. He was also, by then, perfecting the use of humor as a tool to disarm and engage his audiences.

Carver Mead—increasingly recognized as an expert in the field—began to take it on himself to promote Moore's analysis. As time went on, Gordon's numerical predictions came true. The doubling of circuit complexity, every year or so, slowly took on the name "Moore's Law." Even so, most discussions missed his underlying point that the doubling was not itself the fundamental dynamic. His breakthrough insight was that the pursuit of the best, lowest-cost electronics, motivated by economic competition, would necessarily create this doubling through an ongoing, extensive, and expensive social effort.

Failure and Frustrations

Having reached an apotheosis in his *Electronics* piece, it was time to descend from Olympus. There was much to be done. One immediate task was to explore whether microchips could give computers a radically new architecture. Because it was clear that microchips could form computer logic, and microchips might also someday provide main memory, Gordon was convinced that they would reshape computing. Computers would fall in cost and attain far more power. But would computers themselves change?

a few diodes. This allows at least 500 components per linear inch or a quarter million per square inch. Thus, 65,000 components need occupy only about one-fourth a square inch.

On the silicon wafer currently used, usually an inch or more in diameter, there is ample room for such a structure if the components can be closely packed with no space wasted for interconnection patterns. This is realistic, since efforts to achieve a level of complexity above the presently available integrated circuits are already underway using multilayer metalization patterns separated by dielectric films. Such a density of components can be achieved by present optical techniques and does not require the more exotic techniques, such as electron beam operations, which are being studied to make even smaller structures.

Increasing the yield

There is no fundamental obstacle to achieving device yields of 100%. At present, packaging costs so far exceed the cost of the semiconductor structure itself that there is no incentive to improve yields, but they can be raised as high as is economically justified. No barrier exists comparable to the thermodynamic equilibrium considerations that often limit yields in chemical reactions; it is not even necessary to do any fundamental research or to replace present processes. Only the engineering effort is needed.

In the early days of integrated circuitry, when yields were extremely low, there was such incentive. Today ordinary integrated circuits are made with yields comparable with those obtained for individual semiconductor devices. The same pattern will make larger arrays economical, if other considerations make such arrays desirable.

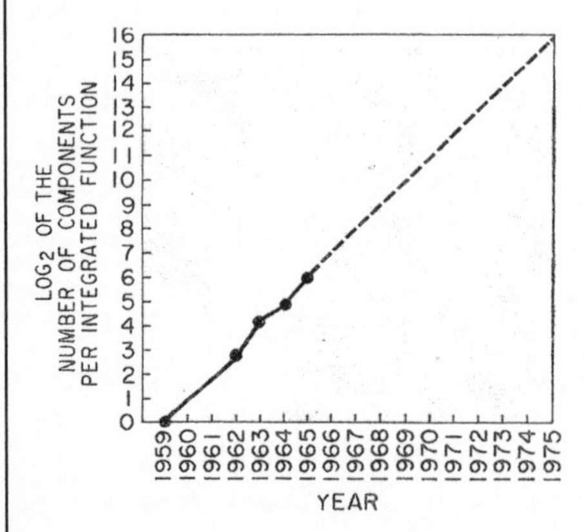

Heat problem

Will it be possible to remove the heat generated by tens of thousands of components in a single silicon chip?

If we could shrink the volume of a standard high-speed digital computer to that required for the components themselves, we would expect it to glow brightly with present power dissipation. But it won't happen with integrated circuits. Since integrated electronic structures are two-dimensional, they have a surface available for cooling close to each center of heat generation. In addition, power is needed primarily to drive the various lines and capacitances associated with the system. As long as a function is confined to a small area on a wafer, the amount of capacitance which must be driven is distinctly limited. In fact, shrinking dimensions on an integrated structure makes it possible to operate the structure at higher speed for the same power per unit area.

Day of reckoning

Clearly, we will be able to build such component-crammed equipment. Next, we ask under what circumstances we should do it. The total cost of making a particular system function must be minimized. To do so, we could amortize the engineering over several identical items, or evolve flexible techniques for the engineering of large functions so that no disproportionate expense need be borne by a particular array. Perhaps newly devised design automation procedures could translate from logic diagram to technological realization without any special engineering.

It may prove to be more economical to build large systems out of smaller functions, which are

116 Electronics | April 19, 1965

Moore's prediction stretched the imagination in 1965. The cartoon shows how the idea of "home computers" was difficult to imagine.

SOURCE: COPYRIGHTED 2015. PENTON MEDIA 115215:115SH.

Would the very character of the computer—its architecture—be transformed by the use of microchips?

To explore this area, Moore hired Rex Rice, a highly experienced computer designer from IBM. Gordon's associate director Vic Grinich had worked on one of the first transistorized computers. Now he and Moore launched an ambitious project, Symbol, to design a novel form of digital computer around the microchip. Gordon wanted the project, yet computer design was more akin to electrical engineering than to chemistry. And electrical engineering was not his specialty. As he had missed the TTL opportunity, so he blundered now with Symbol. The project's eventual failure would reveal the difficulty in understanding exactly *how* microchips would change the nature of computing.

Moore poured resources into Symbol, providing it with a professional staff of more than twenty. The conceit was straightforward: If microchips will be very cheap, why not use them for everything? Why not turn software back into hardware? Typically, computers of the time (and today) have a master software program, or "operating system," that allows other software programs to function (such as Internet Explorer running on Microsoft Windows). Symbol's operating system was made not from digital code but from microchips. It was work that fascinated Moore and went on for years. However, Symbol itself enjoyed little lasting success. A single completed system eventually made its way to the University of Iowa.

In 1966 Moore was suffering other, greater, frustrations than the slow progress on Symbol. As the year got under way, he wrote what he titled "The Plan" in his patent notebook, hinting at problems both within his own R&D lab and in the organization as a whole:

> Projects—Get everything on paper.
> Personnel—Get a program to match needs with ability.
> MOTIVATION TECHNIQUES!!
> Politics—Influence external organization by assigning specific responsibility.

Two sets of psychologists in the 1950s had said Gordon Moore possessed no aptitude for management, but he was now running the industry's most creative research lab. This only proved, he would later demur, that "research doesn't need much management! You hire the right people, give them the right environment, channel them more or less, and you get what you want." As Andy Grove put it, "We did state-of-the-art research from a little lab, running circles around much bigger organizations." One physicist who joined Moore's division in the later 1960s, Herb Kroemer, would even go on to win a Nobel Prize for Physics (coincidentally sharing it with Jack Kilby of Texas Instruments for his development of integrated circuits).

Yet there were serious problems. Along with the rest of Fairchild's semiconductor division, the R&D lab had experienced relentless growth, making management a complex challenge. Gordon now oversaw some seven hundred staff, including sixty PhDs. Congenitally unsuited to "management by walking around," he preferred to attend meetings, make notes, decide on investments, and be available to make suggestions. In the early days, working with peers, he had found it simple to define action points and follow up. Now, his style was ineffective. Each team was in a silo, with workers failing to communicate or collaborate.

Some R&D groups set their own priorities, generating activity that was unfocused. "There was no internal discipline; no external expectations were put on the lab, or on the manufacturing organization to support the lab," recalls Grove. The techniques suited to a small operation lost traction. Failing projects were given latitude by Moore's tendency to avoid conflict, especially over personnel, attributes that Grove saw as passivity. "Every trouble or misgiving I have ever had about Gordon was a case of inaction. Would he interfere in some conflict between X and Y and Z? Not on your life." The recurrent spin-offs also hurt, and camaraderie suffered. The R&D organization was fractured.

Another disturbing issue was the continuing lack of communication between R&D operations in Palo Alto and manufacturing operations in Mountain View, five miles away. Technological competences, fueled by increases in profits, facilities, and staffing, mushroomed independently at each site, enabling Moore and Sporck to operate what became, in effect, two autonomous firms. With the market for silicon devices and microchips incredibly elastic (the lower the price, the greater the market, the opportunity always now), Sporck's greatest challenge was to contend with growth; hence, though he did not interfere with Moore's activities, he was increasingly unreceptive to challenging products and novel technology.

The troubling reality was that Fairchild's semiconductor division had reached the size at which even Bob Noyce's considerable charm was insufficient to conjure the needed magic. Noyce bridged the corporate divide by representing the division to the larger company and by promoting the division's products to the outside world, but as Moore later noted, "Bob wasn't a manager; he was a leader. He set a direction and a tone. He thought that if you suggested good ideas to people, they would naturally do the right thing. He didn't like to manage, because of the follow-up involved." Noyce did not want to alienate either Sporck or Moore. He could not heal the fracture.

Moore was especially concerned by Fairchild's failure to establish a real position in MOS. Despite his conviction that MOS microchips were the future, he could not get the company to seize the opportunity made possible by his lab's pioneering work. Sporck ignored his efforts, and Noyce avoided controversy. Characteristically mild, Moore became galled that he could not get MOS into manufacturing in any serious way. As he later explained, "My laboratory developed the science of how to make MOS microchips stable and the technology of how to make the devices, yet we were unable to transfer that to manufacturing. It was very difficult to get people experienced with the earlier bipolar technology to believe everything that had to be done. They were unwilling to accept that it required special incantations to make it work."

Despite Moore's recommendations and urgings, the manufacturing organization would do only what it wanted to do. With these problematic politics, Moore attempted to get "everything on paper" and have Sporck agree to "specific responsibility" in keeping with his own preference for measurement, analysis, and decision. Yet at the end of December 1966, he recorded bitterly, "Our corporate position in MOS is way, way behind." General Instrument, which by now had recruited Frank Wanlass, was a major producer of MOS microchips, as was General Microelectronics. Incredulous and not a little infuriated, Moore found himself on the outside looking in. "Our spin-off had a spin-off before we got anything happening in Mountain View." At Intel his later bitter boast would be that "we've been able to demonstrate that you really can transfer products from Palo Alto to Mountain View."

Making Strides

Despite these failures, by the late 1960s Fairchild Camera was a $200 million business with some fifteen thousand employees. It remained smaller than its competitors—a dangerous position. In mid-1967, his patience worn thin, Moore took a dramatic step: "I negotiated a deal where I would set up a significant pilot line in the laboratory to try to get MOS going." If Charlie Sporck in Mountain View would not manufacture MOS microchips, Gordon would do it himself. In his notebook he outlined a plan to "establish a facility in which we demonstrate an ability to make P-channel circuits with performance specs at least equal to General Instrument's, including the ability to package them." This pilot line would not differ in kind from Mountain View's factory, except in scale.

Along with his push to establish a manufacturing line, Gordon belatedly reorganized the MOS research effort in his laboratory. Andy Grove

GROWTH OF FAIRCHILD SEMICONDUCTOR, 1957–1967

Year	Employees in semi-conductor	Total employees	Semi-conductor division space (square feet)	FCIC Stock-holders	FCIC shares out-standing	FCIC net earnings
1957	20	1,780	14,400	1,780	17,600	$799,000
1959	1,260	3,580	95,000	3,174	1,037,000	$2,071,000
1961	1,550	5,490	168,000	11,000	2,499,000	$3,819,000
1963		8,100	629,000	10,900	2,551,000	$1,931,000
1965		11,540		9,650	2,576,000	$8,456,000
1967		14,780	1,180,000	12,980	4,305,000	-$7,699,000

Source: Fairchild Camera and Instrument Annual Reports.

and another Hungarian émigré, Les Vadasz, were put in charge. Once more, the race was on. Vadasz sought Gordon's approval to develop a novel "silicon-gate" MOS microchip, recruiting a brilliant physicist, Federico Faggin, from the European joint venture SGS-Fairchild. Over a three-month period in early 1968, Faggin pioneered a process offering speed, size, and power advantages. Gordon watched closely. The renewed push on MOS microchips offered the best route to substantially more complex microchips and, with them, the continued conquest of the digital computer.

Gordon's aim was to create "large-scale integration" (LSI) microchips, each containing several thousand transistors. It became important to find standard functions that could be served by such microchips. Main memory looked to be a very attractive possibility. Since 1962 Gordon had funded explorations of small-scale memory microchips in his lab. Now, with large-scale microchips in his sights, main memory looked like a plausible target, using LSI microchips to compete with magnetic-core arrays that provided digital data at the cost of "a penny a bit." Here was Gordon's breakthrough insight in practice. He was unabashed in his bullishness. Cognizant of manufacturing advances, he took his argument directly to the enemy in the spring of 1968, speaking at the International Conference on Applied Magnetics (Intermag), a gathering of more than nine hundred computer memory experts.

Competitive technologies were invited to "come by" at Intermag so the skeptics "could snicker as we gave our presentations," recalls Moore. Even so, he held nothing back, telling the assembled crowd that the silicon microchip would soon invade magnetic memory's turf. "I come to bury magnetics, not to praise them," he wrote in his speech notes. "The future holds a semiconductor memory for a penny a bit." The gauntlet had been thrown down.

Creative Destruction

As Gordon promoted his vision from 1965 onward, others in the Fairchild Semiconductor Division began to sit up and take notice. His ideas poured fuel on entrepreneurial flames, promising those with vision and appetite a tremendous opportunity. With negative unemployment, an ample market in venture funding, and booming markets in defense and consumer goods, the available technological opportunities meant that fortunes could be made and empires built. However quietly delivered, Moore's message was not lost on his colleague and obstacle Charlie Sporck.

Unlike Moore and Noyce, Sporck lacked a fortune. He had witnessed (several times over) how spin-offs from Fairchild could earn substantial profits. He, too, found the R&D-manufacturing fracture frustrating. Jerry Sanders relates, "Charlie said, 'I can't work here anymore. I've got to be my own man. I'm leaving.'" With his close associate Pierre Lamond (one of Moore's lab engineers since 1962, now head of device development), Sporck began to plot. Fred Bialek (overseas operations) and Roger Smullen (manufacturing engineering) joined them. Marketer Floyd Kvamme threw in his lot with the group.

In one of several spin-offs from Fairchild in the early 1960s, photolithography expert James Nall had created Molectro. National Semiconductor, a struggling Connecticut firm, had bought Nall's start-up in 1965, and two key Fairchild microchip engineers (Bob Widlar and Dave Talbert) had since joined it. Sporck's high opinion of the pair now led him to take a hard look at National Semiconductor itself. By January 1967 he had convinced its board to let him take over the company. He would move the firm to the San Francisco Peninsula, and his microchip experts would establish operations in direct competition with Fairchild.

Bob Noyce depended on Sporck to run the semiconductor division, just as he relied on Moore to advance the technology. Noyce and Sporck got along well, so Noyce was "almost devastated" when Sporck announced his departure that February. In the earliest days of Fairchild Semiconductor, Ed Baldwin's defection to create Rheem Semiconductor had been a body blow, but this was worse: Sporck's position at the heart of the organization made his defection the mother of all spin-offs. Some thirty-five employees went with him.

When Baldwin had quit in 1959, the cofounders had decided not to risk appointing another outsider. Noyce reacted similarly to Sporck's defection, turning to his longest-standing, most trusted associate and partner, Gordon Moore. To help weather the storm, Noyce asked Moore to take the reins and run the division. He must have known it was a vain hope. Gordon, always careful with his time, was conscious enough of his management limitations to doubt that he could fix the division's complex organizational problems, even had he wanted to. His own métier was to drive forward R&D, to take advantage of the vast opportunity in memory (an opportunity he had already signaled to Noyce), and to engineer the microchip's conquest of the digital computer. His plate was more than full.

SELECT GENEALOGY OF FAIRCHILD SPIN-OFFS, 1959–1983

Year	Number of direct spin-offs	Including	Number of spin-offs from direct spin-offs	Including
1959	1	Rheem Semiconductor		
1961	2	Signetics, Amelco	1	Raytheon Semiconductor
1962	1	Molectro		
1963	1	General Microelectronics		
1964			1	Union Carbide Electronics
1966			3	Philco-Ford Microelectronics
1967	1	National Semiconductor	2	
1968	4	Intel	3	Integrated Systems Technology
1969	2	Advanced Micro Devices	4	Signetics Memory Systems
1978	1	California Devices	4	Acrian
1983			7	Cypress Semiconductor
Total direct: 18		**Total indirect:**	**61**	

Sources: Electronic News; SEMI.

Gordon found the decision easy to make, but not to communicate. Awkwardly, he managed to tell Noyce he "really didn't want to." Noyce turned reluctantly to Tom Bay, head of sales and marketing, who had applied for this very job of operations manager back in 1957. Then as now, Gordon judged Bay not up to the challenge. He was "probably not the best choice," says Sanders. "We needed a manufacturing guy in there. We couldn't make this stuff." There were many problems within the semiconductor division and also in its parent firm, Fairchild Camera. Having begun as "perhaps the most extraordinary collection of business talent ever assembled in a start-up company," the Fairchild Semiconductor Division was suffering an inglorious decay. "We had come a long way," comments Moore, but "we had also made a tremendous number of mistakes and had squandered opportunities. Fortunately, good products make up for a lot of problems in an organization."

Rescue and Meltdown

The situation was about to become critical. Pierre Lamond, now at National Semiconductor with Sporck, was courting Moore's protégé and most valuable researcher, Andy Grove. "National Semiconductor was very successful. I was tempted," acknowledges Grove, whose first book, *Physics and Technology of Semiconductor Devices,* was published that year. "When my book came out, it was as if one blew out the light: anticlimax. I was frustrated and bored at Fairchild. Work took nothing. Pierre wooed me; I interviewed. The timing was perfect. I accepted."

Grove had a deep respect for Gordon and went into his boss's office with some trepidation. "I swear to God, Gordon choked up and had tears in his eyes. He said, 'I was always hoping that someday you would succeed me in this lab. I guess it's not to be.' He was very depressed in a quiet way. I felt pretty lousy, but there was nothing to say, so I went back to my office." Gordon had already lost Frank Wanlass. This time, unusually, he took action; he wanted to keep Grove. He called Bob Noyce, seeking help. Noyce rarely appeared at the R&D laboratory, but now he sped over, determined to do whatever it took. Grove recalls that Noyce "sat down across from me, leaned back, and made some comments about my book." Then he cut to the chase: "'We don't want you to leave.' We went back and forth as to why I was leaving. At the end of the conversation—it was a Friday—Noyce gave an assignment, 'Think of the circumstances under which you would change your mind.' 'Are you serious?' 'Absolutely.'"

Grove, wanting "to solve the problem of my becoming useful for the lab," came up with an audacious proposal. He called Moore at home that

Sunday, and the two agreed to meet for breakfast the next morning at Rickey's, where Grove laid out his plan, telling Gordon he wanted to become assistant director of R&D, with technical authority. "I didn't want any dealings with the department heads, who, in my opinion, were useless and who were all substantially older. I wanted to work with section heads and to go to Mountain View to represent key developments" and help resolve the R&D-manufacturing fracture.

Grove saw himself as a successor to associate director Vic Grinich, who had left the lab to lead a new Fairchild division. Grove proposed to report directly to Moore, while also advocating for R&D in Mountain View. It was a jump of several levels on the organizational chart. Moore agreed on the spot. "Gordon made a very gracious announcement of my two-level promotion," recalls Grove, who "chose not to have Vic Grinich's office but to stay near where I was, to make it easy for people to come by and stick their heads into my office." Grove telephoned National to tell them that he was not defecting after all. "I called Pierre from a phone booth, told him I'd changed my mind, and why. He said, 'I knew it!' He had told me, 'We're going to offer you two departments.' When I told him my decision, he said, 'It sounds as if they've given you four.'"

With Grove saved, Bay in place, Noyce in charge at the corporate level, and stability restored, Moore hoped to refocus the leadership of the lab, but it was not to be. Fairchild's semiconductor division continued to disintegrate, a development Andy Grove would later attribute to the weakness of Tom Bay and the inability of Noyce and Moore to remove him. "In retrospect, Sporck had supplied the decisive mass and the operational muscle, knocking heads and correcting behavior. He left a hole, not just in an organizational chart, but in the dynamics of the company."

Other external problems were brewing. Delivery issues and a dip in demand exacerbated poor results from the semiconductor division. In mid-1967 Fairchild Camera, having coasted on the division's success, began to report losses; alarmingly, its stock slid by almost 50 percent. Manufacturing felt the purse strings tighten—"tight as a bull's ass in fly time," says Jerry Sanders. Moore's analysis was milder: "The West Coast tail was not very effective at wagging the East Coast dog." The parent company's board ousted CEO John Carter, replacing him with Dick Hodgson. This sounded like good news in California: Hodgson, who had overseen the creation of Fairchild Semiconductor, had long been a friend to Noyce, Moore, and the division. Yet Hodgson's position was precarious, too, and by January 1968 he was judged to be losing his grip. To Noyce, the situation seemed dire. If Hodgson were ousted, they would have no friend left in top management on Long Island.

That month Noyce confided to Moore that he might leave Fairchild. Noyce planned to start another microchip company, an explosive idea. He knew Gordon to be quiet, trustworthy, and discrete, and he wanted Moore to join him. Why not create a spin-off from their first spin-off? Gordon declined, telling him, "No. I've got the best job in the industry." To a point, this was true: he had built an enormously productive pool of researchers and designed an exquisite facility; he enjoyed good funding and had a sterling reputation. Despite Fairchild's latest woes, Moore's world was still very much intact. He was intent on pursuing a plethora of exciting developments, among which silicon-gate MOS and microchip memory were but the most prominent.

In February a further twist occurred. Hodgson *was* stripped of his title; the CEO function passed to a three-man executive committee. Noyce was on this committee, along with Sherman Fairchild himself. Noyce learned, to his chagrin, that eastern colleagues considered him too young, at forty, to take the top job. In the Valley of Heart's Delight, he was an old hand, and he was the logical internal candidate, but in staid Long Island he was still a youngster and a newcomer. Agreeing to help search for a CEO, Noyce courted C. Lester Hogan, the respected leader of Motorola's semiconductor operations. However, Hogan's appointment would mean there was no future for Noyce at Fairchild. In May 1968 he spoke to Moore again. He really was leaving to start a new company. Would Moore be his partner?

The question left Gordon Moore with a clear choice. As when he asked Betty to marry him, Gordon measured and analyzed the facts before deciding. Without Noyce to provide protection and leadership, and with Tom Bay ill-equipped to solve the division's problems, his situation would worsen: "New management would probably change the nature of the company." He would no longer have the best job in the industry. If he acted now and left with Noyce under optimum conditions to apply their learning to a blank slate, he had a good chance of creating a fresh "best job." Moore took a day to reflect and then called Noyce: "Okay, I'll go too." Fairchild had been a big part of his life. It was not easy to leave, but what's left behind "is usually not worth looking over your shoulder for."

In the spring and summer of 1968, the whole country was in turbulence and disarray. April saw the assassination of Martin Luther King Jr., sparking riots in Chicago, Baltimore, and Washington, DC. In June Robert F. Kennedy was assassinated in Los Angeles, after winning California's Democratic primary. The summer saw an escalation of demonstrations against the Vietnam War, the "establishment," the military-industrial complex, and corporate capitalism. At Columbia University, in high-profile protests, students launched a "sit-in" in the administration building. Several campuses witnessed arson attacks or bombings, and, in California, Stanford's Reserve Officers' Training Corps building was torched.

In contrast, it was business as usual at the Device Research Conference in Boulder, Colorado, that June. The microchip might be closely associated with the military and have significant implications for Cold War politics and the fighting in Vietnam, but the latest announcements of research accomplishments were matter-of-fact, giving no indication of any kind of revolution. It had been a turbulent year for Andy Grove, but he arrived in time to enjoy the entire conference, listening closely to formal talks and informally gleaning useful information. Gordon arrived toward the end of a session, fresh from a Fairchild planning meeting. The pair met up for a walk. Grove was eager to describe conference highlights, but sensed his boss was distracted. He asked about the planning meeting back in California. Moore quipped that the planning had been interesting, but that he himself had not been interested.

It was Moore's turn to return the courtesy that Grove had shown a year earlier in revealing his intent to defect. "I'm leaving Fairchild," he told Grove, who quickly asked, "What are you going to do?" Moore's reply was simple: "I am going to start a new semiconductor company." Grove's reaction was instinctive: "I'm going with you!" Moore quietly accepted the assertion, not saying, "Yes," "Of course," "Certainly," or anything else. They did not shake on it or hug. They simply continued their walk, but now they had a fresh topic: the nature of this future company and what it would do.

7

THE INVENTION OF INTEL

ALWAYS IN A HURRY

The Game Plan

What had been wholly novel in 1957—launching a spin-off focused on silicon electronics—was familiar routine a decade later. The action of the Traitorous Eight had been an improvisation, guided by financiers Arthur Rock and Bud Coyle. Gordon Moore had not been eager to start an enterprise, despite his pioneer heritage. "I'm not the sort who can just say, 'I'm going to start a company.' The accidental entrepreneur like me has to fall into the opportunity or be pushed into it."

In 1968 things were different. Moore had become deeply committed to his vision of the revolution that would be wrought by silicon electronics. He was both the master of chemical printing and the prophet of the microchip's promise. He wanted the best possible vehicle for his self-appointed task. He had become a deliberate entrepreneur. Now, he believed anything less than a breakthrough company would be anticlimactic. "To me the opportunities are few and far between. Things have to line up." He and Noyce planned accordingly. Both understood the stakes were enormous: a successful start-up would not only drive forward the electronics revolution, but also make a very substantial fortune.

Gordon had watched multiple ventures emerge from Fairchild Semiconductor and then give birth to yet further spin-offs, *Fairchildren.* Six of the Traitorous Eight had already played this game, as had members of his original Micrologic team. All the spin-offs stayed within a dozen miles. With the exception of the first (Ed Baldwin's Rheem Semiconductor), each aimed to manufacture microchips. A tacit playbook had developed: A group of technologists and marketers coalesce around a shared enthusiasm for a novel development. Perceiving opportunity, they chafe at constraints in their present company and discuss forming a venture to seize the moment. In secret, they approach backers and lawyers, by now familiar with

Gordon Moore, circa 1968.

SOURCE: INTEL.

technology spin-offs. Once the deal is in place, the group resigns en masse and launches its venture, treading carefully to avoid being sued. The playbook took as its premise the idea that capital was readily available thanks to the expanding market for microchips. Founders would take equity stakes, with potential for personal reward. Since previous experience, even within a failed company or project, was valuable, there was little to lose.

Moore and Noyce's planned start-up was typical in these regards, but it would not be ordinary. Noyce had the zeal, the catalyzing energy, to set off the chain reaction. His business acumen, developed through convincing customers and industry watchers about the value of Fairchild's offerings, could not be bettered. In turn, Moore had a clear idea for what kind of company the venture should become. He provided the strategies for technical and economic achievement. Andy Grove was a skeptical, insightful sounding board. In time, Moore and Noyce would look to him as the enforcer, making good on their schemes.

Because of the prestige Moore and Noyce enjoyed as the two key principals in Fairchild Semiconductor's success, the financing of their second start-up was straightforward. They naturally turned to Arthur Rock, who had remained on good terms with both and had a close personal friendship with Noyce. Rock relates, "They were making 110 percent of Fairchild Camera and Instrument's profits, so they thought they should have a little more voice in how things were run. Bob called and said, 'We're going to leave. We need $2.5 million to get started.' I said, 'Fine, let's do it.' It was that easy." He was also supremely confident in Moore and Noyce. "I thought they could do anything they wanted to. They were that bright and motivated. I was never so sure of an investment in my entire life."

This was a chance to start with a blank slate and do things right: "We saw an opportunity to develop new technologies, oriented to making semiconductor memories. We thought we could compete with established companies by putting cleverness back into processing silicon." With Grove listening in, they talked together in Noyce's study at his home on Loyola Drive in Los Altos. It took Moore only five minutes to drive there from Jabil Lane. The pair had no difficulty in agreeing to repeat the formula through which they had so successfully built Fairchild Semiconductor: "Find a fresh market." Gordon's ideas about the promise of complex memory microchips and their ability to move the main memory market away from magnetic cores appealed to Bob Noyce's passion for the new. He was an easy enthusiast when it came to developments in semiconductor technology. It was pointless to try to compete in an established market with big, diversified firms such as TI, Motorola, Fairchild, or Japan's NEC. Instead, they would build the future.

Gordon's emphasis on having the best silicon printing press would, believed Noyce, "insulate" their start-up from competition. They would use novel manufacturing technology, open fresh markets, seize commanding shares, and front-run the competition by advancing their technology relentlessly. Gordon's focus on microchips and chemical printing techniques stayed true to his philosophy of standard products and his concern with the economics of integration. From the earliest years at Fairchild Semiconductor, Gordon had believed that the best products were standard ones, delivering a function that could be sold in bulk to many customers, offering the potential for large markets and dominant positions.

Moore also believed that his more complex microchips would constitute the cheapest form in which transistors could be deployed at the given moment. Their use would therefore spread widely and rapidly—given one critical socioeconomic proviso. Increasing complexity meant more effort in design, implying much greater initial fixed costs. Ever-growing numbers of engineers would necessarily spend ever-increasing amounts of time on ever more recondite circuits. If Moore were to realize his prediction of a steady doubling of microchip complexity to obtain minimum cost, design costs would need to remain under strict control and be spread across hosts of identical standard products.

Gordon Moore worried that his fledgling electronic revolution was losing momentum. As microchips took on larger functions in the logic of computers, markets for any particular microchip were shrinking. Computer makers each had their own different designs for the logic "brains" of their computers and wanted their own special chips—custom, not standard, products. Increased design costs for so many different logic microchips would throw a wrench into Moore's ideal dynamics that would guarantee the revolutionary future. There was an inherent tension: the more complex the circuits, the more restricted and custom the application. Gordon had to find products that broke this tension: complex microchips that could meet a standard, broadly needed application.

Memory promised a way ahead. Unlike logic chips, memory microchips *could* remain standard products. Computer makers already used more or less standard magnetic-core memory arrays. Memory microchips (the more complex in themselves, the better) aimed to replace magnetic cores. With advances in chemical printing technology, newer microchips would drive down the costs of computer memory, allowing computers to attain vastly greater memory. Customers had a seemingly insatiable appetite for memory, as increased memory made digital computers more useful. Memory microchips would optimize both Moore's philosophy of standard products and his economics of integration. And this virtuous sequence could go on indefinitely.

If memory microchips were to become the new dream product, they would have to be smaller, faster, and more reliable; consume less power; and (most important of all) be cheaper. Magnetic cores had a very low cost of a penny per "bit," per 0 or 1 of digital data. This low cost constituted a formidable barrier. To kill cores, Gordon Moore would require superb chemical printing technology, the ingredient that had given Fairchild its competitive edge.

The combination of the best silicon printing press and a fresh microchip market held other charms for Moore. At Fairchild he had learned the importance of first-mover advantages. The original diffused transistors, planar transistors, and planar microchips—products of advanced chemical printing technology—had broken open fresh markets and scooped commanding positions in established territories. Fairchild had grabbed market share and defended it by front-running the competition, through continued investments in its printing press. With memory microchips Gordon could repeat the exercise. The new firm could put competitors on what he later called the treadmill, to play endless games of catch-up. By constant improvement, Moore and Noyce could keep ahead.

If novelty was to be the firm's basic competitive tactic, it would have to be nimble. At Fairchild both men had become profoundly dissatisfied with the way in which disparate interests, hierarchies, and divisions had hampered their ability to move fast. Failing to transfer MOS technology from his R&D lab into manufacturing, Gordon's frustration had slowly built to a boil. Noyce, heading the corporate division and unable to get key executives to appreciate the semiconductor operation, had butted continually against corporate walls. The pair resolved to avoid these problems by eliminating, as far as possible, any internal divisions. As befitted a start-up, their company would have a flat organization to allow expert opinion to rise and enable their own directions to be heard and followed. While encouraging interaction and discussion, the organization would be neither democratic nor egalitarian. There would be no staff votes about what to do. Moore and Noyce were in charge, but they would be easy to find and to approach.

The most radical consequence of earlier frustrations was his and Noyce's decision to banish the boundary between R&D and manufacturing. At Fairchild, "as the manufacturing people became competent, they had wanted to reengineer everything and start over." Now, there would be just one fabrication line, one silicon printing press. They would use it to wrangle novel microchips into working order, to get them into regular production with decent yields, *and* to churn out products in large volumes. Everything would be focused on innovation and on production, simultaneously. This was something Moore drove "pretty strongly." "I'm convinced that technology in the early stages is very delicate. It only works for true believers! You must have people who are convinced it's going to work. We decided not to set up separate facilities for R&D. We committed a fraction of the production operation to it, taking some inefficiency in manufacturing in order to get efficiency in transferring technology. We were not going to set up a separate central laboratory, as we had lived previously." Moore would later judge this decision to have made Intel "very efficient at evolutionary changes in the technology, allowing it to continue to 'mine the vein' effectively, giving fewer opportunities for spin-offs. Intel has benefited from a much higher capture ratio from our R&D than Fairchild did."

Another important decision concerned the use of stock options: agreements to sell stock to employees at a future date, at a set price. Moore, suffering defections among his Fairchild staff, had blamed the derangement of repeated spin-offs on East Coast management. "We weren't asking for anything except stock options for key technical and managerial employees, to slow down the drain to start-ups," but each request was met with a no. Moore was convinced that options were a critical management tool. "Starting over, we'd get to shake off a lot of our previous mistakes." By industry standards, salaries for fresh hires would be modest, but these would be coupled with stock options that were (for the time) huge. Moore felt this was only fair: "The company creates value. Sharing that between the people who put the money in and the people who did the work made sense. The rule was that 90 percent of the value ought to go to the shareholders and 10 percent to the people who generated it as employees. Options were extremely important in rewarding people for the success of the company."

High levels of employee equity were offset by a no-frills utilitarian approach to the material culture of the enterprise. There would be no mahogany row, executive washrooms, or reserved parking places. Instead, there would be simple desks, chairs, and small offices. "We will provide everyone what they need to get their job done, but it will not be done with flair," joked Noyce.

Rounding out their principles, Moore and Noyce committed to a new mode for the silicon printing press that would be foundational for everything else they planned. Instead of making their own tools, they would—as far as possible—buy commercial equipment. At Shockley Semiconductor Moore and the other researchers had created what they needed; in the early years of Fairchild, it was much the same. Now, the situation had changed. Firms had sprung up to supply silicon wafers, chemicals, and diffusion furnaces. Specialized operations lived or died on their ability to give microchip makers what they needed. To buy equipment was to acquire two additional benefits: true expertise in tools and speed. All that was needed was a purchase order.

Having hashed out and agreed on fundamentals, and with Rock's assurance that funding would be a breeze, Gordon Moore and Bob Noyce cast the die. Noyce resigned from Fairchild on June 25. Moore followed a week later, on July 3.

Obfuscations and Resources

Gordon took Betty, his sons, and his mother-in-law away on a vacation in the Sierra Nevada, to the cabin they often rented. Its deck overlooked the south fork of the Stanislas River, with a majestic view of the great pine trees. Ken and Steve, thirteen and eight, took the bunk beds, while Betty's mother slept on the living room sofa.

The family soon had visitors: Andy Grove and his wife, Eva. It was not a social call. Grove "came up to talk about what might be going to happen in the company." With Gordon having quit Fairchild, Andy wanted to make clear that he was committed to following suit. Grandmother occupied the children while the two couples sat on the deck, enjoying a shared meal and the wonderful view. The men then went off to talk, after which the Groves left. "We didn't have any space for them because it was a tiny cabin," says Betty.

Betty believed her children had little notion that something significant was happening, but older son Ken, on the cusp of adolescence and unaware of the family's wealth, had been acutely concerned: "When Dad went off to start a fresh company, it gave me a shock. I thought, 'Wow. He's leaving his job to go to this unfounded thing.' That was a big deal to me. Dad sounded very confident; my mom didn't seem overly concerned, either. I had the worry. It seemed a large change in direction from a very stable situation. We didn't have a lifestyle that led me to believe we were doing well." Younger son Steve was less perturbed: "I remember seeing Bob Noyce and Andy Grove a lot, and Dad being gone, but I wasn't aware of risk taking. I learned later that it was a big decision for my parents."

Returning from vacation, Moore joined Noyce in an intense round of start-up activities. They commissioned an attorney to draw up documents incorporating not MN ("more noise") but NM (Noyce, Moore) Electronics, while they sought "a nonsense word" as a permanent name. They made four attempts before finding a name that cleared the California and New York secretaries of state. Thus was born Integrated Electronics, abridged to the succinct and "sort of sexy" Intel. The name was not without problems: a midwestern hotel chain was called Intelco. They paid it fifteen thousand dollars for naming rights. Another firm with a similar-sounding name agreed that the two businesses "were sufficiently separate that we didn't have to worry." (When that latter firm went bankrupt, recalls Gordon, "my mother-in-law saw it in the paper and called Betty to commiserate.")

The incorporation of Intel was finalized on July 18, 1968. At this point, Noyce departed for the East Coast and a family vacation, leaving Gordon to talk to a reporter from the *Palo Alto Times*. An article duly appeared, announcing Intel to the world. Moore's descriptions of the company's intentions were deliberately vague. They would engage in extended technological work but had not yet decided what to specialize in, only "product areas that none of the manufacturers are supplying." He saw no reason to go into detail. If investors could back the firm to the tune of several hundred thousand dollars each, without detailed information on their technical plans, then the rest of the world could live without such information, too. Gordon did mention the target of hiring fifty people, including a dozen engineers. As the article included the home addresses of both founders, they were soon receiving inquiries; indeed, "Betty was getting calls on our home phone, from people wanting to put money in."

Because the first funds would come from Moore and Noyce themselves, they could start without delay. "Should we put in half a million or a quarter million? We opted for a quarter of a million each. Bob's idea was we put in at $1 a share and then get a first round of financing outside at $5 a share." Arthur Rock pointed out that it was customary to allow the person arranging the financing to make an investment on the same terms as the founders. He invested $10,000, while Moore and Noyce put in $245,000 each. The three now owned half the stock they intended to issue, half of Intel.

They budgeted $200,000 for equipment and buildings and $100,000 for R&D and reserved the rest for working capital. They also set aside 100,000 shares at $5 each, to constitute stock options for employees. Andy Grove was not an original investor in Intel. Instead, he had been "very concerned that somebody should pay me something, because I was living from

6—PALO ALTO TIMES, FRIDAY, AUG. 2, 1968

2 of founders leave Fairchild, form own electronics firm

By MARGE SCANDLING

Two pioneers in the Midpeninsula integrated circuits industry, Dr. Robert Noyce and Dr. Gordon E. Moore, have resigned from Fairchild Semiconductor to start their own business.

Their new company, to be called Intel Corp., will be in the general area of integrated electronics, according to Moore.

Both men were among the original eight founders of Fairchild Semiconductor in September, 1957. Fairchild is now one of the biggest of the many firms making up the highly competitive integrated circuits industry.

Noyce most recently has served as one of four group vice presidents guiding the operations of the parent Fairchild Camera & Instrument. He has been in charge of both Fairchild Semiconductor and Fairchild Instrumentation, an affiliated company.

Moore has been director of Fairchild Semiconductor's Palo Alto Research Laboratory.

Moore, interviewed at Intel's offices at 365 Middlefield Road in Mountain View, said the new company will get under way with a fairly extensive development period that is to last "many months, probably."

DEVELOPMENT PLAN

"Our strategy is to do some extended technological work before coming out with products," he said. "We particularly are interested in product areas that none of the manufacturers are supplying."

What turns both the founders on most is "the idea of getting back in the laboratory." Moore noted that they have found themselves increasingly absorbed in administrative rather than development work.

They didn't leave Fairchild, however, with the idea of starting a company together. Each made plans to leave separately, and then found their interests were so similar that it would be logical to get together.

A lucky break came when they were able to lease some of the space partly vacated by Union Carbide Corp. when it transferred more than 150 of its Mountain View force to a San Diego plant.

"It's larger than we need at first," Moore explained, "but it's equipped with the installations we need for specialty gas and water supplies."

HIRING PLANS

Moore and Noyce expect to hire a staff of about 50 during Intel's development phase, including about a dozen engineers. A few persons are already on the payroll, and Noyce was interviewing others this week while on a trip to the East Coast.

Financing is being arranged privately by Art Rock, San Francisco, who has specialized in finding capital for a number of well-established companies. Among them are Scientific Data Systems and Teledyne Inc.

Moore said he has to be vague about what areas of technology Intel will specialize in, because it hasn't been decided. In general, the company will avoid direct government business and product areas that others are already active in.

It plans to keep to the industrial rather than the consumer part of the industry.

Moore, 39, lives at 23965 Jabil Lane, Los Altos Hills.

He obtained his Ph.D. degree in chemistry and physics in 1954 from California Institute of Technology.

Before joining Fairchild, he was at Shockley Semiconductor Laboratory in Palo Alto with Noyce.

Noyce, 40, holds a Ph.D. degree in physical electronics and solid state physics from Massachusetts Institute of Technology.

His positions after co-founding Fairchild Semiconductor included research and development director, vice president and general manager, and group vice president.

His home is at 11420 Loyola Drive, Los Altos.

ROBERT NOYCE

GORDON MOORE

Local coverage of the creation of Intel.

SOURCE: INTEL.

paycheck to paycheck," and had elicited from Gordon a guarantee that he would be paid even before the company began operations. Accordingly, he was on the payroll before the end of July, with a generous slice of stock options.

In late September Moore, Noyce, and Rock decided that the time was right to sell the other half of Intel for $2.5 million. Enough time had elapsed that eyebrows would not rise. "We had to assemble an organization, show we had done something, to justify marking the stock up," explains Moore. One potential investor wanted to see a business plan before taking the plunge. Noyce produced a one-page typewritten description. It gave away nothing. "The company will engage in research, development, manufacture and sales of integrated electronic structures to fulfill the needs of electronic systems manufacturers," it began. The following sentences were similarly general, to the point of crafted vapidity.

Rock rapidly sold the additional shares. "Those were the days before faxes and answering machines. By the time we reached all of the people we selected and had answers, it was a day and a half or two days later. It would have gone faster if we could have gotten in touch with them quicker. We didn't get a single no." Rock personally subscribed for a further $300,000 worth of shares at the new price and became the chair of Intel's board and the largest shareholder after Moore and Noyce. "Art saw to it that we did reasonable things," says Moore. Fittingly, the rest of the Traitorous Eight were in this first round of investors.

Fairchild had already spawned nearly a dozen other companies. Now there was a fresh explosion of spin-off activity, as—in the space of twelve months—eight more companies joined Intel, all locating themselves within easy reach of Fairchild's Mountain View headquarters. Among them was Advanced Memory Systems, established by engineers from IBM and Fairchild with a very similar aim as Intel's: to make microchips for computer main memory. Gordon Moore was not alone in seeing the potential of microchips. Keeping vague the details of Intel's strategy was only wise.

Strategy, People, and Premises

Years later, Gordon explained his foundational strategy for Intel:

> In the late sixties, the integrated circuit industry had developed processing technology to the level that fairly complex integrated circuits could be manufactured, if only someone could find a reason to make them. . . . The industry was stuck on small and medium scale circuits. . . . In founding Intel, we thought that we saw a way to change this by focusing on semiconductor memory, a function required in all digital systems, and one where

> it seemed that circuits of unlimitedly high complexity could be used to advantage.

While Gordon's focus on microchip memories was definitional for Intel, he also had an eye out for an additional promising prospect: to find an application that, because of the sheer number of products, gave similar economies. The electronic calculator offered that possibility, so Intel began looking for a manufacturer with whom to develop special-purpose calculator chips."

A regular reader of the *New York Times* or the *Wall Street Journal* in the mid-1960s would quickly have noticed a new and growing market: desktop calculators, priced between one and four thousand dollars (that is, tens of thousands at today's dollar values). Calculator makers partnered with microchip producers to feed this market, booming despite the price. Some of the first commercial MOS microchips, made by Fairchild spin-off General Microelectronics, went into an early electronic desktop calculator. Japanese companies such as Sharp and Sony were in the lead, using microchips made by Hitachi and NEC. The world's giant in microchips, TI, was also active. Gordon's eyes were fixed on memory microchips, but he kept calculator chips in his peripheral vision.

Moore and Noyce wanted to recruit experts, men who had impressed them, to help in the ambitious task of building the best possible silicon printing press. Already they had in place Andy Grove, whose sharp mind and exceptional drive would find full scope at Intel. They began to call others. "We are only going to hire perfect people," Noyce, tongue in cheek, told his children. "A small bunch of people who know what they are doing can accomplish much more than a big group who don't."

Les Vadasz, the talented engineer who had led much of the MOS microchip work within Gordon's lab, was first on the list. At Fairchild this small, skinny Hungarian had been "underappreciated and undervalued," even though his knowledge, smart engineering, and good judgment set him apart. Moore wanted him to spearhead Intel's push into MOS microchips. Vadasz had just returned home from his father's funeral when he learned that "they want to talk to you." He quickly met with Moore, Noyce, and Grove, who all asked him to take on MOS engineering at Intel. He jokes that he had to think about it for five nanoseconds:

> It was a no-brainer. You were going with some of the most brilliant men that you've ever known. You're young and have a strong feeling about your own capabilities. If it doesn't work, you still have those capabilities. If it works, it's a great thing. Their view was that technology had advanced enough that you could make very large-scale integrated circuits with thousands of

> components and that there was a function that was natural for this level of integration: semiconductor memories for computers, which was a zero-dollar business at that time.

Moore's need for a manufacturing expert was equally pressing, and he found the ideal candidate in Eugene Flath, an electrical engineer and former US Navy officer. Flath had risen quickly through the ranks at Fairchild, taking charge of the production line for bipolar microchips. (MOS microchips, to Moore's chagrin, remained stuck in the R&D lab.) In taking Grove's internal course in device physics at the R&D lab, Flath had caught Grove's attention. Grove subsequently signaled his intention to leave Fairchild. Flath took this as "really bad news" and immediately offered his services as "a manufacturing guy" for the new company. Grove arranged a 7:00 a.m. breakfast at the International House of Pancakes. On arrival Flath was surprised to see none other than Gordon Moore and Robert Noyce sitting in the booth opposite Grove. Within an hour he had accepted their invitation. The salary on offer was only two-thirds of his present compensation, but the stock option was generous. He drove straight to Fairchild to quit.

Moore and Noyce were more oblique in securing continued access to other talents, including those of Gordon's friend Carver Mead, the semiconductor expert at Caltech. Intel agreed to finance a small laboratory near Caltech's campus in Pasadena. There, Mead began to explore making a blue light-emitting diode (LED), sought after for use in display screens. He became a business partner of Intel and received an Intel employee badge, one of a growing number of "little blue badges that had our names carved into plastic, kept in a mask box at the front desk." Mead had often traveled to Fairchild to consult for Moore; now the LED lab enabled Gordon to use Mead's services again.

Grove, Vadasz, and Flath were high-profile defections. Further poaching carried the risk of Fairchild deciding to sue the new company. Nonetheless, Ted Jenkins soon received a phone call, inviting him to lunch. Jenkins, who had impressed Moore and Grove with his development of a new fast-switching variation of bipolar microchips in the Fairchild R&D lab, jumped at the opportunity they offered to work in Mead's LED project. More important for Gordon and Grove, they had secured one of the few experts on microchip technology. In January 1969 Jenkins moved to Intel to develop what Gordon believed was the most promising version of the silicon printing press.

Along with talent, Intel needed a home. When Moore and Noyce left Shockley, the only way to create a suitable facility for silicon electronics was to find an empty building and customize it. In 1968, in contrast, they leased an established microchip development and fabrication building: the former home of Union Carbide Electronics, at 365 East Middlefield Road, almost comically close to Fairchild's headquarters and also near the Wagon Wheel, the after-hours favorite. The Union Carbide plant had been set up by Jean Hoerni, Moore's former colleague. Ever restless, he had left Teledyne Amelco to help Union Carbide, a major petrochemical company, move into microchips. When this firm decamped to cheaper land in San Diego, it began to empty the plant that Hoerni had helped to design. Moore and Noyce signed a lease through 1974, paying fifty thousand dollars a year.

"We had the beginnings of an operation," says Gordon, in a facility that was at first "larger than we needed." There were a half-dozen hard-wall offices, a conference room, a small cafeteria with vending machines, and a vast, open space for manufacturing. Noyce took an office by the front door, Grove sat nearer to the heart of the building, and Moore's office was between, a few steps from both. On August 1, 1968, all of them sat around the conference table. Jean Jones, previously secretary to Vic Grinich, was there from the beginning, taking notes. Gordon had convinced her to come on board "temporarily" (she would remain his secretary for a quarter century). For the first year, she was Intel's sole administrative worker.

Gordon Moore was delighted by the custom-designed infrastructure at East Middlefield Road. It was a "good start," buying them time, though some pieces of the facility, such as sewer lines, needed refreshing. "We dug up the pipes that went across the front lawn, connecting into the main drain. The cross-section of the pipes was that of an inverted horseshoe. The bottom was completely eaten out. People had been dumping nitric or hydrochloric acid down there for a long time! Nobody had engineered the fact that quantities would increase more than a hundredfold. I'm appalled at some of the things we did in the beginning, as an industry." Over time, it began to be understood that even pouring acid down leaking pipes was less of a problem than putting organic solvents down the drain. Solvents accumulated underground; eventually, the entire area around Whisman and Middlefield Roads would be designated as an EPA superfund site.

Manufacturing processes still had many primitive elements. Designs and layout were done by hand. Carver Mead recalls "an army of people doing the layout on drafting boards, checking it, putting Rubylith [a plastic film] on tables, and cutting it with razor blades. It was all peeled by hand with tweezers." Even so, Moore's ambitions were sophisticated. Having decided to use commercially available materials wherever possible, he now made a strategic decision about what Intel would buy.

The great majority of firms now used 1.5-inch-diameter wafers, but some vendors were beginning to offer 2-inch wafers. Gordon wanted these

The first Intel facility in Mountain View.

SOURCE: INTEL.

larger wafers. A small increase in diameter meant a big increase in wafer area, allowing each batch to contain more devices and offering the potential for a jump in capacity, together with a fall in the manufacturing cost per transistor. In late August Eugene Flath was on vacation near Los Angeles. Given his proximity to that year's Wescon meeting, he was instructed to purchase the "full complement of semiconductor manufacturing equipment." Flath did so, buying diffusion furnaces and other equipment off the show floor. Gordon worked closely with him, being Intel's ranking expert on diffusion and metallization.

For Moore, the start of Intel was comfortingly similar to that of Fairchild. He was responsible for crafting new approaches in device fabrication into a manufacturing process and for delivering products, both through his own hands-on contribution and by guiding others. As before, he made extensive use of laboratory notebooks. He wrote "Intel #1" on the first page of the first book, along with his signature, and began detailing his efforts at adding oxide or epitaxial layers to a wafer using Intel's new reactor. This was good experimental chemistry, "The purpose," he recorded, is "to try and understand the thing."

Moore routinely lunched with employees in the cafeteria, "sitting down with a group of engineers who had been battling with a problem and suggesting, very quietly, that they might like to take a look at this or try that." With key employees Moore was as clear about his strategies as he had been vague with investors and reporters. When he told recruits that Intel was out to create a big market for memory microchips, they all knew exactly what he was talking about. Fairchild's earlier bipolar microchips were, at this time, powering the NASA computer that was to guide the Apollo 8 moon shot. At Intel Moore was preparing for a moon shot of his own. Memory microchips were a known concept, but his was a fresh level of ambition.

Moore had learned to value Noyce's emphasis on the "principle of minimum information," moving forward rapidly with an idea. "Certain people avoided doing the critical experiment. There was always one test that would tell them if it was going to work or not, and they seemed to want to postpone it. Bob was exactly the opposite. I share his feelings. I like to jump in and do that part first." Noyce remained the "idea factory, the individual people gravitated to, the back-patter, the social bird," Intel's figurehead. With his white Mercury Cougar and INTEL plates, and his penchant for flying planes, skiing, and scuba diving, he was responsible for the notion that Intel was sexy and exciting. Moore was quiet, reserved, the deliberative sifter who pursued ideas he thought had merit. "I was pretty good at listening and then neglecting. Bob didn't mind that."

At Intel their offices were close together. "Most of the time, at the start, both of us were there." Informal interaction led to a division of territory by inclination. "Bob had lots of outside contacts and really enjoyed them. I liked the internal stuff. I would watch what we were spending and look at all the programs internally." In the past Moore had discussed technical problems with Noyce. However, Noyce was becoming more invested in outside work, meeting with potential customers and industry representatives. Eight years away from the lab bench, he could not share Moore's close focus on the manufacturing technology.

It was natural that Gordon should, increasingly, turn to Andy Grove on technical and manufacturing issues. He envisaged that Grove would be Intel's director of R&D "if we came to the point where we needed that." Instead, it was soon clear that Grove was ready and willing not only to ponder technical puzzles, but to take responsibility for the whole internal operation of the company: setting schedules and making decisions to ensure goals were achieved. Moore quickly warmed to the idea. "Grove got over his PhD and became very interested in how organizations worked." Grove, though he "hung on every word of Gordon's," countered his boss's tendency to avoid conflict and instead excelled at action. "He wasn't waiting for [Moore] to deal with the organization."

"Andy really respected Gordon," explains Carver Mead. "I think he even loved Gordon in his own funny way." At Fairchild Semiconductor Grove had come to rely on Gordon, his boss. Grove had been only five when his own father disappeared to the Russian front. At their first meeting in 1963,

Moore and Noyce in consultation, circa 1970.

SOURCE: INTEL.

Gordon showed he could be fully present in a way that mattered to the potential recruit. Only seven years older, Moore was career established; more significantly, his self-assurance and quiet confidence were palpable to the young, insecure Grove. From that first meeting onward, Moore would help him "see what I wanted to be."

Noyce, in contrast, left Grove disillusioned. Early in his time at Fairchild, Grove handed in a piece of work one Friday and received a note the following Monday: "'I just read your report on MOS—it's very nice work,' signed: 'Bob.' I found out who Noyce was: general manager and Gordon's boss. Oh, God! I thought I'd died and gone to heaven. I cherish that note and still have it someplace." When the dazzling Noyce then went silent, to Grove it was another abandonment: "That was the first and the last time I heard from Noyce." He judged Noyce inadequate. "I was not a fan, but Gordon said, 'He's better than you think.'" Grove found Noyce to be "a paradox: very private, very approachable the first inch, and after that you couldn't go any further." Conversely, although Gordon could be detached and passive, Grove experienced him as consistent, dependable, and available. "If you wanted to listen to Gordon, he was there; if you wanted to take his advice, he was there."

As Intel grew, some saw Grove as "the brilliant but truculent son in a perpetual Oedipal battle with Noyce." Admittedly, there were plenty of irritations, jealousies, and attempts at displacement, as Noyce, often traveling, continued to exemplify the absent father. More important, however, was how Grove found in Gordon the authority he needed. Moore himself says, "Andy considered me a kind of father figure," while Moore in turn came to see Grove as his own natural successor. Unsurprisingly, Grove's specialty became reading his boss's mind. "Gordon was a passive, shy guy. You can't figure out what he means in the objective context, let alone the emotional context." In turn, Moore found Grove "easy to talk to," not aggressive but deferential. As their relationship deepened, Grove would become an important proxy, amplifying Moore's thoughts, playing out his strategies, and driving Intel forward. "I'd see things in delicate shades of gray," says Moore. "Andy sees them in black-and-white. We'd have a discussion: 'Maybe we ought to do that.' He'd go out with, 'This is the way we're going to do it!'"

To Jean Jones, Grove's sense of responsibility and its twin—authority—were evident from the start. "I was surprised by his youth and by his obvious sense of command," she reflects. "He knew he was in charge." Grove might act "rough and very businesslike," but this stemmed in part from his concern to make Intel a success. A telling episode occurred late in December. That first Christmas most Intel employees, assembly workers and technicians, wanted a party. Jones remembers that Gordon had decided to put off the party until after December 31, reflecting how seriously he took his work and the breakneck pace adopted to launch the manufacturing technology. "Gordon didn't wish to take the time and had decided to have a holiday party in January instead." She continues, "This didn't sit too well. I was relieving the switchboard at lunchtime, and Andy came out, crouched down next to me in this little booth, and said, 'Did you know about the unhappiness?' I said, 'I know there is some unrest, but I didn't give it too much thought.' He said, 'Why didn't you tell me?' I said, 'It didn't bother me whether we had a party or not.' He said, 'Whenever you hear things like this, Jean, you should come and tell me.'"

Even more revealing was what happened the next day. Grove suspected the "girls in assembly" were holding their own Christmas party in a small lounge connected to the ladies' room and were drinking on the job. He asked Jones to see if he was right. "'Go in there and visit for a little bit. When you come out, either nod or shake your head.' I went in. The first thing they did was hand me a paper cup filled with orange juice and vodka that I could barely swallow. I left, and I nodded my head." Grove, more

Intel's staff in front of the original plant, 1968. Moore is at bottom right, with Andrew Grove behind him to the right.

SOURCE: INTEL.

conscious of immediate human realities and needs than his future-focused boss, gave everybody the rest of the day off, rescuing morale while also saving the "girls" their jobs.

"Goldilocks" and a Christmas Bet

To get ahead, Intel needed to accomplish two things simultaneously: to build the best printing press and to develop memory microchips. The design of any particular chip was highly dependent on the fine details of the manufacturing technology: how closely transistors could be spaced or how aluminum interconnection lines were run across the surface of the chip. The cutting-edge manufacturing technology that most excited Gordon—the silicon-gate MOS approach pursued in his Fairchild R&D lab—promised an advantage in packing transistors much closer together. Its development into a real press, to print microchips, required working with actual chip designs. The printing press and its products needed to be developed together. Intel needed designs.

Intel and Gordon were by no means the first to contemplate memory microchips. By the time Moore and Noyce were launched, there was already a niche market for memory microchips in which Fairchild was active, along with TI, Signetics, IBM, and others. These were small-capacity, "static random-access memory" microchips: SRAMs. Random-access memory was the principal type of memory in computers, built like a wall of boxes stacked in columns and rows. The computer could reach in and get the data from any particular box, through random access.

Intel managed to poach one of the industry's most experienced designers, H. T. Chua, from Fairchild. As Moore puts it, "We didn't want to hire everybody out of Fairchild—in fact, we tried hard not to," but "as with a large passenger ship that was badly damaged, anyone with the chance to move to a more seaworthy vessel understood the advantages of doing so." Arriving at Intel in the opening weeks, Chua set to work on an immediate opportunity: the possibility that the firm could capture an order from Honeywell, then a major producer of computers. Honeywell was seeking a small, fast memory chip for use as a "scratch pad." (At that time, computers were designed with the need to store small amounts of data while performing other operations, much as a Post-it note today can be used to scribble down fleeting thoughts.) Honeywell agreed with Intel that a 64-bit bipolar SRAM would fit the bill. Six other companies were already competing to build prototypes. Chua set to work designing what would become known as the "3101," Intel's first memory microchip.

To invade the memory market, Moore decided he needed to employ a multipronged approach, one he would later dub his "Goldilocks" strategy. He developed three distinct approaches, hedging his bets as to which would prove the most tempting. Bipolar was the most straightforward. High-capacity silicon-gate MOS microchips were a second, untested, possibility. Silicon-gate MOS, recalls Gordon, "would be useful in small memory modules. We could put four times as much memory on the same size chip as bipolar. The chip was a lot slower, but most of the applications were not speed sensitive." The third approach was to hedge the possibility that the economics of complex silicon-gate microchips would fail, namely, a project to build a multichip memory from several less complex silicon-gate chips.

Gordon needed a designer for his silicon-gate MOS approach. Joel Karp was working at General Microelectronics in Santa Clara in 1968. There, he had designed MOS microchips for an early desk calculator, as well as designing shift registers—very simple chips that could be used for both logic and memory. Karp was skeptical about General Microelectronics' fate (following its takeover by Philco, Bob Noyce's old employer in Philadelphia), and on hearing about Intel he immediately sent over his résumé. Grove and Vadasz scooped up Karp and set him to work, to design what they were calling the "1101" memory microchip: a SRAM capable of holding 256 bits of data (zeroes or ones), made using the silicon-gate MOS process. This complex, more capable chip would, Gordon hoped, have a cost per bit that could give magnetic core memories a run for their money. Karp pulled out paper and pencil and got to work.

The ultimate goal was to push magnetic cores entirely out of computer main memory (then the most expensive part of computers). For this, Intel needed its own computer expert. Marcian (Ted) Hoff at Stanford University, a PhD in its computer science laboratory, was duly hired. "Ted was very knowledgeable about the architecture of computers," says Moore. "We were mostly device people. We knew our products would become increasingly complex. He could look in the other direction at what kind of products the market would use." Hoff's job was to locate customers and promote the use of Intel's chips. Max Palevsky, one of Intel's early board members, lobbied for a "computer guru" on the board itself, "somebody who really understood computers. The problem was that Max represented the old computer industry, and the territory was changing dramatically," says Moore. Instead, he hoped Hoff could say "what products we ought to be making," ensuring Intel's memory microchips were appropriate.

Moore's overriding concern was to get manufacturing going for both bipolar and silicon-gate MOS microchips. If he could make them on larger wafers, he could establish dominance in the emerging market before behemoths such as TI, Fairchild, and NEC got in and took over. The stakes were high. To encourage the spirit of competition, he organized a company-wide bet on whether the silicon printing press would be running by the end of 1968. Moore wagered that by midnight on December 31, his staff would have produced both an MOS and a bipolar transistor that met defined specifications. Would they succeed or not? Everyone took sides. The stake was a bottle of Napoleon brandy, signed by all parties, with the winners throwing a party for the losers. "I said we could do it. Andy said we couldn't. I was working on making it happen, and he was cracking the whip! Flath and the process guys were all on my side, but most of management took the other side, to make it a worthwhile bet."

Moore would recall this period of concentrated hard work as "the advantage of the small company, training all our good people on one project, aiming in a narrow direction to solve some particular problems. It was the clean-sheet method of establishing standards: 'Shoot, then paint the bull's-eye.'" With the future of Intel at stake, Gordon returned to the laboratory bench, enjoying the opportunity to tackle directly the material challenges and to test the capabilities and foibles of the firm's equipment. With an assistant, Larry Brown, he ran all the evaporators used to deposit aluminum onto the wafers. Because Moore had been first in the industry to figure out how to use all-aluminum contacts, the metallization would be a piece of cake.

One of the evaporators, he decided, should be devoted to figuring out the tougher problem of how exactly to make silicon-gate MOS microchips. Such novel devices could not leave the laboratory until certain problems were solved. Silicon gate required that a film of polycrystalline silicon be deposited atop the silicon dioxide layer that, in turn, was stuck to the surface of the wafer. No one had yet figured out exactly how to create this last layer. Gordon's experiments were a disaster. On a first attempt, he left the wafers in the evaporator overnight to cool. Next morning, all the wafers were ruined: the polycrystalline silicon films had rolled up into little scrolls, ripping off the oxide layer in random spots. Varying the evaporation procedure did not help. The film "would go down," recalls Gene Flath. "It would look nice. We would sit there and watch; pretty soon the surface would start to roll up, like a sardine can." Then a new hire described how he had, accidentally, grown polycrystalline silicon on top of an oxide in an epitaxial reactor by a different route, thus avoiding the issue.

Even so, problems never seemed to cease. The team had a real challenge with what Moore called "broken metal." When they put down aluminum on top of the oxide layers covering their structures, the aluminum would crack, ruining the devices. On December 20 Gordon came up with a suggestion. In his glassworking experiences at Berkeley and Caltech, Moore had used the common technique of "fire polishing" to smooth out sharp edges: by employing a flame to melt the edge slightly. Gordon saw a connection to the "broken metal" problem with silicon-gate MOS transistors: "The edges were very sharp, and when you tried to run an aluminum film over them, it would cut the film. I thought, 'Why not fire-polish it?'" Putting flame to the chip was out of the question, so Gordon took a chemical approach. He suggested adding a bit of phosphorus to the chip's oxide layer, thereby lowering the melting point. With gentle heating, the oxide layer would reflow, filling in the notches and rounding out the edges. The metal atop it would no longer break.

A different problem surfaced on Christmas Eve. Gordon was among the last to leave. In a concession to the season, he joined assistant Larry Brown on the well-worn track to the Wagon Wheel. Meanwhile, Flath, lingering at East Middlefield Road, walked by the evaporator room and noticed a trickle of water appearing under the door. The evaporator had sprung a leak in its cooling system and was starting to flood the building. He found Bob

The materials-evaporation setup operated by Moore during Intel's early days.
SOURCE: INTEL.

Noyce, himself on the point of leaving. Together they shut off the evaporator and took out the mops, averting disaster. A half hour later, they arrived at the Wagon Wheel to cries of "Where have you been?"

The bet came down to the very last day. On December 31 Moore judged that they had indeed met the technical definition of success. They had the basics in place and could make test batches of both MOS and bipolar microchips. As victor, Gordon kept the signed bottle of Napoleon brandy. It was also up to him to throw the losing side a party, a holiday party, just as he had planned. Bob Noyce had been on the wrong side of the bet, and his wife had, in any case, declared that she wanted no more to do with business socializing. Abandoning their natural, private mode, Gordon and Betty perforce invited everyone to their house on Jabil Lane. Some thirty employees came with their spouses or friends and drank Boston Fish House punch in the Moores' living room. "Everybody at Intel, down to the janitor, fit in our house," recalls Betty, who—if at arm's length—considered herself part of Gordon's new enterprise. "It shows how small we were. From then on, it rolled like a snowball."

INTO PRODUCTION

Pinholes and Progress

From January through April 1969, Gordon divided his efforts between the problem of broken metal and a novel and more vexing issue: pinholes. Small ruinous holes were appearing in the oxide layers, leading to electrical shorts and killing yields. He promptly commandeered the reactors, looking at how to vary settings, steps, and materials. He also instituted an inspection process, examining wafers under a microscope to map where holes appeared. He hoped to solve the problem by his preferred mode of measuring and analyzing.

Thomas Rowe, an experienced production engineer hired by Flath, took up Moore's reflow solution to the broken-metal problem. It worked perfectly. On March 5, 1969, Gordon recorded the details in his lab notebook and had Grove witness the event. Rowe wrote a paper on silicon-gate MOS, with Moore as coauthor, to foster Intel's technical reputation. The paper was deliberately vague on details. Code-named "anneal," reflow became a closely held trade secret and a key part of Intel's operation. "That turned out to be an important step in making silicon-gate work," Gordon recalls. He was issued a patent on the process in 1974.

Using reflow the MOS team succeeded in making its first working memory microchip that month. The whole company squeezed into the cafeteria for champagne, and Gordon called Bob Noyce, incapacitated in an

Aspen hospital, having broken his leg while skiing, to pass on the news. Grove felt it had been "one hell of a month," but with the MOS success he recorded, "General euphoria descended." If they could make one working MOS memory, they could make many; yet unless they could beat the pinholes that killed yield, they would be uncompetitive in the marketplace.

Moore continued to battle pinholes. He ran experiments, putting wafers through the reactor with different variables, examining them under the microscope, plotting the issues, and trying possible solutions. His efforts paid off in June. The incidence of pinholes could be reduced, he discovered, by controlling temperatures, just as he had done at Fairchild while developing all-aluminum contacts. At Tom Rowe's suggestion, they also began to use a double oxide layer. If pinholes were weak spots in the oxide, why not grow a first layer, remove the weak regions with an acid dip, and then grow another oxide layer on top? "Unless the pinholes lined up," explains Moore, "you didn't have a problem." The solution appealed to his pragmatic mind: it was a process that "let you be a bit sloppy and still be okay. The kind of process I liked!"

Moore (*standing*) and Noyce deliver boxes of memory microchips to Intel's first customer, Tony Hamilton of Avnet, 1969.

SOURCE: INTEL.

With pinholes eliminated and broken metal solved by reflow, both by Gordon's hand, the production room at 365 East Middlefield Road began turning out 1101 microchips. Chip designer Joel Karp kept watch during these early runs and remembers listening to the moon landing on July 20, 1969, on a transistor radio. The moment caught imaginations worldwide, but was especially poignant for those like Moore and Noyce who had been involved in the creation of Micrologic microchips at Fairchild, the chips that powered the Apollo guidance computer directing the Lunar Module. As Intel reached silicon-gate MOS milestones, a growing ebullience inside the firm resonated with Americans' wider thrill at reaching the moon ahead of the Soviets. Gordon had launched Fairchild as *Sputnik* went into orbit. Now, as the Americans took the lead and Neil Armstrong walked on the moon, Intel celebrated its first year of success.

Even with Moore's reflow and pinholes controlled, other problems meant yield remained abysmal. Typically, a two-inch diameter wafer produced just two working 1101s. "We threw away a lot more than we kept. The numbers we worked with were terrible." For Intel to really be in business, it would need to produce at least twenty working devices per wafer. At Fairchild Moore had engaged in a major cleanup process to improve yields for the silicon mesa transistor, but at Intel cleanup lagged behind. "The investment needed was pretty hard to swallow. We didn't want to spend more than we had to. Keeping things clean above the benches was important, but we had this idea that dirt below didn't make much difference. We tried to make clean benches, but people walked around in their regular shoes. We were really sloppy."

With the help of Tom Rowe, Gordon embarked on a series of chemical experiments to improve the metallization process, exploring the use of different "dips" to clean the silicon surface before depositing the aluminum. Gordon had Rowe run wafers through the established steps and then change the composition of the dip and run more tests. In early August Rowe processed yet another formulation. When the wafer reached the bench of the man responsible for testing, he yelled, "Holy hell! Look what's going on here!" The wafer had not two good devices, but *twenty-five.* The commotion had everyone running to look. Grove and Vadasz were men whose hopes, careers, and futures were pinned on silicon-gate MOS technology. As they watched wafers coming off the process line, Rowe explained about the formulation of the dip. Moore and his team had hurdled the last barrier.

Quietly thrilled, Gordon Moore showed little overt emotion. "When something really great happened," explains his son Steve, "he'd be in a good mood, but there were no major celebrations." In contrast, Vadasz began shouting, "It's a superdip! A superdip!" Overnight, yield for the silicon-gate 1101 SRAM leaped to meet the goal. Moore had his silicon-gate printing press, the firm was in business, and everyone could be confident about the future. The production line began to issue 1101 chips in profusion. In mid-August 1969, Karp was rewarded with a trip to Boston, at that time a hotbed for minicomputer makers. He introduced the 1101 to a meeting of electronic component distributors. For Intel, it was a successful first step. Moore remembers, "The 1101 returned more than it cost us to make. It wasn't enough to grow a big company on, but it was enough to get us started on the road."

The DRAM and the Microprocessor

Silicon-gate MOS microchips became Intel's primary focus. Gordon had reveled in the opportunity to return to the laboratory bench. To advance the chemical printing technology, safeguard yields, realize a future full of ever more complex microchips, and lower the cost of electronics, there would always be material problems to tackle. Even so, it was time to hand the problems over to others. The year-old firm needed him to step up into leadership and focus on juggling the emerging opportunities. Gordon had played the key role in establishing Intel's printing press. Now, he needed to figure out how best to take advantage of it. "I never really returned to the lab," he says. "You can't do it part-time very effectively."

Bob Noyce, for his part, was looking at how to sell complex microchips to a desktop calculator maker. Noyce was well known within the Japanese electronics community, having traveled there frequently. A Japanese business machine manufacturer, Busicom, was interested. On April 28, 1969, Intel and Busicom signed an agreement to jointly develop microchips for a calculator. It called for Busicom to pay for design work at Intel and to have exclusive rights to the microchips made for it.

To Moore, the agreement made sense. Japanese manufacturers were already delivering fifty thousand calculators each year (more than the total annual production of digital computers). Not only would Busicom pay Intel's design costs, but if the calculator was successful, it would also require a large run of custom chips: a standard product profiting from the economics of integration. In June Busicom engineers traveled to Mountain View. They wanted nine different complex custom MOS chips, a design request that caused immediate heartburn. Ted Hoff knew that Intel did not have the engineering staff to design that many chips, given the push on memory microchips. In desperation, Hoff came up with a fresh approach to the Busicom calculator, inspired by the latest silicon-gate chip designs at Intel.

After designing the 1101 SRAM, Joel Karp had moved immediately to work on a memory chip that would hold four times the amount of data: a 1,000-bit, or "1-kilobit (1K)," memory. Its basic design came from the computer maker Honeywell. There, William (Bill) Regitz had designed a novel type of memory chip called, euphemistically, a "dynamic" memory. What this meant was that the chip could maintain data for only a limited time, so it had to be constantly "refreshed," spending 10 to 20 percent of its time reminding itself of the data it was supposed to store. SRAMs such as the 1101, in contrast, had no need for such reminding. In dynamic RAMs (DRAMs) the drawback of having to refresh data was ameliorated by a simplified design that allowed more capacity to be put on a single chip.

Honeywell approached Intel and several of its competitors about making the DRAM it had designed. When Intel's engineers looked at the designs, the feedback they gave was straight out of Gordon's playbook: at 512 bits, the Honeywell DRAM didn't make economic sense. If, on the other hand, they switched to 1 kilobit (actually 1024 bits)—a more complex design—then the chip would be at the sweet spot of MOS technology and would have the optimal production cost. Honeywell agreed, and Intel set Karp to designing the chip, dubbed the 1102. With their intimate knowledge of design tricks for silicon-gate technology, Karp and his colleagues saw that a simpler design *was* possible. They also began work on a design for their own 1K DRAM, called the 1103.

Ted Hoff had Intel's proposed 1103 DRAM in mind as he rethought the Busicom calculator. If Busicom used more memory, like the 1103 DRAM, then the calculator's logic could be much simpler. Then Hoff realized that he could get rid of Busicom's custom logic altogether, by reimagining the calculator as a simple general-purpose computer that could be programmed by software to *act* as a calculator. Critically, he saw that he could make the central processing unit, the essence of the simple computer, on a single microchip of the same complexity as the 1103. With a "microprocessor," the CPU chip, Busicom's calculator would need just a few chips. With such a shift, Hoff reasoned, Intel might tackle the job. This fresh approach—using a programmed standard microprocessor instead of custom-logic chips—might have great reach. Gordon was quickly enthusiastic; he approved Hoff's pursuit. In September 1969 Hoff began the high-level design, the "architecture," for this microprocessor, known within Intel as the 4004.

Intel and Busicom signed a formal agreement about the 4004's development. Gordon recalls, "The only way we could do this project was to convince the Japanese to throw away their design work and accept our approach. They sent a chief engineer over, and we gave him our pitch. He said, 'All right. We'll do it your way.' I was flabbergasted at how quickly he converted. If they hadn't accepted our way, we would have had to say no, because we couldn't hire people that fast—there weren't that many people capable of designing complex chips."

With the architecture for the 4004 microchip mapped out, the next step was to design the actual circuitry. Gordon supported a decision to let this hang: "Busicom went away and left us for a few months to get organized." Intel, still in Moore's view "a small operation," had a more pressing issue to address: getting the 1102 and 1103 DRAMs out the door. These chips would be cheap enough to compete with magnetic cores. They were key to the market for memory microchips. The 4004 microprocessor could wait. "It required several engineers who had the right background and could be devoted to it," says Gordon. "We were just too busy doing memory chips."

The big bet was on the 1103 DRAM. The fate of the company rested on this larger-capacity chip, with which it hoped to dominate the lucrative market of computer main memory. With a large production volume, and enough customers, Intel could offer a bit of main memory for one cent: the same price as magnetic-core arrays. All this lay in the future. Meanwhile, money worries were pressing. Moore's Law encouraged forward planning and risk taking, but Intel needed actual sales to offset its growing expenses. Carver Mead remembers a "1969 crunch" when "we all swallowed really hard because the bottom went out of everything. There weren't any buyers; Bob Noyce was pacing the halls."

Salvation appeared when a silicon-gate opportunity walked in the door. In 1969 computers had multiple users. Clients interacted with mainframes by dropping off IBM punch cards at a central facility and picking up a printout of the results hours or even days later. However, a new mode of interacting with computers—both mainframes and smaller-scale minicomputers—was on the rise: "time-sharing." Multiple terminals were connected (by direct wiring or telephone) to a single computer. The computer divided its attention among the terminals, switching from one user to another. For the client at his terminal, the interaction was direct and personal, without meaningful delay.

Terminals were little more than glorified typewriters, with the computer printing out its responses on paper. Then, in 1969 computer makers began to offer video terminals, with a cathode-ray tube display and a keyboard. These allowed clients to type in commands and see the results instantly printed as text on a glowing screen. A start-up, Computer Terminals Corporation (CTC), was getting into the business and looking for the microchips it needed. For Moore, this was a welcome opportunity. The necessary microchips, "shift registers," were easy to make, and Intel could crank them out even as it worked on the 1103 DRAM. A deal was quickly signed with CTC.

Mead recalls walking into Gordon's office sometime later: "I said, 'How's it going?' and he said, 'Great!' That wasn't what I was used to hearing. 'Oh, what's going on? Are you selling stuff?' 'Yeah. Shift registers. Before long, the world's going to be coated with a monolayer of shift registers.' It was the first product that made money. Intel was going nowhere in a hurry, and this took them into volume production. It started a real business."

Interlude: Keeping Close

Gordon Moore had put aside his lab work to focus on strategic matters. At home he compensated by sharing his love of practical work and experimentation with his sons. "Gordon had a 'shop' under the house," says Betty. Thrifty as always, "He would go to Sears, buy these great big chests, and fill them with Craftsman tools." Ken recalls making things with his father. "He engineers and builds things super well. He taught me a lot of woodworking. We built a fancy octagonal-shaped table and the shelving to line our family room. Dad built tons of things. I used to watch him in his shop. That was always a kick." Ken also started to take apart old automobiles. Steve, too, became keenly interested in cars.

The year 1969 was a financially ambitious one for the Moores. Having completed their home in Los Altos Hills in the early 1960s, Betty began a "big project" to build a beach house on a lot they had bought at Davenport Landing, on the coast between Santa Cruz and Pescadero. The house featured a huge fireplace and wonderful sea views. The family was soon spending weekends and summertime there. Gordon commuted back and forth to Mountain View, an hour's drive: "I had my little Porsche and could go zipping over the hill. I used to go down with my net on the beach and catch smelt. I never became a surfer in spite of the waves. The water's too cold." Christmas was celebrated at the beach house, with family members. "We would have tree trimming on Christmas Eve, with a huge pot of stew and great crusty bread and a big gooey dessert," says Betty. "It would be storming like crazy outside. Gordon's mother would never stay near the ocean. 'It might roar up and take me out.'"

Both Gordon and Betty are private people who do not enjoy entertaining outside the immediate family. Son Ken notes, "Mom's particular about who comes in her home. Dad's more of a typical guy and isn't lined up on entertaining." They rarely had guests, but would often see Gordon's parents, Walter Harold and Mira Moore, still living in Redwood City. There were regular fishing trips and journeys to visit brother Walt Jr.'s ranch in Pescadero. "If we were together for longer periods, we'd get into poker playing without real money," remembers Ken. "We did a lot of things together as a family. My dad and his brothers had similar interests: fishing and hunting."

Even though Walter Harold Moore was hard of hearing and could be difficult to communicate with, he was still closely involved in his grandsons' lives. Younger son Steve found him "very tall and imposing." Like Gordon, Grandpa enjoyed making things. "We'd run errands in his truck or go to his house," recalls Steve. "We worked with wood, gluing things together, doing stuff where we could help him. I was always interested in seeing the things he had hanging in his garage." Ken and Steve also spent time with the Noyce children, whose backyard was extensively remodeled. Ken remembers "two lakes and a little rapids between. You could kayak."

Betty Moore had little in common with Noyce's first wife, Betty. Socializing between the families gradually dried up. Bob and his wife grew estranged, with Betty Noyce going east to their house in Maine in the summers, leaving Bob in Los Altos Hills. He became a frequent visitor at the Moores' beach house. Ken remembers, "I was building a little meter where you push a button to track a sine wave. He sat down and asked every question about it. He was such an engaging personality. He had a booming, resonant voice. You couldn't help but like him." Betty encouraged Gordon to "bring Bob along to Davenport Landing. We'd put the kids together in the boys' bedroom, and we had a guest bedroom." Other times, "we'd take him out on our boat, or he would meet us at our boat; sometimes he would be at the helm. He'd have a day away from Los Altos, then drive with Gordon back to the office." Moore and Noyce were closer during this period than ever before—"Bob was the only person I was likely to have interaction with, principally because he was left by himself in the summers"—yet Moore, characteristically, maintained a degree of detachment.

Betty remained focused on family responsibilities: planning meals, cooking, and shopping. "She had some housecleaning help," says Steve, "but my parents have never been much into personal help for anything. We're a do-it-ourselves family." Gordon's routine was unchanged: off to work early each morning and home for dinner with the family. After dinner he would rinse the dishes and put them in the dishwasher, before turning back to business concerns.

As the boys grew older, Ken began to enjoy science at school: "I wanted to know how things worked." He experimented with electronics. "Dad was bringing me great things. Though transistors cost around forty dollars, I could have the rejects for free. My grandfather smoked a lot of cigars, and we'd get his boxes. You could build everything in them." Using a piece of cardboard, a ninety-volt battery, and some early integrated circuits, Ken built a binary calculator for his seventh grade science class. "To me this was neat. For Dad it was fun. When I brought the calculator in, the science teacher stopped the class while I presented how it worked. I announced, 'Solid-state is going to wipe vacuum tubes off the face of the planet and is already doing so.'"

Adolescence brought real challenges, as the sixties became an era of youth rebellion. "Even without the Intel side of the story," says Betty, "those years were recipes for stress." Both boys went to the local public schools, but with the Bay Area in the vanguard of student revolt, Ken's progressive high school gained a reputation for delinquency. Gordon believed his elder son "goofed off" and "didn't do anywhere near what he could have." Ken was bored. "I drove Dad crazy. He couldn't understand why I didn't study—I didn't see any value in it." Instead, Ken, like Gordon, enjoyed making a bang. "Dad had taught me how to make nitrogen tri-iodide, a contact explosive. It gives a little purple cloud and leaves a yellow iodine stain on your fingers, as if you've been burned. I brought some into school and put it on the locker bank. 'Boom!' I was suspended. Dad couldn't be too unhappy, because he had thrown some down on the floor in his history class."

As Ken's behavior deteriorated, things reached the point where Gordon and Betty reluctantly took him to a counselor. Ken recalls how the counselor quickly picked up on the family's inexpressive style: "Gee, you guys are all smiling at each other. You're not really communicating your feelings. You're all gussying it up with this facade of niceness." If angry, Betty might "let it out," but Gordon would not. "He would be quiet or disappear" and, as ever, focus on practical tasks. Gordon was keen to pass on his own work ethic. Ken, when in third grade, was promised a longed-for astronomy book if he achieved an A grade. "I remember writing the agreement down on a little card. Of course, I worked extra hard and earned it." Later, Gordon promised Ken three hundred dollars for finishing high school, three thousand dollars for a bachelor's degree, and thirty thousand dollars for a PhD.

Trying to keep Ken out of trouble, Gordon and Betty insisted their rebellious son do chores on weekends. Gordon would stand on the deck railing and clear the gutters, while Ken whacked down weeds with a hoe, dug ditches, chopped wood, and tidied the driveway. "Dad was the taskmaster," he recalls. "No lounging around in bed in the morning. He would come right in and wake me up. 'No lollygagging teenagers. You get up and you do something.' Once, I was ill, and he thought I was trying to slack off. He put me to work, and I came in throwing up. It was the first time he said, 'Gee, I'm sorry,' though I think Mom said it for him."

Steve, five years younger, was studious and quiet like his father, "the opposite of Ken," says Gordon. Allergies delayed his walking, and celiac disease led to bullying at his grade school. Betty moved him to a private academy attended by Bob Noyce's children. Outside of school hours, Steve built car models and set up an electric track in the attic. In the evenings Gordon would sometimes put work aside. "We'd go out and throw the baseball around," remembers Steve. "He let us find what we liked to do."

Ken and Steve received a small weekly allowance: first a quarter, then a dollar. Gordon and Betty were determined not to let wealth ruin their sons. "My father had too much too soon, and it killed his life," says Betty. Ken took on paid cleaning and repair work for a neighbor. "My parents' emphasis on work wasn't so much for money; it was to tie me up timewise. It was a responsibility lesson as well," he says. "As long as I made fourteen to sixteen dollars a week, it allowed me to take my girlfriend out and go to the movies." Working for the woman who ran Common College, he mixed with "residual hippie types" and grew his hair long. "I came from the strait-laced side, so it was absolutely fascinating."

Gordon encouraged Ken's interest in electronics. "My bedroom began to look like an electronics store. I had massive cabling running in all directions." He launched a small television repair business and developed a love of short-wave radio. "I used to listen, send reports, and get cute little cards back. I had one contact in a Communist country. Dad used to joke, 'It's going to be hard for me to get top-secret clearances if you keep having all this stuff sent here.'" Ken also spent time at Intel's offices on East Middlefield Road, where he was allowed to use a minicomputer. "A machine with 8K of memory was huge. I had to be really careful in my coding to save space and be as efficient as possible. I ended up applying that, going into information technology and coding for a living." Gordon was "not much of a circuit man," but to his eldest son, computers were "logical." Gordon says, "If I didn't have the Intel technician available, Ken would be the one I'd call when I had a problem with my computer."

A GOING CONCERN

Funding Success

While Gordon Moore oversaw Intel's technical work, most of Bob Noyce's considerable energy was devoted to business and financial questions. One early issue was that Fairchild's lawyers actively contemplated suing Intel for theft of trade secrets. However, Sherman Fairchild was evidently reluctant to pursue the two men who had done so much to build his firm's success; there was little appetite for redress.

Noyce traveled extensively to see computer makers such as Honeywell and Burroughs, to promote microchip memory. He also oversaw the purchase of a twenty-six-acre pear orchard in Santa Clara, for expansion. Moore and Noyce were thinking big, envisaging the moment when their Union Carbide plant would be outgrown and they could move to a spacious, custom-built headquarters and factory. The purchase mirrored and confirmed the shifting locus for silicon start-ups—from Shockley Semiconductor on the edge of Palo Alto to the fruit groves of the Valley of Heart's Delight, a dozen miles south: cheap, flat, available land in what would become a (metaphorical) valley of silicon.

As anticipated, the start-up months were all about expenses, not sales. Almost $500,000 had gone out the door by year end. In 1969 the pace accelerated. While sales reached $370,000, the year's final result was a loss of almost $2 million. Happily, there were the shift registers and the signs of progress with silicon-gate MOS technology. The reputations of Gordon Moore and Bob Noyce remained compelling, even as the broader US economy began to experience the strains of outgoing president Lyndon B. Johnson's "guns and butter" approach to escalating war in Vietnam. Bob Noyce worked closely with Intel's chairman, Arthur Rock, to pull in fresh funds. The existing stock was split, 1.75 shares for every existing share, while 285,000 additional shares were issued at the significantly enhanced price of $14. The rapid placement of this fresh stock gave Moore and Noyce a further $3.9 million for business. At this point, they launched the planned program to enable employees to purchase the still private stock.

Moore and Noyce knew that Intel required even more money in its coffers. Their plan was to take the company public, to get much-needed cash. Yet 1970 was a horrible time for an initial public offering (IPO). The country was in turmoil, as President Nixon's administration expanded the scope of the conflict raging in Vietnam. When US troops began fighting on the ground in Cambodia, accompanied by significant air raids, America saw a massive revolt at home. Bombs exploded on university campuses, and troops even shot and killed protesting students. The economy languished in recession, and the stock market lost some 20 percent of its value. Technology companies, especially computer makers, saw massive sell-offs. IBM fell by more than 30 percent. TI and others suffered similar declines. The prospects for an Intel IPO did not look bright.

Then one of Intel's larger investors, Fayez Sarofim, stepped in. His confidence in Gordon, semiconductors, and Intel was high. He suggested a private offering through his Houston-based investment firm. Sarofim was a graduate of Berkeley and a classmate of Arthur Rock at the Harvard Business School. He had already managed Gordon Moore's growing personal wealth for years. In 1970 it was a "no-brainer" that private stock was the only option for Intel; Sarofim brought in a welcome $1.5 million.

Second Sourcing

The transistor's early career was built on being an essential ingredient in Cold War technology. Unsurprisingly, it became common practice for military customers to demand a "second source" for any semiconductor product they were buying. A microchip maker needed to convince customers that there was at least one other maker prepared to supply exactly the same device, thereby ensuring that the customer would not be held hostage over price and that supply would remain steady. If Intel's 1103 DRAM was to break open the market for main memory, computer makers were going to demand a second source. Intel located a solution three thousand miles away in Ottawa, Canada. There, a joint venture between the manufacturing arm of Bell Canada and the Canadian government was looking to get into the microchip business. Microsystems International, Ltd. (MIL) planned to make microchips for Bell Canada and for the open market. It was cash rich but had no experience. Intel, needing cash and with leading-edge technology, was the ideal partner.

Confirming the growing awareness of Intel's prowess barely two years after its launch, MIL agreed to make a comprehensive technology purchase, paying $1.5 million to have Intel set up a manufacturing line and get 1103s into production, in Ottawa. MIL would pay a royalty for each 1103 sold and $500,000 in cash if production targets were met. The first installment of MIL funds halved Intel's loss for 1970, to just under $1 million. "Historically, technology transfers have been, 'Here's the recipe. Go do it yourself, and I probably left some things out!'" This time the deal included more than just "the cookbooks on how to make the process work." By paying a sum equal to Intel's net worth, MIL acquired a silicon-gate printing press, the 1103, and even the design for a planned microprocessor—the 4005—similar to Busicom's 4004, but wholly owned by Intel.

Moore says the venture with MIL was "a major decision about what we were willing to do and how we were willing to do it." Andy Grove was not happy. Already, on hearing about earlier ramifying plans, he had been thinking, "Go away, go away, we don't have time for this." The MIL development was worse. Grove was in the midst of trying to raise the 1103 yield, still far too low. Now he would have to send his best workers to Ottawa for a significant period. Despite Grove's objections, Moore and Noyce signed the contracts in July 1970. They were confident that they could pull it off. And they were ruthless, if congenial, in exploiting Andy Grove.

Challenging experiences in early life had made failure intolerable to Grove. By now he had earned a solid reputation for exceeding expectations. He had little sympathy for soul-searching or thumb-twiddling. "Andy was one of those types who gets up at 5:00 a.m. and can't understand why an

Andrew Grove in his early Intel office.

SOURCE: INTEL.

engineer who has worked till 11:00 p.m. the night before might not be in at 8:00 a.m. the next morning." He was committed to executing Moore's push to make Intel's manufacturing technology the most advanced it could be. The insistent drive—with which he channeled Moore's focused determination to avoid the mistakes of Fairchild—made him a less than popular figure. Yet it was Moore, not Grove, who described his desire for Intel to be run "like a marine boot camp."

Gordon was among the few aware that Grove had a vulnerable side. Even Noyce nicknamed him "the whip," commenting, "It is tough for me to do the hard things, but it's not tough for Andy." Grove later told his own biographer, "At Intel, I was scared to death. I had left a very secure job for a brand-new venture in untried territory. It was terrifying. I had nightmares." In one recurring dream, vicious dogs jumped out of a closet: "The dogs have names like Moore and Noyce. They want to know why I'm not getting work done on time and to specification." Grove was risking his all. In contrast, Moore and Noyce had already made their fortunes. For Moore, starting Intel at the age of forty was not "much of a risk." He later described the experience as "fun" and "smooth." "We did things on time and on budget. Everything ended up working." Noyce was sufficiently relaxed to take long weekends skiing. Andy Grove, in contrast, was "fighting off the forces of entropy on a dozen fronts" and found the start-up "pure hell," adding, "I wouldn't want to relive those early years for anything."

Others were even worse off than Grove. In business matters both Moore ("measure, analyze, decide") and Noyce were cold-bloodedly competitive, yet Moore—receptive to personal entreaty and with a growing awareness of human details—found mass firings tolerable only if handled by surrogates and deputies, allowing him to avoid distressing confrontations with affected individuals. As the economy deteriorated and morale plummeted, Intel faced the trauma of major redundancies, with nearly a quarter of the employees laid off between June and October 1970. Grove was also in the hot seat for multiple problems with products. He questioned his own wisdom in believing he could leave stress behind at Fairchild. At times his perpetual near panic led him to be tyrannical in manner. In contrast, Gordon was quietly confident as he guided Intel through its early days, displaying little overt reaction to problems.

A Race for Glory

Massive redundancies, a team of key employees heading to MIL to set up production, issues with the 1103 DRAM and the planned 4004 microprocessor: the period was understandably tense. Intel chairman Arthur Rock pressed Moore and Noyce to get the 1103 into production as fast as possible and launch it, warts and all. The hope was that payments from MIL and widening sales of the 1103 would offer Intel the opportunity to reach profitability and go public.

With the upward trends in yields, Moore and Noyce approved an announcement that the 1103 was ready to ship. Marketers in the semiconductor industry were far from shy when it came to promotion and were already famed for announcing products that did not quite exist; advertisements for transistors and microchips were not modest about their anticipated virtues, either. Intel's ad was in keeping with this culture. In a two-page color spread picturing dozens of black rectangles, the bold text proclaimed, "The End. Cores lose price war to new chip. Ask Intel for proof." A more subdued typeface called the 1103 "a history-making 1024-bit RAM made by our silicon-gate MOS process at such high yields that the cost dips below cores." The 1103 was actually priced at $21 for more than 1,000 bits, around 2 cents a bit. Even so, it spoke to the latent demand among computer makers and system builders. Intel's phones began to ring and ring.

Despite the gratifying early reaction, the success of the 1103 was uncertain. Yield was fragile. Improvement was an endless process, requiring hard, creative work to edge results upward. Grove was acutely aware that "under certain adverse conditions, the thing just couldn't remember." Not only that, but it was hard to use. "The 1103 operated with very small signals, complex circuitry around the edges, all kinds of problems," says Moore. "You had to worry that when you loaded one bit, you didn't disturb the one next door. I once called it 'the most difficult-to-use semiconductor ever created by man.'"

Perversely, the chip's problems worked to its advantage. The "memory engineers" who were the main buyers at computer companies began to accept the 1103, precisely because it was less of a threat than they had imagined. "If to get the memory to function all they had to do was plug in these circuits, their jobs would disappear," explains Moore. "When they started finding problems, they knew they still had a future and so adopted the 1103. If it had been easier to use, it would have been a lot harder to sell." The device triumphantly established itself in a fast-growing market; despite being a "flimsy, flimsy device," it became an industry standard. Hewlett-Packard, a much venerated company, became by far its biggest customer. By 1972 the 1103 was the best-selling semiconductor memory chip in the world, accounting for nearly all of Intel's dramatic increase to $23 million in revenues. Looking back, Moore could call the 1103 Intel's winning technology. "Most of the manufacturers were willing to take a serious look at it for mainframe memory. It was the first commercially successful semiconductor memory device. We made a lot of money off the 1103. It let us grow."

For Moore, the moral of his Goldilocks strategy was clear:

> Had we placed a single bet, it might have been a problem. The bipolar approach was too easy. We got our 64-bit memory out, but we didn't have any edge over the established companies. Everyone picked it up right away. Multichip assembly was way beyond what we were able to do economically, so we abandoned it as too hard. Silicon-gate MOS was just right. We went with the flow, when we found real traction on that one. By focusing our energy on solving the tough process problems, we made it work. Our competitors took a long time to get their equivalent technology onstream. We had a monopoly for seven years.

In 1970 this was all still unknown. Gordon was understandably nervous. The more successful Intel's 1103 DRAM became, the more certain he could be that his competitors would move in. He had already seen firms offer chips to compete with his first product, the 3101 bipolar 64-bit SRAM. Intel was a tiny start-up with a net worth of around $1 million. It was running at a loss. TI had long towered over it and other firms and now sold $200 million worth of microchips a year. Fairchild and Motorola each had microchip sales of around $90 million, while National Semiconductor, Signetics, and the Japanese leader NEC sold more than $40 million apiece. Any of these firms could jump into DRAMs, using superior resources to gain a competitive edge. Earlier, Fairchild itself had regained its advantage by cudgeling Signetics in a costly price war. Intel remained an ant, ripe for squashing. Within the company itself, some felt that "this thing could end in a few years, whether by layoffs or by absorption into some other company. So many companies had fallen by the wayside that it didn't seem reasonable to think of 'the Intel of thirty years from now.'"

Moore's response was to stick to what he knew. The best way to maintain competitive advantage was to design the next generation of microchips,

Moore, with Microma watch and microchips, mid-1970s.

SOURCE: INTEL.

as the present one was leaving the production line, and endlessly push forward the capabilities of chemical printing. In this his strategy for Intel was straightforward and unchanging. Moore would push design and engineering, invest heavily to keep his printing press on the cutting edge, and seek out complex but standard microchip designs that could exploit his manufacturing technology most fully. As the 1103 became a commodity, Intel would not slug it out with well-heeled behemoths in low-margin, high-volume contests. Instead, it would reap the fruits of being first and move along to produce a similar success in the next generation. This front running was predicated on his steadfast belief that silicon technology could stand endless pushing, producing profits all the while. With the competition playing catch-up, Moore would go all out to get a more complex product ready and salable. In November 1970, just as the 1103 arrived in the hands of customers, he asked Joel Karp to design a DRAM with four times the capacity, a 4K, not a 1K, chip, capable of holding four thousand bits of data.

The silicon-gate technology that Gordon had done so much to enable made the 1103 DRAM and the 4004 microprocessor economically "accessible." As a complex microchip with lower cost per transistor and good performance, the 1103 DRAM was competitive with cores. Similarly, the single-chip CPU had become feasible. Gordon was open to any idea that could capitalize on silicon gate, and now he wanted to find other novel microchips that fulfilled a large standard need.

Dov Frohman brought just such a possibility to Moore in 1970. Frohman was a Berkeley electrical engineer who had worked in Moore's Fairchild lab during his days as a graduate student. He had impressed Gordon, Grove, and Vadasz and was hired at Intel early on. Frohman had been assigned to try out the multichip approach in Moore's Goldilocks strategy. After months of work, Moore judged Frohman's results "nice work," but he killed the effort: the 1103 DRAM was better. Gordon and Vadasz now asked Frohman to investigate some puzzling errors on Intel's silicon-gate memories. During the work Frohman imagined how to make a novel memory chip. Unlike Intel's SRAMs and DRAMs, Frohman's chip would be able to hold its data *even when the power was off*. Intel's existing memory chips held data only while they had power.

By now another common form of microchip memory had been created: the ROM, a read-only memory. These chips could hold data permanently, even when the power was off, but that data—being chemically printed into the chip—was fixed and could not be changed. ROMs were good for holding essential programs in computers and other digital systems. Frohman saw a way to use a silicon-gate MOS microchip as a ROM, but with an important advantage: it could be "programmed" with data by electrical signals, but this data could then be "erased" by shining ultraviolet light on the chip and fresh data programmed in.

When Frohman showed a prototype to Moore, he was enthused. This kind of memory chip (an electrically programmable read-only memory, or EPROM) was just what engineers needed. He told Frohman to develop a 2-kilobit EPROM, which Intel introduced very quickly, in January 1971, as the 1601 EPROM. A month later, at the Solid State Circuits Conference in Philadelphia, Frohman and Intel introduced the electronics world to this device, laying out its virtues. At the same time, Joel Karp told listeners that Intel was already at work on its next-generation 4K DRAM—the successor to the 1103—and emphasized its fourfold increase in capacity. Intel was beginning to set the pace and the rules.

While Intel had the wind in its sails, Japanese firm Busicom was in serious financial difficulties. Its desktop calculator, powered by the specially designed Intel microprocessor and three associated support chips, made its debut in April 1971, but its modest success was not enough to stave off competition from other Japanese, US, and European manufacturers. Busicom asked Intel to lower the price of its chips, to help the firm to compete. This gave Intel an opportunity. Hoff, Noyce, and Moore knew that the Busicom "chip set" could do much more than act as a calculator. The 4004 microprocessor and its three support chips actually constituted a general-purpose computer, albeit a very limited one. With the right software, it could run a huge variety of machines and systems, from elevators to traffic lights and beyond. Gordon says, "It was a very intriguing gadget. It had the possibility of being very flexible with a lot of high-volume applications. You could imagine all kind of control functions being done by programming this standard product."

Busicom had the complete rights to the 4004 chip set. Noyce and Moore agreed with Hoff that Intel should win back these rights. Gordon explains, "They came to us in May 1971, saying they needed significantly lower prices to be competitive with their calculators, so we said, 'The way to get the cost down is to increase the volume. The way to do that is to develop further applications.'" Intel gave Busicom the lower price, and Busicom gave Intel the right to sell the 4004 and its support chips to anyone it wanted, with no royalty, as long as they weren't making calculators. The potential for the microprocessor was suddenly much wider.

The Blind Eye: Grove, Graham, and Gelbach

Intel was vindicating Moore's belief in the power of silicon technology and the belief of others in Gordon's superb technological talent. However, that winning combination served only to underline deficiencies that much earlier had made industrial psychologists doubt his management capabilities. As Intel grew, personal conflicts among managers became a recurrent part of the equation. Gordon, by reflex, avoided these conflicts.

For nearly two years, Bob Graham (formerly at Fairchild and coaxed away from a top job at ITT [International Telephone and Telegraph] in May 1968 to become Intel's head of sales and marketing) had been at loggerheads with Andy Grove. Graham was an electrical engineer who had cut his teeth in semiconductor sales at RCA and Raytheon; Noyce had brought him in on the Intel vision from the start, and his "Intel Delivers" slogan would become part of the company's DNA. Moore shared with Graham a love of fishing, and Graham was one of the few people Gordon saw outside of the office. With his wife, Nan, Bob Graham had even been to dinner with Gordon and Betty on Jabil Lane. Unhappily for Gordon, "Bob conflicted with almost everything Andy did." Tensions became palpable. Moore, always an acute observer, knew that part of the problem was the way Graham "had to win arguments and then hit the other guy over the head with it afterwards."

Grove's view of the repeated run-ins was a little different. "Fact: we rapidly began rubbing each other the wrong way. Opinion: he recognized me as a competitor and started putting me down." Grove was already anxious about getting the 1103 DRAM delivered, so the conflict with Graham became a major stressor: "I was utterly miserable." When Grove and Moore attended a technical conference in Washington, DC, in 1970, they went for a walk at the National Zoo. Moore was concerned by what he saw and sought to reassure Grove, telling him in no uncertain terms that he would someday run Intel. Equally candid, Grove said the rancor with Graham was making him miserable. Moore continued to avoid the issue. "He never did a fucking thing about it," says Grove, who was "heading towards leaving" Intel. Grove never directly discussed with Moore the idea of replacing Graham but hoped his mentor would see that the existing situation was "too painful" to continue. "I can't understand that Gordon would have let me go, given what he'd told me about his hopes."

Grove was used to Moore's passivity as a manager. At Fairchild the ongoing issue with Tom Sah's absence had enraged him, with Gordon "looking like he never heard us complaining, yet he heard every word and chose to detach himself." This time, behind the scenes, Moore did go so far as to discuss the situation at length with Noyce. They agreed it was a straightforward choice. "Intel was too small a company for both Graham and Grove, and Grove was clearly more vital to Intel's success." To drive Intel to the stature they desired, their own business and scientific insight would not be sufficient. "They needed a tough manager. And neither of them had ever come across anyone tougher than Andy Grove." Noyce quietly began to look for candidates to replace Graham, soon finding one to Gordon's taste.

Matters deteriorated to the point that Grove and Graham were no longer on speaking terms. Seeking to escape the tension, and with deliberation, Gordon went off on a family vacation. Noyce now made his move. He laid down the gauntlet to Graham: do it Grove's way, or leave. Graham chose to go. Moore was not there to say good-bye and was not involved in Graham's severance terms. He did not speak to his friend again for years. Gordon rationalizes the realities, saying, "Bob Noyce did me a real favor. He fired Graham while I was off on vacation, without telling me. It would have been difficult for me because I had been so friendly with Graham. It was obviously the right way to go." Graham became a successful executive at Applied Materials and later the CEO of Novellus.

Ed Gelbach, already lured away from TI's sales and marketing operation, replaced Graham. From the start he was a strong influence, with a very good sense of the market, and his promotion proved a turning point for Intel's microprocessor efforts, moving them up a gear. Moore and Noyce were deeply interested in the microprocessor, but focused on the immediate promise of the memory microchip. Gelbach, in contrast, was convinced that the microprocessor had as much promise as memories. He knew firsthand that TI was speeding toward both and was convinced that Intel needed to get microprocessors to market: first-mover advantage was needed when competing with giants.

Gelbach believed that the customer would want to customize the microprocessor with software: manuals, application notes, software, and even fresh hardware. He understood the importance of Frohman's new EPROM in this mix. So did Gordon:

> The EPROM turned out to be a very important device. We thought it was going to be a low-volume product, so we priced it high. It turned out the EPROM was the engineer's security blanket. He could always change his program, so he never replaced it with other, cheap, devices. The fact that it came along the same year as the microprocessor was marvelous serendipity. The EPROM was a phenomenally profitable product for us and paid for an awful lot of Intel's expansion. From 1972 clear up to 1985, we were making most of our money on EPROMs.

Vindication: The IPO

Almost three years to the day after Moore and Noyce founded Intel, the company moved into its seventy-eight-thousand-square-foot headquarters in Santa Clara. The physical plant was 300 percent larger and boosted production more than threefold. The facility boasted clean rooms, controlled airflows, and better waste-disposal systems. With its unreserved parking lot and large open floors filled with low-rise, reconfigurable cubicles, the Santa Clara plant reflected Intel's developing culture far more strongly than had the Union Carbide building on East Middlefield Road. Moore explains, "It didn't make sense to have hard-wall offices. If we hired five hundred engineers and put each one in an office, it made a funny-looking building. Engineers did their design work at a desk; offices looked like rows of jail cells, so we went with cubicles. It's easier not to draw a line anyplace than to draw a line and say who gets parking places and who doesn't. We started with as egalitarian an arrangement as we could and institutionalized that as time went on."

Even Moore and Noyce had cubicles. The physical arrangement was functional, but it was also symbolic, reinforcing Intel's image as a no-nonsense manufacturing and engineering company. No frills implied no waste, an important message for employees with stock options. It did not matter where an employee sat. In Intel's meritocracy one's argument, data, and personal performance stood for far more. "Setting up fancy offices for the boss wasn't the way we did things," says Moore.

Moore (*left*) and Noyce walking to the parking lot of Intel's new facility in still-rural Santa Clara, 1971.

SOURCE: INTEL.

Sales of the 1103 were on track, generating millions in revenue: "If silicon gate had been harder, we might have run out of money. If much easier, we would have had competition sooner. Luck was extremely important and helped us off to a great start. In 1971 things were beginning to come together. We were producing the products we'd set out to do. The technology was working well. We had several different directions going."

The success of the 1103, and continuing payments from MIL on the second-sourcing deal, meant profitability was on the horizon. The previous year's hammering of technology stocks had ended, and markets were recovering. Bob Noyce began preparing for the IPO, which occurred on October 15, 1971, with 307,472 shares sold at $23.50 a share. Intel was listed on the new electronic stock exchange, NASDAQ, itself a creation of computers and thus of the microchip. The firm had further funds to fuel its growth, and Moore's Intel shares alone were worth well in excess of $60 million in today's dollars. The self-contained boy from Pescadero, once in danger of being left in the sandbox, was a force to be reckoned with. At forty-two he was seriously rich and leading a revolution.

Exactly a month after Intel's IPO, the firm launched the 4004 to the general market. A big advertisement in the November 15, 1971, issue of *Electronics News* announced a new era of integrated electronics. "A micro-programmable computer on a chip!" would set the stage for the next act in the electronic revolution. The copy explained how the 4004, with three support chips, "gives you a fully functioning micro-programmed computer." Just an eighth of an inch wide, a sixth of an inch long, and with twenty-three hundred MOS transistors, the 4004 had ten times as much ability as ENIAC, the first US electronic general-purpose computer launched twenty-five years before. This monster had weighed thirty tons and required six full-time technicians. The 4004, costing $200, was more than four thousand times cheaper. It sold widely, with Moore getting "a kick out of novel applications," such as the way it was used to automate a chicken house in California's Central Valley. The 4004 made vivid the incredible promise Moore had foreseen for silicon transistors to lower the cost of electronics. This "computer on a chip" was the next step in the digital revolution.

Ed Gelbach had correctly predicted that customers would need tools to customize the microprocessor: manuals, application notes, software, and even additional hardware. Thousands wrote to ask for information. Intel's booth was flooded with visitors at trade fairs. Interest was great, but actual sales were small. More copies of the manual were shipped than of the chip set itself. To Moore, it was clear that many people wanted to explore the capabilities of this device, but to *sell* microprocessors would require Intel to take a different approach. "It was exciting because it was the first product

A close-up photograph of the Intel 4004 microprocessor. Each 4004 die, measuring one-eighth inch by one-sixth inch, contained twenty-three hundred transistors.
SOURCE: GORDON MOORE.

in a series. Not too long after, we started working on next-generation chips that were significantly more powerful."

The commercial microprocessor had arrived—in retrospect, a major moment. Success had a thousand fathers. Ted Hoff, along with Federico Faggin, Stan Mazor, and Masatoshi Shima, deserved the lion's share of the credit. Hiring Faggin away from Fairchild had proved critical. He was a brilliant and dedicated designer, bringing with him technology which he then determinedly developed and which proved critical to microprocessor performance.

What became troubling to Gordon was the widespread impression that he himself cared little for the microprocessor. The perception still rankles:

> My interest never faltered. I didn't necessarily express that interest very strongly, and Hoff may not have known that I was in full support of getting it done. We were at the point we could make a microprocessor. That was the key step.

I always believed in microprocessors, but the market wasn't big enough in the early days. I continued to support it, despite the fact that for a long time the business was smaller than the development systems we sold to implement things. But just when memories were going out, microprocessors were coming into their own. Success came because we always sought to use silicon in unique ways. As boss I had more say than anybody in what happened.

Not even Gordon, with all his insight and vision, could imagine how far-reaching the microprocessor would become. As he reviewed Intel's financial statements and reports for 1971, he was pleased to see that sales had doubled to $9 million and profit was $1 million. Already, the 1103 DRAM had captured half of the existing market for microchip memories, displacing magnetic cores. Moore also took satisfaction in the fact that the 1103 held some 3,500 transistors, an impressive success for his silicon printing press. The first planar transistors at Fairchild in 1960 were sufficiently small that 625 could be squeezed onto a US one-cent coin. By 1970 Intel could place almost 45,000 transistors on that same penny. As he had predicted, the price of transistors also dropped: from an average of $10 each to just 1 cent apiece.

For Gordon Moore, as for others, the beginnings of Intel had been an intense, exciting roller coaster. By late 1971 that start-up ride was over. Not only was the 1103 an unquestioned success, but the EPROM had also been launched commercially. Intel's silicon printing press was a leader in the industry, and, with the IPO under its belt, the company was operating at a

INTEL FINANCIALS, 1968–1972

Year	Revenue	R&D spending	Total costs	Profit/loss
1968	$2,672	$348,000	$449,000	-$446,000
1969	$566,000	$1,293,000	$2,479,000	-$1,913,000
1970	$4,241,000	$1,297,000	$5,692,000	-$970,000
1971	$9,432,000	$1,569,000	$9,945,000	$914,000
1972	$23,417,000	$3,442,000	$19,353,000	$3,084,000

Date	Stock price
10/13/71	$23.50
10/13/72	$58.75

Source: Intel annual reports.

profit. Bob Graham's departure had dissipated some long-standing tensions. The cloud of anxiety began to lift. Looking back, Les Vadasz acknowledged what a calming force Gordon was during these first heady years. "A start-up environment is very tense, lots of anxious moments. Whether things are working or not, you're always on the edge. It was very hard; emotions were high. You had people thrown together, each with their own ego. Of all these people, Gordon had been always the coolest, the most levelheaded. I don't know of one occasion that he lost his cool. He provided the emotional stability for the company."

Like his father, Gordon kept emotion in check. His steady presence offset Noyce's frequent absences, and his calmness counteracted Grove's volatility. His predictable, secure home life grounded him: each morning he bid farewell to his family, and every evening he returned to a home-cooked dinner, followed by quiet reflection and paperwork. Weekends included visits to wider family, fishing trips, or recreation at the beach house, all in the familiar stomping grounds of generations of Moore men.

By betting on new technology, breaking open fresh markets, laboring to establish novel approaches, and relentlessly starting on a 4K DRAM as soon as the first 1K DRAMs went out the door, Gordon was directly responsible for ratcheting up pressure within Intel. At the same time, he was the enterprise's steady rudder. As Intel entered its next phase, Gordon would continue to instigate high-stakes strategies while remaining the cool, calming leader. In language he rarely used with anyone but himself, he wrote in his private notebook that for 1972, "Balls-out expansion is the name of the game."

8

THE REAL REVOLUTION

SILICON VALLEY, USA

Convergence and Competition

The electronics community on the San Francisco Peninsula had by now become self-conscious, aware that it was creating novel realities both social and technological. Industry journalist Donald Hoefler, writing in the trade magazine *Electronic News* in January 1971, presented the one-time Valley of Heart's Delight as the center of the action in microchips and in electronics more broadly. The term *Silicon Valley* was in occasional use, but Hoefler's three-part article "Silicon Valley, USA" provided a formal baptism.

More than a decade had elapsed since the emergence of the first silicon transistors. For the semiconductor industry, these had been years of convergence, both geographic and technological. After Bill Shockley set up shop in Mountain View, spin-off after spin-off had located nearby. The area became home not only to microchip makers but also to tool and equipment suppliers, venture-capital partnerships, and law firms specializing in high-tech issues. Firms headquartered elsewhere in the United States began to open offices and operations in Silicon Valley so as not to miss out.

Not all manufacturers opted to move: for instance, TI was rooted in Dallas, Texas; Motorola's microchip arm remained in Arizona; and RCA, Philco, and AT&T clung to the East Coast. Nevertheless, all began to march to the Silicon Valley drummer and to focus on planar silicon microchips. Competitive pressures, customer demands, and technical realities led management and research staff, wherever located, to adopt similar manufacturing technology, with closely matching chemical printing presses. Each firm reacted rapidly to the success of others and rushed out its own competitive offerings. As the devices became increasingly similar, speed of response became ever more important.

The work of Gordon Moore and his colleagues on the planar silicon microchip convinced and converted an entire industry. Sales of microchips multiplied from $10 million in 1962 to nearly $500 million just seven years later. Learning from this experience, Gordon was determined to tune Intel to its own fresh opportunity, by making the most complex chips through the latest chemical printing technology. His insights guided the company to break open the market for memory microchips and to defeat magnetic cores. Sales jumped from $9 million in 1971 to more than $23 million the next year, as the US market for microchips soared to more than $700 million.

Intel surprised the industry with its successes, proving that silicon-gate MOS held the key to tremendous markets. The 1103 DRAM became the best-selling microchip in the whole industry. During the start-up era, from the summer of 1968 to the beginning of 1972, Moore also opened a second front, the microprocessor approach to digital logic. This second front depended on his belief that the microprocessor could be a standard product for both control and computing, when customized by software.

Intel soon enjoyed not only the leading memory microchip, but also monopoly positions on the profitable EPROM (the rewritable memory that could hold data when the power was off) and the commercial microprocessor. Les Vadasz saw these as "very productive years" with far-reaching implications. "We basically created a manufacturing technology, which fuels a $200 billion business today."

In the mid-1970s, the industry underwent a second intense period of convergence. Microchip production and computer technology in Silicon Valley continued to grow apace, with fresh waves of start-ups and spin-outs. Microchip makers followed Moore's strategic lead in building MOS microchips, particularly silicon-gate MOS microchips, to provide memory and microprocessors. Unusually among his peers, Gordon saw the need for careful guidance and policing to ameliorate two differing aspects of the freshly emerging realities. On one side, there was a need to direct Intel's technological response to products offered by other microchip makers. On the other, he needed to police his company's strategy, and the activities of his own colleagues, in their growing, unavoidable competition with Intel's customers.

As minor differences in manufacturing processes and chip design came to form much of a firm's competitive advantage, the culture of the microchip industry closed in, becoming more secretive. Organizational control and discipline tightened. Companies sought to push into markets ever more quickly. At professional meetings discussions became more formal, centering on already patented processes or on products far from commercial release. Because products from different firms were often strikingly similar in look and price, marketing techniques could make all the difference in securing a multiyear sales contract. Marketing extended to technical and

Chemical printing at Intel, early 1970s. Banks of diffusion furnaces line the wall, with various chemical stations in the center.

SOURCE: GORDON MOORE.

educational efforts, including printed materials and training sessions. Marketing groups enjoyed enhanced power.

A different aspect of the intense competition was less obvious yet for Gordon Moore just as vital. The continuing development of silicon manufacturing technology created a fundamental dilemma. Gordon had long believed that in order to succeed, a microchip maker must pursue products that were "standard"—that is, usable by many customers. Design costs needed to be amortized across long production runs. The snag was that while increasingly complex microchips might offer the cheapest electronics (giving their maker the most competitive advantage), they were steadily more problematic from the viewpoint of Intel's customers.

Memory and logic chips both began to encroach upon the design territory that had belonged to customers: the system makers for digital computers, industrial equipment, military hardware, and consumer electronics. More complex microchips could incorporate increasing aspects of their specialist designs but, being standard, reduced the uniqueness of their particular products. Intel's customers were being required to surrender a valuable part of each one's competitive advantage. "The nature of integrated circuits," explains Moore, was that "we took their architecture, put it on the chip, and gave it back to our customers for free. Design responsibility moved into the semiconductor industry. As chips got more complex, we accumulated more and more of our customers' value."

Microchip suppliers and system makers were two different entities, competing as to which would receive the lion's share of the profits. In the early 1960s, unease about this had led system makers to resist the first integrated circuits. Gordon's writings describing the dynamic through which complex microchips would make electronics profoundly cheaper, arose in that context. Price won out. With standard microchips available for less cost, system makers capitulated. Gordon understood that as he, Intel, and the rest of the industry produced still more complex microchips, a further encroachment would mean competition with customers. This tension was a part of life in the electronic world. The key to success lay in the way the tension was managed.

Like his deputy sheriff father, but in a wholly different context, in the 1970s Gordon began to define the boundary of acceptable practice and police it closely. To compensate system makers for replacing their proprietary designs with Intel's standard chips, he would need to offer high-level products that combined compelling economic and performance advantages. And, while absorbing the extra design costs, Intel must also make profits. These depended upon the ability to sell the resulting standard microchips—DRAMs, EPROMs, and microprocessors—to different competing customers. It was crucial to avoid antagonizing those customers by pushing too far into product territories or teaming up too closely with any one company. For the model to work, Intel had not only to avoid direct competition, but to be seen as doing so. It was a fragile balance. No longer at the lab bench, and increasingly focused on the broader picture, Gordon became quietly efficient in his policing of the boundary, a task he performed with rigor and even, on occasion, ferocity.

Refining the Leadership

By the end of 1972, Intel employed more than a thousand people, with a dominant position in memory microchips and in microprocessors. As the competition regrouped, the company became tougher and more opaque. Other events—the firing of marketing head Bob Graham, the anointing of Andy Grove as heir apparent, and the move to Santa Clara—helped mark the start of a radical shift in Intel's culture. The changed physical location

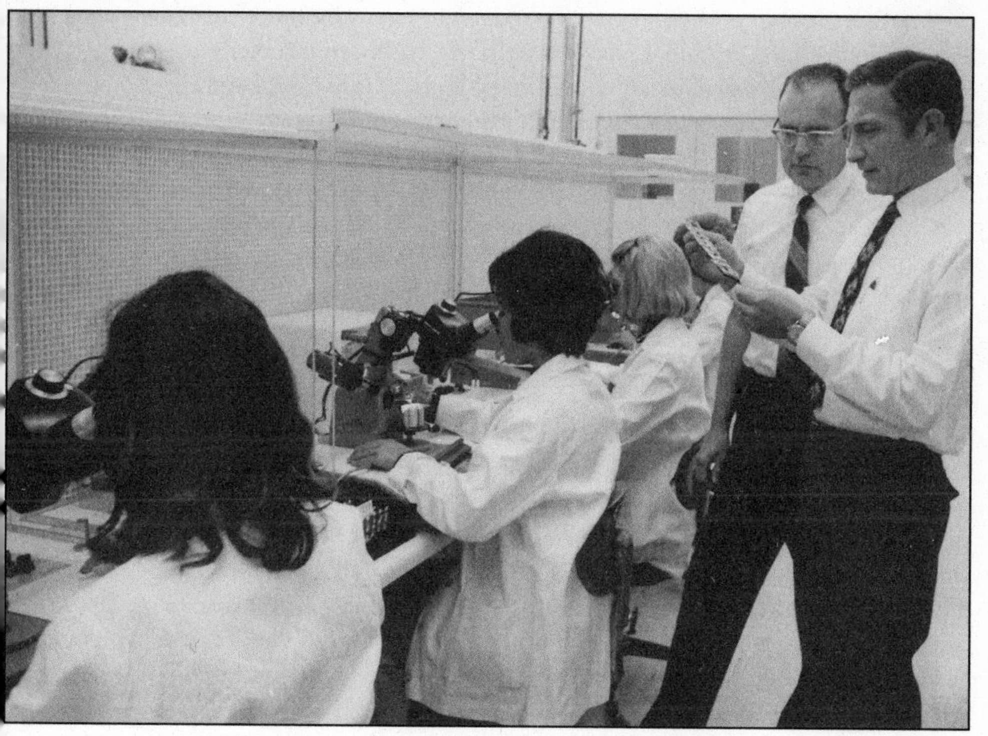

Moore and Noyce inspect early microprocessors, early 1970s.

SOURCE: INTEL.

facilitated the emergence of fresh power dynamics. Noyce, Moore, and Grove were the trio who shared an ability to move from macro to micro. "They all could drop right down to tree line and hammer through the details," explains Gordon's son Ken. "They had an incredible recall combined with large intellectual horsepower. A lot of the top management in some companies doesn't care for detail. They stay up at the fifty-thousand-foot-and-above level."

Early on, Noyce liked to sketch an organization chart that outlined a "reciprocal relationship" among employees. Bill Jordan, a chip designer recruited from Honeywell, recalls that, when he arrived in spring 1971, "I was in Noyce's office, and I asked him, 'What's your organization here?' He went to the blackboard and drew a circle, 'That's you,' and then he drew all these radial lines out, 'That's all the other guys. That's the organization.'" In the early years, Noyce would describe Intel as a community of common interests. Ken Moore, who had often dropped by the earlier Mountain View offices to cadge time on the PDP-8 minicomputer, remembers "seeing guys walk by in their sandals." Moore and Noyce might be "straitlaced types," but they hired "a whole lot of nontraditional types."

Intel's leadership evolved into a triumvirate or "executive office" of Bob Noyce, Gordon Moore, and Andy Grove. Noyce had been the catalyst for change, bailing out of Fairchild, convincing Moore to join him, launching Intel, and, through his interaction with customers, competitors, and Wall Street, establishing the company in the world and the public eye. Moore was at the heart of Intel, shaping its strategies and setting priorities and directions. Grove worked relentlessly to ensure these strategies succeeded. As his importance increased, the balance of power shifted. Moore became the common term in two "dynamite partnerships" that took Intel to a wholly different level.

The complementary styles of each of the trio meshed well. Craig Barrett, a new hire who became CEO in the 1990s, perceived that "Bob saw opportunity under every stone. Gordon saw technology direction under every stone. Andy saw the opportunity for operational efficiency under every stone. If you lined them up, what you got was Bob as CEO, Gordon as the technical brains, and Andy as the operational butt kicker." Noyce possessed what Barrett calls an "angel aura" as the company's public face, meeting investors, interacting with the industry, and selling to key customers. His personal warmth, combined with a "tremendous amount of charisma and great sense of humor," proved key assets. "If things got too serious, he could throw it off with a barbed comment."

Moore, buoyed by the firm's technical successes and the burgeoning fulfillment of his own predictions, continued on his steady trajectory as the technically brilliant introvert who thought deeply about things and never made a lot of hoopla. His strength lay in generating and recognizing important insights and then refining and implementing them, shrewdly and without sentiment. He was the company's "supercompetent engineer and technologist"—as Craig Barrett would characterize him—and his "serene wisdom and capability" nudged others in the right direction. Yet while he described himself as a technologist who never quite outgrew it, he riffed—seemingly without effort and always with success—through major role changes over the decades: from academic chemist to engineer-entrepreneur, from manager to CEO, from CEO to board chair, and from backroom strategist to major philanthropist, statesman, and visionary.

For Andy Grove, the summer of 1971 was a watershed. The hellish production demands and interpersonal conflict of Intel's early days had disillusioned him to the point of resignation, but with Bob Graham vanquished, Noyce often away, and the 1103 DRAM launched successfully, the situation was altogether more promising. The crown was his for the taking. Of three vice presidents under Moore, he rapidly became the dominant one, directing Intel's operations and shaping its culture. Moore was comfortable with this reality. "Very few people are able to make room for somebody else as strong as Andy," says Carver Mead, but Gordon, pragmatic as always, "was smart enough to know pretty early on that he'd reached a point where the only alternatives were to get rid of Andy or have him take over." At Fairchild Semiconductor, a dozen years before, Moore and his cofounders had hired Ed Baldwin to create structure for their growing company. Now, at Intel, Andy Grove began to take on this role. In Santa Clara the power shift became "obvious to all of us," says Mead.

Moore qualifies this assessment. "When was Andy recognized as playing an executive role? I'm not sure that it was as early as 1971, but by the time Bob was backing away, Andy was participating more and more in general management." Bill Davidow, an electrical engineer and marketing executive who came to Intel from Hewlett-Packard in 1973, had the clear impression that Noyce was still running the company, but Ed Gelbach asserts that as early as 1972, "Bob gave advice, but Andy and Gordon ran it." According to Grove himself, it was more gradual. "Gordon was the intellectual power behind Intel, but Bob, being the charmer with the high profile, overshadowed Gordon. Gordon was in charge, but by the time Bob faded on the scene, my light was glowing."

Moore's vision for the coming digital revolution required the cooperation of many players. He knew his own limitations as an actor on the social stage. He could be unflinching and unsentimental when making decisions on paper, but in personal interactions his reluctance to engage in conflict created damaging ambiguity. He had proven psychologists wrong by demonstrating an ability to create a successful organization, yet while he was a shrewd observer of people and what motivated them, his real strength lay in technical leadership and strategy. At Fairchild his R&D lab had pockets of brilliance and enjoyed a superb reputation, but the firm's astonishing, near-uncontrolled growth left it badly tuned, unable to get the most out of Gordon's thinking. His style was inner directed, and he had no deputy charged to elicit his thoughts, make decisions based on his strategies, and force others to follow his hunches. Fairchild's setup and structure simply did not fit Moore's preferred mode of operating and caused him many frustrations.

At Intel things would be different. His strategy was the beating heart of the company, and Noyce and Grove were positioned to capture his insights and take them further, one outside Intel, the other inside. Noyce was a prolific speaker and would habitually purloin Gordon's slides for talks. Often, his details and analysis came directly from Gordon: "Bob was hell for raiding my drawers. He'd give a talk, and he'd come to my office the day before and go through my slide file and take whatever he wanted!" Inside Intel it was Grove who acted on Moore's insights. "You look at the problems, and you try to come up with creative solutions, or you turn them over to Andy,"

Moore explains. Carver Mead says Moore used Grove extremely effectively as his hatchet man.

> Gordon didn't want somebody like himself to be the person working for him. He had great judgment. He knew that in a real organization with real people, you can't be nice all the time. He needed somebody who was not the introverted type. Andy could make it happen. He will jackhammer his way through anything, no matter how solid. He's smart as hell, but not in a nice way. If there was a problem, he'd yell and scream and get it fixed.
>
> That's the part Gordon didn't do. Gordon, being an observer, is never the guy to go out and beat up people to get X, or Y, or Z done, but he was very aware that companies that got things done had somebody like that. Gordon is extremely aware of the kind of energy necessary to be successful. That's why he brought Andy along.

Moore's strategic thinking required continued amplification and implementation if Intel was to succeed. He recognized Grove's capacity to execute in interpersonal situations. Of equal significance was Grove's growing ability to interpret Gordon's subtle facial reactions "better than anybody else." Perhaps this had something to do with Grove's partial deafness since childhood. Grove himself tells how he read Moore's expressions:

> I would be running a meeting and perfectly in charge. People would be bashing each other's heads. I'd look at Gordon—something is wrong. I'd yell, "Stop!" "Gordon, what's bothering you?" "Shut up!" "Gordon, tell us whatever you want to tell us." He usually had the right answer and the right comment, the right concern, but somebody had to stop the traffic. Nobody would have had access to Gordon's insight without my recognizing that it was time to stop the bulls. He was waiting for me; he would give a sheepish little smile and agree. He once said, "You know me better than my wife—not better, but as well as."

Moore in turn recalls that Grove would correctly discern his wishes: "If there was something I did not feel very strongly about, although I thought it might be a better direction for us, Andy would decide that we absolutely had to do it that way, and implement it."

Barrett amplifies:

> Gordon is very much within himself. He might express an opinion on a topic, and that's it. He's not in your face. You recognize his technical brilliance, his experience. You don't immediately launch into a battle. You go off and think about it. I've heard rumors of Gordon getting angry, but you

don't experience it. Gordon used Andy as a surrogate to do all of that. Andy would hit you over the head; you'd have to ask Gordon his opinion on a topic. But he was always there, always accessible, and always very thoughtful. He was interested in technical details. When he asked questions, they were very germane.

On matters he considered important, Moore sometimes intervened. In 1957, during the disintegration of Shockley Semiconductor, he overcame his dislike of conflict and picked up the phone to Arnold Beckman to try to broker a solution. At Intel, in an incident recalled by Carver Mead concerning his protégé, Gerry Parker, Moore stepped in decisively to countermand Grove.

It was Gerry's job to make sure the parts they shipped were reliable. Very early in the Santa Clara period, when Andy was becoming chief operating officer, Gerry told me that when they'd find a bad lot of parts, Andy would tell him to ship them anyway. Gerry would appeal to Gordon, and Gordon would overrule Andy. "No, we're not going to ship parts that aren't reliable." Gordon would intervene in making sure it all came out right, but he didn't do that very often. That was a Supreme Court decision.

The relationship between Moore and Grove remained strong. Grove, in all other ways the tough guy, functioned best when his steadying guide and father figure was physically present. Once, when Gordon was away from the office, Grove wrote to him, "It seems to me that Intel should contract with you not to go on vacation—the world invariably seems to turn into a piece of shit while you're away." In actuality, Grove proved himself more than capable. As competition intensified, Moore saw that for Intel to beat off rivals, Grove needed a free rein.

As the great-grandson of hard-driving California settlers and the son of a deputy sheriff, Moore might be introverted and avoidant, but he was also ideally suited to this risky, macho, unashamedly competitive environment. With much territory unclaimed, and frontiers endlessly shifting, there was all to play for. He continued to apply pressure to Grove, who in turn applied it to the organization. When asked years later if he considered himself a competitive person, Gordon's reply was immediate, "I guess I must be," followed by a laugh and a qualification. "Everybody likes to win. I don't think I'm destructively competitive." Happily, he was in the right place. In the semiconductor industry, the escalating competition was based on technology, products, and manufacturing capability, on measurable merit and performance, not personality. Success was subject to metrics, the kind of competition at which Gordon Moore excelled.

Moore giving a talk in New York City, 1971.
SOURCE: KEN MOORE.

The Challenges of Growth

Gordon's time and energies were increasingly consumed by the strategic challenges of growth. One concerned Intel's ability to meet surging demand for its 1103 DRAM. It was clear the company needed a third factory in addition to existing fabrication plants (or "fabs") on Middlefield Road and in Santa Clara. Gordon quickly agreed that the location for this facility, Fab 3, would be in Livermore, thirty miles northeast of Santa Clara on the eastern side of the San Francisco Bay. "We wanted a place that was close enough so that you could go there in the morning, do a day's work, and come back in the afternoon," he explains. The location enabled Intel to tap into a fresh labor market, and, unlike the other two plants, Livermore was not on the San Andreas Fault. Gordon decided that Fab 3 would use 3-inch wafers, enabling a technological leap that would both drive down cost and increase output. The area of the larger wafer was more than double that of its 2-inch predecessor, allowing for many more 1103s, but the steps in the chemical printing process would be the same. Transistor cost would tumble, as would that of the 1103.

As Moore well knew, the change would be far from simple—temperamental processes would need adjusting, and manufacturing equipment would need altering or upgrading—but the transition was part of the weft and weave of microchip technology and of demonstrating and enforcing Moore's Law. Moves to larger wafer sizes (each requiring careful effort and costing and each paying off handsomely) had punctuated the development of the technology since the early days of Fairchild and would continue to do so through the decades. Fairchild launched with wafers of 3/4-inch diameter; today, 12-inch (300mm) wafers are used, with 18-inch (450mm) set to establish a new norm.

Gordon also determined that Fab 3 would use a fresh iteration of Intel's silicon printing press, producing higher-speed microchips. For Gordon, advances in the chemical printing technology were limited only by the firm's ability to invest massively in engineering all its aspects. There was always a better process out there that could offer smaller transistors, with greater speeds, and lower power demands, resulting in cheaper electronics—embodied in more complex microchips. In 1972 Gordon believed Intel needed to move beyond its original method of making silicon-gate MOS microchips to a novel method that promised smaller, faster transistors and more complex microchips. He directed Intel's engineers to use the new version of chemical printing technology to create a SRAM holding 1,000 bits of data. By the end of the year, Intel hit Moore's goal and took its 1K SRAM to market. With this Intel pioneered a fresh version of its silicon printing press, offering benefits of greater speed and miniaturization for all of Intel's upcoming products.

Intel's success with the 1103 meant that many firms were closed out of the 1K DRAM business. Competitors, including TI, were too far behind to have much hope of capturing market share or making profits. Instead, they eyed the following generation of higher-capacity DRAM chips—holding 4,000 bits of data—as their entry point. Gordon's plan for fending off competition remained focused on the treadmill: keeping others running to catch up. "We kept moving to the next generation. A six-month lead could be very profitable. If you fell behind, prices were falling dramatically. The leader reaped almost all the spoils in those days." Gordon wanted a 4K DRAM as soon as possible.

The design for Intel's first 4K DRAM fell to Joel Karp. In late 1972 it was announced as the 2107 chip. Since Intel's DRAMs were used in both minicomputers and giant mainframes, orders poured in from a wide field of computer makers, yet there remained an important absence: IBM, the dominant player, with an 80 percent share of the US computer market. More than a dozen years earlier, Gordon had taken on the responsibility of making Fairchild's first silicon transistors and delivering them to IBM. Now he sold memories to everyone *except* IBM. "Big Blue," once dependent on outside suppliers, had developed its own semiconductor manufacturing operation to serve its blockbuster System 360 mainframe line. Even for its new line of System 370s, IBM has its own microchip memory supply. "They were doing all their stuff internally," recalls Gordon. He was shut out.

A proposal from two outsiders provided a solution, the chance for Intel to make an end run. Bill Regitz and Bill Jordan, both at Honeywell, had been responsible for initiating Intel's project to make its first DRAM microchip. They now suggested Intel hire them to establish a business selling plug-in memory units—made from many Intel microchips—that could expand the main memory of existing mainframes and minicomputers. Noyce liked the proposal; so did Gordon, who saw how it might enable Intel to sell its DRAMs into the IBM market by creating "big boxes, refrigerator-sized memory systems, that you'd plug into an IBM machine." The only way "to crack that market segment was to sell to their end users," he explains. "We set up a memory systems division with that as a goal."

Elsewhere in the company, Ed Gelbach's enthusiasm propelled the announcement of Intel's *second* microprocessor: the 8008, boasting thirty-five hundred transistors. The firm was now way out ahead and virtually alone in the microprocessor business. It was "taking off in a space where there weren't a lot of people clogging things up." Education and marketing were key challenges. Moore authorized a wide range of efforts to teach customers about the microprocessor approach: seminars, publication of a suite of educational materials, and the production of circuit boards containing microprocessor chip sets (similar to today's "motherboards," providing the essential circuitry for a complete microcomputer system). The aim, again, was to ease customers into using Intel's chips. Ted Hoff took to the road to introduce potential customers to the 8008 and to secure orders.

These customers betrayed a common concern that would come to be a defining feature of the microprocessor business: "What about my investment in software?" To make a microprocessor into a "controller" for any system (whether traffic lights or desktop calculators, industrial machinery or photocopiers), customers needed to spend time, effort, and cash to create software. This software represented a significant customer investment, bringing a lively concern: how long would such software be usable? Intel introduced the 8008 only four months after its 4004, and software was not compatible between the two. Hoff and his colleagues were at pains to assure customers that—despite this—Intel was on top of compatibility issues.

Actually, it had only belatedly dawned on Gordon and his colleagues that backward compatibility of software needed to be a major selling point. Already, Intel was planning a more powerful microprocessor, the 8080. Federico Faggin, the talent behind the 4004 and the 8008, added capabilities while also focusing on the 8080's software compatibility with its predecessor. Improving technology gave him a thousand more transistors to work with. The effort resulted in the most popular and influential of the early microprocessors, expanding the market. Years later Gordon would note, "Compatibility is one of the most important reasons that Intel microprocessors are so broadly used throughout the world. We knew compatibility was important, but none of us appreciated how important."

Gordon had long followed the habit of jotting down his concerns in a plain composition notebook. In the spring of 1972, one or two home-related items appear on his priority list of jobs ("engine for boat" and "burglar alarm"), but most were focused on a handful of Intel topics. The principal concern was to keep on top of exploding demand for the 1103. Intel's microprocessor business was also a priority, needing both a "plan and strategy" and "software" that would enable customers to adopt and use these devices.

Gordon's 1972 "year-end summary of Intel problems"—notes for discussion with Grove, Noyce, and others—included a concern that Intel's "production rate of components" was still inadequate, partly due to "increasing thru-put time" (the lengthening process to create microchips given their increased complexity), but principally because of poor yield. Yield remained a many-headed dragon. Gordon listed a second, related, concern: "Cost is not falling fast enough, something requiring both yield and efficiency." Another concern was "constipation" in novel products and processes. It was also essential to get the right personnel ("We need senior people in the technology area badly"). Moore also listed a major existential concern: "What is the scope of our charter? Is it just semiconductor components, memories, or is it more?" What did it really mean for Intel to make the most out of silicon technology? Where would the economics of integration and the philosophy of standard products take Intel, and Gordon himself? These were questions without clear, quantitative answers.

The $15 Million Wristwatch

A further concern on Moore's list was Microma, a digital wristwatch company that Intel had recently acquired. When starting Intel, Gordon's primary target was the defeat of core memory. Calculators also seemed to offer fertile ground in which standard microchips from Intel's silicon printing press might take root. As Moore looked around for additional uses for high volumes of standard chips, the digital wristwatch seemed promising. Pacemakers and hearing aids had been in the vanguard of wearable electronics. In the spring of 1972 a novel, fully electronic wristwatch, the Hamilton Pulsar, appeared. It deployed power-sipping chips made by East Coast electronics giant RCA. The Pulsar, retailing at more than $12,000 in today's money, featured a digital display made by light-emitting diodes and was a wholly new proposition. To Moore, electronic watches offered, potentially, a large market for sophisticated microchips. In actuality, entering this market would only demonstrate his naïveté about consumer products.

Moore was spurred to exploit the market "opportunity" when Bob Robson—an ex-Fairchild colleague, fishing buddy, and Los Altos Hills neighbor—offered Intel the chance to buy his fledgling digital watch company. Robson, a production foreman in the early days of Fairchild, had gone on to join Intersil, established by Gordon's old colleague Jean Hoerni. In the late 1960s, Intersil developed novel CMOS microchips for a Japanese watchmaker, Seiko. Robson became convinced that by combining liquid crystal displays with these microchips, he could seize the emergent market for electronic wristwatches. In best start-up fashion, Robson and a colleague left Intersil to start Microma, aiming to make the first LCD wristwatch. Eugene Kleiner, another Fairchild cofounder, backed the venture.

Robson enthused to Intel, "Look, we can sell a whole bunch of shit here." He estimated that there was a market for two hundred million wristwatches in the United States, but forecast that within a decade electronic wristwatches would add a further one hundred million. Moore and Noyce went to see Robson's firm in action. Moore was particularly interested in its CMOS microchip techniques and decided that they fitted well with the economics of integration and the philosophy of standard products. "It was intriguing. There were little circuits involved. We thought, 'This is another opportunity to make a complex chip that sells in large volume, like the calculator.' We envisioned a watch to which we kept adding functions." The same reasoning that made microchip memory Gordon's initial target for Intel, and calculator chips a close second, was now applied to—and fit with—the digital wristwatch.

"Predicting costs is the key to our product planning," Moore summarized in Intel's 1972 annual report. "We operate in very fast-changing areas: the cost of a bit of semiconductor memory has dropped a hundredfold since Intel was founded in 1968. The proper choice of products depends upon being able to anticipate such changes." Microma could give Intel the chance to enter "a business with large potential, an excellent example of electronic technology replacing and improving significantly the accuracy of a function previously accomplished mechanically. The engineering and manufacturing requirements for solid-state watches are closely related to our strengths in semiconductor fabrication."

Moore and Noyce decided to buy Microma. The firm was already beginning to ship wristwatches with LCD displays, an industry first. The opportunity so enthused Moore that he spoke to fellow board members individually to convince them that the deal was attractive. Several were skeptical. It would mean shouldering Microma's $1.5 million debt and using nearly seventy thousand Intel shares ($4 million worth) to buy Microma's private stock. This was a big purchase for a newly public, only recently profitable, company, one that necessitated the restatement of Intel's 1971 financials, downgrading a $1 million profit to a $500,000 loss.

Gordon quickly transferred the Microma CMOS chip production process to Intel's fab in Santa Clara. There, Carver Mead's protégé, Ted Jenkins, instrumental in developing Intel's early bipolar technology, got to work on ion implantation. Ion implantation, a technique pioneered by Jim Gibbons—another highly gifted denizen of Shockley's ill-fated enterprise—was by now an important technique within the chemical printing processes of Intel's competitors, but one that Gordon himself had neglected. A virtue of the Microma acquisition was that it brought that technique with it. When paired with diffusion, ion implantation enabled mass production of advanced chips, like CMOS. By 1974 Intel was making a watch chip (the 5810) that combined circuits for timing and for LCD display. Intel's watches had "a beautiful process that ran on low voltage," says Carver Mead, a process "much nicer than their standard one." This was the good news.

Gordon believed that falling manufacturing costs would allow Intel's Microma enterprise to compete on price with other electronic watches. After all, Intel was entering the market early. "When we went in, the one successful digital watch on the market, the Pulsar, had a display you couldn't read in the sunlight. They were expensive machines. In contrast, we were aiming for watches at a tenth of the price." Gordon was far from alone in viewing this consumer market as an exciting opportunity. Fairchild and TI introduced wristwatches in 1975. The latter had an expanding business in pocket calculators, but for Fairchild the move was its first into consumer electronics. Even National Semiconductor got into the game.

Microma quickly found itself in technical difficulties. Intense competition developed with TI: "We had a lot of problems with reliability. Assembly techniques were terrible. Board members bringing in their watches to get them fixed became the first agenda item of our board meetings! We struggled along, but Texas Instruments took the guts—the module with all the displays that go in the case—and said, 'We'll sell them for $19.95.' (We'd been selling them for $75 or $80.) The next year, they said, '$9.95.' It took out all the fun."

The acquisition proved a real thorn in Gordon's side, its troubles a recurrent theme in his notebook. "Microma is not a stable deal," he wrote. "Need business management, but also a long-range technology program." Initially, he saw the digital wristwatch as similar to any other electronic component; in the abstract, it provided a home for the kind of chips he intended to make. Yet to be successful, a consumer product requires far more than the chip that powers it. "The watch business is not a technology business," comments Dick Boucher, the executive brought in to run Microma. This had been the concern of some board members from the start. "They were much wiser than I was," Moore reflects. "It's a different business." He had no experience in either marketing or watches, but soon after the purchase of Microma, he invested his own time in journeying, optimistically, to a major jewelry trade show in Switzerland to learn about the realities of the trade.

Fatefully, both Intel and Microma were started and staffed by old Fairchild buddies, individuals with little or no experience in marketing and distributing luxury goods. Moore downplays the purchase. "We bought what Robson and the others had for almost nothing," he throws off. Nevertheless, Robson did well enough out of it to retire from electronics within the year, cashing out his Intel stock, selling his Los Altos Hills house, and buying a thousand-acre pistachio grove in central California. Like Moore, he was a rural boy, a former high school football player, and a regular huntsman. He and Gordon fished together most years in an arrangement that continued for over three decades.

Microma appealed to the visionary in Moore. He was intrigued by the possibility that "pretty soon you'd have all the things you wanted on your wrist." Electronic devices *would* eventually become multifunctioning, ubiquitous, and wearable—but it would take four more decades to get there. The US military would eventually develop a contact lens to allow a wearer to see an entire battlefield, while Samsung, Apple, and others would pioneer "smart watches" with "killer apps" facilitated through the cloud. A half mile from Intel's Mountain View home, Google would develop its "Google Glass" project. All this lay in the future, one deeply shaped by Moore's Law, but not yet practical in the 1970s. Gordon's imagination had run far ahead of marketing and technological realities.

The wristwatch itself had long been an important item to Moore. Not only did it allow him to measure time and keep careful track of his schedule, but it also served to symbolize his powerful internal drive. Gordon had always been time conscious, sometimes acutely so. From his father, he had learned a habit of checking his watch constantly. "Like Walter, Gordon lives by his watch," Betty explains. In adolescence he had decided not to waste time on activities at which he did not excel.

Intel closed its Microma division in 1977. "They really thought they were going to be in the watch business, which, looking back, seems silly," says Carver Mead. Other US microchip firms also attempted to seize the wristwatch market; all failed, and by the end of the 1970s digital watches imported from Hong Kong and Japan had driven US semiconductor companies (and all other US companies) out. "We misread the way the business was going," Moore says. "We didn't succeed in adding the additional functions we had envisioned. When we got out, the semiconductor content, the chip, cost less than the little pins on the side of the case."

The original Microma wristwatch remained Moore's daily timepiece until 2000, when it ceased to function. He called it "my $15 million watch—the total loss Intel had in the business." The experience brought home to Moore two important principles. First, Intel would do well to steer clear of consumer products; second, effort was needed to "track down the right people" and hire the necessary talent, for any venture to fly. "The watches were really a sidetrack. We took a few sidetracks, but we were successful in keeping the mainstream going, too." For two decades, Gordon Moore's Microma watch testified that, in the ultracompetitive world of semiconductors, time was short and must be wisely invested. The primary danger was to lose focus on what Intel did best: supplying standard chips for everyone who made end products.

Transistors by the Trillion

The number of transistors produced for each person living in the United States rose from 4,000 in 1968 to 90,000 in 1972 (1.9 trillion silicon transistors in all). Most were used as "building bricks" in microchips; the majority were made by US firms, which controlled 80 percent of the $1.3 billion global market. Five firms took most of the spoils: TI ($219 million), Fairchild ($97 million), Motorola ($90 million), and two Fairchild spin-offs, National Semiconductor ($63 million) and Signetics ($48 million). Japan's microchip industry, led by NEC and Hitachi, had a market share that equaled that of TI. European firms, including Dutch electronics giant Philips and German powerhouse Siemens, had a market share worth $100 million. Intel, though growing by leaps and bounds, was still a minnow, with $22 million in sales.

Intel's share might be only a tenth of that of TI, but Gordon was focused on making highly complex MOS microchips, a market worth $250 million in 1972. Here, General Microelectronics spin-off American Micro Systems (AMI) led the way ($29 million), followed by TI ($25 million) and its spin-off Mostek ($18 million). Intel came fourth with $16 million. In this market its share was already worth half that of the market leader and was the same size as that of established Japanese giants NEC and Hitachi.

Not yet four years old, Intel was making its mark. At the same time, the dominance—though not the reality—of Cold War concerns was fading. Only a quarter of the 1.9 trillion transistors produced were destined for military and aerospace systems. A quarter went into the telecommunications network still dominated by the Bell System and a further quarter to industrial machinery and office devices. A third went into computers. The remainder went into microchips that powered consumer products, from televisions to wristwatches.

By reinforcing the appeal of "mass culture," transistors continued to alter the way people engaged with reality. Electrical and electronic appliances were now defined as staples of the "modern" household and facilitated the creation of novel mass markets, as in "hi-fi" music. Radios, home audio sets, phonographs, and tape recorders were everywhere. Of 17 million televisions bought in the United States in 1972, half were color sets; buried in their tuners and cases were multitudes of silicon transistors, which enabled them to beam into US living rooms vivid and disturbing images of the Vietnam War and of urban and student unrest. Transistors also enabled portability. Early transistorized radios were giving way to smaller pocket radios using more powerful chips and boasting more features. Further shrinkage of size and cost would follow, as would new options across the whole field of mobile devices. Meanwhile, in the workplace sophisticated microchip-based pocket calculators were superseding older, chunkier models.

The electronic conquest of the US home continued apace. Already by 1970, nine out of ten households had a telephone, and the average American made more than 800 calls a year (compared to 550 a decade earlier). Ninety-five percent of homes had a television, and a small but growing proportion was switching to paid cable television, a system that relied on microchip electronics.

Information had long been communicated electronically; increasingly, it was used to *control* electronically. Within American workplaces, this reality emerged in the 1960s, with mass computerization of government, military, industrial, and business organizations. Using mainframe computers to control activities through information became routine, and the US computer industry grew at 20 percent per year, with spending increasing to more than $5 billion annually by the end of the decade. Digital computers were used for banking transactions, airline reservations, customer accounts, billing, accounting, inventory control, machine tool operation, oil refinery production, engineering design, and many other functions. Electronic technologies for communication and control were also central to the geopolitics of the thermonuclear Cold War, the space race, and the war in Vietnam. Satellites, reliant on transistors and microchips, provided global communications—radio, television, and telephone—as well as global surveillance.

Digital data (communications between computers and between terminal users and remote computers) now traveled via America's transistor-driven telecommunications network. ARPANET, a network sponsored by the US military, enabled data and messages to pass between computers. By 1972 a third of the transistors produced were going into logic and main memory for mainframes and minicomputers. Two years later there were 165,000 computers in America (more than double the number in 1970). Annual consumption of transistors in computers rose to more than 500 billion.

A New Sparta?

Intel's shift from the open organization of the start-up years to a tighter structure was a response to the company's growth, to an increasingly competitive environment, and to the growing role of Andy Grove. Averse to controversy, Gordon Moore emphasizes Grove's ability to simplify and to create a well-organized company: "I wish I'd had some of Andy's ideas earlier on. He came up with some very good ways of managing. He got out of the technical details. Instead, he became very interested in how organizations work. Andy considered management a science he ought to learn. He took on challenges, attacking them from a scientific point of view."

If Intel became Silicon Valley's Sparta to the gentler Athens of Hewlett-Packard, its ruthlessness lay within the context of fiercer competition in its chosen territory. Other microchip makers such as Texas Instruments and Motorola were equally tough. "Being there first was so valuable; coming in later meant major problems," says Moore. "It's a high-risk business. You bet your company on every generation of the technology. If you take the safe approach, you fall behind. Our tough attitude was what we had to have to really get going." Grove's "tight ship" was uncomfortable for the bottom 10 percent of each division whose jobs were on the line, but not for Moore, "as long as we were winning!"

The perception grew that Intel, like Fairchild, was arrogant. Moore, who came from humble if hardy stock and who had grown up in a rural backwater, explains: "The arrogance is something we kept trying to fight, but it's hard to avoid that image when you're winning. Maybe we were arrogant, but we certainly didn't have that as one of our values. One thing that contributed in the microprocessor days was our unwillingness to negotiate on price. We said, 'This is the price. What else can we do for you?'"

A shift in the behaviors now expected at Intel's Santa Clara headquarters became apparent, as part and parcel of a gradual leadership transition (Bob Noyce retreating, Andy Grove advancing). Grove's growing influence became "part of what was uncomfortable about being there," says Carver Mead, who withdrew from his consultancy.

A minicomputer in use with a teletype terminal, early 1970s.

SOURCE: LICENSED UNDER CREATIVE COMMONS ATTRIBUTION-SHARE ALIKE 2.0 VIA WIKIMEDIA COMMONS, HTTPS://COMMONS.WIKIMEDIA.ORG/WIKI/FILE.

For Mead, Intel changed from congenial camaraderie to a very driven, almost brutal workplace. For Noyce, too, it was ceasing to be fun. In contrast, Moore was less sensitive to the change and continued to be absorbed by strategy, vision, and intriguing manufacturing puzzles. He saw, but avoided the human tangles. Under Grove the company became a ruthless, meritocratic pressure cooker, a corporation with layers of managers and stovepiped divisions, all surveyed by the relentless eyes of the explosive and habitually gruff Hungarian immigrant. The new culture, "way organized," was belatedly joining a long tradition. Organization and discipline were familiar hallmarks of American manufacturing practice. Intense competition and remorseless technological advance simply put a greater premium on them.

One aspect of the shift was Intel's sign-in sheet, the infamous "Late List" instituted in 1971 and used by the company until 1988. Its implementation demonstrates how Grove functioned as Gordon's "amplifier." Moore explains:

> I remember getting frustrated. The normal starting time was 8:00 a.m., but even when we had a meeting, it would be 8:25 a.m. or 8:30 a.m. before everybody was there. I made some comment to Andy. He came up with one of his solutions: "The core hours are 8:00 a.m. to 5:00 p.m. We expect you here during that time!" 8:06 a.m. was the cutoff time—after that you had to sign a sheet. Not much was ever done with it, but it was terrible having to sign your name. The secretaries would go into tirades. If it was getting close, I found myself hurrying.

The Late List caused more trouble, according to Moore, than any other management measure. It divided employees. Many resented the idea of being held accountable. Chip designers Joel Karp and John Reed soon left, citing a new negativity in Intel's culture. Others believed it was necessary to avoid the "slippery slope" of "fab" operators coming in late. "Andy had seen some incredibly sloppy practices at Fairchild and was reacting to this," says Tim May, an engineer who joined Intel in 1974. "Nobody was punished for signing in late too many times." Moore felt the sign-in sheet was worth maintaining. It produced conflict, but that was up to Grove to solve. The list was practical: "It meant we did get there at 8:00 a.m."

With Grove overseeing ground-level operations, Moore spent most of his time on high-level management issues: "All the chores of running a very rapidly growing business: worrying about buildings, staffing, products, technology, and customers. I would watch what we were spending and look at all the programs internally." He no longer had much direct involvement in the lab, but kept abreast through written reports, for him a vital tool to internalize, map, evaluate, and judge technical activities. Gordon seldom responded directly in writing, except through scribbles to himself, but would often offer suggestions and questions later. Other times Grove would pick up his reaction to a report and act upon it.

In Intel's early days, staff meetings had been informal and open: "People got together, and we could all learn what was going on in the various areas." Following the move to Santa Clara, meetings became more restrictive. Moore, aware that his own tendency was to be self-contained, never forgot Bill Shockley's poisonous exclusiveness and made sure to keep employees up-to-date in a larger monthly gathering, "an inclusion meeting as much as anything. We had no secret projects." Yet Intel's growth, and competition, had effects. Many meetings now involved only group leaders and some or all of the leadership trio.

In these years Gordon remembers that "spending half my day in meetings was common." Generally, several people were present, looking at projects or meeting with customers who came in. "That was the way things tended to get done." Moore did try to broaden his informal contacts: "I wasn't very good at 'management by walking around' in the Hewlett-Packard sense—I didn't have the Bill Hewlett approach—but I had certain people that I would see regularly." While Noyce, "open to all comers for the first inch," hosted an unceasing stream of visitors to his office, Moore guarded his privacy. He often sat alone to think, with his notebook or legal pad ("a tablet and a pencil") at hand. Being alone might be classic introvert behavior, but it was a highly effective part of his job as strategist.

Intel's guardedness was evident in technical meetings outside the firm, where Gordon and his colleagues introduced their products and processes. Intel would announce achievements safely made, but it would not reveal active research agendas. Noyce did most of the ambassadorial work, but Moore traveled on occasion to meet customers (typically visiting Europe twice a year) and to present papers or participate in industry panels. Trips afforded a change from the daily grind. "Until the mid-1970s I went to all the device meetings," he explains. "It was a fast-moving field. There were always interesting papers, and you were able to find out what was really going on." In the old days, "we were fairly loose about how much information we gave out, relying on how engineers take information, and instead of saying, 'Look what I can use,' it's 'Boy, look how dumb these guys are, doing it that way!'"

Moore retained a steady focus on developing world-beating manufacturing technology. He also had a strong interest in how best to motivate employees. Andy Grove remembers his "very well-defined ideas" about pay: "low, performance oriented, variable, and equitable as compared to opportunistic." Intel advertised for recent college graduates who would "respond to the challenge of growth and change" and were "capable of the disciplines needed to maintain tight control of operations." Many universities now had semiconductor facilities within their electrical engineering departments. "We were recruiting people trained as electrical engineers who had done the basic processing," says Moore, "but we still hired a few chemists and chemical engineers."

In seeking to hire the brightest and best, Intel faced tough competition, as demand continued to outstrip supply of qualified researchers and engineers. Even Moore's personal qualities, so luminous and appealing to Andy Grove as a fresh-faced PhD in 1963, were not always sufficient to seal the deal. An engineer at AMI, who had previously worked with Moore and Noyce at Fairchild, recounts how Intel missed a trick in failing to recruit "one of the brightest guys in a decade to come out of Stanford."

> I assigned my colleague the job of making sure he came to work for us. My colleague came back week after week, highly discouraged. "What's the problem?" "We'll never get him. Bob Noyce is taking him up to Napa, wine

> tasting, and Gordon takes him out fishing in his boat. There's no way." I explained, "We pay him more money. Intel's going to hire a lot more PhDs than we are this year, and they're going to run into a salary problem. We'll pay whatever it takes." So we hired him, and Intel didn't.

Instead of cash, Gordon relied on Intel's stock-option plan to motivate employees. He saw this "long-term participation" as a means to keep engaged the actors who were necessary to bring his plans to fruition. "My mom might say Dad doesn't have a lot of emotion," says Gordon's son Ken, "but it's totally ingrained in him to think of what motivates people. For compensation, you have to look at some level of emotion. He's done a lot of the compensation work at Intel. That was totally his realm." Andy Grove agrees. "Gordon has strong feelings about a small number of things, compensation being one. To this day, Intel compensation is based on Gordon's philosophy and practices."

Moore and Noyce shared a pragmatic streak. They had refused to set aside parking spaces for executives and were committed to the idea that everyone should work from cubicles that were essentially identical. If everyone was an owner, who would choose to waste money on fancy offices and designated parking instead of boosting profits and the stock price? Stock options, open-plan offices, and the avoidance of a hierarchical parking system gave the company an overtly egalitarian feel, but these elements were rooted in a no-nonsense, hard-nosed approach of meritocratic pragmatism. To become a serious engineering firm, Intel's manufacturing technology required, by its nature, an extraordinary amount of discipline and control. To mitigate conflict, Moore and Noyce had created a single processing line for product development and actual manufacturing. Now, more than ever, the keys to success were rigor, data, and measurement.

Grove took his cue from Moore, picking up on his desire to create transparency, ensure follow-up, and fulfill the promise of the early company slogan, "Intel delivers." Constructive confrontation "was something Andy drove," explains Moore. "It was an approach we tried to have at all meetings. There were no hidden agendas. We brought everything out, and we solved the problem. We lived for problems. We loved to jump in and solve problems." As a management practice, it replicated exactly how Grove wished to operate. Like the Late List, though, it divided opinion. For some, it was liberating. For others, it caused unmanageable stress. "People understood the confrontation part," says Moore, who took no active part in the process, leaving it to Grove and others. "They didn't always understand the constructive part."

Carver Mead, personally disenchanted, could not help but admire in retrospect the way that Grove—acting upon Moore's strategies—created a leading global semiconductor corporation. "If it hadn't been for Andy, it would never have become the Intel we see today. It would have been a less hard-driving, less pugilistic, less monopolistic company. It's extremely important that Andy saw something the rest of us didn't: that we were in a world-competitive situation, and you had to be ruthless to survive, so you'd better get with it."

Staying Grounded

As the demands on Gordon's time, energy, and attention escalated, he remained grounded by his wife and family. With Ken coming into late adolescence and Steve entering his teenage years, Gordon—in the limited time he had at home on Jabil Lane—provided for his sons a steadying hand, a listening ear, companionability, and occasional inspiration for mischief.

Family life had its worries. Steve, at Los Altos High School, came down with a bad case of the flu, exacerbated by lasting joint pain and a sore back. Soon, he was laid up in bed. "The doctors couldn't figure out what it was," he says. Blood tests, X-rays, and bone scans eventually revealed it was likely an immune-system disorder. Initially concerned that he was "going to be paralyzed," Steve was able to return to school after a few months. He became closer to his older brother. "Steve used to help a lot with car activities. We formed a good bond," says Ken.

During Steve's illness Betty drove him to see medical specialists and arranged for him to be tutored at home. Gordon, preoccupied with the demands of Intel, gave Steve math help in the evenings. "Steve worked hard, was very diligent, and did well, but it wasn't easy for him." Steve was hugely grateful. "Dad helped me a lot, all the way from grade school into college. What's always amazed me is that, even now, he remembers the math." Ken attests to the same attribute: "Formulas that I forgot at twenty, he still remembered at seventy. He's always been very precise. If you look at his chemistry notes from college, they are perfectly laid out, with crystal-clear diagrams. He thinks it through and writes it down."

Gordon and Betty encouraged both sons to earn the money necessary for their enthusiasms. Ken's interest in automobiles continued. "He did yard work to support all the vehicles in his mind and the one in the driveway," says Betty. "He didn't wait for a handout." Steve was "belatedly aware of what Intel was, how it was growing, and the significance of it," but neither son had any idea of their parents' real wealth. Ken remembers thinking, "'Gee, my dad's the executive vice president of some company that seems to be doing pretty well. He seems to have some power and influence.' It wasn't until they started taking us on trips to the Galapagos Islands and Ecuador and Peru that I realized my friends weren't doing similar things. I could tell money was really tight for certain of my friends. The power company was shutting their electricity off. That's when I finally started noticing my own very different situation."

Gordon, in his personal habits and purchases, was largely unaffected by his wealth. Success did not change his character. "He's a remarkably non-greedy person," says Carver Mead. "I've never heard him talk about having money. It isn't part of who he is." Unlike more flamboyant personalities in the semiconductor industry, Moore remained careful with money, not wanting to spend unless he had to. Mead says that Gordon once told him about a woodworking shop he was building at home. "He said to me, 'All that tooling is really expensive.' Actually, it would not have made a dent in the fourth digit of anything he had. Personal wealth numbers are different from those that relate to what he's going to spend on a milling machine."

Gordon had passed much of his childhood in outdoor pursuits, growing up in his family's modest bungalow in Pescadero, with little "stuff." Shaped by values such as modesty and frugality, his approach remained utilitarian, focused on what was useful or practical. He was not fancy. He hated waste. Even when it came to his beloved fishing, he was reluctant to replace his old boat and its faulty engine. "Betty kept after him to buy a new boat, but he didn't want another," says Mead. "He loved this old boat. It was going to hurt him to let go of it. Eventually, he conceded, 'I guess I should, because it's getting dangerous.'"

Betty did not spend money lightly, either, but was not slow to support a charity or to invest where she saw genuine merit. Her grandfather, Elijah Whitaker had owned "half of downtown Oakland"; Betty decided to test her own skill in managing money. "Gordon gave me a million to play with. I was running my own little fund." For a time, she was part of a women's investment group but, becoming impatient, decided to make investment choices for herself. "I doubled my money in no time at all." Having satisfied her curiosity and exhausted her interest, she turned the fund over to Fayez Sarofim, Gordon's adviser, to manage.

Gordon also decided to introduce Ken, who was still in high school, to the art of investment. He bought Ken stocks in innovative companies such as Beckman Instruments and offered him ten thousand dollars (the maximum without incurring a gift-tax burden) to invest as he chose. "Dad said he would take a percentage of the upside, but absorb the downside. My dad is a long-term thinker. He used to talk about how putting your money in a market was a good thing to do. I was reluctant to try it, so I didn't. As I look back, my failure was in not taking that opportunity. I was afraid I'd lose the money, and I had a short-term view of what failure meant."

As a young teen, the cautious Ken had been worried about his father starting a fresh, untried business. Gordon, who himself had learned at a very young age how to suppress his own anxiety, now began to realize that his eldest son was unlikely to become an entrepreneur. (Nor did younger son Steve, whose severe health problems caused a significant caution in outlook, see himself as a risk taker.) Nonetheless, Ken absorbed many valuable lessons about honesty, fiscal responsibility, and delayed gratification.

> Every time Dad received a handwritten bill, he added it up. He says that across his life, he would have been better off if he never had, because most of the time they made an error in his favor. He points it out, and the bill goes up. That sends a message about values and honesty. I saw that over and over again. I watched him add up a bill and end up having to pay more. Good lessons.
>
> Dad used to talk about Moore's Law to me. He hadn't distilled Moore's Law to the "better, faster, cheaper" terminology used nowadays, but that's what it was. I learned the time value of compounding. He taught me the classic example, starting with a penny doubling every day for a month. I learned how to have a long-term perspective. To put it in business-school terminology, I slowly understood the "opportunity cost" of spending a dollar now versus not being able to spend it later. It wasn't just Dad telling me that, and he never used that term. It was something I internalized from my parents' actions.

If he was nervous about financial markets, Ken took extreme risks elsewhere. In due course he would own a drag-racing car and, like his father with the silicon printing press, aim to have the most advanced model, in order to dominate the system. "With a race car, you build it any way you want. You design it, develop it, implement it, and test it. It's a feedback loop. It's the full circle. It requires perfection. You win on this race, or you're eliminated. You can't make it up on the next lap. When you're out there at 150 miles an hour and spinning the tires, it's very exciting."

In the early 1970s, the family took its first trip to Hawaii, bringing along Irene Whitaker and Mira Moore. It was "a great two weeks," Betty recalls. "We wanted to bring the grandparents. That was the most wonderful trip of their lives. We all went except Gordon's father, who refused to travel." Walter Harold had undergone a near-death experience sometime earlier, when operated on to repair an obstructed bowel. He now used a colostomy bag and would not travel far, afraid that any problem might land him back in the hospital, an experience he wanted to avoid at all costs. Betty immediately fell in love with Hawaii and would eventually persuade a reluctant Gordon to settle there in retirement, almost forty years later.

Mira, Gordon's mother, who had once yearned to have a daughter of her own, lived a quiet, solitary, occasionally lonely life in the years after her sons moved out of the little house on Westgate Street, in Redwood City.

She was cut off from her roots in Pescadero and traveled rarely, and only with Betty. Her husband had worked long hours until retirement, talked little, and disliked the theater and shows. Mira went to the movies alone. She also attended a local Protestant church alone, her habit since Gordon's childhood. In her late seventies she became increasingly "religious" and "spiritual." Then, not long after the Hawaii holiday, Mira died quite suddenly.

Walter Harold, Gordon and his brothers, and the wider family buried her in Skylawn Memorial Park, ten miles from where she had lived. Walter Harold had chosen the plots in a special section devoted to war veterans. For Mira, too, it was an appropriate resting place, on a ridge of the Santa Cruz Mountains, to the west of Redwood City and San Mateo. These hills had surrounded the Moores over their many years in Pescadero. Two decades later Gordon and Betty would acquire a mansion just five miles down Skyline Boulevard, nestled in the forested slopes.

On losing his mother, Gordon communicated little of his feeling. Now in his midforties, he had achieved remarkable success. Even so, nothing could assuage his fundamental loss. As always he plunged into practical, focused work at Intel as a way to absorb the sorrow and find his place. Betty felt responsible for Gordon's widowed father and made efforts to help him to remain comfortable.

Evidence of Gordon's own mortality now appeared in his partial loss of hearing. A doctor put this down to "rifle ear," common among hunters, but Gordon thought it was more likely due to his adolescent hobby of making bombs and explosions. At least his "$15 million wristwatch" from Microma offered a special, unintended benefit. "It had a very loud low-frequency alarm that could actually wake me up." When this watch finally gave out, a low-cost replacement showed how special the original had been. "My replacement has an alarm, but I can't hear it. At seven o'clock every night it beeps for twenty seconds. Betty hears it, everybody hears it, but I don't hear it at all. I haven't figured out how to shut it off yet!"

THE REVOLUTION TAKES HOLD

War, Weapons, and Winning

Managing Intel's cascading growth was paramount. Shortly after New Year's Day 1973, Moore pulled out his notebook and listed his major concerns, all of them troubling: demand for the 1103 was outstripping Intel's capacity to deliver; to keep competitors on the treadmill, the firm needed to pull ahead with its 4K DRAM; "We are way behind Plan 73A on support personnel"; products lacked dedicated product engineers; no one was in charge of the effort on ROM chips; the quality-assurance department, tasked with seeing that chips coming off the printing press actually worked, had four vacancies. Here, at least, a serendipitous hire soon brought relief.

Craig Barrett was ten years younger than Moore and another native of the San Francisco Peninsula. Barrett had taken the academic path that Gordon had once envisaged for himself, but, as for Gordon at APL, "the luster of basic research" wore off and Barrett began to hanker after work that could be applied. As a materials scientist at Stanford, he consulted in the semiconductor field, using powerful X-ray instruments and electron microscopes to study defects in microchips. When an Intel engineer called to ask for recruitment recommendations, Barrett said, "How about a frustrated associate professor who's looking for something else to do?" Hooked by the "dynamism" of the chip industry, he quit Stanford for Intel, where he would later rise to be CEO. "Academic materials science was always trying to catch up with reality and explain it," he reflects. "The business of silicon is about forecasting what might be and trying to do it, as opposed to catching up on what is. There is an excitement to the industry. Intel, one of a zillion companies that got into the DRAM business, was the most successful." Barrett took up the quality-assurance task that worried Moore.

Another important contributor, Ted Jenkins, had been at Intel since 1969 and now tackled Moore's concern with meeting the demand for the 1103. Jenkins's accomplishments included helping to launch MIL's production of 1103s in Ottawa. He was one of the most skilled hands at pushing chemical printing technology along in the ways that Gordon wanted; Gordon gave him oversight of Intel's gleaming new factory in Livermore, Fab 3, which would produce only 1103s. (Intel's other two fabs, converting from two-inch wafers, were suffering crashes in yield, delays, and major headaches.) It would be Jenkins's responsibility to ensure that Intel kept up with demand for this hugely successful chip.

At Fab 3 equipment and tooling were purposely designed to work with the larger three-inch wafers that, in effect, amplified capacity. Gene Flath, Jenkins's boss, decided to make Fab 3 a "bunny-suit area." In the aerospace and semiconductor industries, so-called clean rooms for high-technology manufacturing—with advanced filters and airflow systems to keep out contaminant dust and chemicals—were becoming widespread. Workers in the most intensive rooms were required to wear "clean-room suits" (white overalls, with hoods, gloves, and foot covers) to prevent microchips from becoming contaminated by human detritus such as skin flakes and hair, a major advance from Gordon's previous attitude that "once dirt got down below the benches, it didn't make much difference."

The clean-room suits were such a novelty that "people used to find excuses to visit the fab labs, so they could put a bunny suit on." Workers

Workers in an Intel fab.

SOURCE: INTEL.

initially found the suits cumbersome, even disorienting. One described immersion in "a sea of white-suited people," the only familiar thing being the "the analog clock mounted to the wall." Bunny suits became another unavoidable, unpopular aspect of what Grove later referred to as his "tight ship." Fab 3 ramped up across 1973. Production capacity was measured by "wafer starts," that is, how many silicon wafers were put into the chemical printing press each week. The launch was one hundred wafer starts a week, with a further one hundred added each week, until by year's end Fab 3 was steaming along at five thousand wafer starts a week. The rising stream of 1103s brought in the great bulk of Intel's $66 million revenue for the year. When the company hit its first $3 million month, bottles of "Domaine d'Intel" champagne were handed out. In the cafeteria so many corks popped that the acoustic ceiling tile had to be replaced.

Carver Mead called the Livermore fab a "foundry for grinding out 1103s." Andy Grove, delving ever deeper into the practice of manufacturing, understood that by being the first to drive the most complex microchips into high-volume production, Intel could reap huge value. "When we make a hundred times as many devices, we typically reduce unit cost tenfold," he wrote. "The advantage of innovation is getting a head start on this experience curve, and making it difficult for the competition to catch up." Moore's Law and Moore's vision of the power of the most advanced chemical printing press were true.

One of the first victims of Intel's 1103 treadmill strategy was its second source, MIL. The costly but comprehensive technology transfer from Intel had at first worked well, with MIL's fab enjoying better yields than Intel. MIL produced versions of Intel's 4005 microprocessor and its popular 8008 8-bit microprocessor. Crucially, however, the MIL-Intel agreement was time limited. It did not cover future shifts in printing technology. Intel now found itself able to produce 1103s far more cost-effectively, having "engineered out" troubles in its move to three-inch wafers. To win sales, MIL also needed to switch to the larger wafers, but "they didn't know how to do it," says Moore. "They fell flat on their face. Yields collapsed. They couldn't get it right." No longer able to look to Intel for help, MIL found itself on the treadmill, where it died a rapid death. In 1975 the company dissolved.

Competition was real, brutal, and unceasing. Gordon, Grove, and others looked on with cold hearts as MIL failed and Intel reaped the spoils, establishing a monopoly on the hugely successful 1103. As Moore remembers, "MIL was a second source while we needed them to get the product adopted by our customers. Then, when they were supposed to deliver volume, they fell apart, and we had the whole thing to ourselves. It was not planned that way, but it worked out beautifully."

With Fab 3 online and the shift to three-inch wafers in Intel's other fabs, capacity was expanding and production was under control. The rising sales of microprocessors and EPROMs, meant that profit margins were becoming "unreal; we basically have what is needed thru 1974." Victory felt good; the destruction of MIL was "beautiful." Moore was not unfeeling, aggressive, or malicious—in person, he could be attentive, genuine, and kind. Among colleagues he would make suggestions, ask insightful questions, and seldom raise his voice—but at one remove, he was unmerciful. Attuned to the smallest details of Intel's financial reports, engineering briefs, and sales memos, he saw the competition circling constantly, always alert to the chance of jerking market share from his grasp. If Intel fell on its face, no competitor would shed a tear; in turn, he would not show his competitors mercy. This was business. In the abstract, Moore could certainly tolerate and even enjoy conflict and combat, born out of analysis, measurement, superior technology, and hard-won trade secrets. Only in interpersonal situations, encountering immediate suffering, did he struggle. And now he had Andy Grove to handle that side of business life.

Even so, it was no time to sit back. Gordon continued to look toward the far horizon, worrying that Intel was "weak in engineering." "I do not feel comfortable long term," he summarized. To consolidate its position in memories, he needed enough engineers to dominate the upcoming 4K DRAM generation and generations beyond. Moore was also struggling to understand the exact direction the memory business would take: What would users want? What would build the market? Was cheapness more important than speed and capacity? As to production itself, "What is most important? Small die size? Very low defect density? High performance?" Each end required the manufacturing technology to move in a slightly differing direction. Moore had to make the right call.

Intel's 1103 DRAM had proven the market for microchip memory, but Gordon did not expect to have the emerging 4K market to himself. It was correspondingly urgent to start shipping its 2107 chip, a 4K DRAM built with fresh technology. The device had twenty-two "pins"—wires to connect it into a product system. More pins meant both more uses and more cost. Intel's choice was confirmed when TI launched its own 4K DRAM soon after, also with twenty-two pins and matching the 2107's specifications. It was a head-to-head competition. As Intel's engineers scrambled to improve the 2107 to outdo the TI chip, both companies were outflanked almost overnight.

Mostek, a TI spin-off, advanced to the market with a 4K DRAM, presented in a cheaper, less bulky, sixteen-pin package. Not only that: Mostek's chip was made using a chemical printing technology even more advanced than Intel's. Mostek's engineers had figured out how to use ion implantation—which Intel had employed in Microma's watch chips—in mainstream technology. This allowed it to squeeze its 4K DRAM into the more competitive package and also get performance boosts. Here was another brutal lesson in competition and change.

"They came along with a different take on the technology," says Moore, and "very good circuit design." He had hoped Intel's chip would set the industry standard, but he had been right to worry. "Mostek took the leadership position with aggressive design." Competition descended like a ton of bricks. Intel's lead quickly evaporated. It had been first to market with its 4K DRAM and in 1974 held 80 percent of all sales. Suddenly, it was on the treadmill itself. In 1975 its share of the DRAM market fell to 45 percent. The 1103 was "the last DRAM we made a lot of money off," says Moore. Mostek consolidated its success with its next generation of chips and by the end of the 1970s sold more DRAMs than anyone.

Microprocessor Rising

Even as Intel established an initial dominance in the DRAM market, Moore was equally committed to the microprocessor. The first ones might not sell "at nearly the volume of memory chips," or be "ridiculously profitable," but they opened the way to novel uses of digital logic, calculation, and control. Moore himself had been quickly intrigued by the possibilities of microprocessors in everything from traffic lights and scoreboards to medical equipment and industrial machines. Envisioning a future in which the microprocessor would take over digital computation and control, Moore was quietly excited about the 4004, as the first of a series of products. Through 1973 the 8080 took shape: it used five thousand transistors, and the main constraint on its design was ensuring backward compatibility to preserve its customers' prior investments in software.

Bob Noyce, traveling widely to speak to customers and competitors about microprocessors, was effervescent in his excitement. He, and other enthusiasts, began to present microprocessors as simply tiny computers—"microcomputers"—that would take their place beside, and encroach upon, mainframes and minicomputers. He envisaged a whole fresh phase of the electronic revolution. Thanks to Moore's Law, increasing numbers of tasks would use digital computation, and there would be a proliferation of data processing. Daniel Bell's *The Coming of the Post-industrial Society* argued that in a large-scale transformation of American society, computers and information would be central. With many books published on the subject, it was becoming common to hear the term *computer revolution.* Noyce, keen for a piece of the action, marketed the microprocessor as a means to spread this revolution.

He drew an analogy between the microprocessor and small electric motors, a comparison that Steve Jobs, of Apple, would later adapt to the personal computer and publicize more widely. In Noyce's original analogy, small-scale electric motors had been a means of mechanization, spreading power—to the factory, to home appliances, to clocks—and becoming indispensable to machines, tools, and products of all kinds. In the computer revolution, the microprocessor was the motor. It would put digital computation into the battlefield, office, home, and everywhere else. It would be the means of mediation between mankind and electronic reality. Today, more than four decades later, we know the truth of this vision, as news reports daily proclaim, "Everything is connected" and as the "internet of things" becomes a larger and larger reality.

Gordon Moore was if anything more convinced than his colleagues of the potential of the microprocessor. At the same time, he was more reserved. For Moore, the microprocessor was a radical innovation. He saw its value to Intel as residing in it being a universal logic component (a standard product, high-volume production) for computing and control, rather than the means to Intel making complete computers (consumer markets, fickle fashion). It could be customized with software to perform any function a customer desired, not simply those within a mainframe or a minicomputer.

Intel's customers could use digital logic for all sorts of control functions, in any manner of products. This versatility was highly attractive.

Moore also understood acutely how Intel's microprocessor "chip set" was an encroachment by Intel into the territory of the firm's main customers, computer makers. On a single circuit board Intel could provide the complete logic and main memory of a general-purpose computer. Add a power supply, housing, and in-out connections (switches, lights, a keyboard, a video screen, and so forth), and this computer would be ready for action. However, computers were consumer products, like watches. He resisted the notion that Intel become a computer maker and thus compete directly with its customers. Standard products, adapted to many large markets and many customers, were what Moore's Law suggested and what he believed would best fuel Intel's success.

Even so, he shared his colleagues' desire to facilitate the growth of Intel's markets. He approved a decision to make circuit boards with the complete microprocessor chip set installed, enabling customers to buy, preassembled, all the hardware needed to use the microprocessor approach. To ease the task of software customization, Moore also approved a plan to have Intel build what were, in essence, special-purpose computers for microprocessor customers to use in building their software. These Intellec-4 and Intellec-8 "development systems" contained boards with 4004 and 8008 chip sets. An engineer could use them to make, test, and debug software, and could then save the software in Intel's EPROMs using an attachment. Bill Davidow, who led this effort, says the first Intellecs were extremely primitive. "I looked at that box, and I was seriously concerned that somebody was going to electrocute themselves if they used it."

Before Davidow's effort, Intel's sales and marketing head Ed Gelbach worked closely with Regis McKenna, one of Silicon Valley's premier marketing and public relations firms, on materials for Intel's development systems to debut at a national computer conference. These display materials presented the Intellecs as complete computers. Shortly before the conference, Gelbach and Hank Smith, who was in charge of microprocessor marketing, set out their "blue boxes" (Intellecs) on tables, along with brochures and posters, and invited Moore, Noyce, and others to attend a presentation. "We were really excited," Gelbach recalls. "We had no doubts that this was the future of Intel."

As Gelbach and Smith introduced the products, they could see "Gordon's face getting longer and longer and redder and redder." Most unusually, he was visibly angry. They paused and asked what was wrong. "You absolutely are *not* going to talk about these systems as computers. *This is not a computer*." Gelbach and Smith, much against their wishes, were forced to remove the more obvious references to computers from their materials. Smith had never seen Gordon "as angry as he was then." For Gordon, he said, the rare show of teeth was necessary. "It was the fact that we were going to be competing against our primary customers, as far as Gordon was concerned." Intel could supply specialized systems to aid customers with their use of components; to say that Intel was making computers was to go too far.

Gordon might oppose positioning Intellecs as computers, but he remained keen on improving these development systems. Not only did they drive sales of microprocessors and Intel's profitable EPROMs and DRAMs, but—with even the simplest model costing more than two thousand dollars—Intellecs were highly profitable products in their own right. Davidow had the needed combination of computer knowledge and marketing savvy to turn this into what Gordon calls "a good business," one growing faster than sales of microprocessors themselves. Davidow told Moore, "Each one is our salesman and will eventually result in the sale of at least fifty thousand microprocessors." For the next decade, Intel would create ever more powerful development systems, increasingly indistinguishable from personal computers. "We were building little computers, really," says Moore. He even gave airtime to Davidow's ideas on how to make Intel's system the "single computer on the engineer's desk."

While Moore recognized that with its microprocessor development system, Intel was "selling a computer for the engineer," he did very little to drive the idea forward. As time went on, he "never put the muscle behind it," preferring to let the venture take its course, rather than kill it off decisively, as a computer. His choice to neglect rather than back this business is another example of his strategy of investment. He invested time in the things he could be excellent at (making explosives) and abandoned things (high school diving) at which he was mediocre. He put effort and resources into businesses he wholly believed in, even as some (such as Symbol) withered and died. He was pleased to have Davidow in development systems, but did not change his mind about the business's potential for conflicts of interest. He enjoyed the profits that came in and the way Intellecs drove sales of Intel's other products, but felt it was not sufficiently compelling to be "a really important stand-alone business."

Gordon remained convinced that Intel's strengths, and his, lay in components. The success of the microchip itself would remain closely linked to developments in the chemical printing technology, which he understood at a visceral level. Systems were inscrutable, thanks to the vagaries of customer desire. As Gordon had painfully learned from the Microma experience, consumer tastes were difficult to comprehend or manipulate.

> I've never understood systems well enough to know when one is complete. You always leave something out or don't make it quite clean enough for the

> user. It takes a different orientation. Emphasis on components was a very important part of Intel's strength. I've been quoted correctly as having said, "I turned down the home computer idea." Even if I hadn't, I don't think we would have done a good job. It took somebody like Apple to get that started. We were always more successful as a component supplier than as a systems supplier.

The truly compelling path lay in the fulfillment of Moore's Law, with its steady increase in the scale of cheaper electronics. Sticking to fundamentals, and focusing on manufacturing technology, was the name of the game. One element of Moore's stance on the microprocessor, and his careful policing of its boundaries, was his growing appreciation of the device's symbiosis with the EPROM, which was itself amazingly profitable. The fact that the two devices "came along the same year" was, says Moore, "marvelous serendipity." The microprocessor required software customization, and the EPROM provided a way to customize it. He explains:

> The EPROM was absolutely the right companion for the microprocessor. Our original idea was that this was a prototyping device that the engineers would use while they were developing programs for their microprocessor-based systems. They would switch then to one of our programmable ROMs, which were much cheaper.
>
> We thought the EPROM was going to be a low-volume product. It would only be used for prototyping, so we priced it high. It turned out to be the engineer's security blanket. He could always change his program, so he never replaced them.
>
> He just bought high-priced EPROMs. We were making most of our money on EPROMs clear up to 1985! We kept this as well hidden as we could.

Where Lies the Future?

On the surface, Gordon rarely showed signs of stress. However, the combination of roles he was playing at Intel—chief strategist, de facto CEO, troubleshooter, and calm counselor—took its toll. Revealingly, his "coping mechanism" was not to have a particular trusted business confidant, or to head to the Wagon Wheel with colleagues, but to seek relief in the outdoors, in family, and in his own counsel, places to which, since early childhood, he had customarily retreated. On the one hand, he found release and reassurance in the familiar natural world of hunting, fishing, and physical activity, as part of the routine of family life. On the other, he counseled himself within his private notebook.

In June 1973, as demand for the 1103 DRAM and competition were both escalating at great speed, Gordon wrote about his concerns for Intel: he felt it was starting to crack under the pressures of growth. He wrote, "For the first time I see signs that we are in danger of losing control." Most concerning was that Andy Grove didn't seem to have his usual firm handle on operations; Grove didn't know why, for example, there had been a slip in revenues to just $1 million in June. Grove might be relentless and even sleepless in demanding information and accountability, but even he could not say why Intel's shipments had stumbled. Over at Microma, careless and unforced errors were causing delays. Disgusted, Moore wrote that "Microma could be clobbered; a complete loss of momentum." Throughout Intel, planning and control problems were bubbling up. One foreman even reported he was "losing control" of raw-materials inventory, the very materials that were required to feed the silicon printing press and keep the line going.

What to do? Gordon wrote a list of "possible responses." The first was to put on the brakes: "Declare a hiatus on growth to swallow what we have." A second was to keep his foot on the gas and ignore warning signs: "Muddle along as far as we can go while the opportunity exists." A third: "Reorganize." Fourth, and most serious: "Resign." There is nothing in Moore's papers or conversational recollections to suggest that leaving Intel had been an option he considered, up until this point. He had long kept a ledger to account for investment and return. From the point of intellectual completeness, resignation was a solution that now seemed both possible and attractive. It would not solve Intel's problems, but the problems would no longer be his.

Bob Robson, his Microma friend, who had already quit for a rural life, was an example of what was possible. Like him, Gordon could simply retire. As a multimillionaire, he did not need to work. He could live very comfortably and even—since he had given his sons Intel's founder stock—provide a wonderful inheritance. Yet no sooner had he written down "Resign" than he took this option off the table. He was at the height of his powers. His job was increasingly exciting. He was doing things he was good at and wanted to do. The company was conquering fresh territories in electronics and beginning to reshape the world. Why give up? What else would he do? In the end, he combined the second and third options on his list: keep his foot on the gas but reorganize. He would refine Intel's focus, police boundaries more closely, and exploit opportunities while muddling along. His counseling session was over.

Intel's sales were following an exponential curve, similar to that of microchip complexity and cost reduction. From 1971 through 1973, sales multiplied sixfold to $66 million, and head count went past 2,500. "The human cost of such growth is high," Andy Grove would write in his notebook that November. "The obvious part of this is that people have to work hard." Staff were expected to "successfully absorb new jobs or increased complexity in old ones." As chips became more complex, so did demands on employees. The start-up years had been unrelenting, but in the face of competition and growth, workers were pressed to continue at the same rate, with Grove writing on Christmas Day 1974, "I don't think I have ever worked this hard before!"

Craig Barrett remembers Grove at that time as being "a rude, outspoken, kick-in-the-butt operations guy," yet his management techniques were proving effective. They included such things as the face-to-face performance reviews with a supervisor that Barrett had found one of the "exotic features associated with Intel." Gordon was very much in favor of Grove maximizing the performance of staff. One of Grove's charts shows employees as falling into one of three categories: failing, "stalled," or capable of growth. As Richard Tetlow, his biographer summarizes, "Grove had both to understand what architecture the organization needed, and master the more subtle and difficult task of figuring out how flexible people were."

Intel had already been involved with watches and memory systems. Now, in his strategic quest to determine the scope of Intel's charter—what did it make sense to pursue, and how should Intel define its business?—Moore explored other chip-based avenues that might lead to sales. Telecommunications offered an interesting opportunity. Much of AT&T's network was reliant on transistors and microchips, with voice and data transmissions increasingly in digital form. Switching stations were themselves specialized digital computers. AT&T and Western Electric also offered small-scale switching systems (private branch exchanges) for businesses and buildings. Western Electric was already buying Intel's 1K and 4K DRAMs. Moore and his colleagues discussed whether to get into making and selling moderate-size switching systems, built from memory microchips, microprocessors, and other standard chips.

The main criterion was cost. "Can we offer an economic advantage?" Moore asked. Could Intel's prowess in making microchips translate into cheaper telephone switches? Even if it could, he was skeptical. Given AT&T's position, the market looked like being worth at most $7.5 million annually. Gordon could not see how to expand this to "as big a market as Intel requires." His own strategy was predicated on making electronic functions ever more cheaply in more complex chips, with price per function falling, enabling markets to grow fast and large. Watches were an elastic market, but telephone switches were not. Just as he had regarded the digital wristwatch from the perspective of the microchip—primarily as a technology business—so he now looked at telephone switching systems. He, and

Intel, decided to pass on the idea. In Gordon's chip-centered view of systems and end products, it lay outside of Intel's scope.

Embargo, Interdependence, and Revolution

Intel was in an enviable position. Moore might struggle with how best to manage the company's explosive growth, but even he marveled at the firm's profitability. Intel remained in its bubble for some time, as a tumultuous series of events in the outside world punctured any general US complacency. As 1973 wore on, Senate hearings on the Watergate scandal, exposing political misdeeds, were televised and beamed into millions of American living rooms. A tenuous cease-fire existed in Vietnam, but in early October 1973 Egypt and Syria launched a joint attack on Israel. With assistance from the United States, Israel beat back incursions into the Sinai and the Golan Heights, yet no enduring resolution was negotiated; instead, Middle Eastern suppliers, within the Organization of Petroleum Exporting Countries (OPEC), responding to America's support of Israel, launched an oil embargo on the United States. The resulting energy crisis exposed Intel's vulnerability to disruptions in the supply chain that fed its silicon printing press. Gordon believed that the embargo would be a long one and began to plan for contingencies. Essentially, the semiconductor industry was a chemical industry, relying on energy and materials. Intel's three Bay Area fabs all used copious electricity for everything from diffusion furnaces to lithography tools, so Gordon's first move was to ensure the company had sufficient electrical power.

Pacific Gas and Electric promised that it had "enough reserve fuel to get to July 1974," even if price rises were inevitable. The embargo brought home Intel's overdependence on a single electrical utility. Unlike its customers, it had no second source. Portland, Oregon (a short plane ride away from the Bay Area), and British Columbia were both leading centers for hydroelectric-power production, immune to OPEC embargoes. Moore wrote these locations in his notebook, and within two years Intel had opened a new fab in Aloha, ten miles from Portland.

Gordon was also concerned about petrochemicals. Intel used several types, including photoresist, the material at the heart of its chemical printing press, and around five hundred gallons a day each of xylene and butyl acetate. Availability and price were core issues. Moore explored how the firm might recover petrochemicals from production waste instead of sending it into the public sewer. Other concerns were that a rise in energy costs could lead to a shortage in the supply of silicon wafers and that dwindling petrochemical supplies could lead to a shortage of plastic parts for chip packages. Either shortage could present a real problem at any time, but with Intel already struggling to manage growth, and an expanding pack of competitors ready to capitalize on missteps, Gordon saw these concerns as especially acute.

A wholly separate issue was the long-term effect of deepening mutual dependence between Intel and its specialist suppliers and between Intel and its customers. In the early days of Shockley and Fairchild, Gordon and his colleagues had built all their own production tools, from crystal pullers to diffusion furnaces. With the rise of specialty producers, this era had passed. On founding Intel Moore and Noyce had made the key decision to rely on specialized outside suppliers, which would enable them to buy time and leverage focused expertise. They would even lease equipment rather than buy it outright. Intel and its suppliers of manufacturing tools and equipment became bound in a deepening mutual dependence.

A parallel interdependence could be seen between computer makers and microchip suppliers. As microchips became more complex, they embodied ever-larger hunks of system design, with lowered cost offsetting a loss of design autonomy. In their turn, computer makers became increasingly dependent on suppliers to deliver chips on time, in quantity, and with

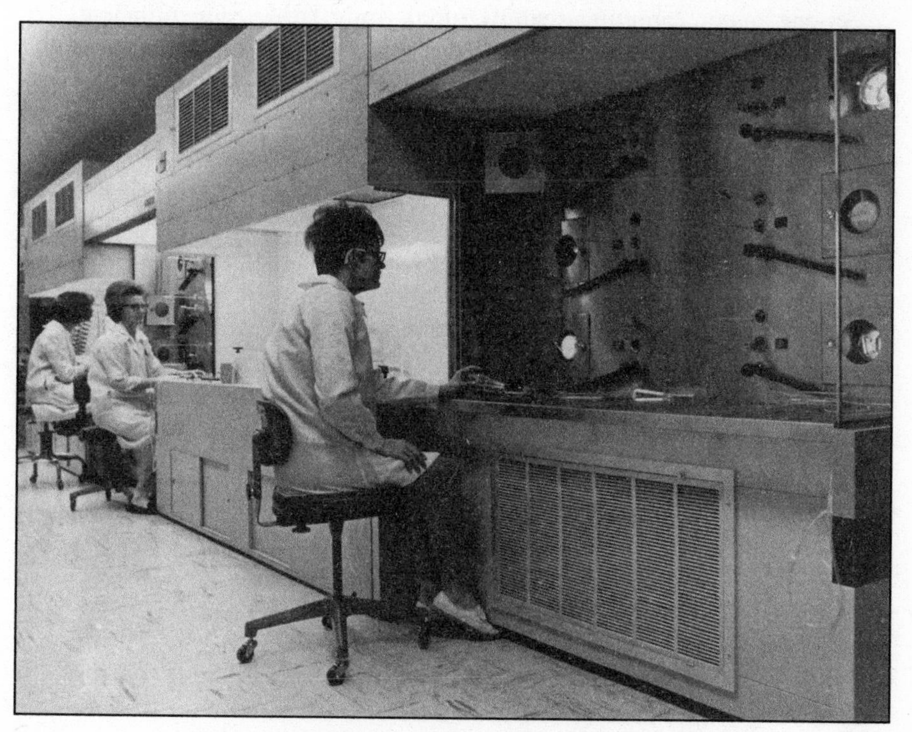

Women technicians operating diffusion furnaces at Intel.

SOURCE: INTEL.

sufficient quality. And suppliers became dependent on computer makers to expand the market for memory and to commit to large multiyear orders, allowing design costs to be adequately amortized. At Fairchild Noyce had proved that—because improvements in the silicon printing press meant higher profits toward the end of the run—an aggressively low price could work.

Not surprisingly, relationships between computer makers and microchip producers went beyond the simple taking of orders and supplying of products. These were carefully negotiated, delicate partnerships, with often-conflicting interests that required both parties to share sensitive information about products and technologies. At Intel, specific executives were assigned responsibility for managing particular relationships. Gordon was the point person for the Burroughs Corporation, a major customer for the 1103 DRAM and, as a computer maker, third only to IBM and Honeywell. (Burroughs had made its big move into computers by buying a Pasadena spin-off of Betty Moore's former employer Consolidated Engineering, shortly after the Moores returned to California in 1956.) If the oil embargo caused Gordon to plan ahead with more urgency than ever, a request from Burroughs also gave him pause.

Among major computer manufacturers, only IBM made sufficient semiconductor memories to meet its own needs. Others, including Burroughs, Honeywell, and Univac, depended on outside suppliers. In the closing weeks of 1973, Burroughs told Moore that it was projecting an annual requirement for 60 billion bits of main memory, in the form of memory chips. It wanted massive supplies of DRAMs each year. Its executives' concern about Intel's ability to deliver led them to demand that Intel guarantee their allocation. They even floated the idea of a collaboration under which Burroughs might, MIL style, make some DRAM chips itself, by setting up an internal second source to meet its semiconductor memory needs. It was a challenging idea, one that required Moore's careful consideration. Eventually, he would approve deals like this, most notably with IBM.

All was not puzzles and stress. Far from it. The November 1973 issue of *Forbes* included a long celebratory article, headlined "How Intel Won Its Bet on Memory Chips." Gene Bylinsky, writing for a national audience, laid out the extent of Intel's successes in its first five years, including its signature achievement: breaking open the market for memory microchips. "When Intel went into business, no market existed for its principal product," he explained. "Today, thanks to the company's trail blazing, no big computer is designed without semiconductor memory components. The market for memory chips has roughly doubled every year for the last five years, Intel's 1103 being employed in computer memories by fifteen of the world's eighteen major computer manufacturers."

Moore might have thwarted his colleagues' attempts to position Intel's microprocessor as a computer, but Bylinsky was operating under no such constraint. He detailed blithely how the company's "microcomputers" were going into "wristwatches, traffic lights, pinball machines and cash registers." He even managed to persuade Moore himself into making a dramatic comment, albeit one delivered in his habitually reserved manner: "We are really the revolutionaries in the world today—not the kinds with the long hair and beards who were wrecking the schools a few years ago." Gordon was not afraid to speak of revolution, having seen its truth.

The technology he had helped to create and develop was changing the basic conditions of life. "It was clear that we were impacting a very broad part of industrialized society. I could see that we were having a big impact. I believed that we were really making a change that was going to be permanent." Moore knew this because he could not only observe but also measure the growth of, and declining cost of buying, electronics, whether as microprocessors or computers. It was harder to quantify the impact of cultural changes in political philosophy, gender roles, artistic expression, and social convention. As Gordon observes, "It wasn't clear to me that the free-speech movement was going to have the same impact!"

Crisis, Cosmic Rays, and Computers

Just after New Year's Day 1974, Gordon pulled out his legal pad and wrote another "State of the Union" analysis. Compared to his turbulent thoughts of six months earlier, his tone was far calmer. Perhaps this was because he had just closed the books on the year. Intel's profit was almost $10 million, and the company was poised to make much greater profits in the coming months. Intel was now the sole source for the highly sought-after 1103. Moreover, the unique EPROM, though selling fewer units, had a stratospheric profit margin. Intel would also gain a lead over other companies by switching its core photolithography process to a new "Micralign" tool produced by PerkinElmer, which would eliminate much of the manual labor needed to align wafers to masks, proving a significant step in automation and precision. By the end of 1976, all of Intel's fabs would have this capability. "That was a huge step in yield," says Moore. "It was high risk, but it worked out very well."

Keeping up with demand remained a challenge. Components would be "production limited all year," Gordon noted, a problem caused not by Intel's ability to churn out naked silicon chips from its chemical printing press, but by the inability to expand assembly capability fast enough. Intel's assembly (packaging and final testing) processes were by now wholly divorced from chip-printing operations. The company flew California-printed chips more than eight thousand miles across the Pacific Ocean to be packaged and assembled in its Penang, Malaysia, factory and then flew them back to the United States. Even as it opened another factory in Manila, taking advantage of low labor costs in the Philippines, Intel still lacked capacity. "Things were really expanding," Moore recalls.

Had Intel been the sole supplier of memory, constrained supply and high demand might have afforded Moore and Noyce the luxury of raising prices. Instead, competition—from low-cost magnetic-core memory producers, on the one side, and from semiconductor companies offering rival 1K and 4K DRAMs, on the other—turned production limitations into a serious concern. By not supplying what "all customers want," wrote Gordon, "we are switching people away from 1103s." Another problem for Gordon was "IBM 145," Intel's attempt to offer a memory system to IBM's customers. Within this giant's vast customer base, there was a market for these "add-ons," but delay in designing and producing the needed DRAMs was holding everything up. Gordon noted these items tersely. His analysis and argument, when presented to Grove and Intel's other senior staff, would be enough to provoke action. Gordon had put himself at the center of an organization purpose built to react quickly to his questions, concerns, and suggestions and take immediate action. Intel's Microma division was still struggling. The picture for commercial microprocessors looked far brighter. Intel dominated this area.

Women assembly workers in Intel's Penang, Malaysia, plant, 1970s.

SOURCE: INTEL.

As 1974 wore on, Intel suffered a major glitch that

Gordon Moore, CEO, visiting Intel Penang plant, 1970s.

SOURCE: INTEL.

illuminated how, when pioneering in unknown territory, a single obstacle could tip the balance from survival toward extinction. The company had only recently won a multimillion-dollar contract from Western Electric to provide 4K DRAM chips for use in telephone exchanges. Western Electric, famously tough in its testing for reliability, noticed a failure in the chips. As Andy Greenberg recalls in *This Machine Kills Secrets,* a single bit of data would flip from a zero to a one or vice versa, seemingly at random. A memory chip that could not remember was no use at all.

Eventually, Intel engineers considered a possibility suggested by Gordon himself: cosmic rays—a form of radiation hitting the Earth from space—might be to blame. Gordon had mentioned this idea in a talk about microchip miniaturization, speculation that there might come a point at which "structures are fine enough that our circuits will be upset by cosmic rays." Quick calculations and experiments ruled out cosmic rays themselves, but not some other source for the radiation. The engineers wondered if it might have a prosaic source. If the packaging material was contaminated with a radioactive isotope, then the silicon chips inside might be bombarded by alpha particles produced by the contamination.

Tim May, one of the engineers, used equipment in his lab to check the hunch. Sure enough, it picked up an excess of alpha particles. The engineers clinched things by tearing out the alpha particle source from a smoke detector and putting it up against an unpackaged DRAM. The DRAM went "bananas." The error's source was revealed. Intel cleaned up its packages. Gordon's initial comments on the perils of radiation inspired a chain of creative work by Intel's engineers that tackled the problem.

What did not disappear, despite Gordon's angry clampdown in mid-1973, were continuing attempts by his senior colleagues to position the microprocessor as a microcomputer and the Intellec development system as a general-purpose computer. Despite the force of his intervention, these attempts intensified. Intel announced a new, faster, and more powerful microprocessor, the 8080, at the Solid State Circuit Conference in 1974. The importance of this device would be reflected for many years in Intel's Santa Clara phone number (765-8080). Ed Gelbach and Regis McKenna were both deeply involved in the 8080's launch. A two-page advertisement in *Electronics* crowed, "From CPU to software, the 8080 microcomputer is here." Whatever one called it, people began building "personal computers" using the 8080.

In New Mexico Ed Roberts led MITS, a small company that made electronics kits for hobbyists, ranging from radio-controlled small toys to calculators and digital voltmeters. These kits provided the customer with parts and instructions on how to solder or screw them together. A "personal minicomputer" hobbyist kit already existed: the Mark-8 custom $50 circuit board, to which a customer had to add parts worth an extra $950, including an Intel 8008, to build a rudimentary computer similar to one of Intel's Intellec-8s. Roberts came up with an improved take. Why not use Intel's 8080 and offer a complete kit of all the parts and instructions required to make a personal computer? Intel agreed to sell Roberts the 8080 for $75 per unit rather than the list price of $360, on the condition that he would buy hundreds. That meant MITS could sell its kit for just shy of $400 or an assembled unit—the Altair 8800—for just over $600. In proportions, layout, and look, the result was very similar to Intel's Intellec-8. The price of the Intellec system, in contrast, was more than $10,000.

News of Roberts's plan for a $400 hobbyist computer reactivated the idea of Intel becoming a computer maker and gave it added impetus. "Bob looked at the Altair, looked at the microprocessor development system, and saw something very similar," recalls Grove. Soon Noyce and Gelbach were jointly enthusing over a plan for Intel to make a $300 computer. Again, Moore's vision for Intel was at stake. He needed to assert control. Grove describes how things came to a head: "Bob Noyce said, 'Now that we are in the computer business, blah, blah, blah,' and Gordon exploded, 'We are not in the computer business! We build computer development systems.' Gordon was more forceful on that than on any other issue—Bob cowered! We did not go into the computer business!" Gelbach, also in the meeting, watched as Gordon's face tightened: "I thought he was going to either faint or hit me." Gordon won the day, and the episode ended further talk of Intel as a straightforward computer maker.

Gordon's intransigence signaled that "fun" had given way to a driven, disciplined approach at Intel. With Gordon Moore and Andy Grove focused on fundamentals and worrying about the competition, Bob Noyce saw "the emphasis shifting to control." It was becoming clear that Intel's entrepreneurial roller-coaster ride—its start-up phase and first glory years—was truly at an end. Moore believed it was time to consolidate. His resistance to Noyce's plan to capture fresh and more distant hills was a response not just to the failures of Microma or to the darkening economic cloud overhanging the nation, but also to an ongoing tension within his own business model.

Gordon Moore, who saw things in "delicate shades of gray," knew that finesse in managing the territory between customers' agendas and Intel's definition of its own business would be central to continued success. Moore's Law, the silicon engine that he thought of as Intel's "great cost-reduction machine," depended on satisfied users of transistors on microchips that Moore would create—first by the hundreds and thousands, then by the millions and billions, and eventually by the trillions and quadrillions. To absorb these chips, the comfort of customers was critical. The success of this stance would later be demonstrated by a notable lack of objection from other customers when IBM bought a significant stake in Intel during challenging times in the early 1980s. Moore was proving himself a deeply thoughtful strategist, maintaining the line and keeping the big picture straight. He remained behind the scenes and out of the public view. In the end, his determination to avoid competing as a maker of low-cost microcomputers proved highly profitable. As Gordon recollects laconically, "It worked fine."

Loosening Ties

If Gordon's life at Intel was becoming more intensely structured, life at home—for his wife, Betty, at least—was the opposite. As her sons grew up, she at last could begin to explore her own interests. For years, she had played the part of "big mother" to everybody. When other parents were absent, she had taken pride in being there for her children and for their friends. "It was a latchkey situation. I took in stray boys. They knew there was always a spot for them and something to eat. One said I was his second mom. Ken's friends would all come after school. We would take them to the beach house, and they would go out surfing. I made a gazillion brownies. They'd all stay; they knew they were welcome."

As her sons moved up into high school, Betty faced a parent's familiar challenge: on to college, and if so, where? Ken knew he was going to get at least a basic degree. "If you didn't have a college degree, you got a lesser job. It was that simple." His mother adds, "Gordon and I had set a precedent; both of us had been. The road you want for your own children goes toward as much as you had, and more." Gordon and Betty were rich, successful, and educated. However, putting pressure on the children was not Betty's way.

These were the years of hippies, Vietnam, the draft, and student sit-ins. "Tuning in and dropping out" were much in vogue, but for the Moores of Pescadero, that was not the way. Ken and Steve both enrolled at Foothill College, a local junior college only a five-minute drive from Jabil Lane. They were commuters, as both of their parents had been, at San Jose State decades earlier. Ken dreaded being drafted to fight in Vietnam, but in the summer of 1973, the threat abated as the war ended. "I was a month away from the physical inspection. They shipped me my 1-A card and were lining people up."

Ken's initial desire was to be an engineer. Gordon believed his son possessed "enough of a technical bent" for an engineering career, but Ken, like his father, did not want to be second best. "Algebra wasn't bad, but geometry was tough. I didn't want to be a crappy engineer." He opted to study business instead, continuing to live at home on Jabil Lane. Betty quietly kept an eye on his progress. "Ken could probably have gone somewhere else and done better, but he didn't have the drive," she recalls. Academic study was "a tough deal," but Ken completed his degree, to his parents' relief. He was the only one in his peer group to graduate—unsurprising given the turmoil of the times, in the Bay Area especially. Several of Ken's friends from that troubled era went off the beaten path, entering cults or spiraling downward into menial jobs. "This is not what you want for your children when you are at that success level. I'm glad our kids did not go away from home. They would have maybe got lost in the cracks. Securing a degree and a job was distinctly passé at that time. It was 'life will come to you.'"

Betty's mother, Irene, in nearby Saratoga, had been closely involved with the upbringing of her grandsons, caring for them when Gordon and Betty went on business trips. She, too, now had more free time. One of

Betty's delights in the 1970s was to pay for her mother to go places. "I sent her on all kinds of trips with her sisters and her neighbor. They went to Europe, to the Holy Land—everywhere that they had any desire." Gordon's father, Walter Harold, adjusting to life as a widower, was finding solitude a challenge, and Betty could see him going downhill. She says, "I had him down to dinner quite often because he was alone." Walter Harold had kept up a property in Pescadero and spent much of his time there. As Betty remembers, "We'd go up in the evening, and Gordon's father would sit there watching some man thing—wrestling or something—and smoking his cigar, and I'd say, 'Oh, God.'"

Gordon's older brother, Walt Jr., still lived and ranched in Pescadero, but the family connection with the little town was dwindling. The most potent symbol of the Moores' influence, the old frame house built by Alexander Moore 120 years earlier, stood derelict and vacant. Fires had taken out interior walls, exposing the redwood flooring and foundations. "The roof and floor of the veranda had collapsed. Nearly all the windows were broken, and there were great holes in the roof." Eventually, vandals set fire to the house and burned it to the ground.

Making Hay, Dodging Rain

In mid-1974, as Bob Noyce lamented the loss of fun at Intel, the fun also disappeared from the US economy. Like the Moore house, the economy was teetering. The toll exacted by the oil embargo was becoming apparent. Inflation spiked to 12 percent. Oil prices jumped to four times their norm. The United States plunged into recession. Drivers of vehicles with even-numbered license plates could buy fuel only on even-numbered days and odd-numbered plates on odd-numbered days. The federal government reduced the speed limit to fifty-five miles per hour, and year-round daylight saving time was introduced. Recession with high inflation became the dread "stagflation." The US stock market tanked. By December 1974 the swooning market had lost about 40 percent of its value. Unemployment rose.

Expenditure on computers had been rising dramatically, by almost 50 percent a year. Microchips had half of the computer memory market, and Gordon envisaged that within a decade, they would take over the territory entirely. Intel's position in microprocessors was no less rosy. This was now a business worth $35 million, of which Intel had three-quarters. Digital logic offered by microprocessors was hailed as revolutionary: the microprocessor was a product with real potential. Intel was booming. "Because MIL had died on the 1103 DRAM, we were the sole source. EPROMs were really growing. And the 1K static RAMs were doing well. We had these three principal products, where we were sole source supplier. We were really enjoying it. Our profit margins were almost 50 percent. That was nice."

Struggling to manage explosive growth in boom times, the whole industry had invested massively in additional capacity. The sudden downturn at the end of 1974 hit hard. As the recession deepened, orders for microchips shrank rapidly. Suddenly, excess capacity confronted vanishing orders. Moore, by now a hardened veteran, knew that capacity gluts leading to price wars were characteristic of the industry's cycle, though this made the experience no less painful:

> It is the natural dynamic of an industry with very high fixed cost. It's like the airlines in a lot of respects. You don't want to have an airplane fly with an empty seat. You're always better off if you can fill it for almost nothing. Similarly, with a wafer plant, once you have the physical structure and the engineering staff in place, whether or not you run it doesn't make an awful lot of difference to the cost. People look at their marginal cost—probably a quarter of the true cost. If you can sell for more than that much lower incremental cost, you keep running your factory. So whenever there's overcapacity, the price spirals down.

The sudden shift in 1974 was not unique, but the severity was without precedent, causing Intel's first really tough period. In July the share price dropped 30 percent. Gordon was still not without a certain business naïveté: "When the market fell off, I told the usual Wall Street analysts, 'Things are changing. The customers aren't shoving orders at us; we are having to ask.' The stock dropped nineteen points the next day. I was absolutely shocked at the impact of keeping the analysts informed about the way business was and the effect it had on the market. The world collapsed."

Growth had been a lot of work, but hid a lot of sins. "Because of supply-limited situations, one can get away with doing fewer things and less perfectly," wrote Grove to himself. The growth mode might be hard on one's physique, he added, "but it was relatively easy on one's mind." Now, Intel no longer needed all its fabs and production workers. Initially, Moore and Noyce tried to ease overcapacity through furloughs. Then it became clear that layoffs were in the cards. Even Moore, the calm, still center of Intel, was shaken. "We did a lot of shutting down. Those decisions were made in agonizing meetings, deciding how much we could do without. The bottom seemed to be falling out of the market. We had to cut capacity to match demand."

In September Gordon sent out a "State of the Union" memo to the entire executive staff, the dozen or so people with whom he ran the company.

In numbered sentences he laid out the challenges of overcapacity and falling prices, using dry humor verging on that of the gallows:

1. We have capacity to bury the world in our standard products.
2. The world doesn't need very many right now.
3. When our competition can match our product capability, they are willing to sell at a much lower mark-up than we are.
4. Since we have had virtual monopolies, others can look at our markets on a differential costing basis, just like we looked at the calculator.

Thus, we have the following:

1. A shrinking (or slowly-growing) market for our main products.
2. Erosion of market share as competitors come in.
3. Rapidly declining prices as our customers pull back.

In the short term, Intel should "develop business in products that are free of competitive pressure," such as the 8080 and the IBM add-ons. Intel might also consider joining the race to the bottom: "Maintain market share at least at steady-state level. The prices are going down anyway. We might as well take them down first." Last, Intel needed to "get down to minimum size to operate at forecast levels of business." This was Moore's analytical, ledger-based, depersonalized way of saying that they needed to fire a significant fraction of the employees.

The following month Intel laid off about 750 of its 2,500 workers. Gordon was at several removes from the messy reality. "Previously, I had to tell employees that they were losing their jobs. During this slowdown, I didn't have to tell anyone directly. Boy, that was a lot easier." Even so, "the layoffs took a tremendous toll on everyone," wrote a commentator. "You can see the sadness in the face of Gordon Moore, a man who does not wear his heart on his sleeve." Moore's partner, Bob Noyce, was himself at both a business and a personal crossroads. He decided to end his long-failing marriage and also to "kick himself upstairs" to chairman. And Andy Grove stepped closer to the heart of the company, joining the board.

Intel itself remained in a position of financial strength. It had no debt and owned its principal facilities and most of its production equipment outright. Funds needed for 1975 would be "well within borrowing capabilities from the banks." Moore felt fortunate that the firm was "exceedingly profitable" at the start of the recession. "Our margins dropped down to around 20 percent. Most companies were starting at less than that, and when they dropped they were in a loss position. We came through pretty well. We had a very good product position, but it was traumatic." The sudden onset of "business winter" was a surprise, but Gordon knew that Intel was fundamentally robust. He also knew that the name of the game was to keep inventing improvements to the manufacturing technology.

The electronic revolution was successfully launched. What mattered was to be ready to make hay when the sun returned.

9

THE GREAT COST-REDUCTION MACHINE

SETTLING IN

Decision Time

For Bob Noyce, the tough business climate and the insistence of Gordon Moore, his partner, that Intel needed to stick to its knitting played into a mounting sense of dissatisfaction. The company was no longer a small, adrenaline-fueled, open-ended start-up: it now employed thousands at home and abroad, with revenue measured in the hundreds of millions. The fun of early days was over. He no longer wanted to be CEO.

Already fabulously wealthy from his involvements in Fairchild and Intel, Noyce also commanded his own substantial private investments from a separate office on Sand Hill Road, the locus of Silicon Valley's growing array of venture-capital firms. He had full financial freedom of choice. What, then, to do? If he no longer wanted to lead Intel, how could the firm continue? Moore had made it clear, when the pair were at Fairchild, that he did not want to shift to a purely management role.

Not short on hubris, Noyce privately approached his former colleague Charlie Sporck at National Semiconductor, four miles up the road. National was thriving and enjoyed revenues nearly twice Intel's. Despite Sporck's defection from Fairchild, he and Noyce had remained friendly. Noyce floated his "no-brainer": that National merge with—in reality, acquire—Intel. The two began to discuss a deal. Belatedly, Noyce laid his cards on the table to Moore. Sporck would be CEO of the merged company, with Moore leading a major division. Unlike Noyce, Moore had retained all his Intel stock, his stake of around 10 percent, making him the largest individual shareholder. He also enjoyed the respect and fealty of the board. Taken off guard, Gordon agreed to hear the pair out. All three knew that no deal could proceed without his support. Never rash, always wanting solid data, he listened carefully to Noyce's proposal: the deal did not make sense. It might offer a short-term boost to investors and stockholders; longer term, there was no advantage in combining. Yet Noyce was clearly on his way out from Intel. If Moore would not let Charlie Sporck lead, who would?

Gordon Moore knew his own answer: "I want to stick around and try to run Intel for a while." Quietly but firmly, he put himself at center stage. *He* would be the CEO. If Arthur Rock consented to step down to vice chairman, Noyce could become chairman and continue to counsel and support the enterprise. Both men agreed that Andy Grove should move up with them, taking Moore's old title of executive vice president. They asked Grove to lunch. He was immediately enthusiastic. All that remained was securing Rock's support.

To Rock, Moore's plan was a natural next step. "Noyce got bored easily. He'd done the CEO job for a while and wanted to try other things. Moore had run research, done a terrific job, and showed a lot of leadership ability. I thought it was great." Rock was confident that Moore and Grove's partnership was solid. "Andy gradually developed into more of a general manager. It was the same evolution Gordon had undergone. They worked fine."

Intel announced the changes in its annual report. Given the widespread business collapse also outlined in the report, the message needed to be positive: "These changes will strengthen the ability of management to respond to the challenges ahead." Rock's postscript, as outgoing chairman, made clear this was no corrective measure. Rather, "Intel has been able to develop breadth and strength of management. Decisions will be made by people who have shown they have the talent and energy to take over more responsibilities." Noyce had long been the charismatic link to customers and competitors, but it was clear to insiders that Moore and Grove were already running the company, together. The fresh titles ratified reality.

In 1967, when Sporck defected and Noyce asked Moore to head Fairchild's semiconductor division, Moore had said no. Now, he wanted to lead. What had changed? The decision to take the reins was less a personal shift than a reflection of the fresh context that Gordon had constructed for himself. In 1967 he had been happy in the R&D laboratory. It was, in his words, the "best job in the world." His vision was clear and his agenda compelling. He knew his own strengths and weaknesses, and given the huge gap left when Sporck—the quintessential production man—departed and the rate at which Fairchild Camera and Instrument was churning through CEOs, he also knew he was not the right choice for the job. In managing a crisis-ridden division within a stumbling company, he doubted he could excel, or even cope. It would be like "reaching into a bag of snakes."

Seven years later, things were different. The immediate problems were external, the result of oil embargo–driven stagflation and intensified competition among microchip makers. Moore was satisfied with Intel's fundamentals, and working with Grove suited him. In 1967, at Fairchild, he personally had felt "on the right track, doing a lot of good work." This, writ large, was how he now felt about Intel itself. Moore had tuned the firm to the opportunities and strategies that would optimize the economics of integration. The year before, he had considered whether to leave and quickly rejected the idea. Intel was exciting, Moore's Law was real, and he wanted to keep moving things along. For him, there was still all to play for.

Running the company afforded Gordon a continuing engagement with the chemical printing technology—the array of chemical and physical steps that transformed slices of silicon crystal into ever more complex, yet miraculously less expensive microchips—microchips that were themselves electrical circuits built from an ever-growing number of ever-shrinking transistor "bricks." Moore was already witnessing the dramatic outcomes of improvements in transistor and microchip technology. Computing cost was down 20 percent year on year, from 1970 through 1974. The computer itself was now dominated by silicon microchips, for number-crunching logic and data-handling memory alike. As Moore had foreseen, microchips were steadily more complex, at prices that fell just as steadily. Clunky though they might be, with their punched cards and temperamental software, massive mainframe computers were reordering the nation's science, technology, business, and government. And silicon transistors, in the form of microchips, were central to their growing role.

Gordon had long looked at his own life as a series of investments and returns. A key satisfaction at Intel was his ability to obtain and use very clear metrics—yield, die size, line width, transistor count, cost per transistor, market share, stock price, revenue, profit, head count, return on investment, and growth—to analyze progress. All fitted easily on the pages of a ledger, like the one used for his personal expenditure. Quantitative financials were useful tools not just to gauge his own success, but also to shape the motivations of others. Compensation directly influenced how people behaved—"operant conditioning." This radical form of behaviorism, promoted by psychologist B. F. Skinner, argued that it was unnecessary to unpack individual emotions and ideations. They were a "black box" in which inputs were converted to outputs. What mattered was to fine-tune a reward system, a code that would deliver the desired input-output conversion. Gordon was comfortable with and preferred such an approach. It was a technical solution to a technical challenge.

With Jay Last's unexpected defection in the early 1960s, he had learned to his cost that personal, empathic communication also mattered, yet he knew, too, that financial compensation (salary, bonus, or stock) was crucial in producing desired behaviors, such as meeting goals or increasing productivity. He had used this approach to coax his son to study. "It's totally ingrained in my father," says Ken, "to think about what motivates people." Quantitative schemes for compensation also meshed well with Grove's developing system of internal objectives, as did stock options and a stock purchase program. With Intel's shares climbing steeply, its senior managerial cadre was both hard driving and secure. Junior employees poured thousands, then millions, into the company through their purchases of stock at advantageous discounts. Interests were aligned. Gordon's approaches were working.

Moore was not the organizational fighter needed at Fairchild in 1967, but at Intel he had an alter ego in Andy Grove. Grove was learning to harness his own punishing drive and brilliant intellect, transforming himself from star researcher into outstanding manager. Grove modeled himself on Sporck—"the operations guy I aspired to become"—and thought endlessly about how to do his job, reading deeply in business school literature. The infrastructure he would slowly build, shaped to execute Moore's technical strategy, would become crucial to Intel's continuing success. With Grove as his key associate, Gordon was confident that the elements were in place to follow Moore's Law and to drive the revolution in which electronics, and especially digital computing, would dramatically reshape every aspect of society.

He had the right team, the right technology, the right strategy, and the right playing field (microchips, deployed in computers) on which to win the game. His decision carried little personal risk. By early 1975, Moore held more than eight hundred thousand shares of stock. Even in the bear market, his stake was worth more than $20 million; by the end of 1975, its value had climbed back to almost $75 million (more than $300 million in today's dollars). He also held significant stakes in the venture-capital partnerships of Art Rock and of Eugene Kleiner and with Fayez Sarofim. Gordon was hugely wealthy—free to do whatever he wanted. And what he wanted was to grow Intel, even as his own stake within it grew.

In April 1975, as CEO, Moore ramped up spending. To expand rather than rein in was counterintuitive, especially in view of his frugal nature. However, in almost two decades of watching boom-bust cycles in the semiconductor industry, Gordon had seen how in times of economic downturn, it was vital to think ahead. As business began to improve, only the newest devices—with increased capability and complexity, made with the most advanced manufacturing technology, and representing the lowest-cost electronics—would take off. Intel must be ready to capture market share and profit. This meant investing in R&D, upgrading fabs, and even hiring production workers ahead of demand. During the rest of 1975, he put these beliefs into practice.

R&D expenditures increased by nearly 40 percent, to $14.5 million. Investment went to the next generation of DRAMs (which would hold

16K of data rather than 4K), and to EPROMs, microprocessors, development systems, and memory systems. Gordon also invested in full conversion of Intel's fabs to the latest technology and to three-inch wafers. He began to counter Mostek at the forefront of chemical printing technology by developing Intel's engagement with ion implantation. In all of this, his aim was to push the manufacturing technology—from materials to photoresist chemistry, and from depositions to optics—to create smaller transistors on more complex microchips at acceptable yields. "Only through such continued investment in products and technology can Intel hope to maintain its progress," Moore explained in the year-end annual report. His investment was reflected in Intel's results, in a time of recession. Revenue was flat at $137 million, while profit fell by $4 million (the size of his increase in R&D spending) to $16 million.

Intel could cope with big investments. The firm had no long-term debt when Gordon took over. "Soft economic conditions," he wrote, "offer exceptional opportunities for well-financed companies to improve their relative technical positions." His spending would both position Intel well for the upturn and challenge competitors when they could least afford to respond. By mid-1975 head count was up to its prelayoff level, and by the end of the year Intel had hired an additional thirteen hundred production workers, giving a total of five thousand employees. Gordon Moore was in charge, driving ahead.

As CEO he asked constantly what could be done, and when. His oversight focused on practicalities and tactics and on balancing competing needs. Intel's most visible success remained the 1103 DRAM. This device had changed the silicon landscape, breaking open the computer memory market and dealing a fatal blow to magnetic cores. Thanks to the 1103, Intel was viewed mainly as a memory-chip maker. For Gordon, however, memory was simply an opportunity that embodied his underlying strategy of finding optimum points of complexity, cost, functionality, competitive position, standardization, and elastic markets. He wanted to make ever-greater numbers of transistors at ever-lower prices for markets that would balloon: to combine the economics of integration with his philosophy of standard products. Microchip memory was the first in a line of possibilities. Calculator chips were another. The microprocessor, with its radically new approach, was a third.

Moore's Law Revisited

Ten years before, when *Electronics* magazine had offered an opportunity to set out his vision, Gordon formulated the argument that would eventually become legendary—"Moore's Law," that is, how a steady doubling of microchip complexity would achieve the lowest manufacturing cost per transistor. Toward the end of 1975, when the Institute of Electrical and Electronics Engineers asked him to talk at the opening session of its Electron Devices Meeting, he decided to update his argument. The meeting, a regular port of call, was the perfect place to reveal how his insights were playing out.

If Silicon Valley was a newfangled name, "Moore's Law" did not yet exist as a phrase. As we saw, little attention had been paid in the wider world to the *Electronics* article of 1965 or the vision it promoted. Caltech professor Carver Mead had been one of the few to grasp its revolutionary potential. In the mid-1960s, consulting at Fairchild, he discussed with Gordon the quantum mechanical tunneling of electrons, an interesting phenomenon whereby electrons blink from one location to another over very short distances, overcoming seemingly insurmountable barriers. Moore was concerned that tunneling might someday present a limit to the microminiaturization of a transistor. Could it overturn his plot of microchip complexity? Mead thought it over and said yes. Tunneling would indeed put a lower bound on how small transistors could be. Moore's follow-up was simple: "How small is that?" Answering Moore's question took several years of intense academic work. In 1972 Mead and his student Bruce Hoeneisen published their results. Transistors could shrink down to one hundred and fifty nanometers, one hundred and fifty billionths of an inch, or much less than one hundredth of the width of a human hair. That was the only ultimate terminus. It was an astonishing conclusion, as extraordinary as the trajectory to which it gave a limit.

Mead increasingly became the spokesman for Moore's insights, taking on a demanding schedule of talks around the United States to the electronics and physics communities. He told his peers that the silicon microchip would change the future. The skepticism and resistance he met only emboldened his personal crusade to show the world that repeated doubling was possible. "Every time I'd go out on the road, I'd come to Gordon and get a fresh version of his plot." Mead helped others to see Moore's belief as credible. Engendering debate was an essential step toward convincing individuals and organizations to make the steep investments required if this "future" was to become real. Moore's vision was a social construct, possible only if others were inspired to take up the cause. In the early days, Carver Mead was principal recruiter and propagandist.

The 1965 article was a forecast and a set of observations, rather than the law it eventually became. With his characteristic lack of self-aggrandizement, Moore points to Mead as the likely originator of the term *Moore's Law,* yet Gordon himself in the published version of his 1975 Electron Devices talk, "Progress in Integrated Electronics," makes the earliest known mention of the "annual doubling law" for microchip complexity. At the meeting itself, he concentrated on "Was I right?" and "How did we do it?" Intel's recent microchips represented a 65,000-fold increase in complexity from the realities of 1959. Three factors were involved: increasing the chip area, decreasing the "geometry" of features, and improving device and circuit "cleverness." The first input, "a factor of approximately twenty," was larger chip size, enabling the inclusion of more transistors. The second—decreasing "geometry" by using advanced photolithography to shrink transistors—gave "a factor of approximately thirty-two." The third and largest factor, "of about 100," lay in the "cleverness" of design advances that enabled more of the area on a chip to be devoted to components, rather than to such features as device isolation and interconnections.

Gordon was acutely aware that it was minimization of cost that drove his strategy. Yet he said nothing about how a perpetual focus on the "sweet spot" of economic optimization was the "why" of increasing complexity. Instead, he again focused on the "how," renewing his earlier doubling prediction and extending it a decade to 1985. Die-size increase and feature shrinking would continue as before, but there would be a slowdown in device and circuit "cleverness." The low-hanging fruit had been harvested. As this last factor fell, the slope of Moore's plot would change. Complexity increase would slow to a doubling every two years rather than every year. "Even at this reduced slope, integrated structures containing several million components can be expected within ten years," Moore announced. These devices, once again reducing the cost of electronic functions, would "extend the utility of digital electronics more broadly throughout society." The revolution—still in its youth—would continue unimpeded. Even to sophisticated audiences, his examples must have seemed startling: "A complete function containing as many as 10,000 fully interconnected components sells today for the price of a single transistor fifteen years ago."

By now Moore was settling into a public role, as he spoke of how "the integration of complex functions in and on a single silicon chip has revolutionized the way that electronic systems are made." As Intel's CEO, he used his presentation as something of a stump speech—for instance, at IEEE's main international convention, "Electro76," in Boston in May 1976. He reprised the talk in Sandia, New Mexico, for the researchers at its national laboratory, dedicated to nuclear realities. Adopting the irreverent, in-crowd joking of the gathered technical assembly, he chose to address "Where Is This Silly Business Going Anyway?" He closed with a self-deprecating joke that made fun of his straightforward approach: "By utilizing the ultra sophisticated technique of extrapolating straight lines on semi-log paper, trends will be predicted for the rest of the decade."

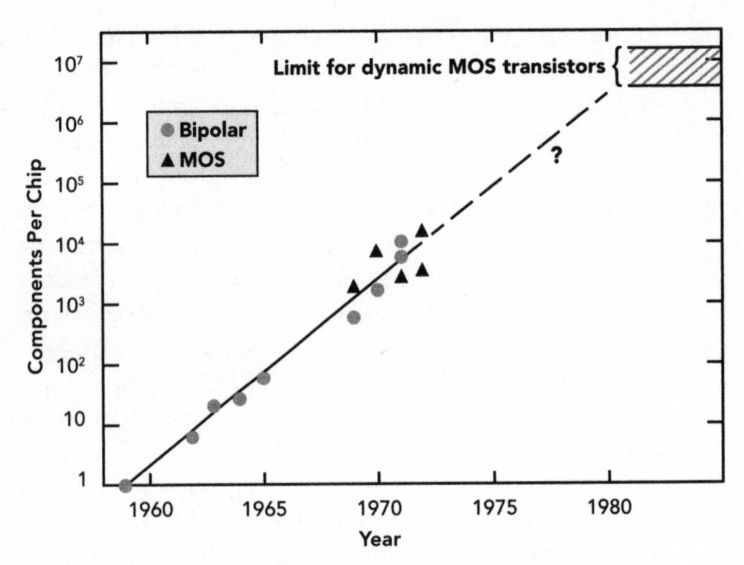

Carver Mead spread awareness of Moore's insight in his 1972 publication with Bruce Hoeneisen.

Source: Solid State Electronics.

Moore's public speaking, combined with Mead's activity to extol his vision, nourished awareness within the semiconductor community, and Moore's Law slowly became an established phrase as well as an operating reality. In 1977 *Science* published a high-profile article by John Linvill (chair of electrical engineering at Stanford) and C. Lester Hogan (vice chairman of Fairchild). It placed "Moore's Law" firmly within the context of revolution. "If the semiconductor industry continues on the curve given by 'Moore's law,' it will achieve a complexity of ten million interconnected components by 1985." The number was stunning, the implications enormous.

Moore's Boom

Intel's 1976 performance vindicated the strategy of investing heavily while the business cycle was in a valley. Revenues exploded to $226 million. Profits swung up more than 50 percent. Head count jumped. With Moore's investment and Grove's drive, Intel was able to take a significant fraction of the early market for 16K DRAMs while pushing its 8080 microprocessor as an industry standard.

As Gordon acted on his wish to "run Intel for a while," spectacular results continued. The company became "a skyrocket, going to the moon." Revenues reached $855 million in 1980, a fivefold increase from his first year as CEO. Profits rose to almost $100 million. The stock price soared. Even after splits in 1978, 1979, and 1980, shares traded at $48. In his office Moore put up a small wooden sign: "This is a profit making organization. That's the way we intended it. And that's the way it is!" His personal stake mushroomed.

INTEL FINANCIALS, 1976–1981

Year	Revenue	R&D spending	Total costs	Profit/loss
1976	$225,979,000	$20,709,000	$174,522,000	$25,214,000
1977	$282,549,000	$27,921,000	$219,403,000	$31,716,000
1978	$399,400,000	$41,360,000	$314,347,000	$44,314,000
1979	$660,984,000	$66,735,000	$511,815,000	$77,804,000
1980	$854,561,000	$92,426,000	$671,441,000	$98,741,000
1981	$788,676,000	$164,960,000	$759,097,000	$27,359,000

Source: Intel annual reports.

During these boom years, Gordon Moore fine-tuned the approach grounded in his fundamental insight about the economics of silicon microchips. While the vision remained unchanged, experience added two corollaries: investing to "cross the valleys" of stagnant periods and downturns and putting competitors on the "treadmill." Moore explains, "You never recover on the old products. The prices never come back. Unless you develop fresh products and keep up with the technology, you don't get well when business gets well."

> You bring out a series of products. From one generation to the next, they sell for about the same price, but offer a lot more capability. You can buy a 64-kilobit memory one year for what you paid for a 16-kilobit memory the year before. You must keep moving to stay with or ahead of the competition. Anybody who wants to compete has to make a huge investment. If we have 80-plus percent market share and our competitor has 20 percent, we spend fast enough that he has a really tough time keeping up. "Get 'em on the treadmill!"

Moore could see no early end to Intel's ability to advance its silicon manufacturing technology. Only the availability of funds limited the company's improvement of yields and its creation of more complex products. There was no fundamental chemical, physical, or electrical impediment to the technology being extended. Advances required manpower, equipment, effort, coordination, and resources and would also depend on the insights of gifted, committed workers, as Jean Hoerni had demonstrated long before at Fairchild. Moore, with his "black box" view of individual desire and his calculus of financial reward, felt confident that innovation could and would occur. He was also confident that his own decisions on when, where, and how to invest in R&D and additional fabs would efficiently control the pace of development.

Intel sold a 16K DRAM in 1976, an improved design in 1979, and a 64K DRAM the following year. It also crafted a more attractive form of EPROM known as the EEPROM. Add-on systems offered customers an increase in memory without needing a computer to be expensively modified. The advantage of making systems was twofold: they drove sales, and they used large numbers of DRAMs (with Intel charging a premium for packaging them). The memory systems business was making more than 20 percent pretax profit and growing at 15 percent a year. From one point of view, all was well. However, major competitors (TI and Mostek as well as NEC in Japan) were bringing forward large production capacities, amid prices that were tumbling. Intel's share of the booming market was contracting from one year to the next; memory microchips were becoming commodities, not a premium product.

Fortunately, the company was no longer solely dependent on the memory business. By 1979 *microprocessors,* not simply *microchips,* were the area of greatest growth, with profit margins rising strongly. Moore, Grove, and their colleagues became increasingly interested in, and focused on, the microprocessor. Grove suggests that this focus was more a case of "strategic recognition" than forward planning. "It's not that I could see the possibilities of microprocessors, going into them," he recalls, "but I could see the horrors of staying in the memory business."

The microprocessor might be the most revolutionary development since the microchip, but it was still novel and little understood. Gordon's awareness of Fairchild's earlier experiences with planar microchips helped inform his stance and his investments in the microprocessor side of the business. He had earlier learned the importance of educating possible purchasers. He therefore increased funding for Intel's development systems that allowed customers (that is, engineers) to create their own software programs for the microprocessor and to store those programs in EPROMs. Intel supplied all an engineer needed to customize the microprocessor for his own desired application.

Moore was heavily involved in educating the technical community about the microprocessor. Already in September 1975 he turned over "Microprocessors and Integrated Electronic Technology" to IEEE's flagship *Proceedings*. His overarching theme was that the microprocessor, "a general-purpose digital electronic block whose function is determined by programming," was "revolutionary." While "its capability and flexibility would have a profound effect on society," a more immediate impact would be "upon the structure of the electronics industry itself."

Users sought increasingly specialized custom microchips, while manufacturers preferred standard parts. The industry was at another crossroads. "The technology could make large quantities of complex parts economically, but the designers could not define many sophisticated functions that required such quantities. Here is where the microprocessor comes in," wrote Moore. This device offers a "completely different" approach to standardization, one with "real power."

When given the proper programming, the microprocessor could generate "any desired logic function." The microprocessor could become a universal part, enormously varied in its functions in particular industrial and consumer products. The addition of "programmability" would require that microprocessors came from their makers with "a new order" of product support. Customers would need design aids, development systems, software, and the like. The semiconductor manufacturer would be "thrown into the software morass, a swamp formerly occupied by the computer industry. Programming manuals, assembler tapes, design aids," and so on might add to costs. Yet the result would be the sale of multitudes of standard microprocessors, with customers performing their own customization and shouldering its expense.

Total cost of microprocessors would fall, with design amortized over massive production runs. The result would be "evolution toward even more complex functions offering improved performance and system reliability, while continuing the rapid decrease in cost per function." Once again, the semiconductor industry would force systems builders to trade customization for lower cost and to adopt standardized microchips in the form of microprocessors. Moore ended his 1975 paper with the observation that the microprocessor had also begun to shape the development of silicon manufacturing technology itself. MOS (metal-oxide-semiconductor) offered the density required for memory microchips. For logic, however, speed was paramount. Intel would now focus upon increasing speed, along with complexity, in its MOS technology. MOS had opened the way to a revolution, but it, too, would be shaped by that revolution.

Chips with Everything

Gordon Moore might routinely make his arguments in written form, but he was also increasingly confident at the podium. His talks became suffused with compelling rhetoric, even as his most effective tool remained dry humor. Preparing for the Mass Storage Device Sales Conference, he used red ink to underline that "I can't imagine a *more exciting* biz, and even beyond sheer excitement, I believe we are *bringing about the next great revolution in the history of mankind*—the transition to the *electronic age*." Underneath, Gordon was as unassuming as ever. Jotted notes for another speech make clear that he had "no comprehensive formula for success." His two start-ups had thrived "at least for the first ten years," but that simply meant, "I'm fortunate."

That December he gave a luncheon address at the Ramada Inn near Chicago's O'Hare Airport to IEEE's Consumer Electronics Group. Because this was a crucial audience, he pulled out all the stops to give a compelling account of the semiconductor industry's history and, more important, its future. The health of the industry will depend, he argued, on "the rapid and broad incorporation of complex digital functions into consumer products." In a deft metaphor, he captured the essence of semiconductor progress: "If the auto industry had made similar advances, we could cruise comfortably in our cars at 100,000 miles an hour, getting 50,000 miles per gallon of gasoline. We would find it cheaper to throw away our Rolls Royce and replace it than to park it downtown for the evening. On the other hand, we could pass it down through several generations without any requirement for repair."

Moore also used another metaphor, one he would repeat frequently in coming years. The semiconductor industry was driving the revolution, he explained, because it was "a very efficient cost reduction machine," decreasing expense more than 10,000-fold over fifteen years, while increasing complexity 64,000-fold. "This is really what is extending the use of electronics in many directions in society today."

This "great cost-reduction machine" was possible only "because of unique features" in semiconductor manufacturing technology:

> Not only do we get a decrease in cost, but by making things smaller in this pleasant world, we also get an improvement in performance. Speed increases and power decreases. Whenever our yield gets in the twenty to thirty percent range, we can decrease cost per function by doubling the complexity and making the die twice as big. Even better than that, everything else isn't usually equal. We learn to improve the process, and we also learn to improve circuit density as time goes on; we pack more and more electronics in the same area.
>
> This is the heart of the cost reduction machine that the semiconductor industry has developed. We put a product of given complexity into production; we work on refining the process, eliminating the defects. We gradually move the yield to higher and higher levels. Then we design a

> still more complex product utilizing all of the improvements, and put that into production. The complexity of our product grows exponentially with time.

On the one hand, exponential growth—Moore's Law, a regular doubling—was good news. On the other, it brought with it an enormous, perpetual, and growing challenge.

Initially, the industry in general and Intel in particular had "lived with the assumption of super-elastic market demand. If we cut the price in half, unit-volume will more than double. It's a basic premise of our existence. So far demand has responded." Yet, the question "So who needs it?" had no abiding answer. "There are always limits to growth; exponential growth of real physical quantities always portends an eventual problem." Moore's acknowledgment points to an issue of growing salience for transistor makers today, some four decades later, but one of only theoretical interest then. Moore's Law will end. In December 1976 in Chicago, the industry was "a long way from any limit," and the central question was simple: could his audience find markets for electronics "at the rate required to maintain this growth"?

The era was one pervaded by broader misplaced worries in the nation's "chattering classes," about the planet's capacity to sustain its population and industrial activity. *Limits to Growth,* a 1972 book full of gloomy assumptions and flawed apocalyptic thinking, provoked a heated and ongoing public debate, while Paul Ehrlich's *Population Bomb* notoriously forecast a future of crisis and doom. Elite researchers, many associated with MIT, used newly available computer modeling to predict rapid collapse in key aspects of civilization. Gordon Moore was altogether more sanguine. However, if the electronics community wanted to maintain the growth rate of its last fifteen years, it faced a very real challenge.

Silicon transistors sold for an average price per transistor of less than a penny in 1976. Where did they go? Remarkably, half went into memory microchips and portable calculators, markets just a few years old. What about the other half? Hundreds of billions went into devices from intercontinental ballistic missiles to color televisions and telephone networks to cash registers. Some 10 billion were in electronic wristwatches. Novel markets for transistors on microchips—such as watches, memory, and pocket calculators—had sustained the steady doubling of complexity, the dramatic drop in price, and the tremendous expansion in production. For the electronics world to stay on track, the industry must find fresh and similarly huge markets to consume the exponentially greater annual production of exponentially cheaper transistors on microchips. As Gordon quipped in his talk, "We only have to find ten such markets in the next six years."

He pointed to two possibilities—automobiles and video games—as applications that might match the size of the memory microchip business. Moore's prescience is impressive. Automobile makers *would* begin to use microprocessors to control ignition and other engine functions. Today, as the driverless car becomes a staple of discussion, the accelerating appetite of automobiles for transistors is obvious. As for video games, these already held consumer appeal in 1976. Even so, Gordon's prediction that games could grow to a market the size of microchip memory seemed far-fetched, but once again he would prove to be correct.

Games had been among the very earliest programs for digital computers: in 1950 Alan Turing, a legendary pioneer, had tried to create a program to play chess. Graphical games followed, like tic-tac-toe and a rudimentary tennis game. In 1961 computer users at MIT created *Space War!,* a program to guide make-believe spaceships that could shoot at one another. Then several companies began producing systems that used television sets for their displays. Fairchild soon debuted a very different, more sophisticated Video Entertainment System, built around its own microprocessor, allowing players to select from games that were sold as cartridges.

Atari's 2600 system brought games into the mainstream. By the early 1980s, industry revenues were around $3 billion per year. Japanese companies came to the forefront with the Nintendo Entertainment System of 1983 and the Sony Playstation of 1995. Ever more powerful microprocessors powered these systems, expanding the range and sophistication of game play. Today, video game systems function as home entertainment computers, integrating television and Internet capability and offering the ability to respond to users' voices and body gestures. Products like *World of Warcraft* and *Minecraft* have become globally popular, boasting millions of players and hundreds of millions of dollars in profit. The video game market is a now $100 billion annual business. Competitions alone have become the fourth-largest category of Internet traffic, with one company, Twitch, changing hands for $1 billion. Gordon himself, now fully retired and seeking activity to echo the strategy, progress, and victory enjoyed in his many years at Intel, turns to *Solitaire* and similar distractions. Betty explains, "He'll be in his study on the computer, playing games. He gets caught up and doesn't know how to get off the treadmill. I become resentful of it, because we are never sitting in the same world."

If Moore was prescient in 1976 with regard to automobiles and video games, he also saw that these two alone could not fully engage the great cost-reduction machine. "The only answer for really large markets beyond the things listed here is in the home," Gordon told his audience that December. "The challenge is to get ten trillion transistors into the household over the next ten years." The transistor was already in devices such as watches, televisions, and video games, but only by getting another "200,000 functions into the average US household" could the semiconductor industry "continue to make its contribution to lower cost electronics." Homes needed to become suffused with microchips. Even so, Gordon—who in 1965 had written of "such wonders as home computers"—had little comprehension as yet of how such a computer might revolutionize everyday life.

Looking back years later, he recalled how, "Long before Apple, one of our engineers came to me with the suggestion that Intel ought to build a computer for the home. I asked him, 'What the heck would anyone want a computer for?' (I still sometimes wonder, in spite of having a few of them.) The only example he could come up with was something for the housewife to put her recipes on. I could imagine Betty at the stove, poking at her computer to read the recipe. It seemed ridiculous!"

Moore and Intel would spend the next three decades grappling with the two central issues he had articulated: playing the microprocessor game, as producers sought to keep their products from being commoditized (described by Andy Grove as "a highly intricate game-theory kind of game"), while building fresh markets to digest the burgeoning cornucopia of transistors produced by the great cost-reduction machine. Transistors would permeate the workplace, the home, the public square, and every other space, real or virtual. Microprocessors, containing myriad millions of transistors, would find their way "by the billions" into every human environment.

CHALLENGES, CHANGES, CONTINUITIES

Managing the Machine

Andy Grove was increasingly settled into his role as both assistant and partner to Gordon Moore. Moore's direction of Grove's fiery intellect and energy took various forms. One was written memos, often peppered with Gordon's trademark dry, ironic humor. Another was face-to-face discussion. Here, the pair built on an earlier formal plan. "For a long time we had tried to have regular meetings, Andy, Bob, and myself. Bob probably made half of them. Andy and I met together, whether he was there or not. Then, when Bob dropped out, Andy and I still kept to regular meetings." In Moore's view, "Andy made sure that everything was planned and done and that all the bits and pieces were in place." Gordon might be orchestrating Intel and the revolution, but Andy Grove, leading the band, was almost always much more visible.

With Grove as his interpreter, enforcer, and hatchet man, Gordon was increasingly efficient in moving Intel in desired directions. It was his to

Gordon Moore and Andy Grove at Moore's office table.

SOURCE: INTEL.

decide how to push forward the manufacturing technology and what to make with it. Profitability was one key metric, in showing whether Intel was succeeding. At Fairchild Moore's use of investment as a strategic tool had been limited to personnel, equipment, and facilities within R&D; at Intel he could deploy investment decisions across the board. The R&D budget and capital expenditures for fabs and tools were especially powerful levers when tuning Intel's response to competitive pressures.

Moore and Grove interacted closely with two small, stable groups—Intel's board and its senior executives—whose memberships had strong continuity and overlap. The comfort afforded by Intel's stellar performance and their support allowed Moore and Grove unfettered control and the ability to chart a course without interference. "We told them what we were doing. We knew a lot more about what ought to go on than they did."

Intel's Board of Directors, 1980s. Moore, CEO, at center flanked by Grove and Noyce.

SOURCE: INTEL.

Moore, Noyce, and Grove were three of the ten board members. Ed Gelbach, now senior vice president, was another. Four members were external but had been associated with Intel from the start: Arthur Rock and Richard Hodgson, who had originally helped set up Fairchild Semiconductor, and James Guzy and Max Palevsky, of Memorex and Scientific Data Systems, respectively, who both became major Intel investors and remained on the board for decades, offering a perspective on the development of the computer industry. Sanford Kaplan (a finance executive at Scientific Data Systems and at Xerox) and Charles Young (a political scientist and chancellor at the University of California at Los Angeles from 1968) completed the lineup. Both served on Intel's board through Moore's tenure as CEO.

On the other side of the Moore-Grove partnership stood Intel's senior executives, a remarkably stable group—as contentious as it was constant—of roughly a dozen men. They were bound to Intel by enticing stock options and a rising share price, as well as by a culture that emphasized loyalty to the team in the cutthroat world of transistor production. Some had worked with Moore and Grove at Fairchild (Leslie Vadasz, Eugene Flath, and attorney Roger Borovoy), while Jack Carsten (brought in from TI to lead marketing) and Larry Hootnick (an MBA recruited to run Intel's finance function) were more recent arrivals. On the rare occasion someone did leave, it was with substantial wealth from Intel stock, more often than not to become an investor in high technology.

Moore spent much of his time in meetings, balancing specialist groups (those who requested his help or in whose work he was particularly interested) with his input at Grove's regular sessions with the various teams. At first the pair met weekly with operating division heads, but as Intel both grew and spread geographically, this changed to a monthly event. "We would review the progress of the company, talk about problems, and make major plans," recalls Moore. "In the 1970s planning was relatively straightforward. As Intel became more complicated, we tried to formalize the process."

Gordon also monitored a tide of correspondence and memos. Legions of monthly reports from engineers and researchers crossed his desk. "I would read, scribble on them a bit, and then throw them away," he recalls. The value, as he saw it, was that the report writer necessarily gauged his own progress, while Gordon was able to skim the vital statistics. His steady discipline in reading these reports reflected his belief that, as Intel's leader and strategist, his most important task was to listen. Only then could he identify and encourage those with the most valuable insights, analyses, and suggestions. Grove, Noyce, and Rock were sounding boards and reliable guides, but he also paid close attention to other colleagues. In directing the company, he explained, "You have to pick up all the bits and pieces that you can. You have to know a bit about the technology, have some idea about how big a step to take, have some idea where the market's going, and know how much money you have. I don't believe there is any magic formula. Knowing who to listen to is a good part of it."

While Moore oversaw Intel's direction, Grove continued to develop the culture that would become Intel's hallmark—toughly competitive; valuing quantification, discipline, and argument; and intolerant of failure. In turn, hourly workers at the fabs spent their days transforming ever more minute amounts of materials into packaged products. The precise chemical complexity of Intel's world, and its ultimate earthly materiality, was in sharp contrast to the transistor's role in making ethereal virtual reality an aspect of everyday life. The invasion of transistors and microchips was everywhere, if often unseen and unheralded. It was still possible, if not easy, to ignore the digital revolution.

Moore developed his own quantitative tools for technology (such as "iso-defect curves," displaying the relationship between die size and yield) and executive compensation (an elaborate system that converted achievements into bonus payments, salary increases, and stock options). Grove led staff sessions on Intel's approach. He stressed quantification and the open discussion of problems, allowing debate on alternative solutions based upon quantitative data and assessments. A disciple of management guru Peter Drucker, Grove honed Intel's "management by objectives," setting quantifiable goals for managers, groups, and individuals. "We had the view, 'If you can't measure it, you don't know what you're doing,'" explains Gordon. He only wished "some of Andy's ideas" had come earlier.

Measurement suffused silicon manufacturing technology. Intel's "analytic form of management" emphasized problem solving based on data. Before any matter was brought to closure, debate was standard. With decisions made in this open and "fair" manner, those on the losing side were expected to fall in line. In reality, macho and combative posturing often preceded "rational judgment." Andy Grove set the tone, frequently shouting and hurling insults at opponents. Craig Barrett describes how evidence-based arguments did not eliminate aggressive confrontation:

> If you know your business, if you know your technology, you don't back down in front of anybody. Jack Carsten—my boss for several years—made Andy look like a pussycat. At his staff meetings, everybody would cower under the table. I would look him in the eye and say, "Jack, you would be really dangerous if you knew what you were talking about, but you obviously don't, so if you want to discuss this in a logical fashion, let's go back to the basics and I'll walk you through them and show you why you're wrong." Neither of us ever shied from a confrontation.

Such socially and personally risky arguments required a strong, stable group of managers, committed to the success of the organization. So long as Intel grew, and the value of its stock rose, options were incredibly lucrative, binding individual futures tightly to the fate of the company. Leaders were personally invested in the success of the firm and felt secure enough in their positions to fight for what mattered.

Gordon Moore was the anchor at the eye of the storm, untouched by the heavy weather. On occasion he raised an eyebrow, but only rarely would he intervene directly. He was at one remove, fine-tuning inputs and outputs as he watched the action unfold. The tiniest of facial reactions was enough for Grove, who was practiced at decoding nonverbal signals. Barrett remembers, "Andy Grove was very deferential to Gordon. If he had to make a decision in the middle of a meeting, he would look at Gordon to see in what direction he was leaning, and then say, 'All right, here's what we're doing.' It was not Gordon saying, 'Here's what we're doing.'" Moore said little, but within Intel's leadership his words carried immense weight. He might express an opinion on a topic, says Barrett, but he was "not in your face. You don't immediately launch into battle. That's not the kind of interaction you would ever dream of having with Gordon." Instead, "you go off and think about it."

Fifteen New Intels?

Because Gordon had not yet connected the growing enthusiasm for "personal computers" with the potential for a market of any size, his presentation on fresh uses for transistors looked elsewhere. He estimated that the market for digital "functions" had risen from 100 million to 100 billion in the fifteen years from 1960. The market would need to rise from 100 billion to 100 *trillion,* perhaps even an annual quadrillion, by 1985, to continue what was, in effect, a perpetual game of musical chairs. Growth in established and emerging markets (computer logic and memory, calculators, watches, video games, and automobiles) might account for much of this, but there remained a gap. With the removal of market expansion, the music would stop and the game of falling prices would be over.

Gordon's concern about this possibility, which would ruin his long-term vision, was mirrored in his concern for Intel itself. On Monday, January 3, 1977, he pulled out his legal pad. "Where does Intel stand; where is it going?" he wrote. Always a man to measure, he translated this question into numbers, dividing the company into its business areas (memory microchips, microprocessors, add-on memory systems, and so on) and adding up the total: $226 million in revenue for 1976. For 1977 he decided revenue should exceed $305 million. Moore next laid out a "reasonable nominal" pattern of growth that continued above 30 percent to 1980, before settling to 20 percent. Calmly displaying the remarkable nature of the enterprise he headed, he noted the implication that revenue would grow more than eightfold in a decade, to $1.9 billion in 1985. (Proving his remarkable feel for figures and knowledge of the microchip industry, Gordon's numbers proved right on target.) With transistors on microchips, his question had been "So who needs it?" Concerning revenue, it became "Where can it come from?"

Intel's memory and microprocessor businesses would reach parity by 1980. However, there would be a $175 million deficit in revenue, unless "contributed by new business segments." Telephone and communications chips might become a $25 million business, but this was far from sufficient. Hence, "A major goal is to identify activities that can contribute $150 million in 1980!" In 1971 Intel had $10 million in sales. Now, in the three years from 1977, he would need to create businesses equivalent to fifteen earlier Intels!

People were listening to what Gordon had to say. As cofounder and CEO of one of the fastest-growing companies in the semiconductor industry, he enjoyed increasing respect. In less than a decade, the firm had risen from nothing to become the world's largest producer of MOS microchips and the fifth-largest producer of microchips worldwide. He had a reputation as one of the industry's most successful technologists. The establishment of Moore's Law gave him further credence. Although his *Electronics* thesis had caused almost no stir, and while he continued to avoid the limelight, his opinions were now sought, valued, and influential. And he was determined to challenge the industry to come up with fresh functions for all the transistors Intel and others were producing.

Mr. Reliable

As CEO Gordon Moore maintained his quiet style. Meetings, private thought, and traveling to speaking engagements and conferences filled his time. As in his personal life, he displayed enormous constancy, maintaining working relationships for decades. His settled nature, so fundamental to his role in the electronic revolution, was witnessed on a daily basis by Jean Jones, Gordon's secretary and the person with whom he interacted most closely.

Jones, a likable woman described by Moore as "a very good interface to the public," had worked for Vic Grinich at Fairchild. Returning to paid employment after having a family, she agreed to join Intel, the new start-up, "temporarily." She stayed more than a quarter century, only retiring when Moore's own role dwindled. As gatekeeper she handled his correspondence, memos, and reports and scheduled his calendar and travel arrangements.

Gordon used longhand on ruled or graph paper, and Jones decoded his jottings. She also took shorthand. "That was something I really missed when she retired," says Moore. "I didn't realize how much easier it is to dictate to somebody than to a machine."

Jones and Moore had little contact outside work, yet their relationship was filled with personal kindnesses. Jones was a widow, and when her son needed to tie a necktie for his senior class photograph, Moore untied his own and patiently walked through the steps by telephone (Jones brought in a photo to show the result). Another time, noticing a pile of unopened boxes, Moore surmised that Jones was intimidated by the prospect of setting up an IBM word processing system. He quietly offered help and then spent several hours unpacking the system and getting down on his hands and knees to run cables and cords around the floor of Jones's cubicle, to the surprise of employees walking by.

Moore and Noyce with Jean Jones.

SOURCE: INTEL.

In what became a habit, Jones would share her lunch with Moore. Noyce often went out to eat, and Grove bought tomato juice from the vending machine to combine with a container of cottage cheese. Moore was often too engrossed to remember lunch. Jones would head to the cafeteria. "It had a make-your-own sandwich bar. I would bring mine back, and he would say, 'That looks good.' I would say, 'You want half?' It didn't matter what kind of sandwich I fixed; he was very happy to share." The common repast each day was a great improvement over the three-martini lunch, the staple of Fairchild a decade earlier. Jones was a kindly and reliable workplace "Betty," ensuring that life remained stable and predictable. To Jones, Gordon was equally dependable. "The main thing was his reliability. You could count on Gordon being there at eight o'clock in the morning. You could count on him attending his meetings and being prepared. You couldn't give him a set of numbers that he could forget. It was absolutely amazing."

On the national scene Bob Noyce continued to overshadow Moore, enjoying fame in his role as Intel's chairman. Andy Grove would inherit Noyce's gilded mantle and in his turn come to personify Intel to attentive publics and the media. Moore's relative invisibility did not trouble him. He had chosen, and preferred, the situation. He was busy and engaged, surveying and settling an uncharted land. In the long term, his achievements would endure, and he would be recognized as the intellectual power behind Intel.

"Both Andy and Bob really enjoyed the limelight. I'm the low-key link in the middle," he explains. "Noyce was the icon of the industry; Grove became the next one. I never had a special need in that direction." His charismatic partners nourished their acclaim by winning over key journalists and "going on all the talk shows," but for Moore this sort of fame took "a certain kind of investment I wasn't willing to make." In 1979, as Noyce became more deeply involved with the microchip industry's trade association, he stepped down to vice chair of Intel. Moore moved up to become board chairman as well as CEO, and Andy Grove added to his chief operating officer role an additional title: Intel president.

Betty; or, The Noncorporate Wife

Betty Moore was equally resistant to the limelight. Though the 1960s and 1970s were the heyday of the corporate wife, she had no wish to "belong." In the late 1960s, she had declined to be interviewed for a book about the wives of Silicon Valley executives. "I said, 'I'm not part of the company or a corporate wife. I am Mrs. Moore, a mother, and my own self. I do what

Gordon and Betty at Caltech in 1973 for Gordon's Alumni Distinguished Service Award.

SOURCE: GORDON MOORE.

I want, if I want.' End of interview! It was great to have evolved along with something this large, but I didn't think that was me completely."

Routine was the key to household happiness. Gordon displayed enormous constancy in his habits; after breakfast with Betty, he would be out of the driveway in his Porsche, around seven, for the short drive to the office. At the end of the day, he would leave around six, eschewing after-work drinks to head home for dinner with the family. He brought work with him, carrying a briefcase back and forth every day. "I probably didn't look at it very much," he jokes. Betty saw it differently. Gordon grew ever busier and less available. "I told him, 'I don't know why you don't take your PJs and your toothbrush and stay down there.' It was so enveloping and engrossing, absorbing all his time and energy. We were on the outside looking

Betty and Gordon, around 1980.

SOURCE: KEN MOORE.

in. They were doing managerial projects and had a ten-year plan. I said, 'How can you go out ten years? You can't even know what's going to happen in two years, let alone in five!'"

Gordon remained as uncommunicative about work as ever. For a while, the family's fishing trips dwindled, as business took precedence. "He was traveling a lot," remembers Steve. "We stopped using the beach house as much. The Forbes 400 list came out, and he was on it. He definitely had more demands on his time, and people wanted him."

At this stage Gordon and Betty "totally ignored" their wealth. Vacations, occasionally focused on more exotic fishing expeditions, were one of the few areas on which they did enjoy spending. "As we got older, when I was mid- to late teens, we took trips to Africa," recalls Ken. "I was in the Galapagos Islands when Nixon resigned. But my parents never took us out of school. Education was paramount. You had to follow the rules, and you didn't yank your kid out for a vacation." In contrast to Noyce, who quickly learned to pilot his own plane, Moore avoided glamorous status symbols. However, when Gordon and Betty celebrated their silver wedding

Betty and her Baja dogtooth snapper catch, 1975.

SOURCE: KEN MOORE.

anniversary in September 1975, Betty decided to treat Gordon to a Turbo Carrera Porsche, one of the first two imported to Silicon Valley. Gordon had initially splashed out on a Porsche in the early 1960s, but since then had bought Buicks and Chryslers (and would go on to buy a much-loved Volkswagen Rabbit diesel).

In the interim between deposit and full purchase, Steve was surprised to see his father's new Porsche flash by "on Stevens Creek Boulevard, in San Jose, quick as lightning." Within the hour, Betty and Steve were at the dealership with a check to pay in full, written on Steve's account to preserve secrecy. Ken recalls, "This was a thirty-thousand-dollar car, in the mid-1970s. People would stop you just to see the car. It was weird, to drive in a fishbowl. I thought, 'Wow, we are on a different curve.'"

Despite the workload at Intel, and the social upheavals of the era, the Moore family maintained their close bonds. Most nights they ate together. When the boys graduated college, Fridays were earmarked for meals out, sometimes including friends and even girlfriends. "This was my parents' way to get to know our friends," says Ken. "It worked well. The parents loved it; the friends loved it; I loved it." Even political arguments were aired without rancor. "My dad's very even-headed, even-minded; he'll understand competing opinions and then have his own." For both Betty and Gordon, family cohesion was a priority. "We tried very hard," says Betty. "By Friday we were all tired, but we could have a quick Chinese dinner and go home. It made the family much closer." Ken, still living at home, completed his bachelor's at San Jose State and went on to study for an MBA. In his teens he had developed a passion for shortwave radio and worked informally with minicomputers at Intel. He took great interest in the emergent Silicon Valley, seeing a career for himself in chips or computing. But his restless side pulled, and he took up drag racing and skydiving. Steve, recovered from illness, returned to Los Altos High School. Gordon dropped him off there each morning before going on to Intel. Betty, with more time available, again looked for opportunities to serve. Through her close friend Rose Kleiner, she began volunteering at a day-care center for seniors in Palo Alto and continued this through the 1970s. "I gave my Wednesdays to that. We went on field trips to county fairs and off to museums. I helped with a ceramics class and learned how to do ceramics. I enjoyed it."

Betty also went back to college, "for enrichment," taking courses in English antiques and watercolor painting. A leading designer at W&J Sloane, visiting the Moore home on Jabil Lane, recognized her artistic eye and asked if she would come on board. Gordon, consulted on the opportunity, opposed it. In his view, having Betty take on additional responsibilities would destabilize their long-standing arrangement. "Gordon said, 'I don't think so. It would take away from the family.' I have some regrets about that, but it was not to be. Mama's got to be there."

Fathers and Sons

Betty kept a close eye not only on her own mother, Irene, in Saratoga, but also on her father-in-law. Walter Harold spent time with his young grandsons. Steve recalls, "He had a big garden. He'd be making lots of different things, gluing things together, doing stuff where we could help him. He'd talk about growing up in San Mateo County and show us his gun and badge collections and things he had hanging in his garage—old saws and implements from earlier in Moore family history."

After the death of Gordon's mother, Mira, Walter Harold continued living on Westgate Street, Redwood City. In early retirement he visited with friends, attended rodeos, and went rock fishing or hunting with his sons. Then in his early eighties he began to go deaf and became isolated. Hunting had long been a Moore tradition, but Walter Harold, increasingly frail, could no longer join in.

With Mira gone, Gordon busy, and his grandsons grown, Walter Harold watched TV and chain-smoked cigars to while away the long evenings. Betty, finding herself the caregiver on both sides of the family, "had him down to dinner quite often because he was alone." Walt Jr., his eldest son, living nearby, was of a solitary temperament and ill-equipped to offer solace. "You couldn't have a great conversation with Walt," says Betty. "He was not going to be forthcoming with much." Hardy, pioneering stock simply did not expect to reach out. The Moore men shared a stoic, determined reserve. In Gordon these qualities nourished Intel's enduring focus. In Walter Harold, lacking a supportive social context, they became pathological. His attachment to mother, sister, and wife had helped him to withstand searing early losses. Later, the autonomy, rootedness, reserve, and habit that characterized his life (much as they did Gordon's) steadied him, but in the 1970s he came undone after suffering a series of losses, including the deaths not only of his wife but also of his sister, Louise.

Walter Harold's sense of purpose and value diminished. Then came a final blow. In the early summer of 1977, he coughed up blood. Reluctantly, he went to see a doctor, who referred him to a hospital for further tests. Walter's brush with the hospital and death in the 1940s had left him with a fear of physical illness and, above all, hospitals, so this was the worst possible news. He had spent much of his life tolerating the humiliations of a colostomy bag and now became concerned that he was seriously ill. Speaking on the phone, he told an old friend that, whatever happened, he was not going back to the hospital.

Betty was brisk and perhaps a little unsympathetic. In her view, Walter needed to face up to whatever was ailing him. "He thought he had cancer, but vomiting blood was the only symptom we knew of." When she and Gordon went to visit in Redwood City one July evening, they found Walter watching television. He was morose and uncommunicative. Betty was troubled. "I knew something was wrong. We went out to the car. I said to Gordon, 'You ought to go back in there; your dad is really hurting.'" Something in Gordon balked. He could not engage with such a difficult personal issue. He had never discussed emotions with his father and, at the point of crisis, was not about to start. "I said, 'Go knock at the door. You need to talk to him alone.' Did he do it? No." The next day, instead, Gordon flew down to Pasadena to fulfill an engagement at Caltech.

For Walter, too, avoidance had been a life motif. "He made up his mind, he didn't want to go back in the hospital," says Gordon. Taking his life was the sole option, "the way to avoid it." As Ken remarks, he "wanted to stay in his house. He thought to himself, 'How can I ease out of this?'" Apart from an oblique hint to a friend, it would have been radically out of character for Walter to speak of his fears or intentions. For Moore men, action defined identity. One did not talk; one did what was needed. Walter's first cousin James Moore, also a veteran, had used a gun to commit suicide two decades before. This method was quick and decisive, but Ken believes his grandfather—a veteran, a deputy sheriff, and all too familiar with the damage a gun could cause—"was never going to dismantle himself. He wouldn't want to mangle anything."

To a man of reserve, carbon monoxide poisoning offered a more fitting end. Two nights after Betty and Gordon's visit, Walter ran a hose from the exhaust pipe into his car's interior, stepped in, rolled the window closed, and started the engine. His plan did not work. He was discovered and rushed to Sequoia Hospital in Redwood City. There, in the very place he so wished to avoid, he languished for a month and then died. Gordon's description of events, given years later, is brief and matter-of-fact: "It was a shock. It took all his family by surprise." Betty, who had long agonized over health issues on behalf of her elders, was openly angry with both Walter and Gordon. "We had a chance to find out what was bothering him, and Gordon didn't take it. It was bizarre that Gordon's father would do something like that to the family, to put us all through this. There was no diagnosis. How can you do that?"

Walter's suicide left her with disturbing thoughts. "Gordon was at Caltech when it happened. He didn't say much about it. He just wraps these things inside somewhere. He has never been able to face straight-out talk about things. He can't face life." The stoic, determined reserve of the Moore men, effective as armor in certain situations, carried the seeds of tragedy. Walter Harold's death was not the first of the line and would not be the last. Gordon buried his father next to his mother in the veterans' section of the Skyline cemetery. Then he went back to work.

Home Computers?

Gordon closely monitored the applications of Intel's microprocessors. In 1976 he had his marketing staff draw up formal lists of actual uses. The data showed that four-fifths of microprocessors went into industrial equipment, business machines, instrumentation, communications, military and medical gear, and transportation. Much smaller percentages went elsewhere, including "consumer" use in such things as video game systems and home appliances. Despite all his tracking of markets, Moore missed the key development. When he spoke about how fresh applications were needed for transistors and microchips, he made no reference to the small but burgeoning use of Intel's 8080 microprocessors by home-hobbyist "techies." He completely failed to see that their amateur computers heralded the most significant shift of the era, the next step in the electronic revolution.

The hobbyist movement was particularly vibrant in Intel's Silicon Valley backyard, where enthusiasts were also constructing video terminals to connect to remote time-sharing computer services. In 1975, as Intel's 8080 microprocessor was offered more cheaply in kits such as MITS' Altair 8800, these enthusiasts began to build their own computers. Such boxes were similar to Intel's development systems, with switches and lights on the front and programs that could be fed in on paper tapes.

Businesses and clubs sprang up to serve the movement. The Homebrew Computer Club, "the crucible for an entire industry," met for the first time in Menlo Park on March 5, 1975. (Steve Wozniak, who attended, says that the meeting inspired him to design the first Apple computer.) There was soon a small market for fully assembled hobbyist computers. Pioneer producers were overwhelmed by the response to their modest advertisements. An astonished MITS shipped ten thousand Altair kits. In all, some forty thousand hobbyist computers were sold, most containing Intel's 8080, yet this new market constituted only a fraction of Intel's microprocessor revenue. Whether a single year's sales might saturate the hobbyist market was, to Gordon, an open question.

Some of the names deeply associated with the personal computer started their businesses during 1976. Steve Wozniak and Steve Jobs began selling a rudimentary kit: the Apple I boasted a typewriter-style keyboard and the ability to display text on a standard television set. A complete integrated version would be marketed as the Apple II. In New Mexico Paul Allen and Bill Gates, two young computer programmers who had worked for MITS on a version of the BASIC programming language, created their own company, Microsoft. From the start, Microsoft software was made to run on Intel microprocessors.

To some, the hobbyist computer was only a first, if thrilling, step in the development of a "personal" computer with wide appeal. These "PC champions" believed that personal computers would suffuse and transform society. While Gordon foresaw the microprocessor working a revolution through an ever-greater number of discrete products and systems, each performing separate functions, the hobbyist's champions believed the power of digital computation would become available through a single vehicle. The personal computer would, in the hands of legions of creative users, perform a multitude of tasks, affecting every aspect of industry, business, home life, and culture.

Moore had long been a components man. He had resisted attempts by Noyce and others to make an "Intel computer." While Intel's microprocessor systems head, Bill Davidow, enthused over the opportunity to create an "engineer's computer," Gordon, always determined to avoid encroaching on the territory of others, insisted the firm stick to its knitting. In the future he would downplay the idea that Intel "missed" the opportunity to manufacture and sell personal computers. "Perhaps we didn't miss that opportunity after all," he explained in 1994, "because we do make a profit out of the PC business; not by being in it, but by serving it. And that may be the best way."

Sales of hobbyist computers took off. Nearly two decades had been needed to sell the first forty thousand mainframes, each costing millions of dollars. More keenly priced minicomputers, still costing hundreds of thousands of dollars, accumulated a similar sales volume somewhat more rapidly. In contrast, the personal computer, priced close to $500, took just a year to hit forty thousand units. The market might be worth only $20 million, but champions of the PC saw this surge as simply the beginning.

Gordon's focus lay elsewhere. In mid-1976, as the Intel spinout Zilog launched its Z-80 microprocessor, Intel was less concerned with small new markets than with the pressure to maintain its microprocessor lead in the face of increased competition. The Z-80, hitting the 8080 in its weak spots, quickly eroded Intel's market share in microprocessors. National Semiconductor and TI were attacking from another side. Their 16-bit microprocessors, with the ability to handle larger chunks of digital data, were inherently more capable than Intel's devices. These various rivals hoped to put the leader itself on the treadmill.

100 Trillion Transistors?

Through his speeches of late 1976 and spring 1977, Gordon urged those in the electronics community to come up with fresh functions to employ the proliferating cascades of transistors. Industry analysts who listened to his words began to conclude that, for semiconductor companies, the game would soon be over. A month after burying his father, and to all appearances unaffected by the tragedy, Gordon decided to change his tune, emphasizing to audiences that he saw a bright future for microchips and microprocessors.

Moore had accepted an invitation to be the closing speaker at a meeting including top financial analysts and leaders of the semiconductor industry, in late September 1977, in Scottsdale, Arizona. On the roster were the CEOs of Fairchild and Signetics and senior vice presidents from Motorola and National. Moore's title was evocative: "Who's Going to Use All Those Functions?" Gordon spoke simply and directly about why he had made the call for fresh trillion-function markets for microchips in his recent talks: "I wanted to stimulate the users to see all this cheap electronic hardware that was coming along, and make them come up with the applications of the order of magnitude we need to keep the semiconductor industry developing in the way it has in the past." Analysts had been scared off, mistakenly believing that the answer to his question "Who is going to use all this stuff?" was "Nobody." He went on, "Unfortunately, I have been cast in the position of being a prophet of doom, which is not the role that I like to play—being naturally optimistic. I guess I have been given this forum to try and extract myself from the problem."

Moore outlined how difficult it was to anticipate the ways in which demand would take off. Pat Haggerty, the legendary leader of TI, had in the early 1960s looked at the potential markets for digital electronics and boldly predicted that a decade later, the world might consume 750 million functions per year. Even with all his insight, experience, and sophistication, Haggerty was "really wrong. He grossly underestimated how demand was going to expand. Usage is growing very much faster than anyone could have envisioned."

In the past Gordon had left his own questions unanswered, hoping others would come up with ideas. Now, to counter the gloom, he offered a vision of two crucial domains of use: the automobile and the home. They would consume transistors in the trillions, through a multitude of microprocessors and memory chips. "What systems can you think of in an automobile that could use a microprocessor?" he asked. His own prophetic answer was "engine control, transmission control, anti-skid systems for braking or acceleration, information display, entertainment systems, safety systems, trip computers, automatic fault monitoring, clocks—and a telephone in the car that looks for empty channels. Each of these will use its own microprocessor."

In sharp contrast he still found the idea of a "home computer" something of a joke. Instead, discrete uses of microprocessors to control specific tasks were plausible and already occurring. Moore took his listeners into the rooms of a fictional house and discussed how microprocessors could improve family life. In the bedroom microprocessors would power clocks, televisions, "television games," and "lighting control." In the kitchen microprocessors would become integral to appliances, including dishwashers, ranges, ovens, microwaves, televisions, refrigerators, freezers, blenders, toasters, washing machines, and dryers. "We see some things already," he noted. "The microwave oven controller, for example, has been a runaway success. Every major appliance and many minor appliances will have their own microprocessor controllers."

All manner of appliances—videotape recorders, heating and cooling controls, telephones, electronic games, security systems, gate and door controls, meters of various kinds, and even "pool controls"—would, in time, employ embedded, invisible controls, performing specialized duties. A single-chip CPU, a masterpiece of electronic sophistication, was not too grand to be used in household objects. It all came back to cost: "These things will be distributed throughout the house as are fractional horsepower motors today."

> It takes a while to realize that it's worth using a digital computer to count the number of times a refrigerator door opens and closes. It's going to turn out to be the cheapest and best way to do it. You have to look at this just as another $2 part rather than a sophisticated computer. A microprocessor is a little piece of logic that does something. It should be viewed as a little chunk of silicon, rather than as a complex microstructure that has to be used with full capability. These things are dedicated controllers that the user never knows are there.

In all of this, he was well ahead of his time. If he did not foresee it as such, the "internet of things" was already implicit in his vision.

Gordon might be prescient, but he had his blind spots. Ironically, he was also correct in predicting that he would duplicate Haggerty's error. The use of microprocessors as dedicated controllers *would* expand greatly in car and home, but first would come the personal computer. With his unique but narrow focus, Moore simply did not see that the fuse had been lit for a revolution in computing, built on microchips. A further reason for his lack of interest in start-ups like Apple was that it was "a small company that used a lot fewer processors than Ford, and quantity was the thing we looked for."

In all, Gordon believed 100 trillion transistors, in microprocessors and memory chips, would pour into US homes in the foreseeable future. The great cost-reduction machine was well established, with Gordon Moore as its champion and Intel as its exemplar. No doubt, this being real life, the way ahead would be strewn with obstacles and booby traps. Even so, he ended his speech on an upbeat note: "We can identify opportunities that are compatible with the growth required to keep the semiconductor industry moving. There's really plenty of life in the old industry, especially if I duplicated Haggerty's error and missed by three orders of magnitude what the actual usage will be."

10

REVOLUTION, STURM UND DRANG

RISING WAVES

Apples and Opportunities

Gordon Moore was correct: he would duplicate Pat Haggerty's error. With his resolute focus on semiconductor manufacturing technology, and his comfort with quantitative measures by which to command and control, he simply could not see the importance of the myriad social interactions that fuel everyday life and that would nourish the sudden rise of personal computers. For once, his instincts were inappropriate. Nevertheless, because PCs were built out of microchips, Intel became a beneficiary.

The first wave of PCs was revolutionary not because of their still basic technical capabilities, but rather because (in accord with Moore's Law) their declining cost put them in the hands of *individuals*—first by hundreds, then by the thousands. At last, a basic computer was sufficiently cheap (six hundred dollars, around twenty-four hundred today) to attract the attention of "techies" and amateur enthusiasts. The resulting explosion of ingenuity would transform ordinary life.

Early "hobbyist" computers held little appeal to Moore. After all, what were they good for? He was well aware of the Altair not because it caused a stir when in 1975 it appeared on the cover of *Popular Economics,* but because it used Intel's 8080. Thanks to the goodwill generated by his Intel policy of generous stock options, he was also given a preview of the Apple II by its financial backer, Mike Markkula. Markkula, an engineer and microchip marketer, had been lured to the fledgling Intel from Fairchild by Bob Graham. Markkula retired as a multimillionaire, when only thirty-four. He decided to back Steve Jobs and Steve Wozniak, the young founders of Apple, buying a third of the company late in 1977.

Some years before Moore had already helped Steve Wozniak, the technical guru of Apple. Wozniak's father was an engineer at Lockheed's vast missile works in Silicon Valley and, like Gordon, a Caltech grad and former football player. He called Gordon, "wanting to know if he could get some memory parts for his son, who had a project." Though Gordon guarded his privacy, he responded to this personal request, "dug up some of our very first 1103s," and sent them off.

To attract funds Markkula, with Wozniak and Jobs, gave the Apple II presentation to Intel's insiders, including his old colleagues Noyce, Moore, Grove, and the rest of Intel's board. In recalling the event, Gordon refers only to Steve Jobs, who evidently and characteristically stole the limelight: "Steve Jobs brought his computer in, but none of us knew what we were looking at—at least I didn't! He had somebody else's processor in the damn thing! He wasn't proposing any kind of an agreement with Intel." The brilliant but socially awkward Wozniak was someone Gordon could relate to, but the mercurial, enthusiastic Jobs was less easy to handle. Moore distrusted his easy, too open approach. "He was the kind of person who, though you liked him the moment you met him," also stirred feelings of ambivalence. "He had that charisma. I don't have that. I'm much colder."

Gordon failed to warm to the Apple. He had already squelched Noyce and Gelbach's plans to go head-to-head with the Altair. On his watch Intel would sell only development systems: microprocessors that controlled functions were his desired future for Intel's "cheap chunks of silicon." Hobbyists assembling their own rudimentary computers were unimportant. Gordon's focus on practical, technological, and financial matters left little room for how personal computers might change the lived experience of individuals, both personally and professionally. At Intel his secretary, Jean Jones, took shorthand dictation and typed up his thoughts. Alone, he used longhand to write in his notebook. He could imagine but not see the point of Betty using a computer to organize her recipes. "I didn't have any feeling for personal computers. The first Apples were not very useful machines."

In his 1965 *Electronics* article, Gordon had mentioned the idea of connecting a video terminal to a computer system, to manage electronic mail. Even so, this use of a microprocessor seemed to him to hold less promise than a command-and-control function that could be deployed in everything from video games to automated garden sprinklers. Social networking was, to him, trivial. "That a mail terminal is going to exist is less certain than that dedicated controllers will," he told an audience in Arizona, about this time.

Intel as a company did not invest in Apple. Andy Grove was intrigued enough to put fifteen thousand dollars of his own money in, and Arthur Rock invested sixty thousand and joined Apple's board. Steve Jobs also targeted Bob Noyce, who at first dismissed him, but in time came to see Jobs almost as a son. Tellingly, Jobs sensed Moore's coldness and made little attempt at cultivation. The two men were polar opposites. Gordon was an old-line Californian, rooted and raised in a close family, self-contained, steady, honest, and quiet. His approach was worlds away from that of the unpredictable Jobs, who had started out by selling "blue boxes"—electronic devices enabling illegal but free long-distance telephone calls. Jobs was the same age as Moore's own son Ken, but deference to an elder was beyond him. Moreover, Apple, an early "lifestyle" company, was making consumer, not technical, products. Moore had had his fingers burned with Microma. There was little common ground.

Microsoft's Bill Gates, a direct contemporary of Jobs, was no better match to Gordon Moore. Gates was somewhat akin to Andy Grove, but whereas Grove acted in support of Moore, Gates acted mostly in opposition. Gordon found Gates "a very bright guy, but very aggressive, too. We were both used to having our own way, and our objectives were somewhat different. He used to say, 'You guys cook sand. We'll do the software.' He wasn't an easy person to deal with."

Personal computers began to evolve from hobbyist designs into fully fledged commercial products. In 1978, Apple sold 25,000 of its Apple II computers, built around an inexpensive microprocessor made by MOS technology. Commodore sold a similar number of its Commodore PET, but Radio Shack outclassed them both, selling 100,000 TRS-80 models. The Byte Shop opened the first of its retail stores, devoted to the PC. Software for personal computers was also surging. Microsoft quickly reached $1 million from operating systems and application programs on paper tapes, magnetic cassettes, and floppy disks. Visicalc, the first spreadsheet—a calculation and database program made initially for the Apple II, bridging the worlds of work and home—helped drive PC sales. Almost 750,000 PCs were sold worldwide in 1980. That December Apple's IPO raised $100 million—fourteen times more than Intel nine years earlier. The electronic computer was becoming domesticated.

Despite the 8080's dominance in the earlier hobbyist computers, none of the leading firms—Apple, Commodore, or Radio Shack—used Intel's microprocessors in their first commercial PCs. Instead, Intel continued on its different heady track with a steady rise in sales of its microprocessors, and associated support chips, for controlling a huge array of products. Revenue from microprocessors and development systems alone grew almost fivefold within three years, reaching $244 million in 1979.

Intel continued to scramble, mixing long-term investment with shortcuts in its bid to get, and stay, ahead. The blue-sky "8816 microprocessor," drawing on the latest ideas from computer science, was already much too late to compete in the advanced microprocessor market. As it inched forward with excruciating slowness, Gordon had to task other engineers to throw together interim products with which to stay in the game. Meanwhile, the 8816 lead designers, who had moved to Intel's new fab near Portland, Oregon, sought Moore's approval to refashion their chip as an even more powerful microprocessor, renamed the 432. Gordon agreed. Costs mounted into the tens of millions. Thousands of man-hours were poured into the project.

The vacuum left by the nonappearing 8816 influenced the microprocessors that Intel did manufacture in the late 1970s. The 8086, Intel's first so-called "16-bit microprocessor," maintained software compatibility with previous chips. Because customers had invested heavily in software when buying Intel's earlier microprocessors, it enjoyed strong sales as an "update" product. In 1979 Intel launched the 8088, whose mandate was to save customers money. Through utilizing cheap 8-bit supporting chips, it enabled its purchasers to get the power of 16-bit computing with less outlay.

Intel faced especially strong competition from Motorola. Its 68000 microprocessor was powerful and sophisticated, with significant performance advantages. Everyone was talking about it. Moore and his colleagues, concerned that Motorola would take the ground from under them, launched an intense, deliberate marketing and sales offensive in response. "Operation Crush" tasked Intel's salesmen to achieve two thousand "design wins" for its 8086 and 8088 microprocessors by 1980. "A design win is a commitment on the part of the buyer to employ your design in a particular application," explains Gordon. "We were looking very broadly at all of the possible ways that you could use microprocessors." Driven by Bill Davidow, a marketing man by inclination, the aim was to entice customers to commit to Intel's 16-bit microprocessors and crush Motorola in the process. Design wins would lock customers into Intel's line through their software investments and would thus have long-term implications.

Family Values

As the recession of the mid-1970s receded, Silicon Valley came alive. Population was growing by leaps and bounds, as the combination of benign climate, distance shrinkage by jetliners, economic opportunity, and media attention made the area an object of attention. Moore recalls that 1978 was a "year of fantastic growth for Intel and a year of trying to cope, not always completely successfully, with the problems of rapid growth." His son Ken

SILICON VALLEY POPULATION, 1950–2010

Year	Population
1950	526,000
1970	1,621,000
1990	2,150,000
2010	2,500,000

Source: Bay Area Census.

was keen for a piece of the action; finishing his MBA could wait. "I pulled the plug: 'The Valley's roaring. It's taking off, I've got to go earn money.' I was really into computers, even though hardly anybody knew what the computer industry was."

Signetics, the microchip spin-off from Gordon's earlier Micrologic team at Fairchild, called Ken to interview. "I knew integrated circuits, which was what they made." He had already discounted working at Intel. "With a very successful parent, you never know if you received your promotion because of some string they pulled. People always assume that string is there, even if it's not. It was emotionally important to me not to go to Intel, even though I really wanted to." Gordon, dispassionate and analytical even about his eldest son, agreed: "I didn't believe bringing him into Intel was a good thing. It's a heck of a shame when the children are brought in like that."

At Signetics Ken still faced the "son of" problem. "At some point I had to let them know I had this connection to the industry." His mother urged him to speak up at the first opportunity. "'You have to tell them who your father is. They will either say, 'Okay, we can go along with that,' or they won't. Be straightforward from the get-go." Ken did spill the beans. Nevertheless, he got the job, beginning in microchip production planning.

While their work was very different, Ken relied on his father's analytic mind when it came to career advice:

> I'd sit down with Dad and say, "I'm in this group and I really love it, but the action is in this other area." I remember him drawing a diagram: "*Here's* your core business, and *here's* how the managers are moving up. Managers tend to get trapped *here.* If you want to move up, you'll have to move yourself out of there." It was like an organizational theory meeting. Over the years, I learned a lot from Dad about higher-level management issues.
>
> Going to Signetics wasn't the best deal financially. It was not a public company and had no real stock, but I had my own career and I was able to do my own thing. The best part was that I was never bored.

The Moore family was and is very close, but now at last it was high time for Ken, at twenty-six, to move out of Jabil Lane, even given his parents' tolerance. Betty elaborates: "Ken had, without a doubt, the weirdest bunch of girlfriends. They would show up very early in the morning, tapping on his window. One even opened a bank account and had mail sent to our address as 'Mr. and Mrs. Kenneth Moore.'" Ambivalent about seeing him go, Betty searched to find the right house for her son. Finding an appropriate possibility in nearby Los Altos, she sent Ken to view the property on his lunch hour. As he recalls, "In under eight minutes, I said, 'I'll take it.'"

Betty followed through. "I said, 'We're not fooling around here.' On the hood of my car, I wrote out a guarantee for the asking price." Intel stock given by Gordon enabled Ken to make a down payment, yet on his $14,200 starting salary he could barely afford the monthly mortgage payments. "I spent the first five years living in the dark and freezing to death. It was tough. My dad saw it as a long-term thing: 'You get in there with a fixed mortgage; as your salary increases, stuff becomes free, right?' It worked exactly that way."

Steve, too, remembers his father's practical advice:

> He wasn't involved in the child-care stuff—he didn't change a lot of diapers—but he was there to help me with my homework and show me how to use tools. We'd work in the yard together. I got my mechanical experience from him. He'd do a lot of fixing around the house. He was a handyman.
>
> He was busy, but we did many family trips together. These became grander; we took trips to Africa and Tahiti and places like that. He showed me fishing and the outdoors. He's still somebody I can talk to if I have a question about financial things or fishing. He's there to help, just as he was before; he's still a hands-on guy.

Betty was less content with her husband's contribution. She felt she was missing opportunities. "I have some regrets," she reflects, albeit with resignation. "I could have become a real estate person or gone into a design firm, but it was not to be." At home she continued to "make it run and make it work." By now Betty was used to a spouse who was uncommunicative, absorbed in his work. "Whenever there was an event with some society, I'd ask, 'Are you giving a lecture? Are you presenting? What's going on?' Gordon would sometimes say no, he wasn't the speaker. Even so, he always had to be there." In his father's defense, Ken says Gordon was busy not from choice, but by accident. "He's a very long-term thinker; he thinks in twenty-year time frames. He's not a planner. Dad gets overcommitted, and he lets the world schedule him. He'll say no to certain things that he doesn't want to do, but that's about it. His calendar fills up rapidly."

Priming the Printing Press

Throughout his professional life, Gordon Moore described himself as a technologist. His central preoccupation was with manufacturing silicon transistors, latterly as microchips, a technology in which he was uniquely steeped. If Intel could maintain the most advanced chemical printing press for making ever-shrinking silicon transistors, it could stay ahead of the pack. With this in mind, he steadily ramped up spending on R&D. From $29 million in 1976, it rose to an annual $96 million in 1979. A huge fraction of funds went into necessary engineering and research. A move to larger silicon wafers (three inch to four inch) enabled Intel to lower production costs again, as it exploited the larger surface area of each batch.

Intel remained the world's leading producer of MOS microchips, now the main line in all of silicon electronics and second only to Texas Instruments in revenue from MOS memories. Nevertheless, Mostek and Motorola were close at its heels, even as Intel continued to create more complex chips through research and process improvement. In manufacturing terms, MOS gave way to NMOS. Then came HMOS and R&D efforts for CHMOS, which was more advanced still.

Much of Gordon's increased spending went for the salaries of engineers, an overwhelmingly male workforce that pushed every aspect of the technology—from chemical etching to projection printing, from ion implantation to aluminum deposition—to achieve ever-greater levels of miniaturization. By continuous, obsessive attention to all of these areas, Intel could print smaller and smaller transistors arrayed in ever more complex microchips and in so doing lower the cost of electronic devices. This was the essential work of Moore and Moore's Law.

Already in 1970 Intel had achieved a "10-micron generation," with features on its microchips as small as ten micrometers (ten millionths of a meter), approximately the size of an amoeba. In theory, Intel could squeeze 44,000 silicon transistors onto a single penny, with each one costing, on average, one cent. By 1980 a "2-micron generation" of the chemical technology could mass-produce features smaller in size than a single red blood cell. Some 630,000 transistors would now fit onto that same penny, each costing only a tenth of one penny. Reducing the size of transistors and cramming more of them onto chips, the firm continued to make Moore's Law a reality.

Intel's manufacturing was fundamentally about R&D rather than about mass production. For Gordon, R&D was the driver, not other disciplines such as statistical quality control, continual quality improvement, disciplined processes, rapid expansion, and the like. Many within and without Intel felt that a more balanced approach was required. "The Intel model was one of absolute technology leadership," remembers Craig Barrett, who was involved with quality assurance during this period.

> The whole industry was one of freelancing new technology and demonstrating it. You make a few devices. You get into the marketplace. If someone else was a bit more efficient at making them, you went on to the next one. Intel kept leapfrogging. As people caught up and become more reliable, Intel would go to the next level.
>
> We were always at the leading edge of technology, never having to look back, never having to sustain a product line, and never having to manufacture in such huge volumes that cost and efficiency and predictability in manufacturing were terribly important. The professionalism was in the technology development, but there was no professionalism to the manufacturing part.

Gordon's strategy was to careen ever forward to fresh technology generations, each with smaller transistors and more complex microchips. As soon as he became CEO, he talked of it openly. For example, in his consumer electronics plea at the end of 1975, he announced and then published his position: "Whenever our yield starts to approach twenty to thirty percent, it becomes very attractive to make something more complex. This is the heart of the cost reduction machine that the semiconductor industry has developed. We put a product of given complexity into production; we refine the process, eliminating the cause of the defects and fine-tuning the design. When yields on this new product approach the twenty to thirty percent range we again design a more complex product, and put it into production."

Late in 1978 Moore, Noyce, and Grove took stock of the company by exchanging their individual "thoughts on how Intel is doing." Candid, serious memos expressed each man's views and dealt with how the firm might transform itself into a truly major corporation. Noyce saw the issues as those of competition, government relations, and industry-wide coherence. With around $850 million in revenue in 1980, Intel was on course to join the Fortune 500. The firm had achieved a leadership position in "a narrow slice of the business world," and "the horizon is expanding." Noyce himself was restless within the confines of Silicon Valley and looked longingly to Washington, DC. He wanted Intel to lobby on matters such as tax and labor law and give guidance to the government. He believed that the American industry's global competitiveness would depend, to an extent, on attitudes and training in its workforce. "We have to shoulder some of the responsibility. No one else is competent to do so."

Of the three men, Noyce was the most concerned about Intel's immediate future. While he knew that growth rates were good, he also maintained that "we are facing a declining market share in some areas which were originally our territory," notably the DRAM market broken open by the 1103. An aggressive drive was needed to maintain market share, not "complacently let it slip away to Japan, Texas Instrument, or Motorola. We should grade ourselves pretty harshly on this one." In the margin of his copy, Gordon penciled a simple "OK."

As this comment might suggest, Gordon's own "assessment of Intel" was calmer in tone: "We are a young, moderately large company on the verge of becoming a big company. We have all the problems and promise of adolescence." On the plus side was technology: "Excellent product line, no dogs. Excellent technical staff." On the minus side were "a lack of systems and procedures, and no clear sense of where we are going in three to five years." Reflecting the limitations of how Andy Grove's aggressive approach played in and out of Gordon's own management style, Intel was "capable of a real fire drill" but was "generally non-responsive and disorganized. We can't even count inventory."

In his usual quiet way, Moore knew that Grove had done much to turn the rapidly growing company into a hard-driving engine. As "the industry's self-appointed numerologist," Gordon was well aware that in 1978, "in spite of everything we do, we could not keep up with demand"; consequently, Intel lost market share "in almost every part of its business." Finding movers and shakers like Grove—even suitably qualified lesser mortals—remained an enormous challenge. "There is truly a shortage of trained people to support industry growth, if we are to retain anything like present market share. The major limitation will be those few individuals who really make it happen. They are always in short supply."

Moore understood that to move to the major leagues, Intel would have to reinforce its technology leadership with stronger manufacturing discipline and much tighter management control. "For the first time, I feel that the original criticism leveled at Intel ('it's an R&D company, not a real manufacturing company') is valid." Intel was at a crossroads. From here "we can become either a junior IBM" (leader in its field) "or a Fairchild" (which lost its grip and became a profitable also-ran). "What we do in the next couple of years will determine which, if either." Andy Grove's "State of the Union" response was far more affirmative in its assessment of the firm and of his own performance. "The strengths of the year are associated first of all with very effective growth. Our product line is in first class shape, and our picture of where we are heading in systems is clearer now than it has ever been." Grove was especially proud of the fabs in Oregon and Arizona. Whereas other companies had stumbled, both of Intel's moves to expand had been executed "literally flawlessly." The glass was half full, even if it was half empty.

Grove's rejoinder to Bob and Gordon that they not forget his "flawless" handling of Intel's dramatic expansion was timely. Between 1977 and 1980 Gordon increased the company's spending on manufacturing plants—fabs—at an even greater rate than the spending on R&D. In 1977 Intel spent some $45 million on its physical plant. In 1978 that doubled to more than $100 million and in 1980 increased to $150 million. With these hundreds of millions, Gordon expanded Intel's manufacturing and R&D facilities outside of Portland, Oregon, and created a new manufacturing and operations campus outside of Phoenix, Arizona.

What did not shift was Moore's, and the broader semiconductor industry's, desire to avoid unionized labor. At the close of the 1970s, Gordon narrowly avoided the unionization of Intel's fab in Livermore. In announcing that the Teamsters' Union was abandoning the effort, Gordon said, "Our non-union status is the direct result of your choice as Intel employees to have open and free communication and to solve our problems together, without outside intervention. We want to keep it that way."

Avoiding unions increased Moore's freedom of action as Intel's CEO. While there were many factors in his and Intel's choice of the Portland and Phoenix areas for Intel's expansion, the lack of strong organized labor was one. In Northern California itself, "We looked at places like Watsonville, which is primarily an agricultural area. They had very, very strong unions, and that was enough to make us decide we ought to look someplace else. Things change so fast in our industry that a union is potentially a disaster. If you lose flexibility in reassigning people, you can't work well at all." If employees faced risk, they also know the possibility of reward. Gordon never took his foot off the gas of the important stock-related programs—aggressive stock options and a stock purchase plan where any employee could purchase Intel stock at a 15 percent discount. He remained convinced that these programs were vitally important to Intel's success.

Another, even more dramatic, example of a developing, changing company was "Intel Israel." The idea arose from Moore's communications with his old friend Dov Frohman, the pioneer of Intel's EPROM—the company's vital hidden cash cow. In 1974 Frohman left to settle in Israel and take up a professorship at the Hebrew University of Jerusalem. He was keen to maintain links with Intel. In conversation he and Gordon agreed that Frohman's students might become useful as fledgling microchip design engineers. Perhaps Intel could establish a small design center in Israel?

As Frohman pressed his case, the Arab-Israeli (Yom Kippur) war broke out, yet to his astonishment, through Moore's agency, Intel took up his idea. Moore, as was his habit, had listened quietly to a man whose opinion he valued and "did not need to be told twice." For Gordon, it all came back to measurement, analysis, and the availability of talent. "That's how we got into Israel. There was a real shortage of engineers. We said, 'Where in the world are there underemployed engineers?' Israel stood out." Soon Gordon was visiting Frohman's fledgling operation and talking with both government and business leaders.

Intel Israel quickly became a key source of manufacturing capability. Early contributions included peripheral support chips for the 8080 and 8086 microprocessors, as well as a mathematics coprocessor (the 8087) that expanded the 8086's computing horsepower: a Jerusalem fab opened in 1984. By the 2010s Intel's subsidiary was Israel's largest private-sector employer, generating a stream of high-tech exports. Frohman gives credit to Gordon Moore as "the key player, at the very beginning. Any other group of people would have decided not to do it. They understood it was not a stable environment, but it offered a very good return on engineering investment."

Killing with Quality: Japan

In his assessment of Intel, Noyce had declared that an "aggressive drive was needed to maintain market share." As mentioned, at the time Moore had simply noted "OK" in the margin. A few months later, however, Moore wrote in his notebook that he was "really getting concerned." In the large, lucrative, and prestigious DRAM market, Japanese products—from the semiconductor manufacturing operations of diversified corporations such as Nippon Electric Company, Toshiba, Fujitsu, and Hitachi—had begun to flood the US market. "It looked like a thunderstorm," he later remarked.

In Pescadero Gordon had spent childhood years with Japanese peers whose parents were coastal farmers, and he had seen them lose their homes and land when packed off to wartime internment camps. Having benefited in 1971 from its association with the financially weak Japanese company Busicom, Intel—and Moore—had dismissed Japanese microchip producers as "not successful competitors." Now they had caught up, with his own dedication to Moore's Law somewhat to blame:

> In the early days, the technology would go one way for a while and then some other way. Japan could not follow that very well. I may have been as guilty as anybody in giving away the direction things were moving: 1K, 4K, 16K. Once they understood, they were very successful in intersecting the trajectory. They put together a major program whose objective was parity with the American industry at the 16K level and then the leadership position at the 64K level. Their 16K product came out as the market was expanding rapidly, in 1979 to 1980. They took over the leadership a generation earlier than we had planned, on the 16K DRAM level.

One clear reason was that Japanese products were superior in quality to those made by US companies. Gordon's strategy at Intel had focused on technological leadership, on front-running the competition and putting it on the treadmill. In sharp contrast, Japanese producers emphasized quality in manufacturing. They engineered and reengineered for improved yields and to ensure chips that performed reliably. This forced Gordon to confront the shortcomings in his own strategy.

> There was a tremendous attention to detail in Japan. In the US, we emphasized that the next generation was much more cost effective than the current one, so we kept rushing to innovation. Our attitude was that our manufacturing was innocent until proven guilty. When guilt was obvious from looking at the final product, we had to go in, find the defect, figure out where it came from, and then go back and eliminate it—all starting by looking at the final product.
>
> Meanwhile, the Japanese kept cleaning up as they moved along. They went from scrubbing the floor, all the way up. Their approach was very effective at the level of sophistication that we had at that time. Their yields were well ahead of ours. They were doing a better job.

The point was brought home in a painful way when Hewlett-Packard, a major consumer of Intel's chips, published findings on disparate producers. HP made no bones about the quality lead enjoyed by Japanese companies. "That was not the kind of thing we wanted to hear," recalls Moore. "Of course, we said, 'Ours are good!'" But, he continues, "the Japanese changed the ground rules. Before then, we would argue with our customers about whether 1 percent defective parts was acceptable or whether they needed only 0.4 percent defective. The idea that customers could get something they did not even have to retest was not considered. All of a sudden, the Japanese started shipping products where one in a million was defective. It took a while for us to understand."

Craig Barrett was dismayed by the HP report, "damning us and praising the other guy." Intel was in danger of losing its DRAM business entirely. Gordon's question, like that of Barrett and many others in the US industry, was simple once he understood: "What are we going to do about it?'" Bob Noyce was the natural leader, as an answer became apparent: "organize an industry-wide response." Noyce had extensive outside contacts and was looking for a fresh challenge. In 1978, along with Charlie Sporck, Jerry Sanders, and the heads of Fairchild and Motorola's semiconductor operations, he helped form the Semiconductor Industry Association, the first US trade association for the industry.

As Noyce raised the war cry, declaring that America would lose the microchip industry in the same way it had conceded the television set industry, the SIA lobbied the US government, on this and on more indirect matters. One initiative was to join business groups pushing for a reduction in the capital gains tax, which would benefit the industry's scientist-entrepreneurs personally. The SIA was successful in this, but to what end? A dramatic reduction in the US capital gains tax did nothing to stem the rising tide of Japanese competition. Intel's share of the DRAM market, once kingly, plummeted to a paltry 2 percent.

FROM MEMORY TO MICROPROCESSORS AND MICROSOFT

CEO and Chairman

For Moore, the challenge of Japan was connected to a larger problem: how could he best set the strategic direction of what had become a much larger company? By 1980 he has been CEO for five years and overseen an extraordinary boom. Sales had risen beyond $850 million and profits to almost $100 million, despite massive spending on R&D, plant, and equipment. The head count had more than tripled, to almost sixteen thousand. Intel was about to join the Fortune 500, a list it would never leave. The company was successful, profitable, established, and, after only twelve short years, a force to be reckoned with. Moore was chairman as well as CEO, in complete control. Andy Grove was president. Bob Noyce was the nominal vice chair of the board, his interests having moved to the national and global politics of the semiconductor industry.

Gordon, deeply rooted in the Bay Area and quietly ambitious, was very much aware not only of Moore's Law but also of Intel's potential to become the dominant force in the microchip industry as it pursued his vision. To continue moving ahead, he must ensure that the firm overcame serious challenges, beating strong competitors not only by advancing the company's products and printing press relentlessly, but also—increasingly difficult for a now large organization—by successfully divining the directions he desired as its leader.

Happily, by now Moore and Grove shared an unacknowledged commitment that Intel's organization must be constructed around Gordon's modes of self-expression, the ways in which he was comfortable giving criticism, making decisions, and providing guidance. The arrangement relied on Grove being a strong and reliable "amplifier," tuning the organization to pick up and act on Moore's signals. Yet despite the strong positive relation between Grove and Moore, to Gordon it often seemed that the size

Moore's formal portrait, 1980s, with his Microma wristwatch visible.

SOURCE: KEN MOORE.

and complexity of Intel interfered with his transmissions. For instance, on July 2, 1980—nearly twelve years to the day since he and Noyce opened the doors on East Middlefield Road—Moore pulled out his legal pad and wrote a lengthy note to himself that began with his sense of discomfort: "My frustration: sometimes I feel that I can't get anybody to listen to me. Clearly I have to talk louder."

This might be any ordinary person's sentiment on occasion, but for Moore, as chairman and CEO, it was an always urgent, and always recurring, issue. (Ten years later he would reveal the problem had worsened, writing in a private note to himself: "I can't effect anything directly anymore.") "We trip all over ourselves," he mused. Worse, he felt utterly impotent. "My present lack of involvement with Intel's general managers is castrating my ability." The critical question was "Where can I best help the company?" His first answer was "protecting other officers from unnecessary outside stuff," but he quickly dismissed this, writing, "Blah!"

Moore was still Intel's inside man. In liaising with Wall Street, customers, competitors, and the government, he did the basics, but the idea of taking on greater burdens in this area was a nonstarter. He moved quickly to another option: defining strategic directions. A third and much more attractive possibility was "organizing and choosing technology, and allocating technological resources." Here, in guiding Intel's great cost-reduction machine, Gordon was entirely comfortable. Having established to his own satisfaction that his best service was continuing as chief architect of the silicon printing press and judge of its microchip products, he graphed on the legal pad the "communication paths" among leadership that so concerned him.

He had no intention of lessening his involvement with technology. His desire to focus on strategy and the long term still burned strongly within him; getting the organization to act was the puzzle. Moore had stepped back from the bench, but he could still both hold and tighten the reins. "My problem is that I want to do Eugene Flath's job [that is, be Intel's manufacturing technology czar], but do it with the CEO's authority. I want to lay out a family of programs that I am sure are then handed off, and that I can also follow over a long time. Mostly these would be worked through E. J. Flath."

That same day, July 2, he distributed to Grove and others his latest "State of the Union" assessment. It gave a stern warning. A storm was approaching, whipped up by competition from Japan as well as domestic US rivals. Nippon Electric Company and Motorola were particular threats. With DRAMs, "we are about to fall off a cliff on prices," which might halve within the year, while direct costs fell only 10 percent. In addition, "based on recent data (IBM, Japan, other rumors), we are running much lower yields than we should. I'm appalled to see that Motorola leads us by nearly 2:1 in first preference in 16-bit microprocessors and is nearly equal in 8-bit. NEC is eating our lunch, and will start on our breakfast and dinner pretty soon." The immediate question was how to face the gathering threats.

Crushing and Crashing

The storm that Gordon Moore faced was blown by the winds of competition. With superior quality, roughly equivalent processing technology, and aggressive pricing, Japanese producers of memory microchips were soaking up global market share. Worse still, Japanese producers had jumped ahead, offering a 64K DRAM for sale well before Intel had any answering product. In 1980 Gordon found himself struggling on the technology treadmill that he had powered so effectively in the past. And the competition was not limited to memories, or to Japan. In microprocessors Motorola debuted its model 68000, a 16-bit machine. Soon, customer sentiment showed that the momentum was in Motorola's favor, not with Intel's 8086.

Gordon had little choice but to reorient toward quality: "I shifted the focus of the company. One nice thing about a technology that evolves as fast as ours is that you have a lot of chances to make changes. It's not like the auto industry, where you have the same production line for years. We were able to change the way we did things fairly quickly."

Quickly, but not easily. "To clean everything up required a cultural change in the company. It was also very expensive." Close control was necessary for every aspect of the chemical printing process, from the purity of etchants to ensuring that photolithography equipment was dust free. Not just Japanese competition was driving Moore to focus on quality, but so were his insights into the previously closed kingdom of the computer industry's dominant giant, IBM. IBM made everything it needed for its computers—from DRAMs to software—itself. Recently, however, it had started to look outside and had purchased some products from Intel in proprietary deals it hoped to keep quiet. Least public of all was that IBM was working with Intel on a custom memory device. IBM's idea was that if it could cram more memory microchips onto a circuit board, it could use fewer boards in its mainframes and thus lower their costs. As Gordon later explained, "They were a big customer for a special memory, which we became involved in building. It was a good-sized business."

Also secret was that Intel had begun to sell add-on memory systems to IBM, which was using them to keep up with demand for supplements to its mainframes. At Intel each major customer was assigned a senior executive, responsible for troubleshooting and overseeing the relationship. Gordon, as CEO and chairman, decided that IBM was so important that he would take on the account himself. On the other side of the relationship was Jack Kuehler. Kuehler, an electrical engineering graduate of Santa

Clara University, had started out in IBM's San Jose research lab and was Big Blue's most senior technologist. Later, Kuehler would become president of IBM from 1989 to 1993. For Gordon, the fit was very comfortable: "He was the highest executive with a strong technical background there, so it was natural that he would be interested in Intel. He was a local guy, and I got along with him pretty well."

IBM began finding uses for Intel's microprocessors in an array of its products. One design win in 1980, as part of Operation Crush, came from a mysterious IBM operation in Florida. Under a cloak of secrecy, the group was attempting to lower the cost of IBM's smallest computers by an order of magnitude, to create IBM's entry into the personal computer market. Breaking with IBM tradition, the team was trying to make a PC by using off-the-shelf components and software, just like every other producer in the market. They wanted a commercial microprocessor for the PC's CPU and for the commercial software.

The IBM team approached Intel, seeking information about various microprocessors and their implications for system design. The inquiry excited Intel's sales staff. They were soon pushing the 8088 microprocessor as the way forward. When the IBM group put in an order, Intel knew they were serious. Negotiations soon turned to price. According to Gordon, "The head of our component manufacturing was arguing that they needed a low price—$10 or so—while it was costing us something closer to $20 to build. He was saying that we could not sell it to them, and I was saying, 'It's not the cost. It's the price! Ten dollars!'" Gordon was "talking louder" to make himself heard. And in demanding that his colleagues secure the win by agreeing to $10, Gordon was drawing on the well-learned lesson of Bob Noyce's huge price cut on silicon transistors at Fairchild, getting them adopted for television tuners. As then, Intel would eventually make money on the deal.

Part of IBM's interest was that they wanted their PC to be free from possible legal entanglements. In effect, IBM was putting its incredible weight behind an open system for its personal computer by accepting Intel's microprocessors and Microsoft's software as IBM grade, laying the basis for "Wintel's" future dominance and profitability. Just as Intel was free to sell its 8088 to PC makers other than IBM, so too Microsoft could sell its operating system software to whomever it wanted. There was nothing stopping any manufacturer from buying Intel microprocessors together with software from Microsoft to make a PC that would perfectly match IBM's offerings in personal computing.

In 1981, at long last, Intel released its "blue-sky" microprocessor that had been in development for six years. Formerly the 8816, and now the iAPX-432, or simply the "432," this was a sophisticated "micromainframe"—that is, a central processing unit for a computer, containing the essential "brains" for a mainframe—all on one microchip. It boasted 32-bit operation, fault tolerance, and the ability to work in tandem with other 432s. Though the chip had cost Intel more than $100 million and untold hours of labor, it proved to be a terrible mistake. Gordon admitted that "to a significant extent," he was personally responsible. "It was a very aggressive shot at a new microprocessor, but we were so aggressive that the performance was way below par."

Computer and system makers wanted nothing to do with the 432. Sales were excruciatingly slow and embarrassingly sparse. Four years after launch, Moore walked away from the venture completely, selling it to Siemens. Like the Symbol project at Fairchild, it was an extraordinary, expensive, unmitigated flop. "The only thing it did was serve as a textbook example for computer science programs around the country to study. When we finally abandoned it, we had some unhappy customers, but not many."

Once again, Intel's success had meant there was a considerable margin for forgiveness that allowed Gordon to experiment; he remained unrepentant about his investment in this particular form of creativity. "The 432 was the wrong concept for the time, with the hardware and the software intimately combined, when the whole market was moving toward open systems. From a product point of view, it was a failure. I do not feel bad about trying big new things and failing." Neither was the 432 project "set up in a manner that would kill us. We had a parallel path going that was conventional." Intel's "conventional" evolutionary line of microprocessors—its lineage tracing back to the video terminal–chip effort in 1969—would remain at center stage. The 8086 would be followed by the 80186 and the fast 80286, or "286," boasting 134,000 transistors.

Adding to Gordon's woes in 1981 was the fact that a severe economic recession took hold in the United States. Intel did indeed "fall over the cliff": with strong Japanese competition and increased capacity, DRAM prices tumbled. Although Intel sold more memory chips than ever, revenue fell by $66 million and profits by $70 million. The company was no more profitable than it had been five years earlier. Gordon kept spending, in keeping with his maxim of "crossing the valleys." And he was still determined to get out in front of the Japanese. Moore believed investment in fresh fab capability, together with intense effort in manufacturing technology and chip design, would help to stem the assault on the DRAM market.

The strategy required Intel's stressed employees to work even harder. This was the unwelcome message of the no-nonsense man from Pescadero, a message he characteristically delivered indirectly, through his chosen implementer: "We decided that the problem wasn't that we had too many people. It was that we had too many things to do to bail ourselves out. Andy came up with his '125 percent solution.' Everybody should put in 25 percent more effort than they had done previously!" Intel's professional staff was requested to work ten extra hours per week, with no extra pay. "We had a big campaign on that. It was very effective. People were already working hard, but they worked harder." Employee stockholding increased the self-interest of those who participated, yet "no amount of effort could help an awful lot, when the problem is you have the capacity to produce much more than the industry needs."

Electronic Reality, 1980

The 1970s had witnessed a massive expansion in the use of electronic information for communication and control. Within US households, telephones, radios, and televisions achieved near-total saturation by 1980; color television was in more than four-fifths of households, while one-fifth now chose to pay for cable television and the expanded access it offered the mind to other places and times—electronic and digital rather than immediate and physical reality. Phonographs and tape recorders were sold in growing volumes, and a quarter of homes had microwave ovens—novel electronic entrants to the home. The video cassette recorder (VCR) made an appearance, allowing US consumers to access virtual reality by retreating into favorite video programs or television programs at times of their choosing.

Personal computers with the electronics unit paired with keyboard, TV screen, and disk or tape storage drive were rapidly making their way into US households. Some adopted the "cartridge" mode of video game makers, with software in the form of ROM microchips; others adapted audiotape recorders and cassettes for the job of software distribution and data storage. Still others, notably Apple Computer, chose the 5¼-inch floppy disk for software and storage. Word processing and spreadsheet software became familiar.

Television, like broadcast radio, had been to this time a passive one-way technology. Information and experiences came from the broadcaster to be consumed by the audience. With video games and personal computers, the television screen now became a medium for the interactive creation of fresh experiences and information. This interaction reflected a broader reality within US workplaces. Electronic communication of information was becoming two-way, especially in engineering and scientific research; in industrial automation; and in information processing of all kinds, from financial transactions to inventory logistics.

For mainframes and minicomputers alike, computer use became associated with time-sharing and networks—means of interaction in addition to communication and control. Initially, time-sharing relied on paper printouts, but these were replaced by video terminals, with text produced on the screen. A user in one location could interact with a computer at a remote location and, through it, still other more remote computer systems. Soon

ELECTRONICS AND PRIVATE LIFE, 1960–1980

	% US Households					
Year	**Telephone**	**Radio**	**TV**	**Cable TV**	**VCR**	**PC**
1960	78	94	87	0	0	0
1970	91	99	95	7	0	0
1980	93	99	98	20	1	8 (1984)

Source: US Census Bureau; US Federal Communications Commission.

users could run programs, generating fresh experiences and information, on any number of computers on the network. Companies were established to provide computer access as a paid service, akin to a utility.

The spread of electronics to the places *between* home and work grew. Portable radios, automobile radios, and two-way radio systems became commonplace. The electronic realm reached the bodies of Americans, as pocket barely became a booming business. By 1980 a simple model cost barely ten dollars. Wristwatches followed a similar pattern. By 1980 they sold for under twenty dollars and by the tens of millions. Hearing aids powered by microchips and transistors became smaller and more powerful. These novel realities of communication, control, and interaction were pervasive at home and work and anywhere there were people. Exchange of electronic information was at once local, national, and global. Through electronic instruments and computers, Americans became connected to new ways of seeing, from satellite transmission of battle scenes to weather forecasting, and from cellular clusters to galaxies.

ELECTRONICS SALES AND EMPLOYMENT, 1940–1980

Year	**US electronics sales**	**US microchip sales**	**Estimated transistors on US microchips**	**Transistor Production per US resident**	**US electronics employees**
1940	$340,000,000	NA			
1950	$2,705,000,000	NA			
1960	$10,677,000,000	$5,000,000	316,000,000	1.75	734,100
1970	$26,580,000,000	$524,000,000	907,000,000,000	4,420	1,200,000
1980	$101,087,000,000	$6,606,000,000	55,500,000,000,000	244,000	1,500,000

Source: Electronics Industries Association Electronic Yearbooks and Market Data Books.

In the Mouth of the Whale

Computers, gauged by sales revenue, remained mainly a matter of mainframes, the military, and the white-collar worker. In 1982 the market for mainframes was ten times and the minicomputer market five times larger in financial terms than the burgeoning personal computer market. Measured by numbers sold, however, the picture was very different. Two and a half million PCs totally eclipsed the traditional large machines. And personal computers were entirely creatures of the microprocessors within them, which determined much of their performance as well as the software they might run.

The success of the IBM PC put things in focus. Despite the broad US economic downturn, it flew off the shelves, quickly selling some 250,000 units. IBM's reliance on the 8088 was one bright spot on the Intel landscape. Low-priced personal computers were spreading throughout the workplace, and IBM was a familiar and trusted brand. Intel stood to gain from IBM's decision to use standard hardware and software. Other manufacturers could now make compatible "clones" that would necessarily also contain Intel microprocessors. For instance, Compaq, a spin-off from Texas Instruments, selected the 8088 for its "portable" computer. As Gordon points out, "Ultimately, Intel's interest is more in the existence of useful standards than in any particular standard. We're the munitions supplier to the electronics revolution. We're here to provide what the competing armies need to fight their battles. It makes our job easier if they all decide to use the same caliber bullet."

Demand was growing on other fronts, too. Intel landed a huge deal with the Ford Motor Company. Every Ford automobile produced for the next fifteen years would use Intel's 8061 microcontroller in its internal operations, allowing greater fuel efficiency and engine reliability to become standard expectations.

With rapid growth in the sales of personal computers, and the expanded use of microprocessors as embedded controllers in automobiles and many other areas, Gordon faced a fresh but not unfamiliar problem: customers making commitments to Intel wanted some sort of guarantee that they would receive reliable supplies of compatible products. He anticipated a growing struggle to meet demand. "We knew how fast we could build capacity and decided we couldn't begin to make what people needed."

Some customers, such as IBM, continued to stipulate that key components must have a second source. Second sourcing, which had originated with the military's dominant position in the early life of the transistor, was common practice. For Intel, as demand mushroomed, the answer was to fall back on other microchip makers. Gordon cut three licensing deals to set up other companies as "second sources." These deals would give royalties on sales; hence, Intel's own return would be disappointingly smaller than had it been able to make the microprocessors itself. For Japanese and European markets, Gordon made agreements with Fujitsu and with Siemens. For the primary market in the United States, he agreed to a more involved and far-reaching agreement with Advanced Micro Devices.

AMD, led by former Fairchild sales star Jerry Sanders, was an obvious choice. Established at the same time as Intel, it too had focused on MOS microchips, initially as a second source for existing products. Gordon now wanted a broader arrangement, so he signed off on a complex, long-term agreement for technology exchange. AMD would get rights to second-source Intel's microprocessors, and Intel would get rights to AMD designs for support chips. IBM's policy of demanding second sources was an important factor in driving this technology-exchange agreement.

If the future was looking bright for Intel's microprocessors, the memory business was in serious trouble, ravaged by recession, global overcapacity, price cutting, and intense Japanese competition. While revenue had jumped back to prerecession levels, profit remained down massively, despite Gordon cutting expenditure on plant and equipment by $20 million and increasing R&D only minimally. The memory part of Intel's business had become a distraction from more profitable territories and fresh frontiers. And this in dangerous times. Terms such as *leveraged buyout, junk bonds, deregulation, mergers and acquisitions,* and *hostile takeovers* were common parlance in US financial circles. The company's stock had taken a pounding, collapsing by half during 1981. The firm was slowly clawing its way back from a couple of "bleak years," but Gordon and the board were very conscious that, at this reduced stock value, the risk of a hostile takeover was substantial.

The DRAM picture had been bleak for some while, yet Gordon had not given up on this enormous market. Neither had others. Through the Semiconductor Industry Association, Bob Noyce and his colleagues continued to lobby the US government, lodging legal complaints about Japanese competitors, railing against Japanese protectionism, and accusing the Japanese of "dumping" memory chips for well below manufacturing cost. The United States had lost its position as the "undisputed world leader" in semiconductor technology, but might yet regain ground. Japan's approach, opined Moore, "may be 'unfair' by our measure—but that's irrelevant. If we are to continue Intel's historic mission of driving the technological revolution and delivering its fruits to the world, we must pursue our strengths and shore up any weaknesses. While technology and new products are important, they alone are not enough. We must deliver products and services of highest quality to the customer when he wants them."

Gordon decided to rush Intel into the next incarnation of its chemical printing technology and push a superior generation of DRAMs to the head of the pack. He would accelerate the rate of technological change by increasing attention on manufacturing technology. "Concentrate on execution," he wrote.

With most microchips now boasting well over 100,000 transistors, complementary MOS increased efficiency by delivering power to transistors only when they actually needed to switch. Intel's engineers had been making CMOS chips for nearly a decade, and the process, previously slow and expensive, had—with tinier transistors and more complex chips—become faster and cheaper. Gordon, long committed to having the best silicon printing press, chose CMOS technology and a move from four-inch to six-inch wafers for Intel's next DRAM, a 256K device containing 250,000 transistors. Further, Gordon decided that CMOS would become the main line of Intel's manufacturing technology overall. Not only would the 256K DRAM be made on a CMOS printing press, but so too would Intel's newest microprocessor.

Gordon had several rival designs under way for Intel's move from 16-bit microprocessors to 32-bit microprocessors following the utter collapse of his blue-sky 432. One of these, eventually known as the 80386, was being designed at Intel's headquarters in Santa Clara. It was intended as the successor to the 80286, fully software compatible with it, but offering features that took advantage of a more expansive 32-bit architecture. The design looked like it might double the number of transistors on the chip, pushing it to a quarter million. For this number of transistors, the lower power requirements of CMOS would be important.

Meanwhile, the risk of another company launching a hostile takeover remained so great that Gordon wrote out and retained a confidential protocol in his office desk drawer. This piece of paper outlined the actions he would take upon learning of an attempt—who to telephone first and so on. Intel's reduced stock price meant that his traditional means for financing fresh manufacturing capability, through retained earnings and sale of stock, was impractical. With these problems in mind, he hatched a rescue plan. To stave off a hostile takeover and obtain a fresh injection of capital under favorable terms to continue fab expansion, he would encourage IBM to take a significant ownership stake in Intel.

Within a very short space of time, IBM had become Intel's largest customer. The two firms' mutual dependence was symbolized by Intel's microprocessor, lodged at the heart of IBM's blockbuster PC. IBM also relied on Intel for microprocessors in its other personal computers and machines and for add-on memory systems. Gordon and his opposite number at IBM, Jack Kuehler, had also signed off on a second sourcing and technology-exchange deal not dissimilar from that between Intel and AMD. The difficulties Gordon faced were not lost on Kuehler. As part of a routine conversation about the relationship between the two companies, he asked Moore what IBM might do to help Intel.

Gordon had an answer ready, albeit somewhat tongue in cheek: "You could lend us half a billion dollars." Kuehler took the request in stride, and seriously. With IBM's profits for 1982 standing at $4 billion (to Intel's $30 million), he had considerably more resources at his disposal. A loan was not out of the question. But as Gordon recounts, "He and the rest of IBM management decided, 'We're not a bank, so we don't think we can lend you the money, but we would be willing to make an equity investment.'"

Initially, Moore was "not quite sure" this was what he wanted. He and Grove mulled over the idea before flying to IBM headquarters for a meeting with top management and board members. There, IBM proposed buying a half-billion dollars' worth of Intel stock and taking an ownership stake of 25 percent. The arrangement included an immediate investment and a warrant to buy more stock at the same price down the road.

"We thought that wasn't a very good deal because our stock was depressed at that time." Instead, Moore and Grove argued for an alternative plan. IBM would make an initial $250 million investment at the current stock price, giving it about 12 percent of the company. And IBM would eventually purchase no more than a 30 percent stake. "Firstly," Gordon says, "we needed the money to expand; secondly, the best shark repellent was to be in the whale's mouth. If IBM owned a significant percentage, the likelihood of anybody trying to acquire us on an unfriendly basis was pretty small." IBM accepted the proposal. The news grabbed headlines nationwide, from the *Wall Street Journal* to the *San Jose Mercury News*. For Intel, it offered respite and forward progress, in what remained an anxious time.

Self-Help: The Family Foundation

Gordon's sons were among the worriers about the company. Steve nervously considered that the Intel stock his father had given him before the 1971 IPO, already depressed, might soon be worthless. "Intel had been growing really quickly. It had made all these advances and was the fastest-growing company in America. Then it started stumbling. IBM had to come in and help bail it out. That was a nervous time: 'Is it going to survive, or is it going to be taken over?'"

Having started out locally at Foothill, Steve was still living at home while studying business at nearby Santa Clara University. Gordon had encouraged him from a young age to pursue science and math, but Steve struggled with calculus. "It wasn't coming. It would take me hours to work out one problem." Business was the obvious alternative, and at Santa Clara classes were small. "I could be a commuter student. That was what made my choice." His conscientious nature, combined with a demanding study schedule, meant social life all but disappeared. "It was survival time. I didn't like slipping from As to Bs. I sure didn't want to get Cs or anything like that. I tried to keep my head above water. I worked myself to death. I'd never go to football games. I didn't even get to go out fishing with my dad."

In his childhood he had enjoyed family trips: "When I was eight or ten, my parents bought all this gear, and we went camping once or twice. I don't think my mom liked camping. She wanted something where they'd have beds. Later we would go up to the cabin, to the Sierras. My grandmother would come along. We'd go hiking and trout fishing on the lakes and enjoy the outdoors."

In June 1984, as he was about to graduate, Steve took a long-awaited trip to Africa with his parents. Illness struck again. "I picked up a bad intestinal bug. My immune system goes into some sort of overdrive that attacks my back. It reinforced my attitude: 'You can't take your health for granted, and no matter how much money you have, you can't necessarily pay to get it fixed, either.'" The illness prevented him from seeking regular employment. "I couldn't get up and do an eight-to-five job." In addition, "Big layoffs hit the valley, and it was a bad time to be looking for a job. The Intel stock was the asset that I might have to live off for a while. 'Is this going to be worth anything?'" Despite his worries, or perhaps in the face of them, Steve also displayed a carefree, passionate side that to his measured, frugal father seemed entirely out of place.

Cars had been Steve's love "as far back as I can remember." He was especially fascinated by the streamlined, technologically advanced cars made by the Auburn Automobile Company under the brand name Cord and at the age of ten began to save up to buy one. To Gordon's initial displeasure, he bought a 1937 Cord, one of only 1,066 made (and worth up to $175,000 today). "This car came along. It was local and I bought it." It was the opposite of his parents' thrifty approach, but it was also a calculated decision. "I knew I could always sell it if I needed the money. I'm not sure my dad was happy. He thought I was impulsive. I don't think he realized how much I liked it."

Steve had no specific plan for employment, other than to follow in his brother's footsteps and work in electronics, perhaps "in the lower end of management at some little local firm." Jobs were hard to come by, and he continued living in the parental home. "For a couple of years I didn't do much. I'm not wild about that, but that's what happened." Slowly, a solution emerged. For years the Moore family had been supporting local causes. "With so many wonderful resources, we wanted to give back to society," remembers Betty. "I could see the need—so much need." Every year through the 1970s, Gordon would put any requests he and Betty received for financial assistance into a drawer. Between Christmas and New Year, "he would pull them all out and spread them out across the family room and see what was there." Ken explains, "It was not strategic. It was what we referred to as gifting or Santa Claus philanthropy. Dad was giving money and not asking for reports or measurable outcomes. It was just coming out of the family checkbook."

In 1986, with Steve at loose ends, the family decided to formalize their giving. "Collectively, we were thinking we had to do something," says Betty. "We could only donate so much through little envelopes at the end of the year." With a family foundation established, it made sense for Steve, with his business skills, to take charge. "Since I wasn't working—my back was still bothering me—it was a good time to start. I could set my own hours, and if necessary I could go and rest."

Self-help, aid to others in the wagon train or settlement, and compassion for those felled by exigency: these were ingredients in the pioneer experience. A sense of community had been tangible in Pescadero and Los Gatos. "My family never had the ability to do anything much," Gordon says. "We used to go to church fairly regularly. That must have been where it started. As Betty and I became a little more financially independent, increasingly it seemed that doing something made sense. It just grew." In fact, both Gordon and Betty had displayed strong charitable impulses from the earliest days of their marriage. Giving at church soon morphed into other things. Gordon's first modest gift to Caltech came as early as 1956, when he joined Shockley Semiconductor Laboratory. By 1975 the couple were establishing the Gordon and Betty Moore Professorship of Engineering at Caltech. In 1983 Gordon joined the Caltech board of trustees. A cascade of gifts followed over the years, such as an undergraduate scholarship fund (1985), the Moore Laboratory for Electronic Materials and Structures (1991), and the Moore Laboratory of Engineering (1991). The cascade culminated in a $600 million pledge in 2011, still today the largest-ever gift to an institution of higher learning.

As the scope of the Moores' giving widened, along with their wealth, Caltech remained at the center of Gordon and Betty's joint affections. At the same time, the family foundation, managed by Steve, channeled more modest funds to good causes, mostly local, and facilitated change in a low-key way. Recipients of unpublicized donations included Cloyne Court co-op in Berkeley, where Gordon had taken his meals as a student. "They never made a big deal about it," explains Bob Naughten, his old roommate.

Betty and Gordon cut the ribbon at the opening of the Moore Laboratory, 1996.

SOURCE: COURTESY OF THE ARCHIVES, CALIFORNIA INSTITUTE OF TECHNOLOGY.

"These contributions help define Gordy. Yes, he has a lot of money, but he gives it away by the ton to people for very good things."

The first year of its operation, the Moore Family Foundation made grants of around $158,000. A decade later, it was giving away more than $1 million annually to causes including conservation, medical research, and children's health. This figure would rise a further sixfold by the year 2000, even as Gordon Moore's own philanthropic ambitions moved to a far-larger scale. While the Moore Family Foundation was the vehicle for philanthropic gifts that Gordon, Betty, and their sons agreed upon, in a fundamental sense the family foundation belonged to Gordon. It was his decision, and his alone, as to how much of his wealth he would place within it. His family may have had opinions, but Gordon was firmly in financial control, as he had been from the day he married Betty.

Betty was gaining practical exposure to nursing, as she cared for family members. It was not only Steve; Gordon's father's health problems highlighted the challenges, as did those of her own elderly mother, Irene Whitaker, suffering from arthritis and narcolepsy. Betty would talk to her mother each day at five o'clock by phone. "She'd have supper, and she'd watch the local teams on TV. She loved the ball games. She even watched basketball: 'I love basketball; it keeps moving. In the wintertime, it's dark too soon, and the evenings are long and lonely.' She depended on me, and I was there for her. When we went to Australia on a fishing trip, I called her from the ship on a satellite link every evening, wherever I was."

When Irene had a fall and became semi-invalid, Betty hired a full-time companion. Betty's mother preferred to remain in her own home and, as time went on, became bedridden. "I would go down early Sunday, so we'd have the rest of the day for the family," Betty recalls. "We'd watch the religious programs on TV."

Treadmill, Tracking, and Transformation

IBM's $250 million infusion came just in time for Intel. Gordon immediately invested in fresh manufacturing fabs capable of using six-inch wafers. Intel's fab outside of Albuquerque, New Mexico, led the leap to these larger wafers in 1983, ahead of the rest of the industry. Moore and Grove also poured money into the completion of the first overseas fab in Israel, and construction was under way for a huge campus in Folsom, California, northeast of Santa Clara headquarters. To this were added expansion and improvements at Intel's sites in Oregon and Arizona.

Undergirding Gordon's willingness to spend big—investing through valleys and taking risks on projects such as the 432 and Symbol—was his characteristically careful approach to spending money. Financial details mattered to this prudent, ever-frugal grandson of a backwoods storekeeper, whose household ledger had once tracked his and Betty's tiniest daily expenditures. He was exceptionally meticulous and desired to inculcate the same care in others. Ken Moore notes, "Dad has the ability to hammer through the details with somebody who needs it and wants it, or doesn't want it but needs it." Intel preferred the "immediate and approximate" to the "precise and historical" in its philosophy, but Moore also valued a "business attitude towards measurable reality." For Gordon, measurability and openness were key. "Open relationship—open discussion—no games," he noted.

Fabs might be expensive overall, but routine cost reductions were always to be found. Intel could buy rubber gloves for $1 a pair, Moore told his financial executives in 1982, rather than the $2.50 it had paid before. (Since even a good fab operator used six pairs a day, such economies appealed to the thrifty Gordon.) He frequently sent memos to address small issues with large implications. Intel was keeping outside callers waiting, he complained: "I recently called in on our general number . . . twice. The first

Fairchild cofounder reunion, 1980s, each with a token of their hobbies: Moore (*seated center*) with fishing rod, Noyce (*standing right*) with ski poles and hat, and Last (*standing center left*) holding a book of his collection of orange box-label art.

SOURCE: GORDON MOORE.

time I gave up after over twenty rings. The second time it was answered after the fourteenth ring. This is much too long for answering outside calls. It is time that we looked at increasing the operator staff, so that we handle all calls by the fifth ring or so."

Nor, he pointed out, was Intel dealing efficiently with its mail: "A letter directed to the Board of Directors sat in the mailroom waiting to be sorted for several weeks. It was actually an important legal notice that should have been handled rapidly. Any mail so addressed to the Board should be sent directly to my office. There should be some procedure for reviewing the distribution of mail that does not fit in any of your standard categories."

Always meticulous, Gordon also learned to pay better attention to human details. Employees told him their workplace was too pressured; he wrote to their manager, "They feel guilty, sitting in their office reading a technical journal. They would like something that told them it is okay to look outside the project they are doing. I believe you should plan for a reading room to contain the appropriate journals and trade magazines." When a young employee visited his office and begged help to overcome a serious drug problem, Gordon gave him a quiet sympathetic response. Three years later the man wrote, "When I left your office that day, I had a little more hope in my life. Today I have been clean and sober for three years. Mr. Moore what you did was listened; it was enough, and for that I thank you with all my heart."

During 1983 Intel began to make a pleasantly sharp recovery. By March Moore was becoming cautiously optimistic. "I expressed my caution in 1982 by mentioning that 'even two robins don't make it spring.' So far this year I'd say we've seen about .75 robins." Revenue for the year rose 25 percent, hitting the $1 billion mark for the first time. Profit also leaped. Intel was once again ahead, with competitors on the treadmill. Just as in the 1970s Gordon had spent heavily to cross the valleys, so in repeating the strategy he saw the firm emerge strongly from this fresh recession. Intel had entered the year implementing pay cuts and aborting projects, but "then the deluge hit! We've been scrambling to try to keep up with this unprecedented increase in demand." March 1983 came "in like a lamb, out like a lion, and the lion continues to roar!"

Intel's 8086 and 8088 microprocessors would soon lie at the heart of more than 1 million personal computers. The lower-performing 8-bit microprocessors made by Zilog and MOS Technology might dominate in cheaper PCs intended for home use, but Intel's chips, able to deliver performance in home and workplace alike, became the industry standard. Through gaining this central role in the PC revolution, Intel finally won a 70 percent share of the 16-bit microprocessor market. Its latest high-performance microprocessors were the 80186 and 80286, with the latter "designed in" as the heart of the next generation of PCs, including IBM's, to be released in 1984. Design of the more powerful 80386 (a 32-bit microprocessor, tuned to Intel's enhanced manufacturing capability) was progressing. Huge numbers of Intel's lower-performance microcontrollers also streamed into industrial goods, scientific instruments, consumer appliances, and automobiles.

With demand surging, and more predicted, Intel was once again booming. IBM alone put in a huge order, for more than 40 million devices, in

1984. The surge was not just about economic recovery. As Gordon noted, "Pervasiveness is here: automobile, home (just beginning), ubiquitous PC and its progeny." Having ensured supply through his broad second-source agreement with AMD, Gordon continued to drive the expansion of capacity to meet anticipated demand. It was crucial to turn the microprocessor treadmill even faster. "More damage can be done by missing an upturn than by missing a downturn," he noted. "We have a real challenge and opportunity," with "very little slack and unrelenting cost pressure," but Intel was "well positioned to respond (although I'd like a couple more plants right now)."

Microprocessors, microcontrollers, and related products were leading the recovery, not memory DRAMS. The EPROM would remain a secret source of profit for a little longer (the company saw the device as "our birthright"), but Intel struggled to regain its competitiveness in DRAMs. Engineers were working at getting a 256K DRAM into manufacturing, but Japanese competitors already had them available for sale. Intel was using the most advanced version of its chemical printing technology to turn out DRAMs in more volume than any other product, but making little money, despite massive effort.

Gordon charged into 1984 at full gallop. Like every other microchip producer, he anticipated an unprecedented demand for personal computers and for the microchips inside them. He also anticipated a continuing exponential rise in sales of microprocessors, now both more sophisticated and much cheaper. Intel also planned continued production of 8-bit and 16-bit microprocessors and microcontrollers in ever-greater numbers for specialized control applications. Such "trailing-edge" devices helped digital computation reach ever further into workplace, home, and government.

Use of more powerful, more expensive personal computers was booming. In the United States alone, sixty new models powered by Intel's microprocessors had emerged in 1983, helping form a $7 billion market. Compaq, which had brought the first IBM PC clone to market, now had sales of $100 million. IBM launched fresh models in 1984: the IBM PC AT ("advanced technology")—along with its clones—became the dominant personal computer in the workplace. Other computer makers such as Hewlett-Packard and Honeywell also used Intel microprocessors in their PCs. PC revenue was now half that of mainframes and three-quarters that of minicomputers.

In 1984 6.5 million PCs were sold. Of these, 2 million were IBM PCs and their clones, all with Intel microprocessors on their motherboards. Yet for microchip makers, all this growth and innovation, pumping expectations sky-high, spelled disaster. Demand for PCs was vastly overestimated and quickly sated. During the second half of 1984, the house of cards tumbled. The industry, including Intel, had tremendous excess capacity. "Everything was booming, and then demand—all of this huge volume we were gearing up to build—disappeared," says Gordon.

Like airlines, microchip makers kept going. Having made huge investments, they sought revenue wherever it could be found. Prices plunged, even for the flagship 80286 microprocessor. The irony was that while Intel itself could fully match actual demand, it had deliberately created even more capacity by licensing the 80286 to others. With little control over pricing, profits leached away. Gordon, despondent, felt Intel was being forced to give away profits "on a whole generation. Demand across the board fell far short of what we had been led to expect." In February 1985 Moore noted, "Considerable turmoil. Outlook (to me) less clear than usual." In a speech that month, he spoke of how "the industry works in nanoseconds. Everything is compressed, like fast-forward on VCR. A 'rapid change, instant riches' environment created excesses we are all trying to recover from." Nor did "blatant appropriation" of US technology by other countries improve matters. "It's much easier to catch up than to lead," he added. "While we in the U.S. semiconductor industry may be our own worst enemy, we are not our only enemy."

Intel's contract with IBM saved it from a nasty crash, as it again fell off the cliff. With the strong mutual dependence of the two firms, IBM agreed to purchase 17 million devices from its original order of 40 million, "taking a considerable hit for a lot of processors it didn't really need," Moore says. "That meant we could delay the impact well into 1986." In other ways, the IBM investment worked as Gordon hoped: he had money to spend on plants and manufacturing technology, and IBM's involvement kept the takeover sharks away. Even so, the relationship began to falter. Moore and John Akers, IBM's president and CEO, were both on Caltech's board of trustees. "Akers approached me, saying they wanted to sell their stock. I said, 'All right, but keep a 5 percent interest long term.' They sold down to the 5 percent, and then, about two meetings later, he wanted to sell the rest. I told him, 'The hell with it, go ahead!'"

Some felt IBM had lost faith, but to Moore the reason was clear: IBM's own finances had started falling apart. When IBM sold the last of its stock in 1987, its profit from Intel stood at $250 million. As Moore points out, by waiting longer, "they could have extracted $100 billion."

The DRAM Decision

IBM's buffering of Intel in the second half of 1984 and across 1985 gave Gordon breathing space as he faced the dispiriting outcome of his bet on the 256K DRAM. That DRAM represented the optimum point of complexity and performance and minimized manufacturing cost per electronic function. Yet rival devices with perfectly acceptable performance were much cheaper. Memory had become a commodity; competition was all about price. Japanese makers were taking a hit of $1 billion a year in the recessionary environment. "They were losing twice as much as the US companies," Moore recalls, but "they had a higher tolerance for pain." Intel gained just 0.1 percent of the world market for 256K DRAMs. Gordon's strategy, using the most advanced chemical printing technology to regain profits and market share, had failed. Intel had "tripped on leadership, choked on investment, and found the competition intense."

With the same quality of production and aggressive pricing with which they had captured the DRAM market, formidable Japanese competitors such as Nippon Electric Company were also advancing on Moore's beloved EPROM. Gordon faced major decisions about Intel's memory business. What should he do next? It was tempting to continue with DRAMs. Intel's "Oregon gang" had recently come up with a next-generation 1-megabit DRAM microchip, which could form features as small as a millionth of a meter and place more than 1 million transistors in the area of a penny. "It was a state-of-the-art device," Gordon recalls. "We had to decide what we were going to do with it."

Looking ahead, Gordon saw two things very clearly: the need for investment and the lack of return. "To become a significant player again, we would have had to build two new fabs at a cost of about $400 million." Yet "nobody was making money in DRAMs. We had the funds to make the investment, but the chances of getting a return on it seemed awfully small, because of the way DRAM prices had collapsed." Memories had been a product that had volume sufficient to drive Intel's business. "We knew it was time to get out of memories when this advantage was lost. The argument was that you had to be in memories because they were the technology driver. But we saw that DRAMs were going off in a different technical direction."

Moore and Grove tossed the arguments back and forth. DRAMs were close to Gordon's heart. Moreover, Intel's marketing leaders felt it was essential to offer customers a full suite of microchips, from memory to microprocessors, and Intel was unique in having the latest CMOS DRAMs. Gordon still could not make the numbers work. There seemed no way that a further investment could pay off. Grove at last brought the matter to a head, posing a question to his boss: "If you were coming in through the door from the outside to run the company, would you stay in DRAMs?" There was a beat. Then Gordon replied, "No." In this "revolving door" moment, the answer emerged with clarity. It was time for Intel to get out of the DRAM market.

The consequences were both profound and immediate. Foregoing the investment meant Intel's future would be bound more tightly than ever to the microprocessor. Gordon now made a significant strategic decision: to refocus his Oregon workforce. The team there would put Intel's powerful 32-bit microprocessor, the 80386, on a version of the 1-micron CMOS technology. In what was "probably the most important strategic decision that we made," Intel's most advanced microprocessor would replace memory as the focus of the continual drive to improve the silicon printing press.

"We focused a very good technology development team on microprocessors," Moore recalls. "Putting the focus there was the most important part of the memory decision." The DRAM had been Intel's "first big winner," but Moore was practiced at pressing on to new pastures and not harboring regrets. "We would not have been able to pursue microprocessors as aggressively had we continued to compete in DRAMs." As he also acknowledges, the matter had to an extent been taken out of his hands: "The single most important decision affecting Intel's business wasn't even Intel's to make. That decision was made by IBM when it decided to base its personal computers on Intel's CPUs." This, more than anything else, "helped to focus Intel's attention on its microprocessor business" and "drove other key decisions that have impacted the company's form and focus to this day." Other US microchip companies were also dropping out of the DRAM market and dying. (At Signetics Gordon's son Ken had been through "a couple of rounds of layoffs" but managed to keep his job.)

Moore had lost his bet on DRAMs. Faced with the need for a capital investment of several hundred million dollars to become a significant participant in the next megabit generation, he chose to drop DRAMs, abandoning the product family with the largest market of any semiconductor. "Fortunately, at Intel, we had another place to go." Many production workers were not so lucky. Intel needed to bring capacity and workforce into line with the realities of demand. Moore approved the permanent closure of the by now ancient Livermore fab, the shuttering of several other facilities, and the layoff of four thousand employees.

Despite deflated expectations, production and sales of PCs continued upward. The 386 was set to enter regular production. Gordon banked on computer makers adopting it for the next generation of IBM-compatible PCs. Compaq, the leading maker of clones, already had half of IBM's market share. It inked an order for the 386 in the PC system it planned for 1986. IBM, which Gordon felt had "abdicated leadership," decided to stick with the 286. They "figured we couldn't make the 386, so they didn't do a design around it."

Sole Sourcing

All eyes were now on the 386 and its prospects. Compatible with the software that worked on Intel's earlier microprocessors, it offered the next generation of IBM PCs and clones a capability that would allow them to rival the computational abilities of advanced minicomputers. The 386, boasted Intel, was "a labor of love, computing's Rosetta Stone," with $100 million expended on its development, this "latest manifestation of Moore's Law" would "ease out mainframes and minis." The 386 could provide users with a graphical user interface (GUI) like that of the Apple Macintosh. Microsoft introduced its rival GUI, Windows 1.0, designed to run on Intel microprocessors. Older "dinosaurs" would now give way to what Moore called the "software eagle. The hardware is in place. Application software is the key. When it comes of age within the next few years, the bird will fly."

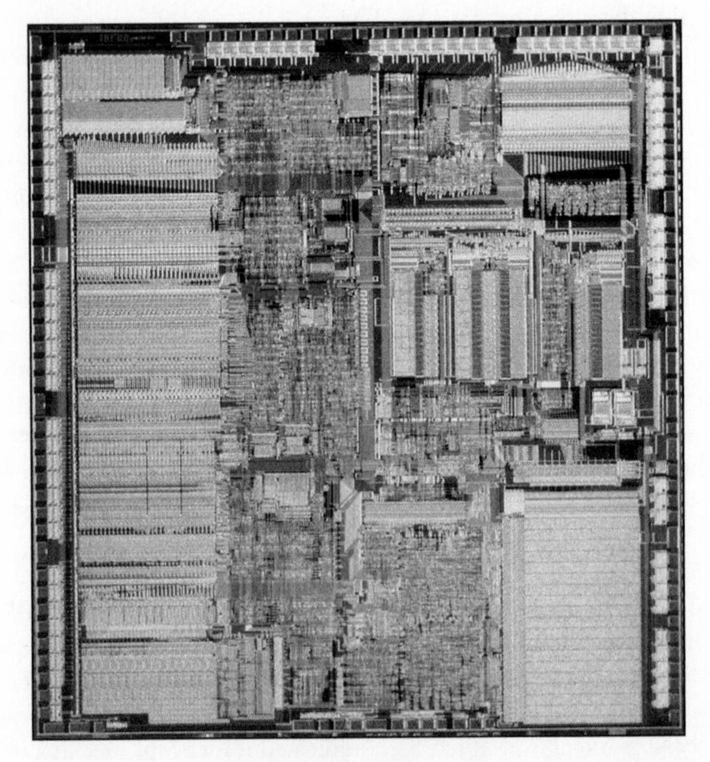

Close-up of the Intel 386 microprocessor. Measuring one-quarter inch by one-quarter inch, this microchip contained some 275,000 transistors.

SOURCE: INTEL.

Moore's senior colleagues were also thinking hard about the way forward. David House was in charge of the microprocessor business and heartily disliked the industry practice of second sourcing. Unwilling to give AMD the 386, and believing Intel would be better served by becoming the only source for its microprocessors, he began to look for a way out of the agreement between the firms. His resolve was bolstered by Intel's recent improvements in manufacturing.

Craig Barrett, who headed this area, had done much to address the quality gap made apparent by the Japanese. To standard Japanese procedures—scrupulously cleaning up all aspects of manufacture and implementing statistical monitoring of quality control—he added his own discipline: "copy exactly." As a team developed a successful fresh version of a technology and brought a microchip to production with good yields, the entire process was documented. Exact copies of the production line were made in other fabs and the exact same procedures used for full, high-volume manufacturing. Such disciplined, scalable production capability gave House confidence and convinced him that Intel could prosper as a sole source.

With assistance from Ted Jenkins, House found an opening in the AMD deal's complicated system of technology credits. If Intel gave AMD's support chips a harshly negative assessment, it would not owe AMD rights to the 386. Gordon was receptive. He had witnessed Fairchild's success with the planar transistor and planar microchip and Intel's own monopoly when its Canadian second source, MIL, dropped out from supplying 1103 DRAMs. He knew these situations could bring bonanzas. Sole-sourcing the 386 would be a complex and ambitious move. On the other hand, if Intel could pull it off, it would lead to a gold mine. Intel would have a monopoly on the microprocessors that dominated in personal computers. Because Microsoft's leading software was specifically tuned to Intel's microprocessors, users of personal computers would have tremendous interest in continuing with Microsoft software and Intel microprocessors. They knew how the software worked. They had millions of dollars invested in it. The momentum afforded by this history of investment in "Wintel" would be powerful.

Gordon argued out the pros and cons with Grove, House, Barrett, and the rest of Intel's top executives and board. The power of the Wintel momentum in personal computers appeared to overwhelm any objection to a lack of second source for the 386. Gordon explains it simply: "We speculated that the software was enough of a lock that they would have to buy the processor from us anyhow, even if we were the sole source." He decided to break the agreement with AMD and sole-source the 386. If the strategy failed, Intel itself might very well fail. If the strategy worked, Intel would have a de facto monopoly on the heart of the personal computer revolution, able to control the price of its microprocessors, and the profit, like never before.

Some of his colleagues believed customers would use the 386 only if a second source existed. It took time to convince them. One, recalls Moore, could not comprehend that "we were not going to second-source to anybody!" At last, Intel's executives all agreed. With the 386, they would go it alone. Gordon recalls, "There were six hundred Taiwanese microprocessor companies trying to get into the PC business, but no company out there that anybody had ever heard of. We decided to be the name brand and do it alone. We thought the fragmentation of the PC market was sufficient that we could get plenty of design wins, even as a sole source. If we couldn't supply all the 386s everybody needed, the industry would just be short!"

Intel aimed to sole-source the 386 for as long as possible, ideally forever. "We had been burned so badly in the previous generation, and the PC business was changing." Gordon had earlier "gone for broke" with his bet on a fresh 256K DRAM and failed. Now, close on its heels, he again proposed to "bet the company," taking a risk that accorded with his technological knowledge. This time success would follow. Moore's decision, backed by the leadership team, would launch nothing less than the reinvention of Intel.

Wintel to the Rescue

As Intel closed its books on 1985, revenue had slipped, from $1.6 to $1.3 billion. Pricing collapses shrank profit to just $2 million, from nearly $200 million. "The interim outlook is the worst that it has been in the thirty years I have been in the business," Gordon wrote to a correspondent. Receiving a request to commission portraits of senior executives, he responded, "With the repressed state of the industry, it is unlikely that we would give serious consideration to any such activity in the foreseeable future."

Further troubles came home to roost. The EPROM price fell from $30 to $3: profit turned to a loss of almost $200 million. Moore later branded 1985–1986 as "the toughest period in our history," with Intel going "to hell and back" as it made thousands redundant and experienced turmoil on an unprecedented scale. In March 1986, he quipped to shareholders, using a *Far Side* cartoon: "The picture's pretty bleak, gentlemen. The world's climates are changing, the mammals are taking over, and we all have a brain about the size of a walnut." Yet Intel was a "master of change," and Gordon remained convinced the next decade would be "an exciting period." His hunch was already being vindicated by the pace at which customers were adopting the 386; buyers brought to market some thirty products containing the 386 in 1986, as Intel recorded two hundred fresh design wins.

An event in New York City on September 9, 1986, brought an iconic moment and hints of a better future. Compaq staged a huge, flashy product debut for its latest PC, the Deskpro 386. As its name suggests, it was built around Intel's 386, and it rapidly became a huge success, selling fifty thousand units in its first six months. Onstage for the unveiling, alongside Compaq's CEO, were Bill Gates—representing Microsoft's Windows software—and Gordon Moore, representing Intel's microprocessor hardware. At that moment, "Wintel" was in control! For the next quarter century, Intel's microprocessors and Microsoft's software would together dominate the PC world, as the PC itself became the ubiquitous enabler of home Internet traffic, the World Wide Web, social media, communications, and the dawning dominance of electronic over physical reality as the preferred locus of human experience. As Moore had often claimed, "We're in the business of revolutionizing society."

The launch also afforded a eureka moment for Gordon Moore, the man who loved to measure and analyze. The self-declared "chiphead" finally became a PC user himself! He explains:

> We announced the 386, and Compaq bet their company on it. I gave a talk at their big press show in New York. For that they gave me a computer. I suddenly had a program that was interesting and a 386-based computer. I thought, "I ought to be learning what the heck these things are good for!" I discovered that spreadsheets are marvelous. I fiddled around with Lotus 1-2-3, which was pretty good. Before too long, I had Excel, which was much easier to use. I still love spreadsheets.

"It was time for me to use a computer personally to understand exactly the contribution that it could make," he explained seven years later. "It takes a little work initially, but yields a stimulating and productive new tool."

If any doubts remained about Gordon's plan to reinvigorate Intel through exiting DRAM production while sole-sourcing the 386, the PC's success blew them away. In 1987 that market blasted past $20 billion, reaching near parity with mainframes. Intel microprocessors provided the core to two-thirds of the 9 million PCs sold that year. Monopoly-based profits gave Intel an astounding recovery to revenue of nearly $2 billion and a record $250 million profit. The stock tripled. Gordon's personal wealth mushroomed.

There was an increasing global realization of "the strategic importance of electronic technology to key future industries." Change brought opportunity. In a speech Gordon noted, "Peter Drucker says, 'The entrepreneur searches for change, responds to it, and exploits it as an opportunity, by shifting resources from areas of low productivity and yield to areas of higher productivity and yield.' We didn't have to search very hard, and we did act as entrepreneurs. The results are beginning to show. For the first time in history, we have some control over pricing. We've had a terrific run, using sole source advantage to get stability in pricing." The decision not to second-source the 386 was the key.

Speaking at Intel's twentieth anniversary celebration in 1988, Gordon tracked how far the company had come. He, Noyce, and its first employees had been like "the immigrants of 1849," following a "direction with promise," focused on survival, while finding "much more than they dreamed." Intel had already made "a major imprint" on the world. "Our products have created completely new industries." Reflecting on the contribution of his own wife, Betty, he acknowledged sacrifices made by "spouses, family and friends—who often can't share in the psychic rewards of our success."

Noyce was now far from Intel, having taken charge of an industry lawsuit against Japanese microchip makers for dumping EPROMs below manufacturing cost. As head of Sematech, the government-funded forum to develop future technology to establish American advantage, he was also seeking to pressure the US government to negotiate with Japan to end the microchip price war. In three short years Noyce would be dead from a heart attack, at sixty-two.

Though it wasn't clear at the time, the glory years were behind Gordon, too. The inevitable strains on a Fortune 500 CEO in a fiercely competitive, innovative industry had become onerous. "I may have been getting a little bit lazy," he later confessed. "Hopping on a plane and going off to talk to a disgruntled customer was less and less appealing. We had come through this terrible period. Things had stabilized and were definitely picking up. We had our new product out there, and it was well accepted in the market." In the business cycle, Intel was at a sweet spot.

11

ONWARD AND OUTWARD

CHAIRMAN OF THE BOARD

Microprocessors and the PC

Gordon Moore turned Intel around, spinning it from its first-ever loss as a public firm, in 1986, to a record income and profit in 1987. The key decisions—to end participation in the DRAM business, continue investing in advanced manufacturing technology, and pursue a sole-source position on the 386 microprocessor—brought stunning results. Moore had seen Intel go through "the toughest period in our history, with turmoil on a scale we have not experienced previously," but his strategic focus on the microprocessor and on leading-edge products for the PC industry, paid off. He re-created the conditions Intel had so enjoyed a decade earlier, when it was the sole source of the 1103 DRAM and the EPROM.

Then, at the 1987 annual dinner for top Intel executives, at the Transamerica building in San Francisco, Moore used the most triumphant moment in his long tenure as CEO to turn over his responsibilities and the full glare of the spotlight—a glare he had never enjoyed—to Andy Grove. Craig Barrett, who was at the dinner, says Moore "stood up and surprised Andy with his new CEO position." However, the handoff was far from being an exit. Instead, by remaining chairman of the board, Gordon aimed to preserve his strategic and technological roles, while shedding those management responsibilities for which his appetite was never strong.

Why change now? Intel was returning to the rewards of monopoly on a popular product, yet this was no rerun of the past but rather a wholly different chapter. Six years later Les Vadasz, one of the few remaining of those who had joined in Intel's first year, would summarize the change this way: "We've transformed from a supplier of memory and logic products to what is essentially a computer company of a different color." Moore had long resisted the idea of Intel as a computer company. Grove explains, "Gordon always placed the emphasis on silicon technology, and he was the leader of the company so long as it was defined by silicon." In contrast, "he was never comfortable with the microprocessor business."

This terse formulation contains an essential truth about the shifting nature of Intel in the late 1980s. Earlier, the central strategy was to advance the manufacturing technology and to put the competition on the treadmill of ever-smaller, cheaper, faster microchips. Now, as the PC assumed center stage, the key to the future would instead lie in the strength of the "Wintel" combination: Microsoft's software plus Intel's hardware. The profitability of Intel's sole-source position was predicated on the notion that the 386 microprocessor would be adopted widely by the dominant IBM, and IBM-compatible, personal computers. This, in turn, depended on a complex of factors that far transcended silicon manufacturing technology. The PC game involved messy, shifting alliances among computer makers, software creators, and customers. And within this game, Wintel was the ace combination.

Important in the success of Wintel was the desire among consumers to retain the value of previous investments, while gaining the improved performance that came through fresh software and hardware. Equally important was the price proposition. With the 386, Intel delivered more "compute for the dollar." Manufacturing cost per transistor had fallen tenfold since the 286 appeared; the 386 harnessed nearly twice as many transistors and offered five times the computing power, for about the same price. To consumers this seemed a giveaway, yet for Intel the 386 was still an extraordinarily profitable product. It was a perfect example of Moore's Law: continuous significant improvements made irresistible by their falling cost.

Negotiation and coordination with Microsoft's Bill Gates became ever more necessary and involved, given Intel's new reality. From his first encounter with Gates in the early 1980s, Moore had concluded that Grove, with his out-there aggression, was better equipped to handle the Microsoft wunderkind. Gordon vividly remembers an early interaction:

> Gates phoned me and chewed me up and down for what we were doing. He said we were stealing their software and made other similar claims. He had lots of highly specific complaints. I couldn't deal with this guy, so I turned him over to Andy. Andy had him for dinner, and they clicked.
>
> Bill used to throw a tirade at the beginning of every meeting. It shocked people so much that he gained complete control of what was going on. Finally, we learned that, after his tirade, we could get down to business.

Moore was far less interested than Grove in the nuances of computer software. For his part, Grove believed that the timing of Moore's decision had everything to do with the nature and evolution of the microprocessor:

"Gordon is the brains behind the first half of Intel. Intel is the brains behind the computer industry. He has everything to do with mass-produced complex integrated circuits, without which microprocessors would not even be a thought. Microprocessors—that's not Gordon; it's not his kind of thing. The microprocessor has changed everything. You have to march it very carefully, with military precision. It's a highly intricate game-theory kind of game. I thrived on it."

With revenue jumping to nearly $3 billion in 1988, and profit almost doubling to more than $450 million, Gordon's decision to bet Intel's reinvention on sole sourcing microprocessors and abandoning memory was affirmed. "When you're a start-up, you have to bet the company on a lot of the programs you do," he explains. "When you get bigger, you only want to bet half the company, if you can. On the other hand, you gotta keep reaching." While IBM's use of Intel's microprocessor and Intel's dropping out of the DRAM market seemed like independent rapid acts, he explains, "They really weren't. They were a series of decisions that ultimately led to a dramatic outcome."

Moore's three decades at the center of semiconductor technology had shown that having an improved offering that performed a standard function, that optimized the chemical printing technology and for which one was the sole source, was to possess a major competitive advantage and the potential for incredible profit. Leading-edge technology, opening fresh possibilities for products and their properties, was the key to success. In contrast, innovative "first-mover advantage" was frequently short lived. A standard microchip pioneered by one firm soon became an interchangeable commodity supplied by many. Gordon's "cheap chunks of silicon"—the concept of using microprocessors as embedded controllers—had so far followed this pattern, but now it emerged that for leading-edge microprocessors within PCs, the pattern no longer held. Instead, Intel had a real chance of disrupting, delaying, or even preventing the process of commoditization and could hope to preserve its advantage.

A cluster of factors was at work in this fresh pattern for marketing Intel's products. One was the rising cost of design. Gone were the days in which chip engineers could lay out a microprocessor on paper or cut designs for photolithography masks from large sheets of Rubylith, a colored plastic film. It now cost upwards of $100 million to design a microprocessor with a quarter-million transistors, such as Intel's 386. It took intensive use of computers and a large team of workers to create a fresh chip to lie at the heart of a personal computer. Such design, tool, and manpower costs deterred many potential competitors from entering the market.

Another factor was increasing fab cost. Leading-edge production facilities required superclean environments, expensive production tools and instruments, ultrapure chemicals, massive volumes of highly purified water, and prodigious amounts of electricity. By 1987 the cost of a fab had risen to $200 million from $1 million less than twenty years earlier. If Intel wanted to succeed in making microchips still smaller and cheaper and in increasing their capacities and performance—and thereby continuing to lower the cost of electronics—it had little option but to invest in fresh, more advanced fabs. Their expense was yet another barrier to entry for other firms.

Further in Intel's favor was the fact that it now possessed the ability to shape the market. By 1984 the microprocessor was the defining feature of any new PC. It sat at the heart of a variegated system, shaping the body of possibilities for memory chips, storage systems, and commercial software. Introducing more advanced microprocessors and more sophisticated personal computers every three years or so, the PC industry was on a brisk product schedule. Intel, setting the pace in microprocessors, was in a position of great power.

A final factor in Intel's newfound ability to resist commoditization was the most direct: historical momentum. During the 1980s, consumers of personal computers collectively invested billions of dollars in software and peripheral equipment (such as printers) that would work only with IBM and IBM-compatible PCs, that is, with Intel's microprocessors. This dependence created a huge incentive to continue to buy PCs that used Intel microprocessors and ran on Microsoft software: the "Wintel" combination.

Risks and RISC

Gordon became convinced that Intel's sole-source position on the 386 would endure for a significant time. The firm enjoyed a unique position, dependent on no single customer. Its fabled manufacturing prowess gave it a tailwind. In addition, with the legal protection afforded by copyrighted microcode, Intel could prevent other manufacturers from making copies of its microprocessors.

When it became clear that he would not be invited to second-source the 386, Advanced Micro Devices' CEO, Jerry Sanders, decided to force Intel's hand. He initiated legal arbitration over disagreements in the comprehensive technology-sharing and second-sourcing agreement. In 1991, after a bitter fight, AMD finally won the right to use Intel's microcode for its own versions of the 386, but because the legal process had dragged on—favoring Intel—AMD was four years behind. Intel had already established a sole-source monopoly for its 486, which, with more than 1 million transistors, could deliver a near-fourfold increase in computing power for about the same price as the 386. IBM took up the 486 in its newest PCs in 1990, and other Wintel PC makers followed suit. AMD launched its 486 version in 1993, but Intel kept its lead.

By the mid-1990s Intel controlled more than 80 percent of the microprocessors-in-PCs market, a position it would retain (with some fluctuation) to the time of this writing. For the next three decades, the firm's competitors would face the daunting prospect created by its position as the sole source for Wintel microprocessors and its leadership in the pursuit of Moore's Law. Every few years Intel released a new, more advanced microprocessor at little or no increase in cost. Each release, based on a fresh incarnation of the chemical printing technology, offered a dramatic increase in computing power for roughly the same price. Under Gordon's chairmanship, Andy Grove used every possible legal and marketing tactic to maintain this highly advantageous position, as did his successors.

Even though there were real barriers to entry into the market that Intel dominated, the company was far from safe. Its historical momentum was a strong advantage, but that advantage was not absolute. The firm had only to make a major stumble in silicon technology or in microprocessor design to become prey to its rivals, which included several lean, agile start-ups. For a period, two would become key players: Sun Microsystems and MIPS Computer Systems. Both emerged from research conducted at Stanford University by their founders, and both offered commercial microprocessors that directly challenged Intel's line and the dominant Wintel mode.

Sun and MIPS introduced "reduced instruction set computing"—RISC for short—which meant, in effect, novel microprocessors that could be run by following relatively simple instructions. Intel's engineers, in developing their line of backward-compatible microprocessors, had used an expanding list of features to make software programs run efficiently and to boost performance. The RISC approach, in contrast, used a smaller number of instructions that were designed to be easier for the microprocessor to perform. To surmount cost barriers, Sun and MIPS outsourced manufacturing to outside fabs, or "foundries," such as those of the Taiwan Semiconductor Manufacturing Company, which would become one of the world's largest semiconductor companies. Established firms such as Fujitsu and Toshiba also offered foundry services and made many of the early RISC microprocessors. The RISC companies aspired to compete with Intel on both performance and cost.

By leaving DRAMs and memory and focusing on sole-sourcing microprocessors for PCs, Gordon greatly diminished the previous threat from Japanese competition, but domestic rivals, especially IBM and Motorola, remained troubling. IBM was developing a powerful, ultrafast 64-bit RISC microprocessor for its own PCs, while Apple Computer, with its small but persistent market share, used Motorola's popular 68000 series in its

Macintosh computers. With all this competition, Andy Grove needed a compelling series of microprocessor offerings.

Under Moore's guidance as chairman, Grove continued to invest in advancing the chemical printing technology, the "great cost-reduction machine." In the four years after 1987, Intel's revenue more than doubled, to nearly $5 billion. Profit more than tripled, to $800 million, and the stock price continued to climb. During these same years, Andy Grove spent an astonishing amount to follow Gordon Moore's technology strategy. By 1991 PC industry revenue was $39 billion, as 20 million systems were sold. Years later Moore would say, "Since 1985, we've had a terrific run, using our sole source position to get stability in pricing, rather than using it aggressively in trying to stop progress. Our average price on a processor has fallen 40 percent, which is discomforting, but that's the nature of competition. We can't find another microprocessor-type radical innovation, but the decision not to second-source the 386 was one of the keys to Intel's success."

Electronics Rule

The US military had funded early computer networks, such as the SAGE air defense system, and the ARPANET. Now PCs began to be connected within workplaces, using telecommunication links. Organizations moved away from minicomputers and even from mainframes. They began to choose a "client-server" mode that could provide high levels of computing resource by networking their PCs with specialized servers.

The year 1991 saw the beginning of the World Wide Web, a system of protocols, languages, and practices for exchanging data over the Internet. Electronic information had steadily altered lived experience across the 1980s. With long-distance and shorter-distance telephone calls dropping in price by more than 60 percent, Americans in the United States used the telephone more than ever before, making five calls each, on average, each day. Microchips provided control over these many calls, thanks to the answering machine, which redefined the nature of the "missed call" and allowed for novel practices such as call screening. Only a very small proportion of the US population (1 percent) as yet used a mobile phone, either as a large portable handset or installed in an automobile.

Within US households television and radio—the now traditional technologies of mass unidirectional electronic communication—had achieved near saturation. Cable television providers served the majority of US households, with most sets showing color. VCR machines, in almost 70 percent of American households by 1990 (up from 1 percent a decade earlier), enabled consumers to control not only what they watched on TV but also when. Other consumer products based on microchips and electronic information proliferated. Four out of five American homes now had a microwave, and stoves and ovens were increasingly supplied with electronic controls and clocks. Video game systems that could be hooked up to the television had become popular in family homes, and annual sales of video game systems approached $4 billion.

Digital clocks and watches were now inexpensive, and electronic calculators, once a luxury, were given away as promotional items. Fully digital sound systems—in particular, the advent of the compact disc (CD)—began to displace vinyl and electronic phonographs and tape players. Digital representations of songs and sounds could now be encoded (as zeroes and ones) on a plastic disc and read by a laser. Home video cameras began to amass material that would supplement film photographs as a means for Americans to preserve their personal histories. Low cost microchips opened up novel opportunities to amateur creators of video programs and projects.

As a consequence of all these new devices, screen time—the hours spent by Americans watching television, viewing a VCR recording, playing video games, or using a personal computer—increased hugely. A growing fraction of the average US citizen's waking hours were spent in contact with electronic reality. Only 15 percent of US households as yet had a personal computer, but as the Apple Macintosh spawned a wider familiarity with the "graphical user interface," or GUI (with its windows and desk metaphors), computers became increasingly "screen centered." Most home computer users relied on commercial software to provide operating systems, games, word processors, and spreadsheets.

In US factories, microchips were used in industrial robotics for manufacturing, controlled by digital computers. Electronic instruments and sensors became key to the operation of chemical processes, making everything from petrochemicals and pharmaceuticals to peanut butter. There were equally profound effects on the US financial and banking sectors. The automatic teller machine, or ATM, became common, replacing teller positions and offering twenty-four-hour banking services. The NASDAQ and other US markets computerized, with securities traded at high speeds through networks of computers rather than on the floor of the New York Stock Exchange or the pit of the Chicago Mercantile Exchange. Electronic information was becoming pervasive.

Road Maps

As chairman of Intel, Gordon did not step back very far. "I didn't do an awful lot differently. I just didn't have to worry about it as much anymore," he says. For years, he had influenced Intel's direction through the interpretative abilities and brusque efficiency of Andy Grove. Asked how he approached his changed role as chairman, Moore gives an oblique response: "Somebody once said to me that the job of a CEO is to do everything he can't find anybody else to do; that was more or less the way I approached it." With characteristic nonchalance, he underplays the brilliance of his strategy. "I had ideas about where the product line ought to go or how the industry ought to develop. I was trying to drive it to grow bigger, become more profitable, and keep customers happy."

Like his father, Gordon paid close attention to the clock, always careful not to waste time on things at which he could not excel. Reflecting on whether his involvement at Intel diminished in the years following 1987, he says, "I still spent as many hours there every week. I had hoped to spend less, but I never figured out how to do anything part-time in this industry." However, with Grove firmly in the driver's seat, Gordon could make a more active contribution to external concerns. He agreed to a one-year term as chairman of the Semiconductor Industry Association.

Bob Noyce had earlier relocated to Austin, Texas, as the founding CEO of Sematech, the organization created to defend US leadership in silicon manufacturing technology. Sematech aimed to develop next-generation production tools and materials. At Intel managers had been in the habit of creating time lines and detailed plans of action. It was an important activity, "a way of looking at things," which Gordon used to push the technology and keep the competition on the treadmill. "We started calling them road maps in the 1980s. We had two or three technology generations laid out and were matching products with technology." Another organization, the National Advisory Committee for Semiconductors, was also compiling an ambitious, long-term plan to enable US producers to move ahead of the Japanese.

When Gordon's twelve-month stint as SIA chairman ended, he took the chair of a new technology committee, charged with developing a road map for chemical printing technology for the industry at large. He saw a clear need for coordination. The complexities of the technology had moved beyond the ability of a single microchip maker, even Intel, to handle on its own. It was second nature to him to focus on alignment and harmonization of strategy, so that longer-term industry goals meshed with the short-term plans of individual companies. He convened a "workshop" in Irving, Texas, involving more than 170 semiconductor experts, with subcommittees meeting for months. The outcome was an SIA report containing a linked series of materials.

The SIA's road map was essentially a spreadsheet, the electronic analog of Gordon's ledger notebook. Into rows and columns of boxes, the constituent specifications for the next generations of chemical printing technology were poured, along with their attendant requirements. The road map defined what needed to be ready, and when, to ensure Moore's Law was observed. It covered silicon crystal growing to final packaging and testing and everything in between. "Gerry Parker, who used to run our manufacturing, is the guy who I first heard of road maps from. He had a pretty good view of how things fit together. When I became chairman of the SIA committee to worry about technology, I adopted the road-map idea."

Road maps would coordinate the activity of the specialized firms that were industry suppliers. Microchip makers had by now adopted a very specific cadence of development: roughly every three years a fresh technology generation emerged. The ability to shrink a microchip's smallest feature by 30 percent—known as scaling—enabled Intel and others to create, and profit from, microchips with at least double the complexity of each prior generation. Moore's report called for this pace of change to continue for the next fifteen years and stressed that long-range technical problems were where the industry needed to focus, while he also called for minimizing unnecessary duplication of effort. "Certain things have to happen on an appropriate time scale and be incorporated." To drive forward the technology, deliver smaller features at high yields, and ensure the status quo of regular exponential change, information exchange had become crucial. Moore knew that the very act of making maps would harmonize the multifarious efforts and products of what by now were hundreds of firms. A map's value as a lasting reference was secondary. Cooperation was the key.

"What the industry wanted to do was well understood. The road map dealt with areas in which Intel wouldn't have the intellectual property. That made us more willing to lay out exactly what ought to be done." SIA took on the collation and publication of this ongoing endeavor, as the National Technology Roadmap for Semiconductors. Its introduction was perfectly direct: "The primary planning assumption is the continuation of Moore's Law. CMOS technology is assumed to be the dominant high-volume high-performance technology throughout the foreseeable future." With the technology road map, Gordon provided an institutional footing for the social construction of Moore's Law.

Broader participation by non-US microchip firms and suppliers prompted a renaming in the late 1990s: the International Technology Roadmap for Semiconductors. Just as US firms find it profitable to collaborate on "precompetitive" matters relating to technical roadblocks, so too does the global industry. Revisions and forward plans provide a central forum for discussion of the ongoing challenges of keeping to Moore's Law. To this day, expert subcommittees update and add detail to the road map to help the world microchip industry and its suppliers carry forward Gordon's global vision.

Losses and Gains

Bob Noyce did not live to see these road maps. He died suddenly, at sixty-two, in June 1990, of a heart attack. The technical and business communities in Silicon Valley mourned loudly. Gordon did so quietly, but with characteristic generosity. "Bob's achievements put him in a class by himself," he wrote in a press release, "inventor of the integrated circuit, founder of two major American companies, and the first chief executive of Sematech."

Noyce's official membership of the Intel board had only recently ended, yet in Santa Clara itself he had long ceased to be a presence. As far back as 1983, Intel's internal magazine remarked humorously on rare sightings of "*Noycensis Robertus,* commonly known as The Bob Noyce." After his death Intel named a building in his honor. Gordon himself was outwardly unaffected. He and Noyce had remained cordial, but no longer as close as earlier on. "When he came up from Texas to Santa Clara, I'm sure I saw him, but my interaction with him was down to a few times a year."

Wherever Noyce had gone, his presence had been felt. In contrast, as Noyce's biographer remarks, Moore remained all but invisible: "Because he was far more reserved than Noyce, he did not mind functioning as Intel's stealth CEO." Moore had no desire to become ambassador; he routinely declined to do "that sort of thing." When he agreed to chair the Semiconductor Industry Association and was obliged to testify before a Senate committee, he found himself "completely blindsided" under tough questioning by senators and contradictory testimony by others. The experience confirmed his reservations about being a public figure.

The partnership of Gordon Moore and Bob Noyce had been among the most productive of the century. First at Shockley Semiconductor, then at Fairchild, and finally at Intel, their work deeply influenced the direction of the industry. They were, as Moore had put it in 1973, "the real revolutionaries." Paying tribute to Noyce more formally, Moore praised his partner's broad interests, ideas, courage ("he was never afraid to try the untried"), and leadership. He also spoke warmly of his personal qualities: "Bob was always willing to take the time to understand."

Yet there had also been a rarely glimpsed, more conflicted vein in the relationship. Noyce's flamboyance and "wild ideas" could be disruptive to his partner. In the early days of Intel, Moore explained, "things would be going so well that you didn't want to be distracted. It was a little frustrating." Also irritating was the fact that Noyce was popularly perceived as the visionary behind Intel. To some, the unassuming Moore was no more than the "researcher-like cofounder," a minor partner, the "other guy." Steve Wozniak at Apple and Paul Allen at Microsoft were both in the shadow of more charismatic cofounders, but unlike them, Gordon Moore quickly became the largest shareholder of his company, remained deeply influential in its work, and ultimately was reluctant to step back. His steady engagement was the organizing principle of both his own life and of the company's; it was Moore who guided Intel from success to success. Indeed, as he would belatedly but truthfully point out to *Fortune* magazine in 2012, "I am still the longest-tenured CEO Intel has had—longer than Noyce, Grove, and Barrett" (and Paul Otellini, Barrett's successor, too).

Moore and Noyce were equal partners, but in the history of technology, Noyce's role as the public face of Intel, and as the early icon, earned him a far larger portion of credit. Gordon's fame was belatedly growing in certain circles (for instance, he received the National Medal of Technology from President George H. W. Bush in 1990), yet outside observers had a tendency to elide him entirely. "[Noyce] wasn't a taskmaster," stated one tech writer in 2013, "and had to bring in Andy Grove to keep order." Few saw how Moore's quiet, brilliant strategizing underlay the electronic revolution and Intel's immense success.

At some level, though he rarely let it bother him, Moore felt cheated. His growing distance from Noyce in the 1980s reflected this. Andy Grove explains, "There's very little emotional affect to Gordon's description of his relations with Bob—positive or negative. It's just a fact of life. 'Bob did this; I did that.' Once, and only once, I saw deep irritation in Gordon when somebody was giving the long-dead Bob credit for something Gordon thought was unfair. He got emotional. That blew me away and suggested to me that lower-level versions of that emotion must have existed."

Changing Scenes

If Bob Noyce's early death was one form of loss for Gordon and Betty, a more significant blow occurred when Irene Whitaker, Betty's mother, died in 1992. She had been bedridden for years. Betty was also involved in caring for other family elders; she would frequently take her mother's brother William to see the former ranch site in Los Gatos where he and his Metzler siblings had lived. "He wanted to go in the week before he died," recalls Betty, but "he was catheterized. He couldn't talk because he'd had a stroke. He never had that last ride."

Betty's own arthritis was worsening. Toward the end of the 1980s, she began to have trouble climbing stairs and applied to the local authorities for permission to extend their driveway on Jabil Lane, to bring it to the second level of the house, behind the carport. Permission was denied. Betty remembers, "I told them, 'I have such bad arthritis that my husband has to carry me down to the car. He's getting older, and I'm ruining his back.' Well, this was like asking for diamonds. 'Maybe you could put in an elevator somewhere.' I said, 'I didn't design this house to put a glass elevator out there as the first thing you see.'"

Gordon, too, was frustrated by red tape. He loved the idea that he could tinker in his home workshop equipped with woodworking tools, among them Walter Harold's cigar box of drill bits with its evocative scent ("I always thought of my father when I opened those drills"). After relinquishing his position as Intel CEO, and with more time at his disposal, Gordon decided to build a larger workshop, but the authorities turned down his application. Los Altos Hills had evolved from a rural enclave into a populous, affluent area. Gordon's plan did not meet its updated, more rigorous planning regulations. At Intel he had built billion-dollar factories, but his home workshop was not a major economic development opportunity for an entire state.

One thing led to another. The couple started to look for an alternative place to call home. Son Steve had never left the nest and now ran the family foundation from Jabil Lane. He "truly wasn't moving out of the house," says his brother, Ken. "My parents had an interest in doing something fresh and getting more land. Both of my dad's brothers had tractors. They could shove dirt around and build barns and put in a fence; that's stuff my dad loves." Gordon began to discuss with Betty the idea of building a home on a larger lot, somewhere in the area.

By this time the Moores were generous supporters of numerous charities. One was the Peninsula Open Space Trust, which bought development rights up and down the mountainous spine of the San Francisco Peninsula and in the coastal regions to the west. In 1990 POST was fund-raising to purchase the Phleger estate, a tract of thirteen hundred acres of pristine woodland in the town of Woodside, on the side of the Santa Cruz Mountains, constituting a significant patch of "natural" environment. It lay four miles toward the Pacific from Gordon's boyhood home in Redwood City and adjacent to Filoli, a historic home and gardens surrounded by six hundred acres of woodland, open to the public. The Phleger estate had its own historic home: Mountain Meadow, designed in the 1920s by the prominent California architect Gardner Dailey.

Mary Elena Phleger, the elderly widow who had occupied Mountain Meadow for many years with her lawyer husband, and lived there still, was determined to keep the estate intact and agreed that POST should raise funds and buy it as a parcel. POST had no special interest in the house and agreed that Gordon and Betty could view it. First impressions were not good. Betty recalls, "It was a shambles. The colors were horrible. The carpet in the dining room was filthy. Gordon said, 'This is dreadful.'" Yet with her eye for design, Betty liked "the old English look" and saw potential. "I said to Gordon, 'See the way it flows. Look at the size of the rooms. Feel the house. It has wonderful bones.' Gordon said, 'I don't like it.'"

Outdoors, he was won over. Mountain Meadow would come with almost twenty-five acres of private grounds, including formal gardens, a pool, an apple orchard, holly hedges, and a hothouse. Beyond this, the woodland environment was largely untamed, filled with second-growth redwoods and oaks and dotted with streams, ponds, and horse trails. For Gordon, it was like stepping back into his earliest years on the far slopes of the very same Santa Cruz Mountains, in Pescadero. "It was marvelous. It's a beautiful piece of property," he later recalled, uncharacteristically enthused.

A new office and workshop could easily be accommodated. Gordon and Betty decided to give POST a considerable gift, to enable the tract of land to be preserved, in exchange for the right of first refusal on the house. Within weeks, unexpectedly, Mrs. Phleger passed away. Soon after, Betty and Gordon became the owners of Mountain Meadow, while POST transferred the surrounding land to the National Park Service. In due course, it would become part of the Golden Gate National Recreation Area.

Betty quickly set to work to renovate the house. At Jabil Lane, with Gordon leaving home by seven thirty, she had called all the shots, decisively declining timber that fell short of the required quality. With the project in Woodside, Gordon had a little time to spare and was more influential. The usual spheres—Betty at home, Gordon at work—lost some definition, with frustrating results. "We employed the wrong contractor and wasted a year with him," says Betty. "I couldn't convince Gordon. I kept saying, 'We're riding a dead horse. Please look at it.' 'Oh, give him a chance.' Gordon finally agreed, 'We do have to get rid of him.' We started over. It took a long time to get it together."

Gordon had a large office built at one end of the house, close to a spacious workshop in the renovated garage. Betty began taking trips to search out antiques, visiting the English countryside with son Steve. "I wanted to look, feel, poke, and pull," she recalls. "We purchased Georgian-period furniture. I found a big old bed that is fabulous, a double poster with wonderful wood." Finally, after three years of renovation, Mountain Meadow became Gordon and Betty's residence.

This was the couple's first move since their early thirties. For Gordon, leaving Jabil Lane was both an enormous change and a mark of constancy. Investing in the Phleger estate was to conserve peninsula wilderness and a historic home. While it trebled his Intel commute to forty-five minutes each way, he now lived just five miles from the cemetery where his parents were buried and in a similar mountain ambience to that of his Pescadero childhood. Betty was also reconnecting with her roots, perhaps especially because her mother had passed away during the renovation. Betty convinced Gordon to take a week's "reunion" cruise to Alaska with her maternal cousins. Joining the Moores were George and Rick (sons of Aunt Ada) and Phil Schiller (the son of Aunt Ruth).

At Intel Gordon had long been in the habit of sitting down with pen and paper to reflect on progress. On this reunion trip, in July 1993, he wrote a detailed account of each day. As the couple left the Bay Area, they first traveled to Portland to join 10,000 of Intel's 29,500 employees—with spouses—for the company's twenty-fifth anniversary party. "The rain held off," noted Gordon, "despite its being the rainiest July on record." His account of his trip bears many similarities to that written by his grandfather Josiah Williamson, on his journey from Massachusetts to the Bay Area in 1869. Like Josiah, Gordon cataloged the ship's dimensions and commented warmly about the entertainment on board ("there was a good pianist"). Both men show great interest in passing geographical features.

When it came to the environment, Gordon's appreciation for the natural world was evident. "We missed the first whales, but saw some in the distance. I think they were humpbacks by the way they occasionally lifted their flukes out of the water when diving. We could see parts of the ice fields in the background with the most rugged mountains." He could be dryly humorous ("the eagle was nice enough to fly into the nest while we were watching, and stick its head over the side to allow pictures") and boyishly eager ("the glaciers were really neat"). He was fond of using his video camera and managed to record a glacier calving: "I took one good sequence." The high point of the trip was a fishing excursion, about which he wrote at length. The low point came with a bout of karaoke singing, after a dinner at the Ritz-Carlton. "That was pretty sad." Overall, Gordon was impressed with cruise-ship traveling. "It all worked out pretty well, and it was a safe way to see the various cities." His warmest comments were reserved for the food. He recorded small details ("the fresh king salmon was barbecued over alder and basted with a sauce consisting of 1/3 cup of butter, 2/3 cup of brown sugar, 2 tablespoons of white wine, and 2 tablespoons of lemon juice") and concluded, "One could eat himself to death, but the food is generally excellent, and the portions are small, so one can try a lot of different things."

The next year, in September 1994, Betty and Gordon had another happy time: the Moore family gained a fresh member when Ken married his longtime companion, Kristen Anderson. Like Ken, Kris worked at Signetics. Betty in particular had grown to love Kristen. "We got to know each other and went on trips together. I thought, 'I'll keep praying hard that this will work out.'" The wedding took place at Mountain Meadow, with an outdoor ceremony in the oak grove. "We set up a church-looking aisle leading to a big well, with a great urn and an iron top," reminisces Betty. "The aisle was spread with rose petals as they came down. The whole day was beautiful." Gordon said little, but took it all in stride.

CHAIRMAN EMERITUS

"Intel Inside"

Gordon remained deeply involved with Intel. He was now one of just eight people who had joined the firm in its first year, 1968. Others included Andy Grove, Jean Jones, Les Vadasz, and Ted Jenkins. At home Gordon and Betty were slowly adjusting to shifting realities. So too, at work, his relationship with Andy Grove was changing. Two episodes in 1994 illustrate the challenges he faced in balancing his position as chairman, rather than CEO, against his own strategic instincts.

The first was far from a do-or-die challenge, but, as Moore says, it made him gulp. While the industry recognized that PC models were defined by the microprocessors within them, consumers were less aware. Andy Grove advocated a radical departure in the form of an advertising campaign: "Intel Inside." The campaign would label every appropriate PC with a distinctive logo to show it was powered by Intel's microprocessors, in an attempt to shape consumer preferences.

As chairman Gordon Moore would weigh in on any major strategic financial decision: in his words, "whether we should do it or not." "Intel Inside," involving a multimillion-dollar consumer-directed advertising expenditure, was just such a decision. To Moore, whose exposure to consumer markets in the form of Microma watches had been less than happy, it looked like a massive risk. He restrained himself. "I gulped quite a bit when we took off to do it, but it worked out very well."

The second episode developed more slowly and was more serious. The result for Moore was much the same: despite his concern, he let the CEO decide. As Grove launched the "Intel Inside" campaign, the company's rivals were beginning to partner. IBM, Apple, and Motorola, desiring a viable alternative to Wintel, announced an alliance to develop the Power PC microprocessor. Meanwhile, Advanced Micro

An early iteration of the Intel Inside logo, 1991.
SOURCE: INTEL.

Devices—continuing to challenge Intel's sole-source position—was edging into the 486 business. Intel's response was to leap to the next level and herald the 486's successor: the Pentium.

As a noun rather than a number, the Pentium represented a definite break with tradition, having a distinct brand of its own: a moniker that could not be reused or confused. Under the hood, it had more than 3 million transistors and depended on technology that could craft features of just six-tenths of a millionth of a meter. For the same price as the 486, the Pentium delivered a hefty tenfold increase in computing power, while maintaining software compatibility.

Intel was now seven years into its sole-source monopoly, enjoying greater power and success than ever before. Anti-Intel sentiment was growing alongside the firm's success, with competitors and customers increasingly perceiving the company as arrogant and brutal. As far back as 1991, a securities analyst had seen Intel's biggest risk as "the tremendous animosity among its customers." Gordon's perception was more nuanced. On the one hand, Silicon Valley could be scrappy and feisty. He knew from the DRAM battles of the 1970s and early 1980s that Intel had to maintain its tough, disciplined culture. If anything broke Intel's stranglehold on the PC market, competitors would be unsparing. On the other hand, many relations were cordial and consensual. "We had our strong arguments with people like AMD. For the rest, the industry came together, looked at common problems, and tried to develop common solutions, like supporting a sales-tax rate increase so we could extend the freeways; like housing cost, education, and infrastructure problems. We could tackle those things area-wide through a big group of companies."

Intel did now stumble, badly. The first huge batches of the Pentium contained a "floating point error" that, occasionally, resulted in incorrect mathematical calculations. In a set of digits buried deep within the microprocessor, some of the necessary numbers were missing. Intel delivered hundreds of thousands of Pentium chips before its own engineers discovered the flaw in May 1994. Fixing the Pentium chips to come was easy: missing numbers could be added to the next, speedier, version that was in development. Less easy to fix were the Pentium chips that had already left the factory.

Grove would later joke that the *S* in his name (Andrew S. Grove) stood for "ship the shit." Shipping a chip on time rather than fully vetted took precedence for Grove. On occasion, Gordon had had to police the line. Now, again, Grove sought to brazen it out. He latched on to a technical argument: the flaw produced errors only in rare circumstances. In the life span of a computer, most users would experience no error. Because Grove had acknowledged the problem was legitimate enough to need fixing in all subsequent Pentiums, this argument was dubious at best.

Whatever its exact consequences, the flaw was not going to go away. A mathematics professor, Thomas Nicely, detected it that October, and—thanks to the novel convenience of e-mail and the emerging popularity of Internet forums—the news began to circulate. The technical press picked it up and forced Intel to acknowledge the problem. At Grove's direction, spokespeople were dismissive, emphasizing the rarity of errors and assuring customers of a fix in the next version. Both Moore and Grove's number two, Craig Barrett, disagreed with this solution. Moore recalls, "When I first heard of it, I thought, 'Oh, God, we've got to replace all those things.'"

CNN soon picked up the story, running an "unpleasant" piece that spread awareness more widely. Intel was painted as a stumbling Goliath. There was a deeper subtext: as PCs permeated modern life, could they be trusted? Intel, in response, reiterated Grove's original brush-off. The company, says Moore, "dug in its heels and said, 'This is ridiculous. It's not a real problem.'" As the press became increasingly negative, Barrett, who shared Gordon's instinct, suggested offering a replacement to anyone with a flawed Pentium. Grove resisted. He relented only when IBM abruptly, and without warning, suspended shipments of PCs containing Pentium microprocessors. Intel's stock price took a hit, and "all hell broke loose."

Grove reluctantly realized that "while it may not have been a real problem, the perception was something that we could not stand." Gordon supported Barrett's plan to replace faulty chips: "We announced a replacement program and defused the situation, which was getting sticky. It cost something like $450 million. That was when we discovered that we were a consumer products company, not an engineering company. We were in a different business." The scandal died off. Intel was powerful and flexible enough to recover. Once again, the Pentium became the microprocessor of choice in personal computing. Moore's instinct from first learning of the problem had been to replace all Pentiums, yet as an individual who hated confrontations he had been disinclined to use his power as chairman and intervene. He explains, "I didn't react strongly enough. Andy decided to take the engineer's approach: 'You'll never see it anyhow. Your computer crashes three times a week, and this problem is statistically very improbable.' I stood back and watched, which was a mistake."

Gordon had, in his time as CEO, allowed board members to go "pretty far down into the operating details" but had always retained the right to turn down "good advice." "Some suggestions we heeded, and some we didn't. Board members told us we were foolish to try to get into the digital watch business, and they were absolutely right! On the other hand, at least one of the directors thought we were foolish to get into the computer business when we started making microprocessors." The value of the board's "outside view" depended, says Moore, on how the company used it. The board could not be allowed to "second-guess" management's decision on preferred actions. As Moore adds, "I don't know any management that wants to entrust the board with real business."

Moore's Law and the Silicocene

The year 1995 was the thirtieth anniversary of Gordon's publication of Moore's Law, in *Electronics*. While few people realized it, the silicon transistor was by now the object most crafted by humans, exclusive of biological products such as grains and human cells or standard chemicals such as polyethylene molecules. Almost 70 quadrillion silicon transistors had so far been made, the overwhelming number as component "bricks" in microchips.

Seventy quadrillion is 70,000,000,000,000,000, or 70 million billion: 1 silicon transistor for every ant on earth, or 600,000 transistors for each human who has ever lived. If those humans made one transistor each, every hour of every day of their lives, the total would not reach this figure. During 1995 alone almost 20 quadrillion transistors were made, and a million could be had for a single dollar. As Gordon had long foreseen, silicon transistors were so cheap that, in microchips, they could be consumed in astonishing quantities. Paradoxically, the cost of the factories used to create those microchips had skyrocketed. It only required around $1 million for Moore to establish silicon manufacturing for Intel at the close of the 1960s; by 1995 the cost of a fresh manufacturing facility was approaching $2 billion.

Intel's new manufacturing plants were now among the most expensive industrial facilities ever created. A fab was more expensive than the largest automobile and aerospace factories, the biggest coal-fired electrical power plants, and the finest hospitals. Gordon explains, "It took me a long time to start saying billions instead of millions. The plants were getting significantly larger as well as each part of them getting more expensive. Production went from a labor-dominated cost to a huge capital investment. That was one of the things that drove us out of the memory business."

By the mid-1990s microprocessors, though minute, had become, in and of themselves, among the most complex machines ever manufactured. Whereas a Boeing 747 jumbo jet comprised 6 million parts, a single state-of-the-art microchip might boast 16 million transistors, formed into dense, intricate patterns and interconnections. Microchips were of unrivaled complexity compared with any other manufactured product.

Across 1994, as Intel ramped up production and delivery of its latest Pentium microprocessors, it began to enjoy the full fruits of Moore's Law and Gordon Moore's work. Revenues jumped to $11.5 billion, with $2.3 billion in profit. "Wintel" PC makers now had a market share close to 90 percent and readily adopted Intel's Pentium; the RISC alliance, conversely, enjoyed only a 3 percent market share. Between 1994 and 1997 the PC industry would double in size. Annual production rose to 80 million PCs. All was set fair—things seemed as close to paradise as may be found on planet Earth. Intel's microprocessors went into four-fifths of PCs, and its revenue climbed to $25 billion, with profit just shy of $7 billion. Intel was the largest semiconductor company on earth by revenue and would remain so to the present day.

Over the decades Gordon consistently communicated his Moore's Law vision through written and spoken words, and his predictions about the future of microchips were consistently realized. In 1995 he spoke of a related but wider future: that of the silicon age, or what one might call the "silicocene." The microchip had matured, and he wanted to communicate a fresh message. The very nature of change in the silicon world was itself about to change. His opening quip in his talk, in Santa Clara, was aimed at insiders: he ruefully reported to them that "Moore's Law" as a label had come to refer to "almost anything related to the semiconductor industry that when plotted on semi-log paper approximates a straight line," that is, anything that doubles and redoubles in a regular fashion. In 1965, he explained, he had been "trying to get across the idea this was a technology that had a future, and would contribute quite a bit in the long run."

Ten years later, in 1975, he had predicted that the rate of doubling would slow to once every two years. Since then the doublings had continued like clockwork. Several different factors had contributed to the overall reality. The "die" size of microchips had increased. Developments in tools, materials, and associated processes allowed the manufacture of features 30 percent smaller, roughly every three years. The smallest features had fallen to four-tenths of a millionth of a meter (400 billionths of a meter). "We could never see more than two or three generations ahead. It always seemed something was going to be a major barrier. Amazingly, Intel's engineers would get by the 'something.' Three or four times, I was really concerned that we were close to the end. We'd talk about alternative strategies, but it's always more effective to spend your energy in solving the obvious, immediate problems."

Gordon knew that "staying on this line gets increasingly difficult." Looming technical challenges concerned him less than the growing cost of finding their solution. Prices for tools and fabs continued to rise. "When Intel was founded in 1968," Moore explained, "a piece of equipment cost about $12,000. You could buy a bank of diffusion furnaces, an evaporator, a lithography exposure machine or whatever for about that amount." By 1997, he suggested, it would be $12 million for a single production tool.

Moore previously perceived a huge margin between the investments required to drive down the cost of transistors and devices and the resulting increase in revenues. Now that gap was narrowing: "Costs are rising exponentially and revenues cannot grow at a commensurate rate. We can no longer make up for increasing costs by improving yields, and by improved equipment utilization. There is little room left in manufacturing efficiency. This is at least as big a problem as the technological challenge of decreasing size."

Moore's Law, as Moore himself understood it, was an economic as well as a technological and social construct. The microchip industry and its suppliers would enjoy a sizable financial return from their investments only if the electronics market expanded at a sufficiently rapid rate. Should the cost of advancing the technology begin to outpace the expansion of markets and revenues, Moore's Law would end. Financial exigencies would trump technological possibilities. "We will not go as fast we would like," Moore explained, "because we cannot afford it."

As far back as 1991, in an interview with *Computer* magazine, he had declared, "The cost component will dominate the rate of evolution." He believed two economic choices could alter the tempo of Moore's Law: the time between generations might be extended, so capital investments were utilized over a longer period, or the price of products would be kept from falling at their accustomed rates. A more complex microchip might be offered without a decrease in price or possibly even with an increase, thereby dampening sales. "Either way," Moore told *Computer,* "the rate of progress slows. We have come to expect the continuation of trends that have brought million-fold cost decreases. The trend will continue, but the rate will slow. At forty, the fires of youth are subsiding. The realities of middle age can no longer be avoided."

On a personal level, Gordon Moore was facing these same realities. The fires of youth had subsided; the realities of middle age had set in. By the late 1990s he was starting to encounter a new and more unsettling transition: the approach of old age.

Exeunt Gordon

In January 1997 Gordon turned sixty-eight. He had served as chairman of Intel for a decade. He had remained instrumental in guiding the strategies that came out of the mid-1980s reinvention. With Andy Grove he had made the firm into the largest, most successful semiconductor company in history. Its sole-source position lay at the center of an expanding PC industry. As the cost of computing steadily plummeted, Intel enjoyed more of the reward accruing from Moore's Law than did any of its competitors.

Business, in Moore's eyes, was "looking pretty bright." The dramatic growth of the World Wide Web, and the entrepreneurial and speculative frenzy it generated, was stimulating even greater demand for PCs and supporting silicon devices. The industry was on the brink of another wave of expansion. Intel, following Moore's Law, would use its microprocessor monopoly to surf this wave. Another change was also coming: Andy Grove, now sixty-one and with a diagnosis of prostate cancer, was ready to retire from the demands of the CEO's position. "It looked like a good time for us both to make a transition," says Gordon. "I'd rather leave on the upswing."

Age and changing interests played into Gordon's next step. For forty years he had been working without significant relief, in the companies he had cofounded, within a brutally competitive field marked by unceasing change. Even as chairman he had arrived at his cubicle every day, ready for work. His dual strategy (sole source and Moore's Law) pointed to tremendous future growth, but Gordon himself was at last ready to move to the role of occasional adviser. He no longer wanted to police Intel's strategies. He and Grove agreed that Craig Barrett should be the next CEO. Barrett, whose respect for Moore's strategic abilities is immense, locates himself "more in the Grove format than the Moore format. I'm a decision maker. You don't have to read my face. I'm much more outwardly active than Gordon."

Craig Barrett, 1990s.

SOURCE: INTEL.

Moore decided to take the title of chairman emeritus, indicative of his special status at Intel. Final authority would reside with Barrett and Grove, but Moore would still be a voting board member, playing a unique role, not least as Intel's largest individual shareholder. His attendance continued at an un-board-like three days a week. As Barrett and tens of thousands of Intel employees rode the Internet boom, continuing Gordon's now traditional strategies, Intel released a flurry of novel microprocessors, all single sourced. The Pentium II, III, and 4 followed the original Pentium, dominating desktop PCs. More power-sipping versions were made for use in laptops. Barrett even expanded the lineup to include low-cost variations to meet the expanding market for cheaper, less-powerful PCs.

The World Wide Web also ran on Intel's microprocessors. The Internet depends on networks of "servers": PC-like computers, but with very powerful microprocessors and large amounts of memory for websites, databases, e-mail, printing, storage, and applications. Intel created devices specially tuned for servers but software compatible with any Wintel computer. As with desktops and laptops, the firm won a near-complete monopoly in server microprocessors—again, something that has endured to the present day.

In all these offerings, Intel hewed to the underlying Moore's Law dynamic on which everything depended. Barrett spent still vaster sums on R&D and on fabs and equipment, embracing the strategy that Gordon had started and Grove continued. Between 1997 and 2001, annual R&D investment climbed to $3.8 billion, to reengineer every aspect of the chemical printing technology to create transistors of exquisitely small size and microprocessors of overwhelming complexity, each transistor roughly the size of a biological virus, with each microprocessor containing more than 40 million transistors. The costs of the highly automated and ultraclean factories, the latest iteration of Gordon's great cost-reduction machine, also continued to climb. Barrett approved capital investments rising to $7.3 billion in 2001. The strategy was that of Gordon as CEO in the mid-1970s, but the scale had changed by orders of magnitude.

In 2001 Gordon Moore reached the mandatory retirement age of seventy-two and, strictly in accordance with the rules he had set, stepped down from membership of the board. "You can end up with collections of fuddy-duddies," he explains. "It doesn't give a good image to the company." (His old friend Arthur Rock had already retired from the board in 1999.) It proved a difficult transition. "I know he didn't like leaving," says son Ken. "All of a sudden you are relegated to the house. It's terrible, but that's what happens." Asked later whether there was "life outside Intel," Gordon's wry response was "Amazingly little."

His retirement coincided with the bursting of the speculative dot-com bubble. With its unique franchise in PCs and servers, Intel's revenues, profits, and stock prices had been swept up in the fervor, as revenue and profits soared—profits reaching $10.5 billion in 2000. Intel's stock price tripled. When the bubble burst, revenue slid back to 1997 levels. Barrett refused to slash Intel's R&D and capital investments. Momentum made his strategy hard to reverse, and—more important—he had fully embraced Gordon's lesson of spending through, and out of, a downturn. Profit for 2001 shrunk to just over $1 billion.

By 2004 revenues had climbed back to their bubble peak of $34 billion, and profits reached $7 billion. As the world and US economies recovered from the collapse of the Internet bubble and the economic fallout from the terror attacks of September 11, 2001, Barrett's investments paid off. Gordon still attended board meetings: Grove, after all, remained chairman, and both men were comfortable in continuing their thirty-year partnership, exchanging thoughts and opinions about what was best for Intel. Gordon came to his office at Intel around a day a week. Barrett had now been CEO for seven years, through boom and bust.

Gordon's further step away coincided with Andy Grove's decision to retire from Intel's board. Grove, having survived his bout with prostate cancer, now faced a diagnosis of Parkinson's disease. He decided to pass the chairman's baton to Barrett, wanting to focus on his own health and that of the US health-care system, full-time. Barrett, for his part, was ready to hand the CEO responsibility over to Paul Otellini, his chief operating officer.

While Grove had been chairman of Intel, he was happy to have Gordon in the board meetings as an adviser, even though Gordon was legally retired from the board. With Grove's departure, and an influx of younger men, it was the end of an era. Moore's visits became less frequent: a day a month. He was no longer the strategic visionary for Intel. While he retained an interest in the technology, it was hard to keep up. "I had all the monthly progress reports e-mailed to me, but I didn't go out of my way to keep informed in detail." He attended board meetings, but his position was becoming unclear. Whether he would have left Intel entirely on his own is impossible to say. Instead, the decision was made for him.

Intel's new leadership grew uncomfortable at making certain business decisions in Gordon's presence. In 2006, during a particularly sensitive discussion, Otellini and Barrett asked him to leave the room. "It was, 'This is beyond what your ears need to hear,'" recalls Betty, "'so therefore, adios.'" Gordon left quietly, but deeply hurt. "He told me he was not going to go to board meetings any longer," says Betty. His relationship with Intel was never the same again.

Jean Jones had long since retired. Gordon's space at the company—a small square room, defined by cubicle partitions and a window wall overlooking the main entrance—remained as he left it. Folders lay unopened on the desk. An outdated IBM PC sat next to a Busicom calculator powered by Intel's first 4004 microprocessor, even as Intel itself announced its newest "Core Duo Microprocessor" built from more than a quarter of a billion transistor "bricks." Silicon "tombstone" wafers, commemorating long-ago triumphs, hung like hunting trophies, along with poster-size schematics of Intel's microprocessor designs. Gordon even left in situ his National Medal of Technology and other prestigious awards. Yet this was not a shrine to Intel's cofounder, merely a dusty reminder of a different epoch.

In late 2007 Paul Otellini made the decision to pack up these modest personal effects and hand them back to Gordon. Moore's time was truly over. Intel had been central to most of his working life. Now the center was gone. Some staff continued to remember him fondly, but he had created no cult of personality to perpetuate his name. Instead, Moore's Law, the strategy through which Intel continued to enjoy massive successes, had become his enduring testament.

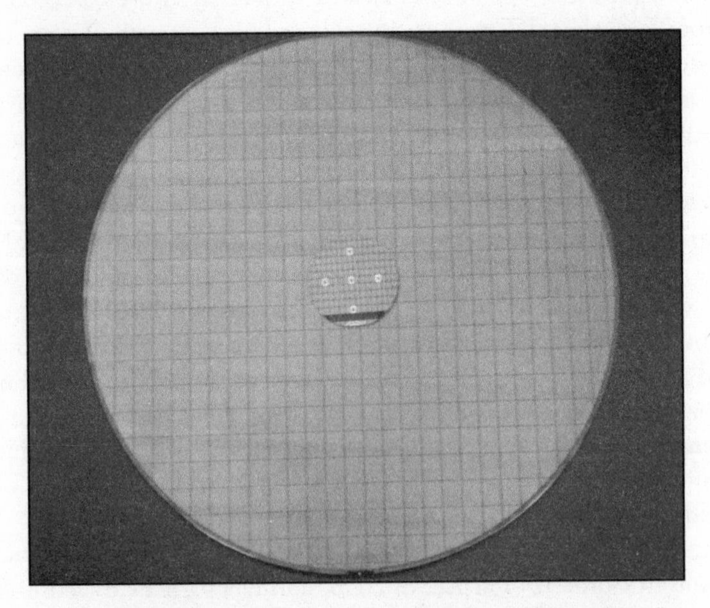

Picturing Moore's Law. A two-inch wafer printed with 4004 microprocessors sits atop a twelve-inch wafer printed with Core 2 Duo microprocessors. The two-inch wafer holds two hundred 4004s, each built of 2,300 transistors. The twelve-inch wafer holds more than five hundred Core 2 Duos, each built of 290 million transistors.

SOURCE: INTEL.

PUBLIC, PRIVATE, PHILANTHROPIC

California to Hawaii

As Gordon's role at Intel diminished, other options and challenges came to the fore. The silicon transistor had taken him from being merely very rich to being, at his apogee, one of the twenty richest people in the world. The challenge of how best to handle such wealth was not easily resolved.

What drove further changes was Betty's arthritis. Betty had lived on the San Francisco Peninsula for almost seventy years: she had been born and grew up and studied there, met Gordon there, and raised her family there. Now, she could no longer be comfortable at home in Mountain Meadow. The cold and mists might resemble those of Gordon's Pescadero, but for Betty they proved too much. "The winter of 1997–1998 was my worst winter ever. I couldn't get in and out of the car to go in to town to shop." Hawaii, where she had enjoyed many vacations, provided the obvious solution. The Kona coast of the Big Island offered a dry, warm climate. Betty made plans to move. She need never experience winter again.

According to Ken, Gordon had "no interest in Hawaii, absolutely none." To be sure, he had been there on family trips and even attended Fairchild sales meetings there. To live halfway across the Pacific Ocean was something else. The San Francisco Peninsula was home in a deep sense: the place not only of his roots, but also of Silicon Valley and the activities he most enjoyed. However, as Ken explains, "Mom wanted a house in Hawaii, and she was an irresistible force." Gordon gave way. Betty found a suitable furnished residence: Lava House. They bought it for a winter refuge, late in 1998. "That was when our pattern of residence to-and-fro began," says Gordon. Mountain Meadow remained their—or at least Gordon's—"real" home.

Betty had chosen a relatively modest dwelling within a larger planned development with a golf course, a clubhouse, and all the usual Hawaiian amenities. With neighbors close at hand on a flat lava-flow formation close to the sea, the house offered a complete contrast to Gordon and Betty's previous lives in Woodside and Los Altos Hills. It was a world away from Gordon's professional milieu, not just in its setting but also in Hawaii's slow, erratic, and casual pace.

By the late 1990s Gordon had developed several active commitments in California aside from Intel and had spent a decade on the board of Transamerica Corporation. In addition to enjoying "good views and great lunches" at Transamerica Pyramid in San Francisco, his involvement helped broaden his business experience, giving him exposure to "a different bunch of problems." And at Transamerica he mixed on familiar terms with "an interesting collection of people," including Condoleezza Rice, the Stanford provost. He also served on the board of Varian Associates, an early jewel in the crown of Silicon Valley. Moore had only reluctantly agreed to be a director of Varian, which was not in good shape. His quiet and astute inputs did little to change how it operated: "They weren't doing well financially and were overpaying management."

Further afield, he had joined Caltech's board in 1983 and served as its chairman from 1994 to 2001, then remained as a life trustee. Discussions ranged from finance to construction and issues from the annual $1.5 billion Jet Propulsion Laboratory (a federally financed R&D center) to increasing the institute's racial and ethnic diversity. Moore was also instrumental in inviting Arthur Rock and Intel's Bill Davidow to become Caltech trustees and had heavy involvement in appointing a Caltech president in 1997. Another long-term commitment was to Gilead Sciences, a biotech company on whose board he served with such luminaries as Donald Rumsfeld, George Shultz, and Paul Berg. "In semiconductors we never see a price go up; in the drug business, they never see one go down."

Unlike Intel, Gilead struggled for ten years before it became profitable. Moore enjoyed the "new scientific stuff" and only reluctantly retired in 2013, as he moved toward the age of eighty-five, by which time Gilead was a multibillion-dollar behemoth. Other obligations included membership on the President's Council of Advisors on Science and Technology, but Moore saw clear-eyed that his own effectiveness, and that of the council, was limited. "These committees have guys who used to be important, but who don't really know what the situation is. I find myself in that category often. Many reports never have any impact."

Meanwhile, in Hawaii, Betty carved out a fresh life for herself. She joined an informal women-only group, the "Dragon Ladies" club. "People seem to be more open," she noticed. Thriftily, she and Gordon relied on a very part-time helper in California to sift, triage, and forward their mail from Woodside, a "care package" they received every Friday. Ramping up their philanthropic efforts had led to an increase in mail: "enough to drive us crazy," says Betty. "That's the reason I like hiding in Hawaii. Then they found us over here, too. We receive more requests than ever."

For years the couple had shared a tacit "leave Intel outside the door when you get home" understanding. In Hawaii, with no conveniently located technology, science, or business to attract him, Gordon was home far more, yet he was slow to settle. Still focused on California, he described himself to colleagues there as "gone a good portion of the year." He continued to prefer his own company but lacked the focused interaction, intellectual stimulation, and steady rhythm of his working life. There was both nothing to leave at the door and nothing outside of it. According to Betty:

> Soon after we bought our first place, we were at the local refuse dump. One man made an overture to Gordon about golf. I could tell he was local, because he had palm fronds to drop off in the back of his sport utility vehicle. Gordon turned him off. I said, "That man was reaching out to you; he wanted to be friends. You put him down. You didn't even ask him to come over, to see how he played." Gordon was looking at his watch, thinking, "Where am I supposed to be in the next half hour?" I mentioned this to Steven, who said, "He really has to stop running."

This was easier said than done. With Moore's Law, Gordon had set himself and millions of others to move at a punishing pace. For decades he cranked the treadmill and was on it himself. Intel was renowned for being always in a hurry; Gordon's personal sense of purpose was predicated on the need to stay ahead and in control. To stop "running" entirely was too much of a shift and might feel perilously like ceasing to exist. He would also have to confront things he had spent his life avoiding.

Soon after the couple's partial move to Hawaii, Gordon's last significant tie to his ancestral home in Pescadero was suddenly and brutally severed. In early August 1999 his older brother, Walt Jr., went missing one Tuesday. Later that night his son discovered Walt lying dead underneath his all-terrain vehicle (ATV) at the bottom of a remote ravine on his ranch. Walt lived alone as a widower. In 1946 longtime Pescadero resident Joseph Cabral had bought the 1,513-acre Rancho Pescadero, once owned by Walt and Gordon's great-great-uncle Thomas Moore. By the time of the sale, "dozens of people who were heirs of the Moores were on the title." Cabral cleaned things up and took possession. Within two years Walt married Cabral's daughter Darlene and became an employee of Cabral's.

Unlike his forebears, and brothers, Walt did not own a business or work independently in the manner of the old middle class. His wife, Darlene, had been his mainstay, but in 1984 she died of cancer. Gordon, running hard at Intel, left his brother to rely on his immediate family for support. The two never discussed their feelings, and while they appreciated each other's company, they had always been self-sufficient. "When Walt was ill over on the coast, or when Darlene was, we could have tried to go over and see them. We didn't," says Betty briskly. "It was, 'Well, they can take care of themselves.' You couldn't have a great conversation with Walt; he was not going to be forthcoming. Fran was the outgoing one; the two older brothers, Walt and Gordon, were very secretive."

The eventual death of Joseph Cabral in the mid-1990s meant that Walt inherited the ranch and with it considerable responsibilities for which he was ill-prepared. The good times were past. Betty is sure his death was no accident: "Walt took his own life. He rolled that thing on purpose." Gordon concedes, "My brother may have lost some of his enthusiasm for living. He was very competent about everything he did and, I thought, pretty careful. For him to kill himself on one of those ATVs was really a surprise." The family buried Walt in the Skylawn Memorial Park, alongside Gordon's parents.

Betty and Gordon hardly discussed their feelings about the event. From an early age, Gordon had tended to retreat into what attachment theorist Robert Karen describes as a protective state of indifference. Precociously independent as a child, he had long employed "an emotional cutoff" to disguise hurt and anger even from himself. It was a pattern of avoidant response that had first become evident to Betty many years before, in the shock and pain of her first miscarriage.

Among far happier moments was the September 1999 birth of Ken and Kristen's second son, David, complementing the arrival of his older brother, Alexander, in June 1997, and Gordon and Betty's golden wedding anniversary in September 2000. This last saw a celebratory dinner and a pear-drop diamond brooch on a platinum chain, in contrast to the initial missed engagement ring. That month also Steve married Kathleen Justice, a San Francisco attorney. Gordon saw her as "a go-getter, a very bright girl." Betty felt that "she fit in so beautifully." Gordon reflects, "Getting Steve married was very nice. We ended up with two marvelous daughters-in-law."

In August 2001, two years after the death of his older brother, Gordon lost his younger brother to cancer. Whereas the deaths of Walter Harold and Walt Jr. had been sudden and traumatic, this time Gordon was able to spend time reestablishing early bonds with Fran; he was "choked up" to lose him at the end. Gordon himself remained fit and trim. (At Intel, conscious of the need to keep a healthy weight, he had instructed his secretary, Jean Jones, to fetch only yogurt for dessert.) He was a little past seventy, but far from ready for retreat.

Exponential Assets

Gordon had always taken great care over his personal finances. He tracked income, expenses, and investments meticulously. In an era less familiar with massive wealth accruing to innovators, *Forbes* magazine in 1987 had already placed him at number 155 on its list of the 400 wealthiest Americans, with more than $450 million. Around that time Gordon decided—with an eye to both tax and estate planning—to put his annual giving on a fresh footing. He and Betty created the Moore Family Foundation, and he transferred some $20 million to its endowment. His wealth and position also tugged him into other philanthropic orbits of the mid-1980s, such as the high-profile Search for Extra Terrestrial Intelligence through the SETI Institute, a project he viewed as "intriguing, relatively inexpensive, and with potentially a very high payoff."

The family foundation focused on themes closer to home, with environmental preservation and conservation, especially in the Bay Area, a special concern. "At the local level, you could really see the difference you could make," Gordon explains. In the early 1950s the Valley of Heart's Delight consisted of farms and orchards, with fewer than eight hundred manufacturing positions, including seasonal jobs in the canneries. Three decades later, "Every place I liked was changing. All the naturalness was disappearing." He became passionate about the quality of life in the Bay Area. "I sure wouldn't want to see the coastline get like Malibu," he told an interviewer. On a wider scale, "The places that we used to go to for fishing trips were wild. Now you go and you find high-rise hotels and golf courses. Resorts and golf courses are nice, but they shouldn't be everyplace. Seeing all of these things disappearing really bothered me. Trying to save some of what was left became attractive for me."

Gordon supported the Peninsula Open Spaces Trust and in 1987 sent a $100 check to Conservation International (CI), a spinout from the Nature Conservancy. Peter Seligmann, its founder and chair, was quick to scent opportunity, following up with a note and a call. "Moore asked if I contacted everybody who sent in a hundred bucks. I said, 'No, but they're not all Gordon Moore.'" Whereas the Nature Conservancy focused on US wilderness, habitat, and wildlife, CI wanted to concentrate on preserving biodiversity, the rich array of plants and animals found in particular regions around the globe. "There was a fight about international versus domestic conservation," explains Gordon. "The people who wanted to do things internationally ended up leaving in a big uproar. It sounded a lot like Shockley and Fairchild."

CI went on to partner with Lewis Coleman, a prominent figure in San Francisco finance and a top executive with Bank of America, in one of the first "debt-for-nature" swaps. Coleman arranged to sell $650,000 of Bolivia's national debt (held by Bank of America), at the vastly discounted price of $100,000, to CI. In turn, CI exchanged the debt for an agreement from Bolivia to protect more than five hundred square miles of sensitive biodiversity habitat and provide Bolivian conservation projects with funds. Through this ingenious loop, debt forgiveness could be converted into protecting biodiversity "hot spots." Gordon liked this innovative, direct use of finance for conservation. Through the intervention of John Young, CEO of Hewlett-Packard, he agreed to join CI's small advisory council and subsequently its board. With its combination of science, direct financial action, and conservation, CI's approach appealed to Moore: it was a nonconfrontational organization that "took a science-based approach to what was worth saving and what the priorities ought to be," and it showed on a larger scale "how it's possible to achieve significant conservation results."

Gordon's commercial activities and technologies had been leading agents in the disappearance of "naturalness" from the Valley of Heart's Delight. His own attitude was congruent with a 1998 description of Silicon Valley as "a business phenomenon that includes more than 2,600 high-tech companies, employing 300,000 people. It's also a social phenomenon that includes drugs, industrial spies, instant millionaires, crowded freeways, and a population that has burdened the infrastructure." In his professional life Gordon was dedicated to unsentimental, explicit revolution and to fast-paced industrial and economic development. He was far from agreeing with *San Francisco* magazine that, "The underside of Intel is a nation of landfills, bulging with obsolescent computers, leaking a whole soup of chemicals." For Gordon, giving time and money to conservation of biodiversity and open space helped to balance the equation and exert a counterpressure.

Separately, Caltech became a prime focus for Gordon and Betty's public, personal philanthropy. From an initial gift of $1,000, their generosity climbed upward. As Intel became an obvious success, Caltech sought the endowment of a chair, and in 1977 the Gordon and Betty Moore Professorship of Computer Science was launched (Gordon being a self-professed "sucker" for novel computer architectures). Eager to continue the momentum, Caltech invited Gordon to its board of trustees, and with a scaling up of Gordon's philanthropic investments in the late 1980s, Caltech and the University of California, Berkeley—the institutions that confirmed his calling as a chemist and gave him an entrée into the world of research—became major beneficiaries.

The Berkeley trajectory was similar to, but more modest, than that at Caltech—reflecting in part how Gordon's undergraduate two years there were less defining of his and Betty's world. Even so, a first gift of $20 in 1973 had translated into $300,000 for environmental research within a decade, and morphed into gifts in the millions by the 1990s.

Gordon had put $245,000 of his own money into the launch of Intel. He held on to the bulk of his investment, stock that, over the years, followed its own exponential trajectory. In keeping with the example set by his storekeeper grandfather, and his own parents' modest lifestyle, he approached personal finance not simply with meticulous care, but also with a strong sense of responsibility. He wanted his decisions to result in an improvement of his balance sheet, and have his savings go into sound investments. Retaining Intel stock also seemed appropriate, given that Gordon was more invested than anyone, literally and figuratively, in making the company work.

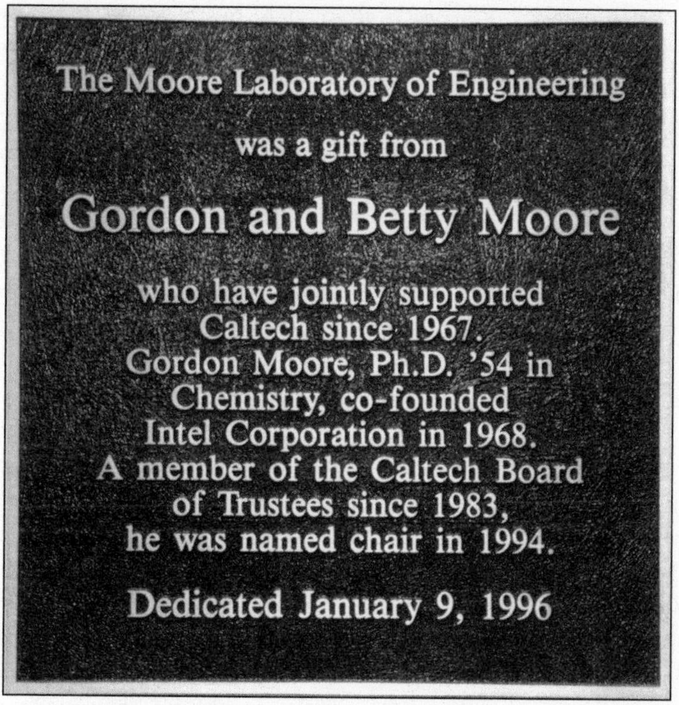

Dedication plaque, Moore Laboratory.

SOURCE: COURTESY OF THE ARCHIVES, CALIFORNIA INSTITUTE OF TECHNOLOGY.

Gordon rose to ever-higher positions on the annual *Forbes* 400 list of the richest Americans. By 1991 he was a billionaire. As the Internet bubble expanded, and as Intel's microprocessors dominated in the servers undergirding the World Wide Web and the PCs connecting to it, sales and stock continued to climb. In 1999 Gordon saw his wealth grow to $15 billion. In early 2000 it ballooned to $26 billion, an impressive return on investment—of time, money, and knowledge—making him America's fifth wealthiest man and the richest person in California. Since he and Betty had no plans to change their lifestyle, and as both personal morality and taxation rules deterred them from transferring major wealth to their sons, Gordon decided that he must start making much larger philanthropic investments for the common good.

Board experience at Caltech and CI gave him insight, but a more immediate inspiration was in Silicon Valley itself, where David Packard, cofounder of Hewlett-Packard, had created a foundation with his wife, Lucile, in the 1960s. It supported conservation, education, and research, causes close to Gordon's heart. William Hewlett had founded a foundation with similar aims. Together, they provided a model for how pioneering Silicon

Valley entrepreneurs could go beyond mere business success to establish philanthropic enterprises of lasting value.

In January 1999 Gordon turned seventy. For at least five years, he had contemplated—and procrastinated over—the idea of a much larger philanthropic venture. Now, more than ever, he wanted to "take advantage" of his "ridiculous level" of wealth. Betty pushed him: "She'd say, 'Let's see it,'" says Ken, also in favor of the idea. "We asked Dad, 'Don't you want to get a chance to see it working?'" Giving away a large portion of personal wealth was a big decision. "When you have resources of a major scale, do you do major-scale philanthropy?" asks Ken. "It's not a given."

Serious Giving

In the world of the silicon microchip, Moore was a master strategist and risk taker. Even so, he was not especially a self-starter. As an accidental entrepreneur, he said he had to "fall into an opportunity or be pushed." He had turned to Fairchild Semiconductor only as a last resort after losing faith in Bill Shockley and as one of a group of equal partners. It was Bob Noyce who provided the catalyst to leave Fairchild, where Moore was settled, and start Intel. Now, in preparing to launch his own foundation, Gordon was at a loss. He wanted to see things happen, but could not identify a partner—an Andy Grove—with the skills necessary to put his ideas into action. "Giving away money effectively isn't easy," he explains. "I had to find somebody I was happy with to run it. There was no clear way to make a first move. I wasn't willing to take it on myself, so I let the idea languish."

Gordon now had time available, but deeper conflicts were at work. For decades, at Intel and at home, he had lived in a world of high-level strategy and analysis. His deep, long-term partnerships with Andy Grove and with Betty were highly effective and had allowed him to remain in strategic control while indulging his desire to delegate operational management and its attendant conflicts. His preferred mode was to create situations that could exploit his insights and leave the actual doing to others. For the family foundation, he replicated this model with son Steve.

For major philanthropy—a foundation dealing with billions of dollars, not millions—Gordon needed a partner experienced in complex financial transactions. The foundation, as a public and prudent entity, would be obliged to sell all the Intel stock he intended to donate; to do this without upsetting the market would be no small challenge. Gordon desired an operational partner with the ability not only to "set up something that could handle this large amount of money," but also to interpret his desires, turn them into a solid strategy, and lead a major philanthropic organization, a multitasking manager equivalent or even superior to Grove. The diverse and high-level skill sets required were rarely to be found in a single individual, as would become painfully clear.

With the dot-com boom of the late 1990s, the urging of others for Gordon to think about making big gifts became more intense. One suitor was CI, which believed—as did Gordon—that "thinking big" could engender important change. If Moore's Law symbolized the "never-ending human drive to create newer and better things," could not the same apply to philanthropy? In the mid- to late 1990s, Lewis Coleman began to woo Gordon for a major donation to CI. The two men were both old-stock Californians and took part in fishing expeditions to remote spots arranged by CI's Seligmann. Gordon opened up to Coleman, revealing his desire to create a large foundation and explaining that his wealth was not cash, but Intel stock.

Caltech, too, was gearing up for a major fund-raising campaign. Gordon, as chairman of its trustees, was central to the discussions. Digging into the financial realities, he saw that to continue to excel at cutting-edge research, the school would need a much larger endowment. "It looked to me like they needed another billion and a half." Caltech expected its board chair to make a significant lead gift. Moore suggested a stunning $300 million "to get the campaign started," even though endowment went against his principles—though not against his faith in assessing impacts by their longevity: "I don't like endowments (I guess the idea is that I can invest the money better than they can!), but in this case it seemed the most important thing to do. Endowment comes closest to my idea of ten-thousand-year goals. To keep Caltech playing its unique role was one of the most valuable things I could do."

A gift of Intel stock and its subsequent sale might well damage the share price. Gordon did not want to flood the market or signal that he had lost confidence in the company. Coleman remembers his being "as nervous about how the money was going to get handled as he was about how grants would be made." As they talked further, Gordon became convinced he had found his partner. Coleman was involved in conservation issues and committed to science-based efforts, but he also saw the importance of finance, business perspectives, analysis, and action. Because Gordon believed—optimistically—that it would be easier to bring a financial expert up to speed in philanthropy than vice versa, Lew Coleman's solid reputation in the financial community was what was needed. Fortuitously, Coleman was looking for a career change: "We had some standing-in-the-fishing-stream conversations. He found out that I was thinking about leaving banking. I had him up for lunch in my office, to figure out how to pry a large gift out of him for CI. I asked him for a lot of money. He said, 'If you'd run my foundation, I'd probably figure out how to make that gift.'" At the close of 2000 Moore and Coleman shook hands on the deal. Coleman would set up the Gordon and Betty Moore Foundation, using a small initial gift, and when ready would then handle the transfer of billions of dollars' worth of Intel stock, half of all Gordon and Betty's holdings.

The realities of that transfer were quintessentially Gordon. Incredibly, all the shares were in the form of paper stock—stacks and stacks of certificates. "Gordon lived a very simple life," comments Ed Penhoet, Coleman's successor at the foundation. "All the shares he kept getting over the years, he put metaphorically under his pillow." For Coleman, the transfer was "not the easiest thing to handle." "We had to go get the certificates. I sent my chief financial officer down in a car with a banker, knowing that the banker's blanket bond would cover any automobile accidents. They met Gordon at the bank, walked in, and carried out 192 million shares. Intel had a rule that you could not issue more than 100,000 shares per certificate, so there was a big wad of paper. To put it into electronic form was one massive paper shuffle."

Coleman slowly and deftly sold the stock, creating a diversified asset base for the foundation. During this transfer the dot-com bubble burst, Intel's stock plunged, and Gordon and Betty's fortune deflated. Even so, the Gordon and Betty Moore Foundation leapfrogged into the ranks of the largest US charitable organizations. In 2005 *Forbes* named him the year's most charitable citizen, also ranking him fourth in a longer-term global list of "billion-dollar givers," after Bill Gates, Warren Buffett, and George Soros. Remarkably, he was the only one to have donated more to philanthropy than the total of his remaining net worth.

In 2001 Caltech announced the receipt from Moore of a transforming pledge—$600 million—to its campaign. Under some pressure, Gordon had decided to give not just shares of Intel, to the tune of $300 million over five years, but also an additional $300 million from his new foundation, over ten years. Gordon's gift was the largest ever (in current dollars) to an institution of higher learning. (A challenging example of largesse had been set for Gordon in 1986 by the then-record gift of Arnold Beckman, an earlier Caltech chemist turned board chair, who all those years ago at Shockley Semiconductor had been the unwitting catalyst for Moore's success.) Gordon and Betty also gave the single largest gift ever to an environmental organization: $261 million to CI, over ten years, to enable scientific research into, and action within, biodiversity hot spots.

Gordon had clear aspirations for his new foundation. He wanted it to tackle "big problems; do things with a big impact, long term; and make a difference way down the road." He believed targeted philanthropy could make a meaningful change in domains of significance. To Gordon, making good grants equaled "accomplishing important things." Just as he had revolutionized society through his unerring focus on transistors, microchips, and microprocessors, in his philanthropy he wanted to avoid spreading money "all over, willy-nilly. Governments have far more resources than individuals like me and Bill Gates, but they don't seem to be able to initiate much," he told an interviewer. "Private money can get something going. It can take a lot more chances. We're venture capitalists in these conservation efforts."

As Silicon Valley wealth began to translate into major charitable ambitions, Moore's thinking and his venture style were very much part of a fresh West Coast reality. California was outpacing the rest of the country, with a big leap in the number of charities created between 1999 and 2004. Fastest growing was Silicon Valley itself, with some three hundred additional foundations. What made Gordon's approach distinctive was his interest in method. He wanted to apply the "logic of the ledger" that had guided his personal and professional life. "We insisted on being able to measure the results," he explains. "Historically, that hasn't been part of philanthropy, but is a major part of engineering and business. If you can't measure it, how do you know if you're doing any good?" Measuring, analyzing, and retaining control were as important now as they had been to Moore as an adolescent boy, creating explosives in his parents' garage.

Bringing measurement into philanthropy was also to bring a business approach, not just to conduct philanthropy as a business, but also to use finance as a tool (as in debt-for-nature swaps and other large-scale conservation deals). Gordon's foundation would set its own agenda, quantify goals to be tracked, and then identify grantees to work with it. Control was key. Steve McCormick, who was then at the Nature Conservancy but later became the Moore Foundation's third president, recalls its early attitude: "'As a foundation, we have our own outcomes, our own strategies. We'll buy outcomes from you'—as if we were contractors." Moore's "businessman's view of getting economic incentives right" was recognized by Carver Mead, who was intrigued to "hear the echo" of an Intel style ("saying, 'You're going to do this,' then expecting it to get done, by applying the metrics and methods of for-profit businesses").

Perhaps most distinctive was Gordon's insistence that his philanthropy be science centered. He believed science was key to addressing fundamental social issues, such as conservation, and wanted to make a difference "not as a philanthropist but as a scientist," says Coleman. The environmental movement "could use more science and maybe a little less emotion," he felt, "a little more head and a little less heart."

At Fairchild and Intel, Moore had invested in research where he felt return would match risk. He strove to compete in only those areas of silicon electronics where opportunity existed to become a leading player in an area of potential significance. This emphasis passed directly from his professional life into the foundation. He would target particular areas of science to generate results on which to base effective solutions. Using a measurement-driven method, he would innovate just as he had done in silicon electronics. At Intel hardheaded research had been the bedrock of his success; so too, he hoped, would it be at his new foundation.

Making a Difference

If Gordon was clear about method, he was equally determined to focus on well-defined concerns: environmental conservation, scientific research and higher education, and to a lesser extent opportunities in the Bay Area itself. All were core interests in his life, if also deeply conventional. The Packard and Hewlett foundations supported just these areas; so had great philanthropists such as Carnegie, Rockefeller, and Ford. Yet Gordon did not simply want his foundation to "walk into something and add its two bits' worth."

An "early, easy hit" was a $9 million donation enabling the fledgling Public Library of Science to launch open-access biomedical journals. Such online journals were themselves a fruit of the electronic revolution. In other higher education–linked areas, it was less easy to make a difference. For example, Moore wanted to help productive scientists in academia who faced structural impediments when they desired to change fields. Coleman, tasked with investigating, talked to "half a dozen major university presidents," but "we heard nothing for two years. Hell, no. It just upsets their system. As I understood university structures more, I could see why."

The foundation also began to pour millions into two major initiatives: to preserve salmon in the North Pacific Rim and to protect the Amazon rain forest. Maintaining a healthy salmon ecosystem was a cause especially close to Gordon's heart, as a fisherman, but Coleman admitted, "The real reason we picked salmon is that we can count them." The second program, through CI, worked to maintain biodiversity in the Amazon Basin and also funded centers in the Andes, Brazil, Madagascar, and Melanesia. Within two years it had helped protect nearly 50 million acres of biologically importance. Asked what difference he hoped to make, Gordon responded, "We're making a better world for people to live in, for the long term. By long term I mean 10,000 years. If all we do is prevent the forest from getting cut down for twenty years, we've wasted our money as far as I'm concerned. The difference has to be permanent."

The foundation's principal beneficiary, unsurprisingly, was Caltech, but other academic institutions were eager to get into the game. Cambridge University in England had earlier and cleverly invited Gordon to join celebrations of the centennial of J. J. Thompson's discovery of the electron and, no slouch at fund-raising, soon elicited a grant to construct a Betty and Gordon Moore Science Library. Cambridge then invited the whole Moore family to return to England for a special luncheon with HRH Prince Philip, the university's chancellor (Betty, especially, enjoyed this "outstanding" trip). Amusingly, the follow-up—with Stephen Hawking and other Cambridge stars tasked with traveling to Gordon's home in Mountain Meadow to plant an apple sapling seeded from the legendary tree in Isaac Newton's childhood garden in Lincolnshire—had to be abandoned after US agricultural inspectors impounded the sapling.

One event that did go with a bang was in 2002, when Gordon received the Presidential Medal of Freedom, the United States' highest civilian honor. This award, given simultaneously to management guru Peter Drucker and to Nelson Mandela, Bill Cosby, Placido Domingo, and Nancy Reagan, came as a "complete surprise" and provided Gordon with the greatest pleasure of all his honors. His family was invited to the ceremony. "We had everybody's picture with the president and his wife." Gordon enjoyed other one-off events as his stature and significance were increasingly recognized, both in the US and abroad.

Gordon was scrupulous in his attendance at his new foundation's board meetings, but he declined involvement in its operations. He had long left such matters to others. While Microsoft's Bill Gates turned to philanthropy as a new career (becoming his foundation's principal strategist and spokesperson), Gordon, a generation older and living an ocean away from California, chose to remain at one remove. Lew Coleman persuaded him to locate the foundation not in expensive Silicon Valley (Sand Hill Road real estate was "going for eye-popping amounts," Coleman argued) but in San Francisco. In short measure, Gordon approved the lease and renovation of part of a building in San Francisco's Presidio. This longtime US Army base, released for civilian use, was where Walter Harold Moore had been discharged from wartime duty in 1919. The location enjoyed extraordinary views of the San Francisco Bay and was convenient for Coleman's own commute from Marin County, across the Golden Gate Bridge northward. However, from Gordon's home in Woodside to foundation headquarters was thirty traffic-congested miles.

Coleman and his recruits had free rein. Almost entirely without experience in nonprofits, they saw an opportunity to break with tradition and practice philanthropy in what they felt would be a more businesslike mode. They modeled their office with scrupulous attention to ecological features and sustainable materials, spending some $8.5 million. "Fashioning the digs with resolutely green building materials wasn't cheap," sniped local magazine *San Francisco* of the "fanciful office." It was a far cry from Gordon's thrifty, hands-on approach at Fairchild, when he outfitted its first lab with kitchen furniture at the cheapest possible price.

The foundation made attempts to elicit direction from Moore. For instance, Coleman's aide Sherry Bartolucci met with him in early 2001. "I

SELECTED LIST OF HONORS RECEIVED

Date	Institution	Award
1967	IEEE	Fellow
1974	National Electronics Conference	Best Paper
1975	California Institute of Technology	Distinguished Alumnus
1982	Financial magazine	CEO of the Year
1984	National Academy of Engineering	Elected to membership
1985	American Society for Metals	Medal for Advancement of Research
1985	University of Beijing	Honorary professor
1986	Michigan Technical University	Honorary doctorate
1988	National Academy of Engineering	Founders Medal
1990	American Academy of Arts and Sciences	Fellow
1990	President George H. W. Bush	National Medal of Technology
1992	Rensselaer Polytechnic Institute	Honorary doctorate
1993	American Association of Engineering Societies	John Fritz Medal
1996	Horatio Alger Association	Horatio Alger Award
1996	University of California	UC Berkeley Medal
1997	IEEE	Founders Medal
1999	Cambridge University	Honorary doctorate
2000	Princeton University	Honorary doctorate
2002	Franklin Institute	Bower Medal for Business Leadership
2002	President George W. Bush	Medal of Freedom
2002	Yale University	Honorary doctorate
2003	Royal Society of Engineering	Elected foreign fellow
2004	Society of Chemical Industry	Perkin Medal
2005	American Philosophical Society	Elected to membership
2008	IEEE	Medal of Honor
2009	National Inventors Hall of Fame	Inducted
2009	Sequoia High School Community Hall of Fame	Inducted
2009	Carnegie Corporation	Medal of Philanthropy
2013	University of California	Presidential Medal

Source: Gordon Moore.

said, 'Gordon, we need to understand what drives you. We want you to be proud of us. You represent certain things, and when people work at this foundation a hundred years from now . . . ' He said, 'I'm going to be dead.' 'I know, Gordon, but let's assume you could look back.' 'I'll be dead; I'm not going to look.' We really had to prompt him. At last he said, 'Okay, all right,' agreeing to an inconclusive session with staff."

Gordon, never comfortable with "touchy-feely" discussion, wanted his foundation to be serious and professional. He wanted it to reflect his desires in its methodology and areas of focus, but he did not have an answer for what exactly he wanted it to *do.* The great silicon revolutionary could provide no specifics on how to spend his money. Coleman and his executives floundered. "Is this what he wants?" they asked. Steve McCormick, a dozen years later and with his own presidency of the foundation behind him, would explain: "Gordon is a hard person to connect with. It's hard to get real feedback; it was difficult for me to know where he stood and where I stood. He never said, 'This is what I want.' In a lot of respects he's not decisive at all. He just lets things go."

Coleman says, "Gordon has been fairly explicit in what he doesn't want to do. He doesn't want to do political or policy things. That comes from the traditional Silicon Valley, iconoclastic independence—not a lot of faith in government. What Gordon has asked us to do is to try to make grants where results can be measured." With formal board meetings only once a quarter, Coleman called his boss "every other week, to see whether or not what we were doing made some sense to him."

Coleman correctly saw his job as guiding the foundation down the road envisaged by Moore, but that road was only vaguely sketched. With no road map, no relevant experience of his own, and plenty of egos to manage, he began to stumble. Uncertainty and ambiguity percolated, and the foundation suffered a considerable turnover of staff.

Cracks Appear

By 2002 the foundation had more than forty employees, including Ken Moore. In his semiconductor career Ken had been "running out of things to do. I didn't want my boss's job, and I'd been doing mine for too long." He became the foundation's research director and a member of the growing staff. Already a trustee, he was well aware of the foundation's teething troubles and chaotic state. Some in the nonprofit world interpreted its ambitions as tantamount to hubris, feeling that established practice was negated by an emphasis on research, the scientific targeting of approaches, and a focus on metrics. In January 2004 *San Francisco* magazine published an in-depth profile, alleging the foundation had "generated a rash of critics" in its short existence. "Few environmental groups, in the Bay Area and beyond, have struck gold at the Moore Foundation. Today, countless environmentalists are not only green with envy but pale with exasperation at trying to figure out what it takes to crack the Moore safe."

At the core of this situation lay a vacuum: Gordon's paucity of instruction to and engagement with the foundation and Lew Coleman's inability—borne of his lack of experience as an executive in the nonprofit arena—to fill the gap. Staff looked to Gordon for vision and guidance, but he was retreating. "I didn't want the foundation to take an awful lot of my time." Unsurprisingly, there were stumbles. "Because we didn't know what the hell we were doing," says Coleman, "we were quite willing to have unsolicited conversations with eager, assertive prospective grantees." Gordon's son Steve believes it was a wider failing: "As a family, we weren't able to define what we wanted to support. It took four or five years to focus in." Internal contention developed within the foundation, and feelings ran high about priorities for funding. "You get very passionate people attracted to foundation work, who believe that their passion is slightly more important than the institution they're working for," explains Coleman.

Staff received little training, and their confident approach (like that of "a high-powered group of graduate students") was at odds with their inability to deliver the goods. Expectant development officers, especially of Bay Area institutions, participated in meetings with as many as ten or fifteen foundation staff, each apparently able to grant money but in the end all without the needed authority. Sources reported "a new species of disdain," as calls to the foundation went unreturned: "Their attitude is, 'Why are you bothering us?'" reported *San Francisco.* Among would-be grantees frustration ran deep "because expectations run so high." Gordon admits, "We hired mostly people without previous foundation experience, who had a lot of on-the-job learning to do. I'm sure we upset several potential grantees, because we kept changing our methodology and made them jump through hoops to get an award from us."

Direction by indirection was Gordon's style, but as in the early days of Fairchild and Intel, his policy of allowing others to explore ideas meant they would sometimes drift off course. As then, the one essential ingredient for success was a talented, driven operator with a highly similar education and employment experience, and a skill at reading his mind. While Gordon was comfortable with Lew Coleman, the latter was not a scientist but a financial genius, and neither Coleman nor Moore was versed in the culture of nonprofit management. Steve McCormick, Coleman's successor-but-one, later recalled how early staff enthusiasms had far outrun Coleman's ability to manage, and how Moore was finally forced to pull the plug on a major initiative: "Lew had gotten way ahead. It caught Gordon off guard. We had an organization expecting a $300 million ten-year commitment, based on miscommunication. It compromised us."

Others expressed their concerns much earlier. "Ardent enviros," said *San Francisco,* were doubting that "the business titans at the helm of the foundation are willing to back nonprofit groups determined to force American corporations to clean up their act." Could the foundation's leaders, from corporate backgrounds, make the big changes they were promising, without conflicts of interest? No, said the magazine, suggesting the difficulties began with "the Henry Ford of Silicon Valley" (Moore himself) and singling out Coleman's recruitment of John Seidl for special opprobrium. As the first chief program officer for the environment, Seidl was a "corporate type"—formerly at Enron—who shared Coleman's view that "the path to enlightened environmentalism runs through corporate America."

The magazine alleged that the foundation was "locked into funding only groups that don't challenge the business values of Moore, Coleman and Seidl" and was "little more than a glorified fishing club for ex-business guys enjoying one last challenge in the twilight of their careers." If Gordon's foundation did fund only its own narrow business-based interests, it would fail to uphold public responsibilities, said *San Francisco.* "Foundations get a huge tax break because they are supposed to fill some of the social potholes that government leaves behind. Once Moore stashed his billions in a foundation, he and those hired to run the foundation became public stewards."

"Coleman and company" stood accused of "marching only to their own tune," an idea made plausible by the continuation of both Coleman and Moore on the board of Conservation International, Gordon's presence on the board of Caltech, and their closeness to other groups funded by the foundation. Meanwhile, there were fears that the foundation had "lost sight" of indigenous people and their rights. Rebecca Adamson, president of the First Nations Development Institute, met with Lew Coleman and told *San Francisco,* "There's a whole new priesthood in conservation that believes the way to preserve endangered lands is to set up protected areas for science and keep them laboratory-pristine. These science-based conservationists have less use for people." The magazine also quoted the Silicon Valley Toxics Coalition: "Moore is putting huge amounts of money into the rain forest. And that's fine. But that does nothing for the problems they've caused at home." Of course, for Gordon, measurability was everything. Scientists working in pristine wilderness were better able, as one analyst suggested, "to make empirical calls on the success or failure of conservation."

The foundation's insistence on measurement met resistance. "When you push other people to be accountable and they're used to just getting gifts, they don't like it," says Ed Penhoet, the respected University of California biochemist, biotechnology entrepreneur, and academic administrator who succeeded Coleman as foundation president. Gordon's "sharp focus, top quality, and accountability"—so successful at Intel, which did "one thing and did it supremely well"—did not easily translate, and was not quickly accepted, in this very different arena. According to McCormick, the foundation quickly developed a reputation for being "absolutely sure of its own opinions," arrogant and condescending. "It was very much a closed-door thing and very hard to deal with." The environmental program was run by "people who had no experience in the field, businesspeople. You couldn't talk to them; they were uninterested in different parts of the discussion. They would say, 'This is what we do. This is what you do. Can you perform or not?'"

Intel had faced accusations of arrogance, but it could afford to. Its focus on measurement fitted the aim of creating honed exactitude in its manufacturing technologies. Philanthropy was a messier, more human endeavor. In 2010 a *Stanford Social Innovation Review* article concluded that the "huge push toward measurability skews the work of nonprofits through its narrow strictures and highly directive requirements." Gordon, by thinking in broad strategic and quantitative terms, was learning that—in philanthropy at least—no good deed goes unpunished.

Ken Moore believed the combination of Lew Coleman and his father (the one, as he saw it, ill-equipped to manage the daily operations of a foundation, the other unwilling to step in) meant problems went unaddressed. Ken distrusted Coleman's "pure democracy," his "totally egalitarian" approach to organizational matters. Ken's brother, Steve, at one remove from the action, was more forgiving, seeing Coleman as "a really strong businessman" who "could have used a little more training on how to run a foundation." In 2004 things came to a head when Coleman transgressed long-established conventions on workplace norms of conduct. Senior foundation staff petitioned Gordon to intervene, and with great reluctance he did. Coleman left.

Ed Penhoet, already a senior staff member and at home in both business and the world of ideas, took on the role of president. He had long admired Gordon for the way he combined a quiet demeanor with powerful insight. "When Gordon says something, everybody listens," he explains. "He has an amazing ability to distill a complex subject into a very succinct summary." Penhoet refocused the foundation on its basic themes. In addition, he chose to launch an initiative in the much smaller but potentially very fruitful area of marine microbiology. Here, there was "a lot of science to be done." The foundation made grants to J. Craig Venter, the academic entrepreneur who in 2000—thanks to electronic instrumentation and the calculating power unleashed by digital computers—had completed the mapping of the human genome. It also supported Venter's Global Ocean Sampling Expedition with a $150 million investment over ten years, which was a resounding success.

Caltech became an indirect beneficiary of large pledges made by the foundation to create the world's most powerful optical telescope atop the extinct Mauna Kea volcano, close to Gordon and Betty's home in Hawaii. This "Thirty Meter Telescope" promised to address questions about the nature of the cosmos that had long intrigued Moore. "Every time we get a bigger telescope, we look further back in time and discover new things," he explains. "You get back as close to the big bang as you can." The foundation pledged $50 million to its design and, later, $200 million toward construction. This was in line with Gordon's wishes for the foundation: to act as the early-stage investor in a major science initiative having measurable results, with the hope to attract larger sums from other investors in due course. The telescope, with its promise of enabling astronomers to look back another billion years into the past of the universe, is almost literally in Gordon's backyard. The groundbreaking, following years of preparation, took place in November 2014.

More slowly, Betty has added her imprint to the Gordon and Betty Moore Foundation. California's "community property" laws mean that half

Gordon Moore atop Mauna Kea, Hawaii,
at the groundbreaking for the Thirty Meter Telescope.
SOURCE: GORDON AND BETTY MOORE FOUNDATION.

of Gordon's wealth is Betty's, and from the start their philanthropy has been a joint endeavor. If slow to articulate philanthropic goals for the foundation, Betty has been "a real presence on the board, not just a rubber stamp for Gordon." Like her husband, she is shrewd, results-focused, and "quiet, but very clear on what she's interested in." Nursing and patient care in the United States are the big subjects that concern Betty. She had had close involvement in caring for her son Steve, for her mother, and for other elders, and she had experienced brief hospitalizations herself. During one, a nurse came into Betty's room and, despite her objections, gave her a shot. Ken explains, "Mom got the insulin shot that should have gone to the patient in the neighboring bed. They nearly had two deaths out of one medical error. That was the start of her being really interested in nursing care."

In 2003 the foundation made nursing and related education into a formal theme: the Betty Irene Moore Nursing Initiative allocated $123 million over ten years to improve standards and nursing-related patient outcomes. An additional $100 million followed in 2009 to establish the Betty Irene Moore School of Nursing at the University of California, Davis. And Betty and Gordon personally gave $50 million in 2014 to fund the Betty Irene Moore Women's Hospital of the University of California, San Francisco. Patient care, on a broader national basis, is a natural next step, in Betty's view.

Betty Moore with nursing leaders.

SOURCE: GORDON AND BETTY MOORE FOUNDATION.

Under Penhoet the Moore Foundation moved on from the failure of Coleman's attempts to pursue radical innovation. It steadied itself, becoming more incremental and more professional in expectation and practice. Yet Penhoet had many other commitments and calls on his time. His real love lay in biomedicine, public health, and entrepreneurship: he became restless. Fundamental questions about the foundation's receipt of further assets, and the family's ongoing role, went unaddressed. Gordon was the boss, and the boss kept his own counsel.

Family Business

Gordon and Betty were spending more time than ever in Hawaii, though Gordon remained "signed on for too many things" in California. "He thinks he has to be there," as Betty sees it. Gordon traveled to Silicon Valley frequently on short trips to attend a variety of board meetings, tend his financial holdings, and meet supplicants, but his sons now had their own families and agendas, and the large house in Woodside lacked Betty's comforting presence. He also traveled widely on fishing trips or to give short humorous speeches while collecting a few of the awards and honorary degrees offered in all directions. Betty rarely returned to California. "I'm happy not to be there. I will only go when it's not cold."

Gordon remained involved in more than a dozen venture-capital partnerships. Most were in Silicon Valley, with several set up by his former colleague Eugene Kleiner. "I put the money in and get reports, and they send me stock certificates after the IPO. It's a very inactive kind of investment, a limited partnership, which is all I have time and energy for." Occasionally, in discussions with Betty, he invests in a company directly, "usually because some friend has reason to think this is an especially good opportunity." "It keeps him connected to the market," says Betty, "knowing how the world is going and when to jump." At home he manages the couple's finances, as he has done since the first days of their marriage. Betty says the papers relating to their very considerable wealth are "all on Gordon's desk. He's always been very secretive. At breakfast time we're watching the news and how the market's doing, the Down and Jones as he calls it, but he doesn't talk to me about things."

Gordon and Betty changed little in their interests and values, but by 2005 a decline in Betty's health prevented her from taking adventurous trips, whereas Gordon, still quite active, "could hike and jump on and off boats." He remained an avid saltwater fisherman, traveling far overseas with his sons or with long-term associates. Craig Barrett recalls with pleasure, "Here's a guy worth billions of dollars, and the only thing that he's interested in is that brown trout over there. If you put Gordon on the river, he's just your average fly-fishing buddy. He enjoys every minute of it. Fly-fishing forces you to forget everything else and concentrate on what's in front of you."

Betty was settled in Hawaii, active in the Dragon Ladies group who would lunch together and help each other out of difficult patches. She enjoyed these lunches, with their companionable talk "about the world's problems. I never had that satisfaction before. I was always tending someone else." Like Gordon, she too could not entirely settle, so she started looking for a larger, more permanent home, a single-level house more suited to visitors and to the late stages of life. Soon she found "the choice lot, a big spread along the ocean," still on the Kona coast. Gordon was summoned back from a California visit to help secure the deal.

The house on Jabil Lane had been contemporary, the house in Woodside was an English country manor, and their first Hawaiian home, Lava House, had been refurbished in Mediterranean Tuscan style. For this new project, to be designed by architect Warren Sunnland, Betty wanted something different: an Asian "less is more" feel. She envisaged three homes in one big U-shaped complex, with a pavilion at the front, a lotus pond with bridge, a swimming pool and spa, and in the main house an octagonal breakfast room and a study for Gordon projecting out to the cliff above the ocean. There would be a wing for visiting children and grandchildren, separate caretaker's quarter for future use, and electronic controls and security.

Approaching eighty, Betty became project manager: obtaining permits, reviewing materials, managing workers, and making key decisions. "Everything had to be 'click, click, click' or we were months behind, given the tempo of Hawaii. We were crazy to start to build a house at this age, but it was a wonderful challenge, and I needed a focus." Gordon was often absent on the mainland or on fishing trips. The stunning complex—planned for completion by Betty's eightieth birthday in 2008—was finally ready two years later. Gordon and Betty Moore have lived there since. (Betty officially became a resident of Hawaii in 2009.)

Though Gordon was slowly moving away from most professional engagements, he did not establish fresh friendships. "He doesn't really care for Hawaii," explains son Ken, "because there's not a lot to do. He's not a social type. Let's just say my e-mails get answered fast when he's over there." Betty became frustrated by Gordon's unwillingness to engage: "He'd have calls backed up. I'd say, 'You have to return calls,' but he didn't get at it. Yet he would drop everything to go to a meeting in California." Betty's arthritis prevented her from painting and playing the piano, but she still loves to listen to music and maintains her passion for reading. Once regular and even assiduous churchgoers, Gordon and Betty Moore are now detached from organized religion, though Betty retains a fondness for religious programs on television, a habit picked up when she was caring for her elderly mother. "Religion has become a very private matter," she says.

Intel and the electronic revolution had been the major themes in Gordon Moore's life, but that phase was over now. Betty was the sole remaining constant. She set the terms of family life, decreeing the move to Hawaii and eschewing domestic help. Gordon, deeply committed to Betty, shows her an unwavering loyalty and deference in domestic arrangements, with the result that he spends long days in Hawaii with little on his agenda. It is his choice, or lack thereof. He often passes his leisure hours alone in his study, using his computer to compile a collection of pictures or send e-mails. One interviewer noticed that he did not push his PC to the limit of its performance. Moore told him, "I get very upset about the software sometimes. I hate how long it takes for my computer to reboot."

Betty is less passive; in fact, she's "not very tolerant of things that don't work well," says Gordon. "One day at home, she picked up her laptop, took it outside, and threw it in the garbage can. Her Earthlink connection was crashing more than she found acceptable. I went out and retrieved it. I couldn't bear to see it go to the dump." In Pasadena Betty had taken her book to read in her husband's basement lab. In Hawaii she would try to draw him out of his study. "I'll find he's not doing work; he's playing solitaire on the computer. I'll say, 'Let's watch the baseball or football game on TV together.' Sometimes I'll just go and read in there with him at his desk."

When upset, Betty acts. In contrast, Gordon retreats, but he has characteristic ways to express his impatience. "He says 'Oh, hell' a lot," explains Betty. "His father used to do that. The older Gordon gets, the more he looks like his dad. He even has that same mole on his forehead." Shopping with Betty can trigger a bout of irritation, as Gordon himself explains. "I can only shop so long. All of sudden something switches inside, and I have to get out of there. It's a male thing. I don't shop; I buy. I know exactly what I want. I buy it and leave. Not like walking through Costco with Betty. 'We're going to go look for something.' Pretty soon she's found three other things to look at."

Despite their vast wealth, neither Gordon nor Betty ever became used to the idea of "help." They simply did not wish to employ others to do menial chores, beyond an absolute minimum. Even in Hawaii, Gordon still vacuums parts of the house. Ken explains that "they like it that way. It's never been a Moore ambition to have lots of personal attention." They remain careful with money, with Betty price shopping for groceries and both refusing to buy gas at the expensive gas station. "They would be living even more modestly today if it wasn't for us, the kids. My brother, Steve, and I play 'good cop, bad cop.' We started to push them into bigger expenditures. 'Spend it. What are you waiting for?'"

The brothers urged their father to update his car and even to invest for a period in a one-eighth share of a private jet for fishing trips and visits to Caltech. More recently, he has reluctantly subscribed to a regular shared-ownership jet service between Kona and San Francisco. "Dad thought, 'Private aircraft make no financial sense. No amount of calculating will get you there.' We said, 'It saves you time. You have to think of it as other than money.' Private plane owners in his peer group said, 'What's wrong with you?' After a while, with family pressure and peer pressure, he said, 'Oh, okay.' It took input from a lot of people before he decided this was a worthwhile expense."

Ken and Steve, not reared as trust-fund kids, nevertheless possess an ease in spending that eludes their father. "You buy one, you keep it" is Gordon's lifelong maxim. Even where his own philanthropy is concerned, he is careful to avoid talk that might be seen as boasting. Carver Mead gives an example.

> Once in a while, in California, Gordon comes to the local restaurant down the hill, and he'll say hi. Last time I saw him, he was there with three young ladies. I said, "Is Betty here?" He said, "No, she's in Hawaii." Then he said, "Oh, this is my daughter-in-law and her two friends, who are starting a school." He was worried I thought he was stepping out on Betty! That's as close as he gets to anything around money, talking to these people about a school they want to start. He doesn't say "that we're going to fund," but it's obviously a foundation thing. He doesn't talk about that.

Gordon and Betty show exceptional generosity to others, directly through private and often year-end gifts, but at home Gordon remains frugal. Some years ago, Betty described wanting a microwave:

> Gordon thinks, "We already have something that works. Who needs a new one? I'm in no rush." Or if I'm on my computer and Gordon's on his, neither of us get much bandwidth. I say, "How can you live this way? There's better stuff out there." He will eventually sign the papers. It just takes time and a little bit of, you know—push, push. It's a big shock for Gordon to watch our grandkids. They're living on a different level, even though they are not doing anything differently from their peers.

Ken and Kristen are parents to two teenage sons, Alex and David, while Steve and Kathleen have two younger daughters, Laura-Ann and Sarah. The extended Moore family no longer celebrates Christmases at the beach house in Davenport Landing—used now only by Ken and Steve and their families—or at Woodside; instead, in their early Hawaii days, Gordon and Betty would travel to their sons' California homes. Gordon enjoyed the "hands-on" element of being a grandfather, "but I don't think he likes the connotation that he's getting old," says Betty. She herself was "thrilled at being a grandparent," and today she delights in having her grandchildren visit in Hawaii. Like Gordon's mother, Mira, Betty raised only sons; Mira would have been "pleased to know I finally got my girls."

Betty remains especially close to Steve and Kathleen, whose first daughter was born with a hole in her heart, requiring surgery at the Lucile Packard Children's Hospital. The situation so concerned Betty that she came down with shingles. "I told Gordon, 'You're going to have to go in and be there for me.' I don't think I've ever seen him more shaken than after he saw his little granddaughter with a gazillion tubes coming out of her body. They lose one out of ten. But she did well, and she's stayed doing well."

The stairs of Chef Chu's, the Moore family's favorite Los Altos restaurant, had long been a problem for Betty when attending Friday-night family dinners. After she moved to Hawaii, her attendance naturally lapsed. When Gordon does meet with his sons and their families in Silicon Valley, it is there or at some other child-friendly Chinese or Mexican restaurant. The Moore Family Foundation has become the main vehicle for family communication, with extensive quarterly meetings scheduled on a weekday evening. Betty participates by telephone. Gordon relates, "Steve orders the pizza and I bring the wine, and we spend an evening deciding how much money we can afford to give away."

Some years after his marriage, Steve moved out of his parents' old home on Jabil Lane, while continuing as the sole staff for the family foundation. He monitors investments, visits grantees and applicants, and handles accounts and administration. Gordon, in clear contrast to his detachment from the larger foundation, plays a defining role within this smaller family philanthropy. "He takes it seriously," says Steve. Betty, too, is vocal, with projects often driven by her passion to change unjust or difficult situations. "She lets you know if she likes something or not," says Steve. His wife and Ken's are also active, as the family foundation receives more than a hundred (mostly local) requests each year and funds about a third. "We get a proposal and say, 'That sounds neat.' It's whatever interests one and the others agree to," says Gordon. "If anybody's adamant about funding a proposal," adds Steve, "the rest of us say, 'Okay.'" Modest grants support an eclectic array of projects close to the hearts of the family.

Whose Foundation?

Ken works for the larger foundation, responsible for trying to execute his father's "logic of the ledger." Aware that research ought to be part of, not separate from, grant making, he moved to focus on evaluation of the foundation's work, eventually adding the role of facilities manager. In 2008 he successfully lobbied to move operations from San Francisco—distant both geographically and culturally from the Moore family's sphere—to a vacant building on Page Mill Road in Palo Alto, close to the birthplace of Shockley Semiconductor and to the family homes. Gordon sees his eldest son's employment at the foundation less as an opportunity to apprentice him to leadership and more as a serendipitous escape route. "About the time a normal midlife crisis might develop for Ken, the foundation was getting set up," Gordon explained. "Ken thought he'd like to take a shot at that."

Gordon sees no enduring or necessary connection between the larger Moore Foundation and his family. The family disagrees. Betty and her sons, from the start, wanted the big foundation to remain under the direction of the family, pursuing both Gordon's aims and the family's values. In 2005—thanks to direct and unilateral action by Betty who, attending by telephone from Hawaii, presented Gordon with a fait accompli in the middle of a regular board meeting—the couple's daughters-in-law were appointed to its board, a move that strengthened the family's voice.

Ken is the eldest son of the founders, a board member, and a program director. His multiple positions have sometimes created uncertainty and fed intrigue. Not only was it hard for the foundation's CEO to know if he was doing what Gordon wanted (or would be satisfied with), he also needed to consider whether Ken imagined it was what Gordon wanted. In 2008 Ed Penhoet returned full-time to the biomedical arena. In the manner of a closely held corporation, Gordon chose Steve McCormick, previously CEO of the Nature Conservancy, to become the third president. The foundation "had calmed down," recalls McCormick, but "it had no experience of true management at the center."

The new president promised to continue the focus on science, environment, and the Bay Area. A lifelong activist, McCormick had strong ambitions for philanthropy in general and for the Moore Foundation in particular. Gordon also has deep aspirations to make lasting impacts in important arenas and to practice a fresh kind of philanthropy based on meaningful measurement. Gradually, it emerged that there was a mismatch between the two sets of ambitions and aspirations. Gordon, "middle of the road" Republican by temperament and conviction, was firmly in support of individual action, personal responsibility, and limited government.

Although Moore's Law was fundamentally a product of human cooperation, he was suspicious of grand social schemes. Steve McCormick, conversely, was persuaded that the foundation should take necessarily political means to encourage broad government actions on environmental issues. In addition, he believed that "foundations should be using their asset base much more imaginatively." It grew clear to him that "Gordon had no interest in this sort of thing, none at all."

McCormick published his vision in a booklet, *Game Changers,* arguing that philanthropy could turn "bold ideas into lasting change, across the spectrum of critical social issues." Philanthropy needed to shift from doing good to asserting leadership; only this would bring the necessary "breakthrough solutions." Philanthropists should act fast, striking "when the iron is hot." They should take risks, "supporting unorthodox but well-conceived ideas." Philanthropy was not a donation but an investment. Philanthropists should "be persistent, staying with an issue for the long haul." The approach valued both agility and lasting results and saw "failure" as the inability to take well-calculated risks. Thinking of this kind was not foreign to Gordon, who at Intel had backed such ideas as Hoff's microprocessor and Frohman's EPROM. And Gordon certainly championed a persistent core strategy for Intel, both during his active leadership and subsequently. However, he had little enthusiasm for McCormick's activist social views.

In March 2010 Betty was jolted into a new reality when Gordon, now eighty-one, suffered a serious health scare. Due to medical errors, a routine checkup, followed by surgery for a minor colon tumor, mutated into three months in the intensive care unit (ICU) at Stanford Hospital. With Gordon suddenly incapacitated and in critical condition, Betty—forced to stay in California, where she was supported by Ken and Steve—began to realize the enormity of her ignorance about their financial affairs. By August, with Gordon making progress (though on dialysis), the couple returned together to Hawaii, and Betty turned her attention to improving the equipment at their local hospital. The episode also made her acutely aware of the value of maintaining "a base on the mainland" in case either of them became seriously ill again.

Gordon himself has strong views on death—at least in the abstract. "I'm concerned about the pressure of the human population on everything on Earth. If you look at some projections, there's a good chance that the world's population will peak in about fifty years and move down after that. That will be the period of maximum pressure on the earth's resources—a bottleneck, if you will. If we extend everybody's life span, the bottleneck is going to be longer and more severe. So I think our obligation to die—and make room for everything else—isn't something that should be given up casually."

In Hawaii Betty was thankful for her foresight in building a one-level home. "It's flat all the way, so Gordon was able to get up and be ambulatory. We worked with a trainer a couple times a week, with exercise, walking, and the pool, to make sure that he regained his equilibrium." Gordon was eager to get back to his business and foundation-related activities; Betty remained concerned that "there was too much on his plate, too soon." Gordon took back the financial reins, but Betty had learned a lesson. "I don't manage any particular thing now, but I do watch all the bills that come through. I'm the watching bird." She also became determined to push for change, to "sit down with Gordon and talk about our financial situation" and "get our paperwork together." As always, Betty is the initiator of action in the home. "I like to see things get done."

At the large Moore Foundation, change is afoot. Steve McCormick wished to disengage from Conservation International and—since original initiatives were coming to fulfillment—was keen to plan the next steps. This unsettled Gordon's family. "They didn't want to see anything change in terms of the programs: 'We like these initiatives. They're set up.'" Gordon himself was open to change, says McCormick. "He said, 'It's probably time to think of what to do next.'" The gap grew between Steve McCormick's vision and the values of Gordon's family, with McCormick concluding his was "a philosophy of philanthropy which the family doesn't share."

Gordon's serious illness in 2010 proved a catalyst. "Gordon was withdrawing. As the family saw him fading physically, they were becoming more engaged. They wanted to assert that it was a family foundation." Board meetings were already far from relaxed affairs. "The staff would come in and, with high anxiety, give these complicated PowerPoint presentations," explains McCormick. "Gordon's not a person who delivers praise easily or often, and so people are hungry for that. People would hang on his every word."

Tensions became more pronounced, as Gordon worked through a prolonged period of recovery in health. McCormick asked James C. Gaither, a mutual friend, to join the board and to work with Gordon to establish his desires for the foundation's future. "Gordon was clear he wanted it to be an independent foundation, with significant family involvement. That led to Jim creating a set of bylaws saying that the family would never constitute the majority of the board. That stirred up the family a lot, but Gordon wanted it to go through." Lew Coleman agrees that Gordon was concerned that mission drift—the bane of many family foundations—might set in: "The landscape is littered with people who haven't quite figured out how to pass on their foundations. Take a highly successful guy like Gordon. He's ultimately going to pass some of this work to his kids. Their status in life becomes more tied to the foundation, which they inherit, than their careers. Will they change things when Gordon and Betty are gone? Probably. Will they change it too much? I hope not, but it will become their source of public identity."

Matters came to a head when Steve's wife, Kathleen, "took on Jim Gaither in a board meeting over the proper definition and role of a governance committee." Amid family resistance to his urgency about a half-thought-out agenda for major change, McCormick found Gordon resolutely avoidant: "Candidly, the family did not want me as president, but Gordon will avoid any kind of tension around a difference of opinion. There was no clarity or resolution." McCormick pushed Gordon to decide on governance, with Moore at last making plain to him, "I have to run it by the family. I have to talk with Betty." To McCormick, this signaled, "This is the way it is going to be." Separately, Kathleen told him, "You're not doing what the family wants, and Gordon is too scared to tell you." It was a Mexican standoff, says McCormick. "We were stuck in this awkward situation."

At Intel Moore had proved himself a master of boundaries and balance, a strategist who never competed directly with others but cooperated in just the right measure. He sounds almost wistful in reflecting on his activities as a philanthropist: "We accomplish so much more if we cooperate than if we compete." By 2014 Gordon had recovered his own health and continued to pursue a busy—though more limited—schedule, including regular travel. At his foundation, however, the situation was far from resolved.

CODA

A LEGACY ABORNING

Our tale is mainly told. For Gordon, as for all of us eventually, one final question remains—that of legacy. What will we remember of the life of Gordon Moore, Silicon Valley's quiet revolutionary? The outline of an answer, if not the detail, is already clear. The question may best be addressed on several levels, widening from the most personal to the most general: immediate family, philanthropy, the big foundation, and lastly, what of Silicon Valley and the electronic revolution?

Family Man

The saga of the Moore family is one robust element in the story of the American West. The Moores were pioneers by wagon train in a far and unknown land, original settlers of a remote community, and ambivalent witnesses of the Gold Rush. The family, through succeeding generations, created a rooted, secure context for the young Gordon Moore—and ultimately for his and Betty Moore's children and grandchildren. Gordon and Betty in turn have built a strong, stable marriage, over six decades.

Gordon Moore's private life is a laudable success, of a kind too rarely acknowledged—to be an upright family man, respected and honored in one's locality, is something precious in our fragmented, frenetic world. In itself, this achievement and legacy are cause for celebration and a model for emulation. Focus, study, work, constancy, and commitment: these themes are timeless and honorable. Put simply, Gordon Moore is a good man. Paradoxically and fittingly, the success of Silicon Valley's greatest revolutionary depends not on flashes of inspiration, unbridled ego, or moments of upheaval—of personal angst and turmoil—but rather upon rootedness and consistency as the underpinning for enduring progress.

The Moore family in Hawaii.

SOURCE: KATHLEEN JUSTICE-MOORE.

Philanthropy and Ambiguity

The legacy of Gordon and Betty Moore is written in larger letters in their philanthropy, most especially in the creation, development, and evolving path of the Gordon and Betty Moore Foundation (GBMF)—the big foundation, as they call it. Once again, the story is quintessentially American. From the first days of industrial, technological innovation—from the times of Andrew Carnegie and John D. Rockefeller—down through the exploits of an Arnold Beckman or a Mark Zuckerberg—economic progress has been accompanied not simply by personal wealth but also by the creation of enduring philanthropies. This is the socially approved model of success. In the United States, there is no "fourth Lord Carnegie" or "Sir Henry Ford, descendant of the automobile pioneer," no "Earl of Seattle" or "Duke of California." Rather, society casts its eye approvingly on the good works of a Bill and Melinda Gates, or a William R. Hewlett, Foundation. These and other similar philanthropies are the preferred legacy of well-lived lives.

Moore presents a certificate to Arnold Beckman for his one hundredth birthday at a Caltech celebration. Beckman, like Moore, served as Caltech's chairman of the Board of Trustees. Nearly a half century earlier, Moore phoned Beckman to alert him to the troubles at Shockley Semiconductor.

SOURCE: COURTESY OF THE ARCHIVES, CALIFORNIA INSTITUTE OF TECHNOLOGY.

Interestingly, the Gordon and Betty Moore Foundation is far from the only vehicle for the couple's generosity, documented or quietly private. Thanks not least to many unpublicized acts, the Moores belong in that very small group of individuals who have given away more than half their wealth within their lifetimes. The GBMF alone is one of the top-ten US public foundations—and one slated to grow substantially from already arranged estate gifts. If GBMF is today regrouping in the midst of ambiguity, it is in good company. Other family-based philanthropies, in their early days, have displayed similar uncertainty; it is one that reflects Gordon's own learning experiences and the extent to which the institution is but the lengthened shadow of the man.

Initially, the big foundation had been a vehicle to convert Moore's Intel stock into capital to fulfill his megagifts to Caltech and Conservation International. These gifts fitted his logic of the ledger. Like transistors, or like dollars invested in manufacturing technology versus the resulting profit on microchip products, dollars invested versus rain-forest acres conserved and dollars invested against research results can be measured. Gordon wanted to take his Intel model, especially the core element of measurement linked to

Gordon Moore (*center*) with the staff of the Gordon and Betty Moore Foundation, 2004.

SOURCE: GORDON AND BETTY MOORE FOUNDATION.

a strategic approach, and transfer it to philanthropy. As the GBMF moved beyond stock conversion and initial megagifts, a mismatch emerged.

At Intel much of Gordon's success had come from keeping to his fundamental understanding of the business, remaining steady, and doing a single important thing in the right way. Above all, he used measurement to manage both the technology and the business. In 2014 he reflected, "Keeping the foundation doing the things it was set up to do is an interesting challenge. Every time you have a management change, you get incremental directional change. My beliefs haven't altered, but people who do philanthropy as a living have a tough time thinking about measurement. If you don't measure what you've done, you have no idea if it's been useful or not. It's harder work than I imagined, but I'm still a believer."

However, the practice of philanthropy remains radically and essentially different from that of semiconductor manufacturing. Many issues treated by philanthropy, while economic, are not competitive, capitalist market issues. Dollar investments can be converted into activities that address issues, but the activities themselves are embedded in complex social and political circumstance, rendering problematic any measures of return on investment, revenue, or profit. Counting salmon is a possible but in the end a partial term in a complex set of human and environmental equations.

Most for-profit start-ups, pursuing innovation in competitive commercial markets, fail. Gordon's large foundation faces a quite different reality. Its capital resources mean the worst that can happen is inefficiency, controversy, and waste. If failure is not an option, success is also hard to conjure. Gordon had already admitted as much in 2005, when he reflected that while changing the world was "what I've been doing for forty years," it was "less as a philanthropist, and more in my professional life. My involvement at Intel is really the high point of my career."

From the point of view of financial stewardship, the big foundation has been a great success. Its endowment has grown more than most, while in its first dozen years it has given away around $3 billion. Its environmental conservation work has settled on efforts amenable to measurement: acres conserved, biodiversity protected, fish populations. In science it has chosen fields (marine microbiology, plant science, quantum materials, the Thirty Meter Telescope) in which it can be a significant or initiating funder. In the Bay Area, land conservation and science-education activity, such as museum support, have dominated. In the nursing and patient-care activities initiated at Betty's prompting, the foundation has also gravitated to areas that can be measured: nurses trained, ICU patient outcomes.

Because the philanthropic sector is more conservative than the semiconductor industry or even banking, it has been difficult, and costly, for the

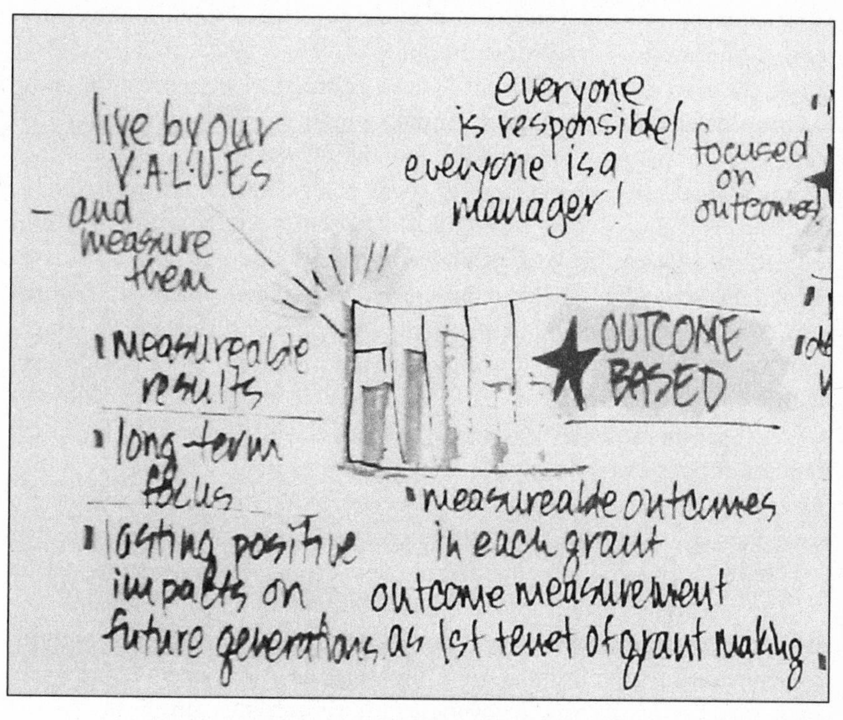

Portion of a Moore Foundation planning document, stressing measurement, 2001.

SOURCE: GORDON AND BETTY MOORE FOUNDATION.

foundation to find areas that fit—to a degree—with Gordon's desire to impose method. Each area chosen has some coherence, but is in itself mixed. In science little connects quantum materials to plant science or an optical telescope to marine microbiology. In environment little connects Pacific salmon with Andes rain forest. The foundation has found pockets of themes, but it is hard to see an overall "portfolio effect," says Steve McCormick, who believes that, as a result, the foundation has not yet fulfilled its potential.

The core issues, as seen from within and without, are ones of identity and direction. Gordon could have chosen to lead the big foundation, but, in his seventies and mainly in Hawaii, the idea did not appeal. In any case, his style of quietly focusing on strategy requires an Andy Grove or Betty Moore who shares a common background with him, as a partner executing the vision. He listens discerningly at board meetings and is present without fail, but he is the founder, funder, and chairman. The president, program officers, and leadership team, who come from a quite different universe of experiences, must first perceive or intuit and then apply his aims—and those of the family—to the organization.

At the start of 2014 Gordon Moore, restored to better health, chose to abruptly terminate Steve McCormick's presidency. The immediate challenge of calming and refocusing the foundation was handed to Paul Gray, already a foundation board member. Gray is an electrical engineer, first at Fairchild, and then with a distinguished Berkeley career—culminating in the provostship—in his background. He has ably pursued this task, over the ten months of his interim presidency. He points out how "Moore's commitment to strategic philanthropy is ambitious and audacious. Achieving measurable long-term outcomes that are important is a difficult task, requiring both sophisticated and knowledgeable program leadership and strong engagement with partners, to execute effective strategies. It's much more ambitious than simply funding worthy causes." Gray correctly observes how "If the success rate is 100 percent, then the goals were set too conservatively," and how "much more time will be needed to assess whether GBMF lives up to the full potential of the strategic philanthropy model."

The physical relocation of the foundation typifies its growing pains. Initially, Gordon let Lew Coleman site GBMF in San Francisco for his own convenience, but the foundation is now in Palo Alto, where it has again spent millions on a green renovation of a large facility. Ironically, as the foundation moved geographically closer to the family's traditional territory, Gordon and Betty were mainly far away. By 2009 Betty was almost always in Hawaii, rarely attending board meetings, and would soon resign. Gordon flew in only for board sessions or in an emergency.

Gordon wants his family to have only limited involvement. He prefers the GBMF to be overseen by a qualified board, on which "the family has significant representation, but not control. I want enough senior outside people to give continuity if something happens to me; broad people with good thoughts." That the actions of the board meet with the long-term approval of his descendants is, in his view, of secondary importance. His descendants' involvement in leadership will not guarantee that the foundation fulfills his agenda. Betty, Ken, and Steve disagree with Gordon's position. And Ken Moore holds dual roles as paid executive and as board member. At times, some have perceived him as a shadow president. In actuality, Gordon keeps his own counsel and makes his own decisions, most recently when choosing GBMF's fifth president.

Harvey Fineberg, a deeply respected academic and medical administrator from the East Coast, became the president of GBMF on January 1, 2015. Reflecting the stature, challenge and promise of the foundation, and Gordon Moore's instinctive reach for excellence, Fineberg brings impressive credentials and fresh perspectives to his new position. Entirely educated at

Harvard (AB, MD, MPP, PhD), he rose within its faculty ranks to the position of provost. From there he moved to become president of the nation's Institute of Medicine, in Washington, DC, from 2002 to 2014. Like Penhoet and Gray, he is well-practiced in the art of fashioning excellence and coherence from strong intellects and conflicting agendas.

Meanwhile Steve has made the management of the Moore Family Foundation the whole of his career. By 1990 it had $20 million in assets, mainly Intel stock. In the dot-com bubble, the value rose to $70 million, but by 2005 it was back to $40 million and today it hovers around $50 million. The family hopes to see an increase, but Gordon has declined to add more funds. Even Betty is not privy to the details of his estate plan. In the last years of his life, Gordon's avoidance—perhaps outright denial of the need to act—is once more in the ascendant. Crucially, he remains in control.

At the same time, it is already apparent that his and Betty's natural and unforced generosity places them among the greatest of philanthropic benefactors in a hallowed American tradition.

Transistors Triumphant

In contrast, Gordon Moore's legacy in silicon electronics is clear and unchallenged. Today, the electronic revolution is palpable. Its home, its undisputed heartland, is Silicon Valley—nestling close to Gordon's native Pescadero. The very name (a chemical element, modifying a geographical term) enshrines the role of chemistry, of silicon transistors, and of Moore in creating a revolutionary way of doing business on an exponential curve of growth and change. Thanks to Moore and his Fairchild colleagues, "start-ups" and "spin-offs" are no longer the locus of any "Traitorous Eight" but rather hallowed terms of canonization for the "Fairchildren" and their progeny. "Moore's Law" is at once a vogue phrase for rapid, cumulative, revolutionary growth in technological capabilities and an exact descriptor of the particular forward pathway of silicon electronics for the past six decades, as envisaged and enacted by Gordon Moore. And Intel itself continues to dominate its territory, with Moore's Law as its mantra.

We clutch our smartphones, marvel at Silicon Valley's latest IPO, and salute with envy its growing legion of multimillionaires and billionaires, still in their twenties and thirties. Yet the two key actors remain invisible to all but the most interested observers. Moore's Law may receive routine obeisance (Google shows around 4 million citations at any one time), but Gordon himself is unknown outside a small cohort of admirers. Meanwhile, the transistor—the most manufactured object in the world—is surprisingly little talked about and has signally failed to capture the public

Intel's past CEOs. *From left:* Moore, Barrett, Grove, and Otellini.

SOURCE: INTEL.

imagination. Even in 1948 news of its invention was so underwhelming that the *New York Times* saw fit only to mention on page 46, in the "News of Radio" column, that the transistor might one day replace vacuum tubes in hearing aids.

One reason for the transistor's low profile is its literal and metaphorical invisibility. Initially fingernail size, it has shrunk far beyond the limits of sight, becoming encapsulated in microchips that are themselves concealed in business and consumer products deployed all around us. Unlike visibly arresting technologies, the transistor operates silently and unobserved. Yet like the air we breathe, transistors pervade our environment and are essential to our lives. They have brought profound change in areas from warfare to medical treatment, financial planning, entertainment, politics, communication, and transportation. Gordon Moore declared in the 1970s that "cheap chunks of silicon" would come to control all manner of devices. By 2020 billions and billions of such chunks will form the backbone of the "internet of things," further transforming daily life. It is hard to imagine the world without the transistor. As one commentator has remarked, "Our servers would be three stories high, and laptops would only be props on *Star Trek*. Our televisions would still use vacuum tubes, and our cars couldn't guide us to the nearest Indian restaurant. Heck, without the transistor, what would the digital economy look like? Would Microsoft Corp. and Google Inc. have become giants? Would geeks have become cool, rich guys driving BMWs? Probably not."

The digital revolution has become a matter of everyday impact, of transformed lives. E-mail became a primary form of two-way communication, along with instant messaging and online chat. Napster, a service that allowed users to share digital music, made apparent the capacity for the Web to disrupt entire industries. "File sharing" became both an opportunity and a concern for all types of electronic information. The global positioning system, emerging from military investment, employed satellite signals to determine precise geographical location. Automobiles began to be equipped with navigation, communication, and computing systems.

Microchips—and the transistors within them—have changed how people do their jobs, pay their bills, become educated, and buy everything from books to used toaster ovens. One casualty is the landline telephone. By 2010, with nine out of ten Americans having a mobile phone, landline use had fallen considerably. Cable TV providers joined telephone companies in offering high-speed Internet access and Web-based telephone systems. Skype provides free user-to-user voice calls and free Web-based video calling, too. Text messaging has helped make mobile phones predominantly screen devices: Americans send trillions of messages every year. The launch of the category-defining iPhone—rapidly preferred by one in three Americans with a mobile phone—has increased access to electronic information via the use of "apps." Connection to the Web is now possible from almost any location, while the Web itself is an essential tool for commercial, governmental, and military entities.

Today, global financial markets are interlinked by computer networks, with traders employing ultra-high-speed computers to execute a vast number of trades in microseconds. Their algorithms function more quickly than human perception. One early result was the "dot-com" investment boom. With the ability of ordinary citizens to create electronic information—and to disseminate it cheaply and globally through the Web—have come exhilarating fresh forms of expression. The rise of blogs and Twitter has changed the face of reporting and analysis, overturning business models and modes. YouTube's video platform has proved a turning point for broadcasting. Netflix, providing movies over the Web, has superseded the video rental business. Aiding, abetting, and aggregating the creation and dissemination of electronic information are social networks such as Facebook, which has well over a billion users around the globe. Meanwhile, the privacy practices of social networks—selling fine-grained information about users, behaviors, habits, interests, and contacts, to generate revenue and profits—remain a matter for debate. Bill Davidow, Gordon's colleague at Intel in the 1970s, has written about "the dark side of Moore's Law": "Moore's Law has made mass automated surveillance dirt-cheap. In the past, surveillance was labor intensive, but when it became automated, its cost declined exponentially. We have yet to fully grasp the implications of mass surveillance. The only thing that is certain is that we will be seeing a great deal more—of ordinary citizens, potential terrorists, and heads of state—and that it will have major consequences."

The wars and unrest in the Middle East have demonstrated how military use of electronics, computing, and the Web has become indispensable. From drones to stealth bombers, satellites to tanks, and smart bombs to infantry equipment, electronics and computing are central. The "global war on terror" involves unprecedented surveillance, with the National Security Agency in the United States, and its partners overseas, compiling huge quantities of electronic information. A new term, *cyber warfare,* describes how governments engage in both familiar and innovative espionage activities—the United States deployed the Stuxnet code to derail control systems in Iranian nuclear centrifuges. Meanwhile, on a brighter note, autonomous vehicles are already on the public highways of Northern California, promising early relief from the drudgery of driving.

The Metronome of Moore's Law

As annual sales of smartphones (1 billion worldwide in 2014) outstrip those of desktop and laptop PCs (300 million combined), computing and connecting are ubiquitous in space and time. Manufacturers eye the potential of inexpensive computers and Web connections incorporated in products, increasing their functionality and value. "Wearable" digital electronics are heralded by the Apple Watch, while cochlear implants, medicine pumps, pacemakers, and artificial retinas, among others, are becoming integral to the human body. With the advent of sensors that pick up on electromagnetic brain signals, humans may soon control their own prostheses and wheelchairs through thought alone. Some speculate about a forthcoming "singularity" when human and electronic brains merge.

Moore has long seen it all in prospect. "The real benefits of what we have done are yet to come," he told an interviewer in 2001. "I sure wish I could be here in a hundred years, to see how it all plays out." To another interviewer, he spoke of his hope for "a technology that lets a computer understand what we're saying." This idea holds special appeal for Gordon, who in school found it easier to diagram sentences than to perceive and interpret shades of meaning: conversing with a computer might prove more comfortable than navigating the nuances of verbal, interpersonal exchange. Already, in 1997, he had enthused, "It has been exciting to watch semiconductor technology enable one application after another. The impact of low-cost computing is just beginning to be felt in our society. I believe that we are near the time when natural speech recognition will make the computer accessible to many more people throughout the world. Eventually, it will be possible to converse with a computer." Today, Siri wholeheartedly agrees.

Early in his career Gordon Moore realized that silicon electronics combined the economic and the technical. Moore's Law has two dimensions: the shrinking cost of and the technological extensibility of silicon manufacturing. To achieve cost and performance advantage, microchip makers needed to advance technology and double complexity every year or two. Initially, there was a long, open field for both financial investment and technical improvement. The conviction about the expanding future of digital electronics took hold. Since that time the proliferation of novel devices has been underpinned by the certainty that cost will plummet, as power and performance rise. The world has become accustomed to, and dependent upon, this state of perpetual progress.

Through the efforts of myriad researchers and the investment of hundreds of billions of dollars, the semiconductor industry has maintained this dynamic, this social construction, for more than a half century. At the head of the pack is Intel. Its CEO Brian Krzanich, a chemist who—like Gordon—began at San Jose State, unabashedly proclaims that the company's relentless pursuit of Moore's Law is "our driving force." Development of microchips continues at a steady, metronomic pace, as the digital world itself changes rapidly. Electronic influence is everywhere: in social media, Alibaba, and the Arab Spring; in Google, Facebook, and Wikileaks; and in Microsoft, IBM, and Oracle as today's tech underdogs.

The digital revolution—proving Gordon Moore correct in his belief that electronics would suffuse every aspect of society—has introduced a dazzling menagerie of transformations and possibilities: "Nothing else we've ever discovered has scaled like semiconductor design. From mud huts to skyscrapers, we've never built a structure that's thousands of times smaller, thousands of times faster, and thousands of times more power efficient within a handful of decades." As the average adult today spends around half of waking time immersed in electronic interactions, does this alter us fundamentally? With the silicon transistor impinging upon every facet of our material existence, how are we being shaped in expectation and action? Moore's Law has been a singularly important enabler of our lives, revolunizing the realities of being human.

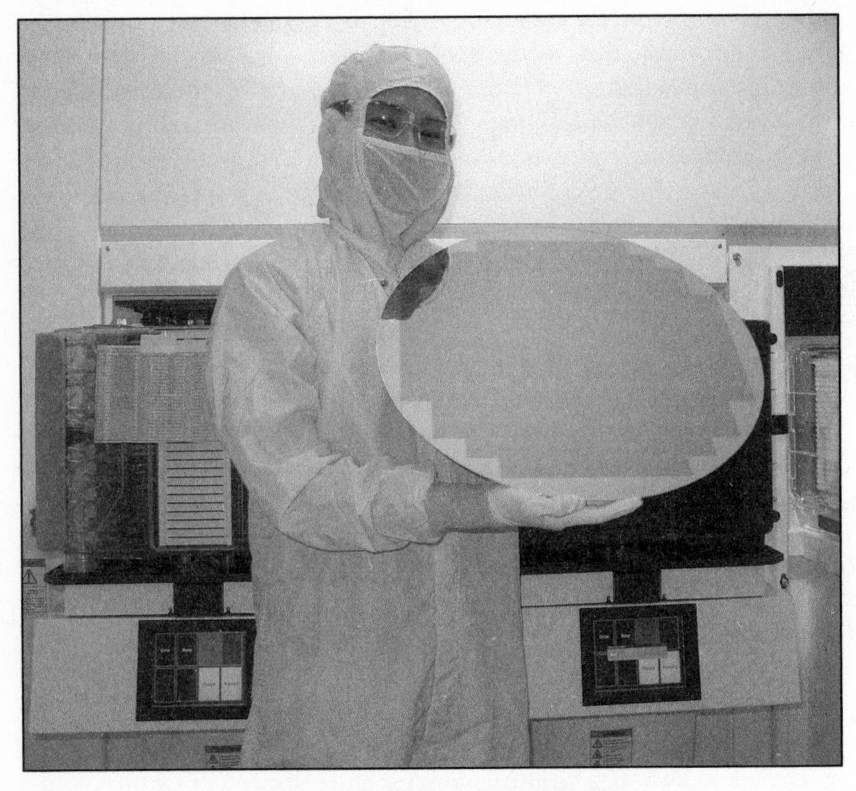

An experimental 450mm (18-inch) wafer,
patterned by advanced chemical printing, 2013.
SOURCE: MOLECULAR IMPRINTS, INC.

All Good Exponentials End

A shared conviction about the future of electronics has shaped our purchase of goods and services. As consumers we are content to wait to buy all manner of gadgets, in the expectation that fresh versions, at virtually the same prices, will deliver more bang for the buck. On a larger scale, decisions about government purchases and military weapons systems are grounded in Moore's Law and its expectation of change in forthcoming realities. How long can this continue? Commentators from across the semiconductor and financial communities have correctly begun to voice concerns about the cost and extensibility of silicon manufacturing. There is a decade or less to go before transistor size shrinks toward that of the silicon atom itself, a fundamental physical limit, now fast approaching.

The expert community involved with the International Technology Roadmap for Semiconductors, contemplating the prospects for microchip technology in the early 2020s, has added a fresh dimension to its planning: "More than Moore" is its plaintive subject. Gordon himself has long known that the end is at hand. "I can only see the next couple of generations. There always seems to be an insurmountable barrier there, but as we get closer, the engineers have so far come up with a solution. However, we shall soon reach truly fundamental limits, within a couple of semiconductor generations, I think."

What will happen when electronics no longer become cheaper and faster, when computing begins to cost more, not less? How will industries and enterprises dependent upon and accustomed to Moore's Law continue to thrive? What might this disruption mean for society? And will the certainty of continued exponential change enshrined in Moore's Law ever become standard again? Former Intel fellow Robert Colwell, the architect of the Pentium processors, says that while technological developments have created an illusion of inevitability for decades, the game is up. "I pick about 2020 as the earliest we could call Moore's Law dead. You could talk me into 2022." That is only a short seven—even five—years away.

Colwell, like Moore himself, predicts that economics will be the deciding factor. "Everybody concentrates on how many atoms. Those things matter, but my suspicion is there is so much expense that this is what will break first." John Gustafson, another former architect of Intel's chips, points out, "The original statement of Moore's Law is that the number of transistors it is more economical to produce will double every two years. It has become warped into all these other forms, but that is what he originally said." Moore's Law hinges on "the economics of integration." Just as costs and practicalities dampened our abstract enthusiasm to travel ever more quickly over longer distances (by supersonic air flight, for example), so too with the transistor: chip makers "won't be able to get a return on their $4 billion–plus chip factories, because they won't be able to sell their chips for high-enough prices to make up for huge start-up costs." As transistors shrink, "new technologies required to keep the show going are getting prohibitively expensive. So, we may choose not to pursue them."

In the short term, the end of Moore's Law could spur better use of what we have. Chip designers will have to exercise more ingenuity. Some commentators emphasize that, post Moore's Law, not all roads are blocked. "There is plenty of room to improve on software. As transistor counts stagnate, a combination of clever parallel programming techniques and engineering tricks will become more important. These tweaks will keep the industry moving forward until a postsilicon computing era can take root. There are several prospects on the radar to replace traditional silicon chips, including graphene, light-based logic circuits, and quantum processors. The next big thing beyond silicon is anybody's guess."

As the machine grinds on, and the promise of Moore's Law begins to dissolve, there is a sense of loss. "This isn't something that we're going to just find a way around," shrugs another commentator: "The old way—the old promise—of a perpetually improving technology stretching into infinity? That's gone. No one seriously thinks graphene, III–V semiconductors, or carbon nanotubes are going to bring it back, even if those technologies eventually become common."

So how might the end of Moore's Law affect daily life? Companies that depend on consumers replacing their electronic devices every few years, and those whose future plans depend heavily on fast computers, better bandwidth, and cheaper storage, may be worst affected. Gains in corporate productivity may shrivel, as computerization spreads more slowly to fresh domains. Theoretical physicist Michio Kaku has ventured that if Silicon Valley can find no effective replacement for silicon, its once lush orchard lands might eventually turn into a rust belt. Others predict a "Moorepocalypse." "Will the mantra of 'good enough,' coupled with incremental improvements over time, be sufficient to stave off a meltdown in the tech sector? Will device upgrade cycles lengthen, or will users purchase new toys at the same rate, even though they aren't much faster than their predecessors?" What will happen to Silicon Valley, the "secret home of the future tense"? Because no one active today can remember a reality not governed by Moore's Law, no one has definitive answers to the plethora of questions arising as that law reaches its final stages. "It's over folks. Get over it" is about all that can be said.

The very novelty of this situation is testament to Gordon Moore's fundamental contribution, his legacy. The quiet revolutionary of Silicon Valley has provided the world with a paradox: the past certainty of exponential advance and the present certainty of its ending. As his own life enters its last chapter, so does the story of Moore's Law and of the electronic revolution as we have known it.

ESSAY ON SOURCES

Notes, attributions, glossary of technical terms used, and bibliography may be found online at www.mooreslawbook.com. Here, we offer a brief overview of the rich resources we have relied on for this biography, resources developed by dedicated writers, researchers, and archivists. Without their prior labors, this book would not have been possible. The faults and deficiencies of this account are entirely our own creation.

The most important resource for this work has been oral history interview transcripts. In 2001 David Brock and Arnold Thackray conducted what proved to be the first of many sessions with Gordon Moore. Thirteen multihour interviews for the Chemical Heritage Foundation became one oral history transcript, running for more than 650 pages. At that stage, interviews switched in character from oral history to discussions in support of a planned biography. Audio-recorded sessions continued into January 2015, the time of this writing. Most of the quotations of Moore in this book are from the interviews. Gordon Moore is a reliable narrator. When he could not recall something, he did not hazard a guess. The events he recounted were routinely corroborated by the memories of others and often by documentary evidence. Additionally, with the benefit of time and familiarity, the very private Moore became forthcoming about many deeply personal events across his lifetime.

As our interviews with Gordon Moore turned toward this biography, we also began to sit with Betty Moore. A half-dozen multihour interviews in the mid-2000s led to an oral history of her for the Chemical Heritage Foundation. From then into 2014, Arnold Thackray and Betty Moore have had nearly another dozen recorded discussions about her and her husband's lives. Again, the great preponderance of quotes from Betty Moore are from these interviews. Those who know her will not be surprised by the cogency and candor of her insights. Indispensable to filling in our portrait were interviews with Gordon and Betty's two sons, Ken and Steve Moore, and with Ron Duarte and Robert Naughten, friends from Gordon's youth. To document Gordon and Betty Moore's philanthropic endeavors, we created oral histories with Sherry Bartolucci, Lew Coleman, Thomas Everhart,

John Martin, Steve McCormick, Russell Mittermeier, Ed Penhoet, Audrey Rust, and Kenneth Siebel.

We also had the privilege of creating a set of oral histories that focus on the origins of Silicon Valley, especially the stories of Shockley Semiconductor, of Fairchild Semiconductor, and of Intel—including Gordon Moore's place within each. Oral interviewees included Craig Barrett, Julius Blank, Roger Borovoy, William Davidow, Eugene Flath, Dov Frohman, Andrew Grove, Ted Jenkins, Victor Jones, Jay Last, Carver Mead, Robert Robson, Arthur Rock, Chitang Sah, Harry Sello, and Les Vadasz. All provided essential details and perspective; particularly vital for constructing our account were those with Andy Grove, Jay Last, Carver Mead, and Les Vadasz. Indeed, our oral history with Andy Grove provided a core resource for Grove's biographer Richard Tedlow and for Michael Malone's recent book on Intel.

We, in turn, benefited substantially from oral histories created by others. We have used many transcripts from the collections of the Computer History Museum. Equally helpful were the oral histories in the "Silicon Genesis" and SEMI-sponsored collections now part of the Silicon Valley Archives in the Stanford University Libraries. Additionally, we have drawn on the resources made available by the IEEE History Center.

While subject to the foibles of human memory and sociability, oral history is an invaluable tool in contemporary biography in contexts marked by a lack of available documentation—such as the inner workings of highly competitive for-profit corporations. That said, we went to great lengths to find, and use, documents—both for the vital information they contain and to check against oral histories and interviews. Our greatest good fortune was to acquire a very substantial collection of Gordon Moore's professional papers—notes, reports, correspondence, and the like—from his nearly four decades at the Intel Corporation.

Gordon Moore granted us permission to look through his professional papers at Intel, where a very substantial collection of documents slumbered, following the end of his active engagement with the firm. We advocated for a permanent archival home for this collection, and, with support from Gordon Moore, Sasha Abrams of the Moore Foundation, then CEO Paul Otellini, and Intel's recently retired archivist-historian Jodelle French, it has found one. The Moore Papers are now in the Silicon Valley Archive of the Stanford University Libraries, where archivist-historian Henry Lowood has been our key partner over the years. The Chemical Heritage Foundation, which processed and cataloged the papers, will possess a digital facsimile of the collection, as will the Intel Corporation. Our narration of Gordon Moore at Intel depends heavily on these documents.

Another crucial collection is the professional papers of Arnold O. Beckman and the records of Beckman Instruments, now at the Chemical Heritage Foundation. Much of our story of the origins of Shockley Semiconductor and of its short history is in these materials. Further details about Shockley Semiconductor and Fairchild Semiconductor can be found in the William Shockley Papers and the Fairchild Semiconductor Collection, in the Special Collections of Stanford University Libraries. The Computer History Museum includes the Fairchild notebooks of Jean Hoerni and Robert Noyce, recording their seminal contributions, and also Gordon Moore's Fairchild laboratory records, patent disclosures, personal aide-mémoire, and journal.

Also in the Computer History Museum is a large set of semiconductor industry market reports created by Dataquest, from which we drew facts and figures. For our estimates of transistor production, the work of G. Dan Hutcheson of VLSI Research, which he so kindly shared with us, was definitive. Early information on electronic markets and production in the United States was drawn from the Electronic Industry Association's data yearbooks. The archives of the Intel Corporation, through Jodelle French, provided us with useful interview transcripts and photographs.

At the Chemical Heritage Foundation, with the support of many colleagues, including most prominently Amanda Antonucci and Patrick Shea, the Moore Papers were supplemented with a large collection of documents about Gordon and Betty Moore. This research collection includes materials gathered from the San Mateo History Museum Archives, San Jose State University, Santa Clara County Library, University of California–Santa Cruz Special Collections, Genealogical Society of Santa Cruz County, University of California–Berkeley Bancroft Library, California Historical Society, California Genealogical Society, San Francisco Library, Redwood City Public Library, Pescadero Historical Society, Santa Cruz Historical Society, Los Gatos Public Library, Sequoia Union High School, Los Gatos High School, California Institute of Technology, and the personal papers of both Gordon and Betty Moore.

In addition to unpublished and ephemeral archival materials, we have also drawn on a considerable corpus of published work. The masterful work of Kevin Starr was an essential guide: *Americans and the California Dream, 1850–1915* (1986), *Embattled Dreams: California in War and Peace, 1940–1950* (2002), and *Golden Dreams: California in an Age of Abundance, 1950–1963* (2009). An earlier evocative perspective is in Carey McWilliams's *California: The Great Exception* (1999) and *The California Revolution* (1968). Robert Hine and John Faragher's *The American West: A New Interpretive History* (2000), C. W. Mills's *White Collar: The American Middle Classes* (1951), Jane Morgan's *Electronics in the West* (1967), and Alan

Hynding's *From Frontier to Suburb: The Story of the San Mateo Peninsula* (1982) also contributed to our work.

Gordon Moore's educational and professional life up to, and after, his encounter with the silicon transistor was focused on chemistry. William Brock's *The Norton History of Chemistry* (1993) provides a backdrop, while Peter J. T. Morris's edited volume *From Classical Chemistry to Modern Chemistry: The Instrumental Revolution* (2002), Carsten Reinhardt's *Shifting and Rearranging: Physical Methods and the Transformation of Modern Chemistry* (2006), several articles by Davis Baird, and Arnold Thackray and Minor Myers's *Arnold O. Beckman: One Hundred Years of Excellence* (2000) were our guides to the instrumentation revolution that Gordon Moore joined. William Klingaman's *APL—Fifty Years of Service to the Nation* (1993) was also helpful.

From his PhD thesis at Caltech through his scientific writings and issued patents to several reflective historical articles in the early 2000s, Moore's own publications were another critical resource. We also owe a debt of gratitude to historian Christophe Lécuyer: his advice and perspective were invaluable, as was his *Making Silicon Valley: Innovation and the Growth of High Tech, 1930–1970* (2006). Lécuyer's collaborations with David Brock provided further backbone to our story, notably their *Makers of the Microchip: A Documentary History of Fairchild Semiconductor* (2010) and "Digital Foundations: The Making of Silicon-Gate Manufacturing Technology" (2012). Brock's edited volume *Understanding Moore's Law* (2006) was another trusted guide.

Leslie Berlin's exemplary biography of Robert Noyce, *The Man Behind the Microchip* (2005), was indispensable. So too was Ross Bassett's *To the Digital Age* (2002), the best single account of the development of silicon electronics. Michael Riordan's articles on Fairchild Semiconductor, Jean Hoerni, and the planar process were extremely helpful, as was his seminal book with Lillian Hoddeson on the beginning of transistor electronics, *Crystal Fire* (1997). Richard Tedlow's *Andy Grove* (2006) and Joel Shurkin's biography of Shockley, *Broken Genius* (2006), were also relevant and instructive.

On Intel Robert Burgelman's *Strategy Is Destiny* (2002) was essential reading. Tim Jackson's *Inside Intel* (1997) offers useful detail, if from an overly negative perspective. Michael Malone's *Intel Trinity* (2014) also proved useful, from an equally positive perspective. Hector Ruiz's *Slingshot* (2013) paints a dour portrait of Intel from the perspective of the former CEO of its rival AMD. Andy Grove's *Only the Paranoid Survive* (1996) is comparatively well balanced. Michael Malone's earlier work provides interesting perspectives on Silicon Valley, including his *The Big Score: The Billion Dollar Story of Silicon Valley* (1985), *The Microprocessor: A Biography* (1995), and his history of HP, *Bill and Dave* (2007). Chuck House and Raymond Price's *The HP Phenomenon* (2009) is a wonderfully detailed history of that seminal Silicon Valley firm.

Ernest Braun and Stuart McDonald's *Revolution in Miniature* (1982) remains a valuable history of microelectronics. Hyungsub Choi's 2007 PhD thesis on RCA and the rise of the Japanese industry was useful, as was his article with Cyrus Mody, "The Long History of Molecular Electronics" (2009). Dan Holbrook's PhD thesis on the semiconductor industry (1999) and his "Government Support of the Semiconductor Industry" (1995) were instructive. Bo Lojeck covers a great deal of ground in his *History of Semiconductor Engineering* (2006). Thomas Misa's exemplary article on the early story of the transistor in Merritt Roe Smith's edited volume *Military Enterprise and Technological Change* (1985) is required reading. Charles Sporck's engaging *Spin-off* (2001) is a compendium of insider accounts and reflections.

Robert Colwell's *The Pentium Chronicles* (2006) offers a rare inside glimpse into creating microprocessors, as does Albert Yu's *Creating the Digital Future* (1998). Glenna Matthews's *Silicon Valley, Women, and the California Dream* (2003) provides an entry into the gendered labor history of the semiconductor industry, while Ted Smith et al.'s edited work *Challenging the Chip* (2006) considers environmental and labor issues. Finally, David Morton and Joseph Gabriel's *Electronics: The Story of a Technology* (2004) provides an able overview of the entire terrain.

On the history of computing, which becomes so inextricably intertwined with that of Gordon Moore and semiconductor electronics, Paul Ceruzzi's *A History of Modern Computing* (1998) and Martin Campbell-Kelly and William Aspray's *Computer: A History of the Information Machine* (1996) are the crucial duo. Jeff Yost's *The Computer Industry* (2005) gives a wonderful overview, and Paul Freiberger and Michael Swaine's *Fire in the Valley: The Making of the Personal Computer* (1984) is a powerful treatment of its subject. Janet Abbate's *Inventing the Internet* (2000) is the essential study of the development of networking. William Aspray's article on the Intel 4004 microprocessor (1997) was most helpful. Our concept of electronic reality and its development and spread was inspired in part by the writings of Manuel Castells, particularly *The Internet Galaxy* (2001).

James Cortada's *The Digital Hand* (2003) was essential to our explication of the computerization of US industry. Stan Mazor's participant histories of Fairchild and Intel, especially his article on the history of the microprocessor (1995), were of direct use. Arthur Norberg and Judy O'Neil's *Transforming Computer Technology* (1996) reveals much of the government and military context for computing. The work of Emerson Pugh, especially his, Lyle Johnson, and John Palmer's *IBM's 360 and Early 370*

Systems (1991), is a core reference. James Chposky and Ted Leonsis's *Blue Magic* (1998) details the IBM PC story. Nathan Ensmenger's *The Computer Boys Take Over* (2010) was essential for our understanding of the history of computer programs and software. JoAnne Yates's *Structuring the Information Age* (2005) provided an enlightening case study of computerization.

On the story of Silicon Valley, great rivers of ink (and pixels) have been spilled by journalists and bloggers but less by historians and scholars. Stephen Adams has written several instructive articles on the subject. Udayan Gupta's edited volume *Done Deals* (2000) gives insight into the Silicon Valley venture-capital story, along with John Wilson's *The New Venturers* (1985) and Tom Perkin's *Valley Boy* (2007). Martin Kenney's edited work *Understanding Silicon Valley* (2000) contains essential academic studies, as does Chong-Moon Lee et al.'s *The Silicon Valley Edge* (2000). Leslie Stewart's "How the West Was Won: The Military and the Making of Silicon Valley" (1993) is important. Rebecca Slayton's *Creating the Cold War University* (1997) is illuminating on Stanford's history. Last, Annalee Saxenian's *Regional Advantage* (1994) and *New Argonauts* (2006) remain indispensable.

For fuller information, the reader is again directed to visit the website at www.mooreslawbook.com.

ACKNOWLEDGMENTS

This is an authorized biography, made with the cooperation of its subject but free of review or oversight by him. It is fitting, then, that we begin our acknowledgments by expressing our deep debt to Gordon Moore himself. Without his opening of his life to us, through conversations, documentation, and introductions to others, this book would have been impossible. The generosity of Betty Moore with her time and insights was also vital to our appreciation of the full dimensions of this story, in which she plays such a central role. Ken and Steve Moore both were bravely candid in reflecting on their lives and their parents. Many other individuals shared their experiences with us, through documents, conversations, and oral histories, for which we are most grateful: here we must especially single out Craig Barrett, Andy Grove, Jay Last, Carver Mead, Art Rock, and Les Vadasz.

The Chemical Heritage Foundation (CHF)—the leading global center for the heritage of the chemical sciences, technologies, and industries—was the institutional home for our work. We thank its leaders—Kevin Cavanaugh, Denise Creedon, Carsten Reinhardt, Miriam Schaefer, and Thomas Tritton—for their support. Staff members from its library, archives, oral history center, and Institute for Research gave our efforts valued help. We particularly salute Hyungsub Choi, Christophe Lécuyer, Patrick Shea, and Amanda Shields for their contributions. Della Keyser deserves especial recognition and thanks. The Gordon and Betty Moore Foundation was a strong supporter for the project. Sasha Abrams, Genny Biggs, George Bo-Linn, Paul Gray, Steve McCormick, and Ed Penhoet all gave encouragement.

The Intel Corporation was generous, facilitating our access to Gordon Moore's personal papers and working with us to preserve those records at Stanford, CHF, and Intel. Our very special thanks go to Jodelle French, Jennifer Lee, Terri Murphy, Paul Otellini, Young-ae Park, and Tom Waldrop. The Special Collections and University Archives of the Stanford University Libraries was a kind host as we consulted their invaluable holdings. Henry Lowood and Leslie Berlin gave generously of their time and expert knowledge. John Hollar at the Computer History Museum was most helpful, as were Dag Spicer, David Laws, Paula Jabloner, and Sara Lott. As we developed our manuscript, we benefited from the readings and careful eyes of Ross Bassett, Jim Gibbons, Chuck House, Henry Lowood, and Kathleen Justice-Moore, and Ken, Kristen, and Steve Moore. We thank Melissa Chinchillo and Donald Lamm of Fletcher and Company, our agents, and our editor, T. J. Kelleher, and our project editor, Melissa Veronesi, at Basic Books.

As is not wholly uncommon with biographies, this work has been a long time gestating. Above all, we thank our families for their understanding of and tolerance for our preoccupation with *the book*. Arnold Thackray offers special gratitude to his wife and helpmeet, Diana, for her patience with his perpetual preoccupation; to his children, Helen, Gillian, and Timothy (Timbo), for their interest in Silicon Valley stories; to Harris Dienstfrey for many excellent suggestions; to Nikki Myoraku for research assistance; and to Marthenia Perrin for enduring support. David Brock wishes to thank Jennifer Stromsten, Vivian Brock, and Lucinda Brock for their love, support, and patience; Greenfield, Massachusetts, for being a great place to write; and Ben Gross, Dan Holbrook, Ann Johnson, Christophe Lécuyer, Patrick McCray, and Cyrus Mody for their collegiality. Rachel Jones is indebted to Andrew, Molly, and Pippa Jones who were patient through the long hours of writing, and grew increasingly interested in the life and times of Gordon Moore as the project progressed. She would also like to thank Hayley McGregor, Estelle Williams, and Dixie Wills for their encouragement and hospitality.

The defects of the work are wholly of the authors' creation.

INDEX